The STM32F103 Arm Microcontroller

and Embedded Systems

Using Assembly and C

First Edition

Muhammad Ali Mazidi

Sepehr Naimi

Sarmad Naimi

Naimi & Mazidi books

Copyright © 2019-2022

Arm, Cortex, Keil, and uVision are registered trade mark of Arm Limited.

To contact authors, use the following email addresses:

Sepehr.Naimi@gmail.com

mazidibooks@gmail.com

Visit our website at

https://NicerLand.com

ISBN-13: 978-1-970054-01-9

ISBN-10: 1-970054-01-8

"Regard man as a mine

rich in gems of inestimable value.

Education can, alone,

cause it to reveal its treasures,

and enable mankind to benefit therefrom."

Baha'u'llah

Dedication

To the faculty, staff, and students of BIHE university for their dedication and steadfastness.

Table of Contents

See the following website to download the chapters which are labeled as "(Web)":

https://NicerLand.com

Preface

The Arm processor is becoming the dominant CPU architecture in the computer industry. It is already the leading architecture in cell phones and tablet computers. With such a large number of companies producing Arm chips, it is certain that the architecture will move to the laptop, desktop and high-performance computers presently dominated by x86 architecture from Intel and AMD. Currently the PIC and AVR microcontrollers dominate the 8-bit microcontroller market. The Arm architecture will have a major impact in this area too as designers become more familiar with its architecture. This book is intended as an introduction to STM32F103 Arm programming. To write programs for Arm microcontrollers, you need to know both Assembly and C languages. Chapters 2 to 6 cover the Arm Assembly language. However, Chapter 6 contains more advanced topics and you can skip it if you like. Some general topics about embedded C programming are covered in Chapter 7. Then, the peripheral programming of the STM32F103 chip is discussed in Chapters 7 to 19. You will learn interfacing to some real-world devices such as LCD, Keypad, Motor, 7-segment, relay, and sensors, as well.

Prerequisites

We assume no prior background in assembly language programming with other CPUs. But a basic knowledge of C programming is required. We also urge you to study Chapter 0 covering the fundamentals of digital systems such as hexadecimal numbers, various types of memory, memory and I/O interfacing, bus designing, and memory address decoding. Chapter 0 is available free of charge on our website (https://NicerLand.com).

Trainer board

This book covers the STM32F103 microcontrollers and you can use any trainer boards with STM32F10x chips. But the Blue Pill board can be a very good choice. Since they are low priced and widely available around the world. The pins are labeled on the board, as well.

Keil tutorials

We have used the Keil Compiler for the programs throughout this book. See our website (www.NicerLand.com) for the Keil step-by-step tutorial. You can freely download the Keil IDE and use it for programs which are less than 32KB.

Power Point, Source codes, and other materials

The source codes, lab manuals, and Power points of the book are available on the website (https://NicerLand.com). If you are a professor using this book for a university course you can contact us to receive the solutions to the end-of-chapter problems.

Chapter 1: The History of Arm and Microcontrollers

In Section 1.1 we look at the history of microcontrollers then we introduce some of the available microcontrollers. The history of Arm is provided in Section 1.2.

Section 1.1: Introduction to Microcontrollers

The evolution of Microprocessors and Microcontrollers

In early computers, CPUs were designed using a number of vacuum tubes. The vacuum tube was bulky and consumed a lot of electricity. The invention of transistors, followed by the IC (Integrated Circuit), provided the means to put a CPU on printed circuit boards. The advances in IC technology allowed putting the entire CPU on a single IC chip. This IC was called a *microprocessor*. Some of the microprocessors are the x86 family of Intel used widely in desktop computers, and the 68000 of Motorola. The microprocessors do not contain RAM, ROM, or I/O peripherals. As a result, they must be connected externally to RAM, ROM and I/O, as shown in Figure 1-1.

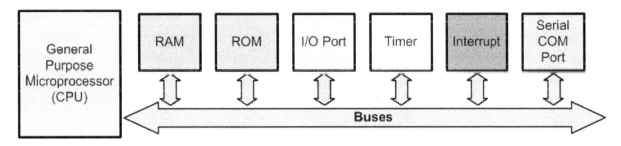

Figure 1-1: A Computer Made by General Purpose Microprocessor

In the next step, the different parts of a system, including CPU, RAM, ROM, and I/Os, were put together on a single IC chip and it was called *microcontroller*. MCU (Micro Controller Unit) is another name used to refer to microcontrollers. Figure 1-2 shows the simplified view of the internal parts of microcontrollers.

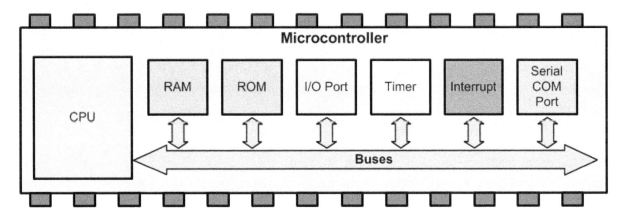

Figure 1-2: Simplified View of the Internal Parts of Microcontrollers (SOC)

Since the microcontrollers are cheap and small, they are widely used in many devices.

Types of Computers

Typically, computers are categorized into 3 groups: desktop computers, servers, and embedded systems.

Desktop computers, including PCs, tablets, and laptops, are general purpose computers. They can be used to play games, read and edit articles, and do any other task just by running the proper application programs. The desktop computers use microprocessors.

In contrast, embedded systems are special-purpose computers. In embedded system devices, the software application and hardware are embedded together and are designed to do a specific task. For example, digital camera, vacuum cleaner, mp3 player, mouse, keyboard, and printer, are some examples of embedded systems. It is interesting to note that embedded systems are the largest class of computers though they are not normally considered as computers by the general public. In most cases embedded systems run a fixed program and contain a microcontroller. But sometimes microcontrollers are inadequate for a task. For this reason, sometimes general-purpose microprocessors are used to design embedded systems. In recent years many manufacturers of general-purpose microprocessors such as Intel, NXP (formerly Motorola), and AMD (Advanced Micro Devices, Inc.) have targeted their microprocessors for the high end of the embedded market. Currently, because of Linux and Windows standardization, in these embedded systems Linux and Windows operating systems are widely used. In many cases, using the operating systems shortens development time because a vast library of software already exists for the Linux and Windows platforms. The fact that Windows and Linux are widely used and well-understood platforms means that developing a Windows-based or Linux-based embedded product reduces the cost and shortens the development time considerably.

Servers are the fast computers which might be used as web hosts, database servers, and in any application in which we need to process a huge amount of data such as weather forecasting. Similar to desktop computers, servers are made of microprocessors but, multiple processors are usually used in each server. Both servers and desktop computers are connected to a number of embedded system devices such as mouse, keyboard, disk controller, Flash stick memory and so on.

Making computers using SoCs

It is becoming common to integrate the processor with the most parts of the system to make a single chip. Such a chip is called an SoC (System on Chip). In recent years, companies have begun to sell Field-Programmable Gate Array (FPGA) and Application-Specific Integrated Circuit (ASIC) libraries for their processors. This makes the production of the new chips easier.

A Brief History of the Processors

In the late 1970s the first processor chips were introduced. In the beginning years of 1980s IBM used the x86 (8088/86, 80286, 80386, 80486, and Pentium) to make their Personal Computers and Apple used the 68xxx (68000, 68010, 68020, etc.) to make their Macintosh PC. Consequently, Intel and Motorola became the dominated the field of microprocessors and also microcontrollers in the 1980s and 1990s. Many embedded systems used Intel's 32-bit chips of x86 (386, 486, Pentium) and Motorola's 32-bit 68xxx for high-end embedded products such as routers. For example, Cisco routers used 68xxx for the CPU. At the low end, the 8051 from Intel and 68HC11 from Motorola were the dominant 8-bit microcontrollers. With the introduction of PIC from Microchip and AVR from Atmel, they became major players in the 8-bit market for microcontroller. At the time of this writing, PIC and AVR are the leaders in terms of volume for 8-bit microcontrollers. In the late 1990s, the Arm microcontroller started to challenge the dominance of Intel and Motorola in the 32-bit market. Although both Intel and Motorola used RISC features to enhance the performance of their microprocessors, due to the need to maintain compatibility with legacy software, they could not make a clean break and start over. Intel used massive amounts of gates to keep up the performance of x86 architecture and that in turn increased the power consumption of the x86 to a level unacceptable for battery-powered embedded products. Meanwhile Motorola streamlined the instructions of the 68xxx CPU and created a new line of microprocessors called ColdFire, while at the same time worked with IBM to design a new RISC processor called PowerPC. While both PowerPC and Coldfire are still alive and being used in the 32-bit market, it is Arm which has become the leading microcontroller in the 32-bit market.

Introduction to some 32-bit microprocessors and microcontrollers

x86: The x86 and Pentium processors are based on the 32-bit architecture of the 386. Although both Intel and AMD are pushing the x86 into the embedded market, due to the high power consumption of these chips, the embedded market has not embraced the x86. Intel is working hard to make a low-power version of the 386 called Atom available for the embedded market.

PIC32: It is based on the MIPS architecture and is getting some attention due to the fact it shares some of the peripherals with the PIC24/PIC18 chips and also using the MPLAB for IDE. Microchip hopes the free MPLAB IDE and engineers' knowledge of the 8-bit PIC will attract embedded developers to the PIC32 as they move to 32-bit systems for their high-end embedded products.

ColdFire: The NXP (formerly Freescale, Motorola) is based on the venerable 680x0 (68000, 68010) popular in the 1980s and 1990s. They streamlined the 68000 instructions to make it more RISC-type architecture and is the top seller of 32-bit processors from the Freescale. In recent years Freescale revamped and redesigned the 8-bit HCS08 (from the 6808) to share some of the peripherals with ColdFire and are pushing them under the name Flexis. They hope engineers use the HCS08 at the low-end and move to Coldfire for high-end of the embedded products with minimum learning curve.

PowerPC: This was developed jointly by IBM and Motorola. It was used in the Apple Mac for a few years. Then Apple switched to x86 for a while and currently is using Arm in all their products. Nowadays, both Freescale and IBM market the PowerPC for the high-end of the embedded systems.

How to choose a microcontroller
The following two factors can be important in choosing a microcontroller:

- *Chip characteristics:* Some of the factors in choosing a microcontroller chip are clock speed, power consumption, price, and on-chip memories and peripherals.
- *Available resources:* Other factors in choosing a microcontroller include the IDE compiler, legacy software, and multiple sources of production.

Review Questions

1. True or false. Microcontrollers are normally less expensive than microprocessors.
2. When comparing a system board based on a microcontroller and a general- purpose microprocessor, which one is cheaper?
3. A microcontroller normally has which of the following devices on-chip?
   (a) RAM   (b) ROM   (c) I/O   (d) all of the above
4. A general-purpose microprocessor normally needs which of the following devices to be attached to it?
   (a) RAM   (b) ROM   (c) I/O   (d) all of the above
5. An embedded system is also called a dedicated system. Why?
6. What does the term "embedded system" mean?
7. Why does having multiple sources of a given product matter?

Section 1.2: The Arm Family History

In this section, we look at the Arm and its history.

A brief history of the Arm

The Arm came out of a company called Acorn Computers in United Kingdom in the 1980s. Professor Steve Furber of Manchester University worked with Sophie Wilson to define the Arm architecture and instructions. The VLSI Technology Corp. produced the first Arm chip in 1985 for Acorn Computers and was designated as Acorn RISC Machine (Arm). Unable to compete with x86 (8088, 80286, 80386, …) PCs from IBM and other personal computer makers, the Acorn was forced to push the Arm chip into the single-chip microcontroller market for embedded products. That is when Apple Corp. got interested in using the Arm chip for the PDA (personal digital assistants) products. This renewed interest in the chip led to the creation of a new company called Arm (Advanced RISC Machine). This new company bet its entire fortune on selling the rights to this new CPU to other silicon manufacturers and design houses. Since the early 1990s, an ever increasing number of companies have licensed the right to make the Arm chip. See Table 1-1 for the major milestones of the Arm.

Table 1-1: Arm Company milestones (www.Arm.com)
1982
■ Acorn produced a computer for BBC named BBC micro. Good sales of the computer motivated Acorn to decide to make its own microprocessor.
1983
■ Acorn and VLSI began designing the Arm microprocessor.
1985
■ Acorn Computer Group developed the world's first commercial RISC processor. The ARMv1 had 25,000 transistors, and worked with a frequency of 4MHz.

_1987_

- Acorn's Arm processor debuts as the first RISC processor for low-cost PCs

_1989_

- Acorn introduced ARMv3 with a frequency of 25MHz. It had a 4KB cache as well.

_1990_

- Advanced RISC Machines (Arm) spins out of Acorn and Apple Computer's collaboration efforts with a charter to create a new microprocessor standard. VLSI Technology becomes an investor and the first licensee.

_1991_

- Arm introduced its first embeddable RISC core, the ARM6 solution using ARMv3 architecture.

_1992_

- GEC Plessey and Sharp licensed Arm technology

_1993_

- Texas Instruments licensed Arm technology
- Arm introduced the ARM7 core.

_1995_

- Arm announced the Thumb architecture extension, which gives 32-bit RISC performance at 16-bit system cost and offers industry-leading code density
- Arm launched Software Development Toolkit

_1996_

- Arm and VLSI Technology introduced the ARM810 microprocessor
- Arm and Microsoft worked together to extend Windows CE to the Arm architecture

_1997_

- Hyundai, Lucent, Philips, Rockwell and Sony licensed Arm technology

- ARM9TDMI family announced

1998

- HP, IBM, Matsushita, Seiko Epson and Qualcomm licensed Arm technology
- Arm developed synthesizable version of the ARM7TDMI core
- Arm Partners shipped more than 50 million Arm-powered products

1999

- LSI Logic, STMicroelectronics and Fujitsu licensed Arm technology
- Arm announced synthesizable ARM9E processor with enhanced signal processing

2000

- Agilent, Altera, Micronas, Mitsubishi, Motorola, Sanyo, Triscend and ZTEIC licensed Arm technology
- Arm launched SecurCore family for smartcards
- TSMC and UMC became members of Arm Foundry Program

2001

- Arm's share of the 32-bit embedded RISC microprocessor market grew to 76.8 per cent
- Arm announced new ARMv6 architecture
- Fujitsu, Global UniChip, Samsung and Zeevo licensed Arm technology
- Arm acquired key technologies and an embedded debug design team from Noral Micrologics Ltd

2002

- Arm announced that it had shipped over one billion of its microprocessor cores to date
- Arm technology licensed to Seagate, Broadcom, Philips, Matsushita, Micrel, eSilicon, Chip Express and ITRI
- Arm launched the ARM11 micro-architecture
- Arm launches its RealView family of development tools
- Flextronics became the first Arm Licensing Partner program member, allowing it to sub-license Arm technology to its own customers

2004

- The Arm Cortex family of processors, based on the ARMv7 architecture, is announced. The Arm Cortex-M3 is announced in conjunction, as the first of the new family of processors
- Arm Cortex-M3 processor announced, the first of a new Cortex family of processor cores
- MPCore multiprocessor launched, the first integrated multiprocessor
- OptimoDE technology launched, the groundbreaking embedded signal processing core

_2005_

- Arm acquired Keil Software
- Arm Cortex-A8 processor announced

_2007_

- Five billionth Arm Powered processor shipped to the mobile device market
- Arm Cortex-M1 processor launched – the first Arm processor designed specifically for implementation on FPGAs
- RealView Profiler for Embedded Software Analysis introduced
- Arm unveils Cortex-A9 processors for scalable performance and low-power designs

_2008_

- Arm announces 10 billionth processors shipment
- Arm Mali-200 GPU Worlds First to achieve Khronos Open GL ES 2.0 conformance at 1080p HDTV resolution

_2009_

- Arm announces 2GHz capable Cortex-A9 dual core processor implementation
- Arm launches its smallest, lowest power, most energy efficient processor, Cortex-M0

_2010_

- Arm launches Cortex-M4 processor for high performance digital signal control
- Arm together with key Partners form Linaro to speed rollout of Linux-based devices
- Microsoft becomes an Arm Architecture Licensee
- Arm & TSMC sign long-term agreement to achieve optimized Systems-on-Chip based on Arm processors, extending down to 20nm
- Arm extends performance range of processor offering with the Cortex-A15 MPCore processor
- Arm Mali becomes the most widely licensed embedded GPU architecture
- Arm Mali-T604 Graphics Processing Unit introduced providing industry-leading graphics performance with an energy-efficient profile

_2011_

- Microsoft unveils Windows on Arm at CES 2011
- IBM and Arm collaborate to provide comprehensive design platforms down to 14nm
- Arm and UMC extend partnership into 28nm
- Cortex-A7 processor launched
- Big-Little processing announced, linking Cortex-A15 and Cortex-A7 processors
- ARMv8 architecture unveiled at TechCon
- AMP announce license and plans for first ARMv8-based processor
- Arm Mali-T658 GPU launched

- Arm expands R&D presence in Taiwan with Hsinchu Design Center
- Arm and Avnet launch Embedded Software Store (ESS)
- Arm, Cadence and TSMC tape out first 20nm Cortex-A15 multicore processor

2012

- Arm, Gemalto and G&D form joint venture to deliver next-generation mobile security
- First Windows RT (Windows on Arm) devices revealed
- Arm, AMD, Imagination, MediaTek and Texas Instruments founding members of Heterogeneous System Architecture (HAS) Foundation
- Arm and TSMC work together on FinFET process technology for next-generation 64-bit Arm processors
- Arm forms first UK forum to create technology blueprint "Internet of Things" devices
- Arm named one of Britain's Top Employers
- MIT Technology Review named Arm in its list of 50 Most Innovative Companies

Currently the Arm Corp. receives its entire revenue from licensing the Arm to other companies since it does not own state of the art chip fabrication facility. This business model of making money from selling IP (intellectual property) has made Arm one of the most widely used CPU architectures in the world. Unlike Intel or Freescale who define the architecture and fabricate the chip, hundreds of companies who have licensed the Arm IP feel a level playing field when it comes to competing with the originator of the chip.

Arm and Apple

When Steve Jobs came back to run the Apple in 1996, the company was in decline. It had lost the personal computer race that had started 20 years earlier. The introduction of iPod in 2001 changed the fortune of that company more than anything else. Apple had tried to sell a PDA called Newton in the 1990s but was not successful. The Newton was using the Arm processor and it was too early for its time. The iPod used an enhanced version of Arm called ARM7 and became an instant success. iPod brought the attention to the Arm chip that it deserved. Since then Apple has been using the Arm chip in iPhones and iPads. Today, the Arm microcontroller is the CPU of choice for designing cell phone and other hand-held devices. In the future, Arm will make further in-roads into the tablet and laptop PC market now that Microsoft Corp has introduced the Arm version of its Windows operating system.

Arm family variations

Although the ARM7 family is the most widely used version, Arm is determined to push the architecture into the low end of the microcontroller market where 8- and 16-bit microcontrollers have been traditionally dominating. For this reason, they have come up with a microcontroller version of Arm called Cortex. As we will see in future chapters, the Cortex family of Arm microcontrollers maintains compatibility with the ARM7 without sacrificing performance. The Arm architecture is also being pushed into high-performance systems where multicore chips such as Intel Xeon dominate.

Figure 1-3 shows some of the most widely used Arm processors. It should be emphasized that we cannot use the terms Arm family and Arm architecture interchangeably. For example, ARM11 family is based on ARMv6 architecture and ARMv7A is the architecture of Cortex-A family.

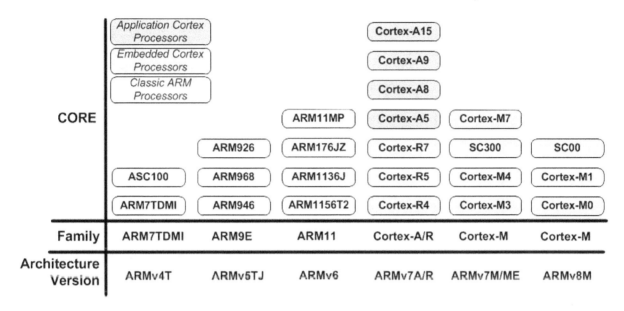

Figure 1-3: Arm Family and Architecture

One CPU, many peripherals

Arm has defined the details of architecture, registers, instruction set, memory map, and timing of the Arm CPU and holds the copyright to it. The various design houses and semiconductor manufacturers license the IP (intellectual property) for the CPU and can add their own peripherals as they please. It is up to the licensee (design houses and semiconductor manufactures) to define the details of peripherals such as I/O ports, serial port UART, timer, ADC, SPI, DAC, I2C, and so on. As a result, while the CPU instructions and architecture are the same across all the Arm chips made by different vendors, their peripherals are not compatible. That means if you write a program for the serial port of an Arm chip made by TI (Texas Instrument), the program might not necessarily run on an Arm chip sold by NXP. This is the only drawback of the Arm microcontroller. The good news is that the manufacturers do provide peripheral libraries or tools for their chips and make the job of programming the peripherals much easier. For example, ST Micro has the Cube, TI has the TivaWare for Tiva series devices, and Freescale (now part of NXP) has Processor Expert. Figure 1-4 shows the Arm simplified block diagram and Table 1-2 provides a list of some Arm vendors.

Actel	Analog Devices	Atmel (now Microchip)
Broadcom	Cypress	Ember
Dust Networks	Energy	Freescale
Fujitso	Nuvoton	NXP
Renesas	Samsung	ST
Toshiba	Texas Instruments	Triad Semiconductor

Table 1-2: Arm Vendors

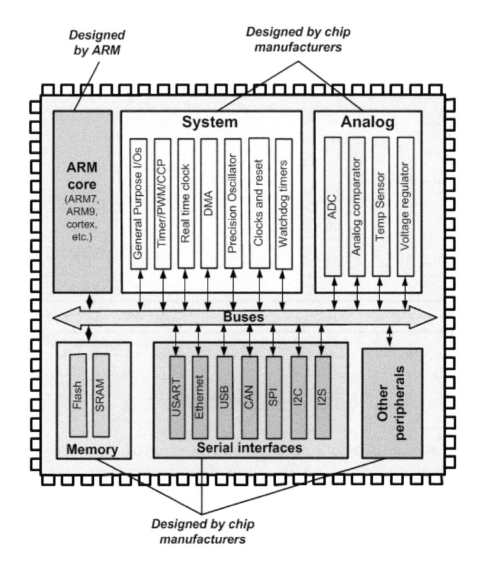

Figure 1-4: Arm Simplified Block Diagram

Review Questions

1. True or false. The Arm CPU instructions are universal regardless of who makes the chip.
2. True or false. The peripherals of Arm microcontroller are standardized regardless of who makes the chip.
3. An Arm microcontroller normally has which of the following devices on-chip?
   (a) RAM   (b) Timer   (c) I/O   (d) all of the above

4. For which of the followings, Arm has defined standard?
   (a) RAM size   (b) ROM size   (c) instruction set   (d) all of the above

Section 1.3: STM32 Family

Figure 1-5 shows the series of STM32 microcontroller.

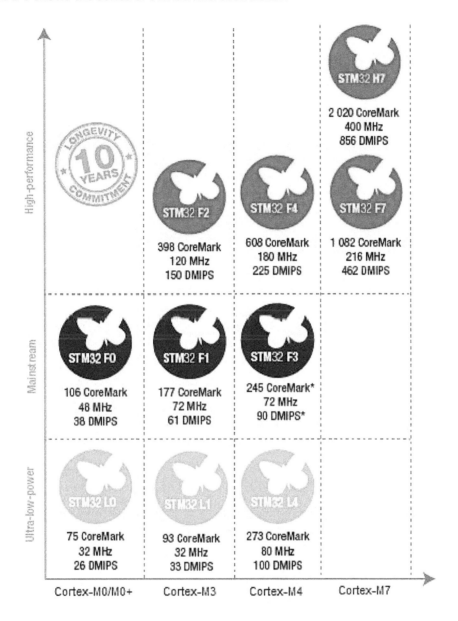

Figure 1-5: STM32 Microcontrollers

Introduction to some of the series

STM32Lxxx Series

The series have a low power consumption. So, they are well suited for devices which use battery.

STM32F0xx Series

Their prices are low and they are well suited for simple works.

STM32F1xx Series

Their prices and capabilities make them suitable for most of the projects. They have a complete Thumb-2 instruction set and there are very different trainer boards for it. That is why, in this book, we concentrate on the series.

STM32F2xx, STM32F4xx, and STM32F7xx Series

They are the high performance series of STM32. They can be used in the projects which need high processing power. Table 1-3 summarizes some of the features of the STM32 series.

Series	Max RAM	Max Flash	Speed	Some features
STM32W	256KB	1MB	64MHz	IEEE802.15.4 Wireless transceiver
STM32L0	20KB	192KB	32MHz	EEPROM
STM32L1	48KB	384KB	32MHz	USB device, EEPROM, AES
STM32L4	640KB	2MB	80MHz	DSP, MPU, FPU, capacity sensing, camera interface, FSMC
STM32F0	32KB	256KB	48MHz	
STM32F1	96KB	1MB	72MHz	USB OTG, CAN, SDIO, HDMI, Ethernet, FSMC, DAC
STM32F2	128KB	1MB	120MHz	USB OTG, CAN, SDIO, Ethernet, Crypto/hash processor, camera interface, FSMC, DAC
STM32F3	80KB	512KB	72MHz	USB, CAN, 16-bit ADC, DSP, FPU, DAC, capacity sensing
STM32F4	384KB	2MB	180MHz	USB OTG, CAN, SDIO, Ethernet, Crypto, DSP, FPU, LCD controller, camera interface
STM32F7	512KB	2MB	216MHz	DSP, FPU, FMC, Crypto, LCD controller, HDMI, camera interface

Table 1-3: Some Series of STM32

STM32 chips naming convention

Names of the new Arm products of ST begin with STM32. The following figure shows their naming convention. Table 1-4 lists some of the STM32F103 chips together with their memory sizes and packages.

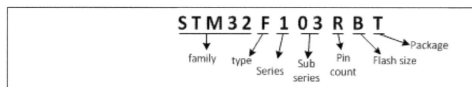

Family: Names of the new Arm products of ST begin with STM32.

Type:

Type	L	F	H	W
	Low power	Mainstream (Foundation)	High performance	Wireless

Series:

CPU Core	0	1	2	3	4	7
	Cortex-M0	Cortex-M3		Cortex-M4		Cortex-M7

Sub series:

The peripheral features of chips are dependant to the sub series. Chips with higher sub-series numbers most probably have richer configurations.

Pin count:

	F	G	K	T	S	C	R	V	Z	A
Number of pins	20	28	32	36	44	48	64	100	144	169

Flash memory Size:

	4	6	8	B	C	D	E	F	G	H	I
Group	Low density		Medium density		High density						
Flash	16KB	32KB	64KB	128KB	256KB	384KB	512KB	768KB	1MB	1.5MB	2MB

Package:

	H	T	U	Y
Package	BGA	LQFP	QFN	WLCSP

Figure 1-6: STM32 Naming Conventions

Part number	CPU Core	Pin count	Flash size	RAM	Max. Freq.
STM32F103RB	Cortex-M3	64 pins	128KB	20KB	72MHz
STM32F103C8	Cortex-M3	48 pins	64KB	20KB	72MHz
STM32F030K6	Cortex-M0	32 pins	32KB	4KB	48MHz

Table 1-4: Some STM32 Chips

Review Questions

1. Give the number of pins in each of the following chips:
   (a) STM32F103VC   (b) STM32F103R8
2. Give the size of Flash in each of the following chips:
   (a) STM32F103VC   (b) STM32F103R8

Problems

Section 1.1: Introduction to Microcontrollers

1. True or False. A general-purpose microprocessor has on-chip ROM.
2. True or False. Generally, a microcontroller has on-chip ROM.
3. True or False. A microcontroller has on-chip I/O ports.
4. True or False. A microcontroller has a fixed amount of RAM on the chip.
5. What components are usually put together with the microcontroller onto a single chip?
6. List three embedded products attached to a PC.
7. Give the name and the manufacturer of some of the most widely used 8-bit microcontrollers.
8. In Question 7, which of them sell most?
9. Name some 32-bit microcontrollers.
10. In a battery-based embedded product, what is the most important factor in choosing a microcontroller?
11. In an embedded controller with on-chip ROM, why does the size of the ROM matter?

Section 1.2: The Arm Family History

12. What does Arm stand for?
13. True or false. In Arm, architectures have the same names as families.
14. True or false. In 1990s, Arm was widely used in microprocessor world.
15. True or false. Arm is widely used in Apple products, like iPhone and iPod.
16. True or false. All Arm chips have standard instructions.
17. True or false. All Arm chips have the same peripherals
18. True or false. The Arm corp. also manufactures the Arm chip.
19. True or false. The Arm IP must be licensed from Arm corp.
20. True or false. A given serial communication program is written for TI Arm chip. It should work without any modification on NXP Arm chip
21. True or false. At the present time, Arm has just one manufacturer.
22. What is the difference between the Arm products of different manufacturers?

Section 1.3: STM32 Family

23. Give the Flash size in: (a) STM32F401RC (b) STM32F100C6T
24. Give the number of pins in: (a) STM32F103R6T (b) STM32F050C6T

Answers to Review Questions

Section 1.1

1. True
2. A microcontroller-based system
3. d
4. d
5. It is dedicated because it does only one type of job.
6. Embedded system means that the application (software) and the processor (hardware such as CPU and memory) are embedded together into a single system.
7. Having multiple sources for a given part means you are not hostage to one supplier. More importantly, competition among suppliers brings about lower cost for that product.

Section 1.2

1. True
2. False
3. d
4. c

Section 1.3

1. (a) 100 pins   (b) 64 pins
2. (a) 256KB   (b) 64KB

Chapter 2: Arm Architecture and Assembly Language Programming

CPUs use registers to store data temporarily and most of the operations involve the registers. To program in assembly language, we must understand the registers of a given CPU and the role they play in processing data. In Section 2.1 we look at the registers of the Arm CPU. We demonstrate the use of registers with simple instructions such as MOV and ADD. Memory map and memory access of the Arm are discussed in Sections 2.2 and 2.3, respectively. In Section 2.4 we discuss the status register's flag bits and how they are affected by arithmetic instructions. In Section 2.5 we look at some widely used assembly language directives, pseudo-instruction, and data types related to the Arm. Section 2.6 discusses the memory allocation. In Section 2.7 we examine assembly language and machine language programming. The process of assembling and creating a ready-to-run program for the Arm is discussed in Section 2.8. Step-by-step execution of an Arm program and the role of the program counter are examined in Section 2.9. Section 2.10 examines some Arm addressing modes. The Pipeline and RISC architecture are examined in Sections 2.11 and 2.12.

Section 2.1: The General Purpose Registers in the Arm

In the CPU, registers are used to store information temporarily. That information could be a piece of data to be processed, or an address pointing to the data to be fetched. Arm microcontrollers have 16 registers for arithmetic and logic operations. See Figure 2-2. All of the registers are 32-bit wide. The 32 bits of a register are shown in Figure 2-1. These range from the MSB (most-significant bit) D31 to the LSB (least-significant bit) D0. With a 32-bit data type, any data larger than 32 bits must be broken into 32-bit chunks before it is processed. Although the Arm default data size is 32-bit, some instructions also support the single bit, 8-bit, and 16-bit data types, as we will see in future chapters. In Arm, the 32-bit data size is often referred as "word" and the 16-bit data is referred to as half-word. Therefore, Arm supports byte, half-word (two bytes), and word (four bytes) data types.

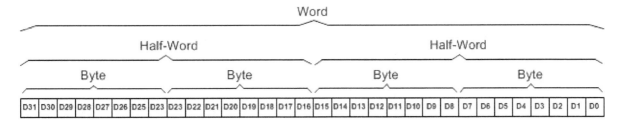

Figure 2-1: Arm Registers Data Size

To understand the use of the registers, we will show it in the context of some simple instructions.

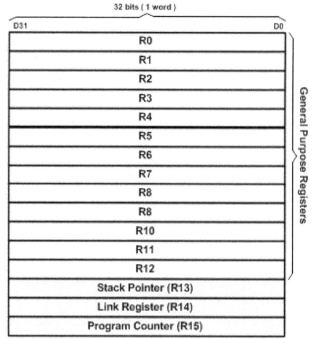

Figure 2-2: Arm Registers

MOV instruction

Simply stated, the MOV instruction copies data into register from register to register or from an immediate value. It has the following formats:

```
MOV     Rn, Op2        ; load Rn register with Op2 (Operand2)
                       ; Op2 can be an immediate value
```

Op2 can be a register Rm. Rn or Rm are any of the registers R0 to R15. Op2 can also be an immediate value.

Immediate value is a literal constant encoded in the instruction. In the Arm data processing instructions, the immediate value is an 8-bit value that can be 0–255 in decimal, (00–FF in hex). An immediate value is preceded by a '#' in the instruction.

The following instruction loads R5 with the value of R7.

```
MOV R5, R7             ; copy contents of R7 into R5 (R5 = R7)
```

The following instruction loads the R2 register with a value of 25 (decimal).

```
MOV R2, #25            ; load R2 with 25 (R2 = 25)
```

The following instruction loads the R1 register with the value 0x87 (87 in hex).

```
MOV R1, #0x87          ; copy 0x87 into R1 (R1 = 0x87)
```

Notice the order of the source and destination operands. As you can see, the MOV loads the right operand into the left operand. In other words, the destination register is written first in the instruction.

To write a comment in assembly language we use '; '. It is similar to the use of '//' in C language, which causes the remainder of the line to be ignored by the assembler. For instance, in the above examples the words after '; ' were written to explain the functionality of the instructions to the human reader, and do not have any effects on the execution of the instructions.

When programming the registers of the Arm microcontroller with an immediate value, the following points should be noted:

1. A '#' sign is written in front of an immediate value.
2. If we want to specify an immediate number in hexadecimal, a '0x' is put between '#' and the number, otherwise the number is treated as decimal. For example, in "MOV R1, #50", R1 is loaded with 50 in decimal, whereas in "MOV R1, #0x50", R1 is loaded with 50 in hex (80 in decimal).
3. Eight bits are moved into a 32-bit register, and the remaining 24 bits are loaded with all zeros. For example, in "MOV R1, #0xA5" the result will be R1 = 0x000000A5; that is, R1 = 00000000000000000000000010100101 in binary.
4. If an immediate value cannot be represented by an 8-bit value with even number bits of right rotate, the assembler will flag it as a syntax error.

LDR pseudo-instruction

We stated loading a register with MOV immediate value is limited to an 8-bit value. So the valid immediate values are limited. What do we do if we need to load a value that is not a legal immediate value of the MOV instruction? The Arm assembler provides us a pseudo-instruction of "LDR Rd, =32-bit_immediate_value" to load any 32-bit value into a register. We will examine how this pseudo-instruction works in Chapter 6. For now, just notice the '=' sign used in the syntax. The following pseudo-instruction loads R7 with 0x11223344.

```
LDR    R7, =0x11223344
```

We will use this pseudo-instruction to load 32-bit value into register extensively throughout the book. Some assembler such as Keil Arm Assembler, will replace the LDR pseudo-instruction with a MOV instruction if the immediate value fits a MOV instruction.

ADD instruction

The ADD instruction has the following format:

```
ADD    Rd, Rn, Op2 ; ADD Op2 to Rn and store the result in Rd
                    ; Op2 can be immediate value or Register Rm
```

The ADD instruction tells the CPU to add the value of Op2 to Rn and put the result into the Rd (destination) register. As we mentioned before, Op2 can be an immediate value or a register Rm. To add two numbers such as 0x25 and 0x34, one can do any of the following:

```
MOV    R1, #0x25        ; copy 0x25 into R1 (R1 = 0x25)
MOV    R7, #0x34        ; copy 0x34 into R1 (R7 = 0x34)
ADD    R5, R1, R7       ; add value R7 to R1 and put it in R5
                        ; (R5 = R1 + R7)
```

or

```
MOV    R1, #0x25         ; load (copy) 0x25 into R1 (R1 = 0x25)
ADD    R5, R1, #0x34     ; add 0x34 to R1 and put it in R5
                         ; (R5 = R1 + 0x34)
```

Executing the above lines results in R5 = 0x59 (0x59 = 0x25 + 0x34).

Figure 2-3 shows the general-purpose registers (GPRs) and the ALU in Arm. The effect of arithmetic and logic operations on the status register will be discussed in Section 2.4. In Table 2-1 you see some of the Arm ALU instructions.

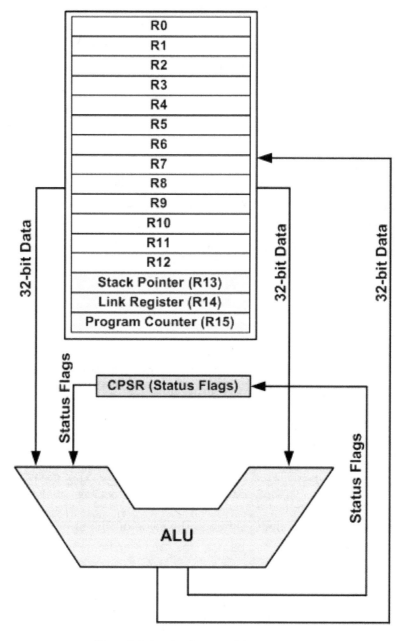

Figure 2-3: Arm Registers and ALU

Instruction		Description
ADD	Rd, Rn,Op2*	ADD Rn to Op2 and place the result in Rd
ADC	Rd, Rn,Op2	ADD Rn to Op2 with Carry and place the result in Rd
AND	Rd, Rn,Op2	AND Rn with Op2 and place the result in Rd
BIC	Rd, Rn,Op2	AND Rn with NOT of Op2 and place the result in Rd
CMP	Rn,Op2	Compare Rn with Op2 and set the status bits of CPSR**
CMN	Rn,Op2	Compare Rn with negative of Op2 and set the status bits
EOR	Rd, Rn,Op2	Exclusive OR Rn with Op2 and place the result in Rd
MVN	Rd,Op2	Store the negative of Op2 in Rd
MOV	Rd,Op2	Move (Copy) Op2 to Rd
ORR	Rd, Rn,Op2	OR Rn with Op2 and place the result in Rd
RSB	Rd, Rn,Op2	Subtract Rn from Op2 and place the result in Rd
RSC	Rd, Rn,Op2	Subtract Rn from Op2 with carry and place the result in Rd
SBC	Rd, Rn,Op2	Subtract Op2 from Rn with carry and place the result in Rd
SUB	Rd, Rn,Op2	Subtract Op2 from Rn and place the result in Rd
TEQ	Rn,Op2	Exclusive-OR Rn with Op2 and set the status bits of CPSR
TST	Rn,Op2	AND Rn with Op2 and set the status bits of CPSR
* *Op2 can be an immediate 8-bit value #K which can be 0–255 in decimal, (00–FF in hex). Op2 can also be a register Rm. Rd, Rn and Rm are any of the general-purpose registers*		
** *CPSR is discussed later in this chapter*		
*** *The instructions are discussed in detail in the next chapters*		

Table 2-1: ALU Instructions Using GPRs

SUB instruction

The SUB instruction is like ADD instruction format. It subtracts Op2 from Rn and put the result in Rd (destination).

```
SUB    Rd, Rn, Op2     ; Rd = Rn - Op2
```

To subtract two numbers such as 0x34 and 0x25, one can do the following:

```
MOV    R1, #0x34        ; load 0x34 into R1 (R1 = 0x34)
SUB    R5, R1, #0x25    ; R5 = R1 - 0x25 (R5 = 0x34 - 0x25)
```

The Special Function Registers in Arm

In Arm the R13, R14, R15, and CPSR (current program status register) registers are called *SFRs (special function registers)* since each one is dedicated to a specific function. The function of each SFR is fixed by the CPU designer at the time of design because it is used for control of the microcontroller or keeping track of specific CPU status. The four SFRs of R13, R14, R15, and CPSR play extremely important roles in the systems with Arm CPU. The R13 is set aside for stack pointer. The R14 is designated as link register which holds the return address when the CPU calls a subroutine and the R15 is the program counter (PC). The CPSR (current program status register) is used for keeping condition flags among other things, as we will see in Section 2.4. In contrast to SFRs, the General-Purpose Registers (R0-R12) do not have any specific function and are used for storing data or as a pointer to the memory.

29

Program Counter in the Arm

One of the most important register in the Arm CPU is the PC (program counter). As we mentioned earlier, the R15 is the program counter. The program counter is used by the CPU to point to the address of the next instruction to be executed. As the CPU fetches the opcode from the program memory, the program counter is incremented automatically to point to the next instruction. The more bits the program counter has, the more memory locations a CPU can access. A 32-bit program counter can access a maximum of 4 gigabytes (2^{32} = 4G) of program memory locations.

Review Questions
1. Write instructions to move the value 0x34 into the R2 register.
2. Write instructions to add the values 0x16 and 0xCD. Place the result in the R1 register.
3. True or false. No value can be moved directly into the GPRs.
4. The GPR registers in Arm are _____-bit.
5. The R13-R15 registers are called _____.
6. The SFR registers in Arm are _____ -bit.

Section 2.2: The Arm Memory Map

In this section we discuss the memory map for Arm family members.

Memory mapped I/O in the Arm

Some of the CPU designs have two distinct spaces: the I/O space and memory space. In the Arm CPU we have only one space and it is memory space and it can be as high as 4 gigabytes. The Arm uses these 4 gigabytes for both memory and I/O space. This mapping of the I/O ports to memory space is called memory mapped I/O and was discussed in Chapter 0 on the website. This 4 gigabytes of memory space can be allocated to on-chip or off-chip memory.

Memory space allocation in Arm Microcontrollers

See Figure 2-4; the memory spaces of most Arm microcontrollers have 3 on-chip sections:

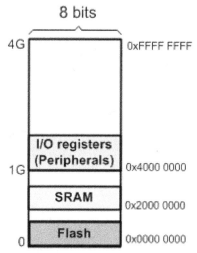

Figure 2-4: Memory Map in most Arm Microcontrollers

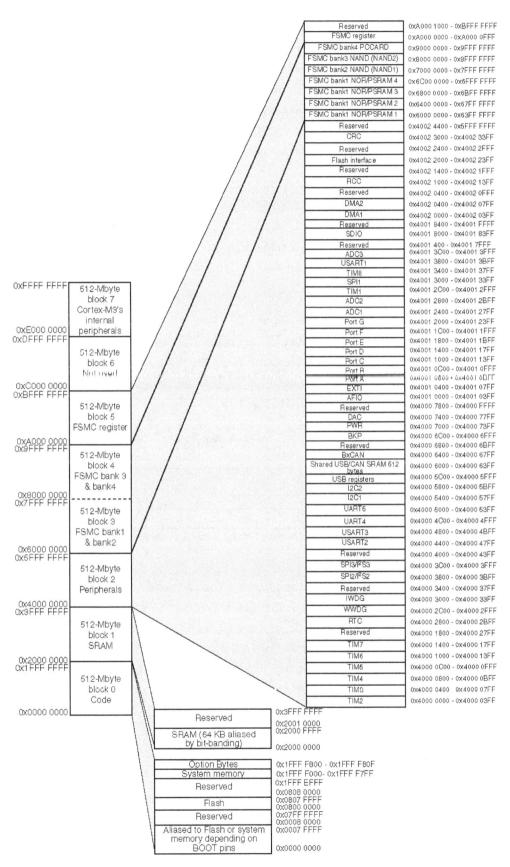

Figure 2-4B: Memory Map in STM32F103 (Copied from STM32F103xC Datasheet)

1. **I/O registers (Peripherals):** This area is dedicated to registers of peripherals such as timers, serial communication, ADC, and so on. The function and address location of each register is fixed by the chip vendor at the time of design. The number of locations set aside for registers depend on the pin numbers and peripheral functions supported by that chip. That number can vary from chip to chip even among members of the same family from the same vendor. Due to the fact that Arm does not define the type and number of I/O peripherals one must not expect to have same address locations for the peripheral registers among various devices.

2. **On-chip SRAM:** The SRAM space is used for data variables and stack and is accessed by the microcontroller instructions. The Arm microcontrollers' SRAM size ranges from 2K bytes to several thousand kilobytes depending on the chip. Even within the same family, the size of the SRAM space varies from chip to chip. Although in many of the Arm microcontrollers embedded systems the SRAM is used only for data; one can also design an Arm-based system in which the RAM is used for both data and program codes.

3. **On-chip Flash ROM:** A block of memory from a few kilobytes to megabytes is set aside for flash ROM. The flash ROM is used to store the program code. The memory can also be used for storage of static data such as text strings and look-up tables. The amount and the location of the Flash ROM space vary from chip to chip in the Arm products. See Table 2-2 and Examples 2-1 and 2-2. Figure 2-4B shows the memory map for STM32F103.

Company	Device	Flash (K Bytes)	RAM (K Bytes)
Atmel	AT91SAM7X512	512	128
ST	STM32F103RB	128	20
TI	MSP432P401RIPZ	256	64
NXP (Freescale)	KL25Z128VLK4	128	16

Table 2-2: On-chip Memory Size for some Arm Chips

Arm-based Motherboards (Case Study)

In Arm systems for Microsoft Windows, Unix, and Android operating systems the Arm motherboards use DRAM for the RAM memory, just like the x86 and Pentium PCs. As the Arm CPU is pushed into the laptop, desktop, and tablets PCs, and the high end of embedded systems products such as routers, we will see the use of DRAM as primary memory to store both the operating systems and the applications. In such systems, the Flash memory will be holding the POST (power on self-test), BIOS (basic Input/output systems) and boot programs. Just like x86 system, such systems have both on-chip and off-chip high speed SRAM for cache. Currently, there are Arm chips on the market with some on-chip Flash ROM, SRAM, and memory decoding circuitry for connection to external (off-chip) memory. This off-chip memory can be SRAM, Flash, or DRAM. The datasheets for such Arm chips provide the details of memory map for both on-chip and off-chip memories. Next, we examine the Arm buses and memory access.

Example 2-1

A given Arm chip has the following address assignments. Calculate the space and the amount of memory given to each section.

(a) Address range of 0x20000000 – 0x20007FFF for SRAM

(b) Address range of 0x00000000 – 0x0007FFFF for Flash

(c) Address range of 0xFFFC0000 – 0xFFFFFFFF for peripherals

Solution:

(a) With address space of 0x20000000 to 0x20007FFF, we have 20007FFF – 20000000 + 1 = 8000 bytes. Converting 8000 hex to decimal, we get 32,768, which is equal to 32K bytes.

(b) With address space of 0x00000000 to 0x0007FFFF, we have 7FFFF – 0 + 1= 80000 bytes. Converting 80000 hex to decimal, we get 524,288, which is equal to 512K bytes.

(c) With address space of 0xFFFC0000 to 0xFFFFFFFF, we have FFFFFFFF–FFFC0000 + 1 = 40000 bytes. Converting 40000 hex to decimal, we get 262,144, which is equal to 256K bytes.

Example 2-2

Find the address space range of each of the following memory of an Arm chip:
(a) 2 KB of peripherals starting at address 0x50000000
(b) 16 KB of SRAM starting at address 0x20000000
(c) 64 KB of Flash ROM starting at address 0x00000000

Solution:

(a) With 2K bytes of peripheral memory, we have 2048 bytes (2 × 1024 = 2048). This maps to address locations of 0x50000000 to 0x500007FF.
(b) With 16K bytes of on-chip SRAM memory, we have 16,384 bytes (16 × 1024 = 16,384), and 16,384 locations gives 0x20000000–0x20003FFF.
(c) With 64K we have 65,536 bytes (64 × 1024 = 65,536), therefore, the memory space is 0x00000000 to 0x0000FFFF.

Review Questions
   1.   True or false. I/O registers are located in memory space in Arm microcontrollers.
   2.   What is the use of Flash memory in Arm microcontrollers?

Section 2.3: Load and Store Instructions in Arm

The instructions we have used so far worked with the immediate value and the content of registers. They also used the registers as their destination. We saw simple examples of using MOV, ADD, and SUB earlier in Section 2.1. This section discusses the instructions for accessing the data memory. Since these instructions either load the register with data from memory or store the data in the register to the memory, they are called the load/store instructions.

LDR Rd, [Rx] instruction

```
LDR     Rd, [Rx]         ; load Rd with the contents of location pointed
                         ; to by Rx register. Rx contains an address between
                         ; 0x00000000 to 0xFFFFFFFF
```

The LDR instruction tells the CPU to load (read in) one word (32-bit or 4 bytes) of data from a memory location pointed to by Rx to the register Rd. Since each memory location can hold only one byte (Arm is a byte addressable CPU), and the CPU registers are 32-bit wide, the LDR will bring in 4 bytes of data from 4 consecutive memory locations. The locations can be in the SRAM, a Flash memory or I/O registers. For example, the "LDR R2, [R5]" instruction copies the contents of memory locations pointed to by R5 into register R2. Since the R2 register is 32-bit wide, it expects a 32-bit operand in the range of 0x00000000 to 0xFFFFFFFF. That means the R5 register gives the base address of the memory in which it holds the data. Therefore, if R5=0x80000, the CPU will fetch into register R2 the contents of memory locations 0x80000, 0x80001,0x80002, and 0x80003.

The following instructions loads R7 with the contents of location 0x20000200. See Figure 2-5.

```
LDR    R5,=0x40000200    ; R5 = 0x40000200
LDR    R7, [R5]          ; load R7 with the contents of memory locations
                         ; 0x20000200-0x20000203
```

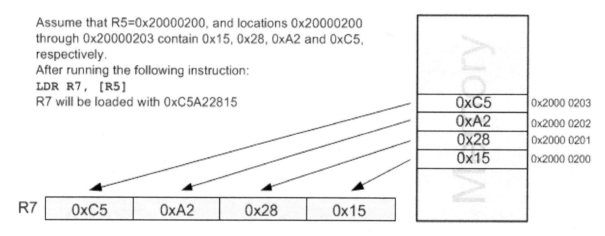

Figure 2-5: Executing the LDR Instruction

```
STR    Rx, [Rd]          ; store register Rx into locations pointed to by Rd
```

The STR instruction tells the CPU to store (copy) the contents of a CPU register to a memory location pointed to by the Rd register. Notice that the source register of STR instruction is placed before the destination register. Obviously since CPU registers are 32-bit wide (4-byte) we need four consecutive memory locations to store the contents of the register. The memory locations must be writable such as SRAM or I/O registers. See Figure 2-6. The "STR R3, [R6]" instruction will copy the contents of R3 into locations pointed to by R6, the locations 0x20000200 through 0x20000203 in the SRAM memory.

The following instruction stores the contents of R5 into locations pointed to by R1. Assume 0x40000340 is held by register R1.

```
                        ; assume R1 = 0x40000340
      STR   R5, [R1]    ; store R5 into locations pointed to by R1.
```

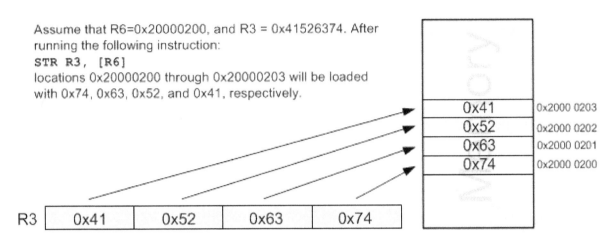

Figure 2-6: Executing the STR Instruction

LDRB Rd, [Rx] instruction

The load/store instructions can also operate on smaller data sizes by appending 'B' or 'H' to the opcode.

```
      LDRB   Rd, [Rx]   ; load Rd with the contents of the location
                        ; pointed to by Rx register.
```

The LDRB instruction tells the CPU to load (copy) one byte from a memory location pointed to by Rx into the least significant byte of Rd. After this instruction is executed, the least significant byte of Rd will have the same value as the memory location pointed to by Rx. It must be noted that the unused portion (the upper 24 bits) of the Rd register will be filled by all zeros, as shown in Figure 2-7.

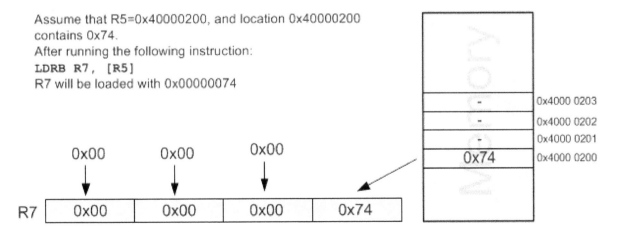

Figure 2-7: Executing the LDRB Instruction

LDR vs. LDRB

As we mentioned earlier, we can use the LDR instruction to copy the contents of four consecutive memory locations into a 32-bit register. There are situations that we do not need to bring in all 4 bytes of data. A UART register is such a case. The UART registers are generally 8-bit and take only one memory space location (memory mapped I/O). Using LDRB, we can bring into CPU register a single byte of data from UART registers. This is a widely used instruction for accessing the 8-bit peripheral ports.

STRB Rx, [Rd] instruction

```
STRB   Rx, [Rd]      ; store the byte in register Rx into
                     ; memory location pointed to by Rd
```

The STRB instruction tells the CPU to store (copy) the least significant byte of Rx to a memory location pointed to by the Rd register. After this instruction is executed, the memory locations pointed to by the Rd will have the same byte as the lower byte of the Rx, as shown in Figure 2-8.

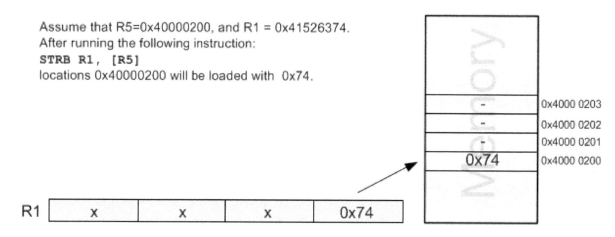

Figure 2-8: Executing the STRB Instruction

The following program first loads the R1 register with value 0x55, then stores this value into location 0x40000100:

```
LDR    R5, =0x40000100   ; R5 = 0x40000100
MOV    R1, #0x55    ; R1 = 0x55 (in hex)
STRB   R1, [R5]     ; copy R1 to location pointed to by R5
```

Example 2-3

State the contents of RAM locations 0x20000092 to 0x20000096 after the following program is executed:

```
LDR    R6, =0x20000092   ; R6 = 0x20000092
MOV    R1, #0x99    ; R1 = 0x99
STRB   R1, [R6]     ; store R1 into location pointed to by R6
                    ; (location 0x20000092)
ADD    R6, R6, #1   ; R6 = R6 + 1
MOV    R1, #0x85    ; R1 = 0x85
STRB   R1, [R6]     ; store R1 into location pointed to by R6
```

```
                       ; (location 0x20000093)
ADD    R6, R6, #1   ; R6 = R6 + 1
MOV    R1, #0x3F    ; R1 = 0x3F
STRB   R1, [R6]     ; store R1 into location pointed to by R6

ADD    R6, R6, #1   ; R6 = R6 + 1
MOV    R1, #0x63    ; R1 = 0x63
STRB   R1, [R6]     ; store R1 into location pointed to by R6

ADD    R6, R6, #1   ; R6 = R6 + 1
MOV    R1, #0x12    ; R1 = 0x12
STRB   R1, [R6]
```

Solution:

After the execution of STRB R1, [R6] data memory location 0x20000092 has value 0x99.

After the execution of STRB R1, [R6] data memory location 0x20000093 has value 0x85.

After the execution of STRB R1, [R6] data memory location 0x20000094 has value 0x3F; and so on, as shown in the chart.

Address	Data
0x20000092	0x99
0x20000093	0x85
0x20000094	0x3F
0x20000095	0x63
0x20000096	0x12

Example 2-4

State the contents of R2, R1, and memory location 0x20000020 after the following program:

```
MOV    R2, #0x5      ; load R2 with 5   (R2 = 0x05)
MOV    R1, #0x2      ; load R1 with 2  (R1 = 0x02)
ADD    R2, R1, R2    ; R2 = R1 + R2
ADD    R2, R1, R2    ; R2 = R1 + R2
LDR    R5, =0x20000020   ; R5 = 0x20000020
STRB   R2, [R5]      ; store R2 into location pointed to by R5
```

Solution:

The program loads R2 with value 5. Then it loads R1 with value 2. Then it adds the R1 register to R2 twice. At the end, it stores the result in location 0x20000020 of memory.

After MOV R2, #0x05

Location	Data
R2	5
R1	
0x20000020	

After MOV R1, #0x02

Location	Data
R2	5
R1	2
0x20000020	

After ADD R2, R1, R2

Location	Data
R2	7
R1	2
0x20000020	

After ADD R2, R1, R2

Location	Data
R2	9
R1	2
0x20000020	

After STRB [R5], R2

Location	Data
R2	9
R1	2
0x20000020	9

STR vs. STRB

As we mentioned earlier, we can use the STR instruction to copy the content of a 32-bit register into four consecutive memory locations. Some of the peripheral registers are 8-bit and take only one memory space location (memory mapped I/O). Using STRB, we can send a byte of data from register to memory location such as a peripheral register.

LDRH Rd, [Rx] instruction

```
LDRH  Rd, [Rx]      ; load Rd with the half-word pointed
                    ; to by Rx register
```

The LDRH instruction tells the CPU to load (copy) half-word (16-bit or 2 bytes) from a memory location pointed to by Rx into the lower 16-bits of Rd Register. After this instruction is executed, the lower 16-bit of Rd will have the same value as two consecutive locations in the memory pointed to by base address of Rx. It must be noted that the unused portion (the upper 16 bits) of the Rd register will be filled with all zeros, as shown in Figure 2-9.

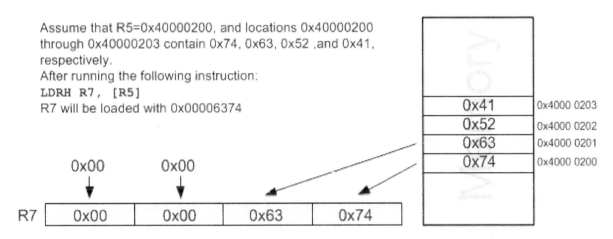

Figure 2-9: Executing the LDRH Instruction

Table 2-3 compares LDRB, LDRH, and LDR.

Data Size	Bits	Decimal	Hexadecimal	Load instruction used
Byte	8	0 – 255	0 - 0xFF	LDRB
Half-word	16	0 – 65535	0 - 0xFFFF	LDRH
Word	32	$0 - 2^{32}-1$	0 - 0xFFFFFFFF	LDR

Table 2-3: Unsigned Data Range in Arm and associated Load Instructions

STRH Rx,[Rd] instruction

```
STRH  Rx, [Rd]      ; store half-word (2-byte) in register Rx
                    ; into locations pointed to by Rd
```

The STRH instruction tells the CPU to store (copy) the lower 16-bit contents of the Rx to an address location pointed to by the Rd register. After this instruction is executed, the memory locations pointed to by the Rd will have the same value as the lower 16-bit of Rx Register. The locations are part of the data read/write memory space such as on-chip SRAM. For example, the "STRH R3,[R6]" instruction will copy the 16-bit lower contents of R3 into two consecutive locations pointed to by R6. As you can see in Figure 2-10, locations 0x20002000 and 0x20002001 of the SRAM memory will have the contents of the lower half word of R3 since R6 = 0x20002000.

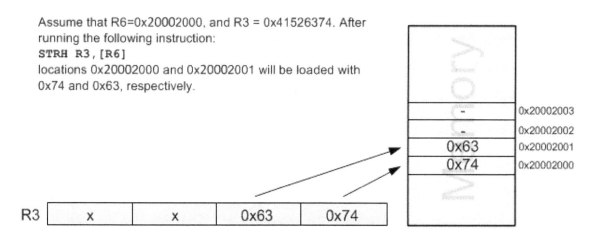

Assume that R6=0x20002000, and R3 = 0x41526374. After running the following instruction:
```
STRH R3, [R6]
```
locations 0x20002000 and 0x20002001 will be loaded with 0x74 and 0x63, respectively.

Figure 2-10: Executing the STRH Instruction

In Table 2-4 you see a comparison between STRB, STRH, and STR.

Data Size	Bits	Decimal	Hexadecimal	Load instruction used
Byte	8	0 – 255	0 - 0xFF	STRB
Half-word	16	0 – 65535	0 - 0xFFFF	STRH
Word	32	$0 – 2^{32}-1$	0 - 0xFFFFFFFF	STR

Table 2-4: Unsigned Data Range in Arm and associated Store Instructions

Review Questions
1. True or false. You can't store an immediate value directly into a memory location.
2. Write instructions to store byte value 0x95 into memory location with address 0x20.
3. Write instructions to store the content of R2 to memory location pointed to by R8.
4. Write instructions to load values from memory locations 0x20–0x23 into R4 register.
5. What is the largest hex value that can be stored in a single byte location in the data memory? What is the decimal equivalent of this value?
6. "LDR R6, [R3]" puts the result in _____.
7. What does "STRB R1, [R2]" do?
8. What is the largest hex value that can be moved into four consecutive locations in the data memory? What is the decimal equivalent of this value?

Section 2.4: Arm CPSR (Current Program Status Register)

Like all other microprocessors, the Arm has a flag register to indicate arithmetic conditions such as the carry bit. The flag register in the Arm is called the *current program status register (CPSR)*. In this section, we discuss various bits of this register and provide some examples of how it is altered. Chapters 3 and 4 show how the flag bits of the status register are used.

Arm current program status register

The status register is a 32-bit register. See Figure 2-11 for the bits of the status register. The bits N, Z, C, and V are called conditional flags, meaning that they indicate some conditions that resulted after an instruction is executed. Each of the conditional flags or the combinations of them can be used to perform a conditional execution, as we will see in Chapter 4.

31	30	29	28	27	26	25	24		19	16	15	10	9	8	7	6	5	4	3	2	1	0
N	Z	C	V	Q	IT1	IT0	J	Reser.	GE[3:0]		IT[7:2]		E	A	I	F	T	M4	M3	M2	M1	M0

Figure 2-11: CPSR (Current Program Status Register)

The following is a brief explanation of the flag bits of the current program status register (CPSR). The impact of instructions on this register is then discussed.

N, the negative flag

Binary representation of signed numbers uses D31 as the sign bit. The negative flag reflects the result of an arithmetic operation. If the D31 bit of the result is zero, then N = 0 and the result is positive. If the D31 bit is one, then N = 1 and the result is negative. The negative and V flag bits are used for the signed number arithmetic operations and are discussed in Chapter 5.

Z, the zero flag

The zero flag reflects the result of an arithmetic or logic operation. If the result is zero, then Z = 1. Therefore, Z = 0 if the result is not zero. See Chapter 4 to see how we use the Z flag for looping.

C, the carry flag

This flag is set whenever there is a carry out from the D31 bit. This flag bit is affected after a 32-bit addition or subtraction. Chapter 4 shows how the carry flag is used.

V, the overflow flag

This flag is set whenever the result of a signed number operation is too large, causing the high-order bit to overflow into the sign bit. In general, the carry flag is used to detect errors in unsigned arithmetic operations while the overflow flag is used to detect errors in signed arithmetic operations. The C and V flag bits are used for signed number arithmetic operations and are discussed in Chapter 5.

The T flag bit is used to indicate the Arm is in Thumb state. The I and F flags are used to enable or disable the interrupt.

S suffix and the status register

Most of Arm data processing instructions generate the status flags according to the result. But by default, the status flags of CPSR are not updated. If we need an instruction to update the value of status bits in CPSR, we have to put the 'S' suffix at the end of the opcode. That means, for example, ADDS instead of ADD is used.

ADD instruction and the status register

Next we examine the impact of the SUBS and ADDS instructions on the flag bits C and Z of the status register. Some examples should clarify their meanings. Although all the flag bits C, Z, V, and N are affected by the ADDS and SUBS instruction, we will focus on flags C and Z for now. The other flag bits are discussed in Chapter 5, because they relate only to signed number operations. Examine Example 2-5 to see the impact of the ADDS instruction on selected flag bits. See also Example 2-6 to see the impact of the SUBS instruction on selected flag bits.

41

Example 2-5

Show the status of the C and Z flags after the addition of

a) 0x0000009C and 0xFFFFFF64 in the following instruction:

```
; assume R1 = 0x0000009C and R2 = 0xFFFFFF64
ADDS   R2, R1, R2    ; add R1 to R2 and place the result in R2
```

b) 0x0000009C and 0xFFFFFF69 in the following instruction:

```
; assume R1 = 0x0000009C and R2 = 0xFFFFFF69
ADDS   R2, R1, R2    ; add R1 to R2 and place the result in R2
```

Solution:

a)

```
    0x0000009C        0000 0000 0000 0000 0000 0000 1001 1100
+   0xFFFFFF64      + 1111 1111 1111 1111 1111 1111 0110 0100
    0x100000000     1 0000 0000 0000 0000 0000 0000 0000 0000
```

C = 1 because there is a carry beyond the D31 bit.

Z = 1 because the R2 (the result) has value 0 in it after the addition.

b)

```
    0x0000009C      0000 0000 0000 0000 0000 0000 1001 1100
+   0xFFFFFF69    + 1111 1111 1111 1111 1111 1111 0110 1001
    0x100000005   1 0000 0000 0000 0000 0000 0000 0000 0101
```

C = 1 because there is a carry beyond the D31 bit.

Z = 0 because the R2 (the result) does not have value 0 in it after the addition. (R2=0x00000005)

Example 2-6

Show the status of the Z flag during the execution of the following program:

```
MOV    R2, #4          ; R2 = 4
MOV    R3, #2          ; R3 = 2
MOV    R4, #4          ; R4 = 4
SUBS   R5, R2, R3      ; R5 = R2 - R3 (R5 = 4 - 2 = 2)
SUBS   R5, R2, R4      ; R5 = R2 - R4 (R5 = 4 - 4 = 0)
```

Solution:

The Z flag is raised when the result is zero. Otherwise, it is cleared (zero). Thus:

After	Value of R5	Z flag
SUBS R5,R2,R3	2	0
SUBS R5,R2,R4	0	1

Not all instructions affect the flags

Some instructions affect all the four flag bits C, Z, V, and N (e.g. ADDS). But some instructions affect no flag bits at all. The branch instructions are in this category. Some instructions affect only some of the flag bits. The logic instructions (e.g. ANDS) are in this category. In general, only data processing instructions affect the status flags.

Table 2-5 shows some instructions and the flag bits affected by them. Appendix A provides a complete list of all the instructions and their associated flag bits.

Instruction	Flags Affected
ANDS	C, Z, N
ORRS	C, Z, N
MOVS	C, Z, N
ADDS	C, Z, N, V
SUBS	C, Z, N, V
B	No flags
Note that we cannot put S after B instruction.	

Table 2-5: Flag Bits Affected by Different Instructions

Flag bits and decision making

There are instructions that will make a conditional jump (branch) based on the status of the flag bits. Table 2-6 provides some of these instructions. Chapter 4 discusses the conditional branch instructions and how they are used.

Instruction	Flags Affecting the branch
BCS	Branch if C = 1
BCC	Branch if C = 0
BEQ	Branch if Z = 1
BNE	Branch if Z = 0
BMI	Branch if N = 1
BPL	Branch if N = 0
BVS	Branch if V = 1
BVC	Branch if V = 0

Table 2-6: Arm Branch (Jump) Instructions Using Flag Bits

Review Questions

1. The register holding the status flags in the Arm CPU is called the _____.

2. What is the size of the status register in the Arm?

3. Find the C and Z flag bits for the following code:

```
; assume R2 = 0xFFFFFF9F
; assume R1 = 0x00000061
ADDS   R2, R1, R2
```

4. Find the Z flag bit for the following code:

```
; assume R7 = 0x22
; assume R3 = 0x22
ADDS   R7, R3, R7
```

5. Find the C and Z flag bits for the following code:

```
; assume R2 = 0x67
; assume R1 = 0x99
ADDS   R2, R1, R2
```

Section 2.5: Arm Data Formats and Assembler Directives

In this section we look at some commonly used data formats and directives supported by the Arm assembler.

Data format representation

There are several ways to represent literal data in the Arm assembly source code. The numbers can be in hex, binary, decimal, ASCII or other formats. The following are examples of how each works using Keil Arm Assembler.

Hexadecimal numbers

To represent Hex numbers in Keil Arm assembler we put 0x (or 0X) in front of the number like this:

```
MOV   R1, #0x99
```

Here are a few lines of code that use the hex format:

```
MOV   R2, #0x75        ; R2 = 0x75
ADD   R1, R2, #0x11    ; R2 = R2 + 0x11
```

Decimal numbers

To indicate decimal numbers in some Arm assemblers such as Keil we simply use the decimal (e.g., 12) and nothing before or after it. Here are some examples of how to use it:

```
MOV   R7, #12        ; R7 = 00001100 or 0C in hex
MOV   R1, #32        ; R1 = 32 = 0x20
```

Binary numbers

To represent binary numbers in Keil Arm Assembler we put 2_ in front of the number. It is as follows:

```
MOV    R6, #2_10011001  ; R6 = 10011001 in binary or 99 in hex
```

Numbers in any base between 2 and 9

To indicate a number in any base n between 2 and 9 in Keil Arm Assembler we simply use the n_ in front of it. Here are some examples of how to use it:

```
MOV    R7, #8_33  ; R7 = 33 in base 8 or 011011 in binary format
MOV    R6, #2_10011001  ; R6 = 10011001 in base 2 or 99 in hex
```

ASCII characters

To represent ASCII data in Keil Arm Assembler we use single quotes as follows:

```
MOV    R3, #'2'   ; R3 = 00110010 or 32 in hex (See Appendix F)
```

This is the same as other assemblers such as the 8051 and x86. Here is another example:

```
MOV    R2, #'9'   ; R2 = 0x39, which is hex number for ASCII '9'
```

To represent a string, double quotes are used; and for defining ASCII strings (more than one character), we use the DCB directive which will be discussed next.

Assembler directives

In this section we look at some commonly used assembler directives supported by the Arm assembler. While instructions tell the CPU what to do, directives give directions to the assembler. For example, the MOV and ADD instructions are commands to the CPU, but EQU, END, and ENTRY are directives to the assembler. Table 2-7 shows some assembler directives.

Directive	Description
AREA	Instructs the assembler to assemble a new code or data section
END	Informs the assembler that it has reached the end of a source code.
EQU	Associate a symbolic name to a numeric constant.
INCLUDE	It adds the contents of a file to the current program.

Table 2-7: Some Widely Used Arm Directive

Note
Traditionally, pseudo-instruction and directive are treated as synonyms. But with Arm, the pseudo-instructions are translated to real instructions for CPU while directives are not.

AREA

The AREA directive tells the assembler to define a new section of memory. The memory can be code (instructions) or data and can have attributes such as READONLY, READWRITE, and so on. This is used to define one or more blocks of indivisible memory for code or data to be used by the linker. Every assembly language program has at least one AREA. The following is the format:

```
AREA            sectionname, attribute, attribute, ...
```

The following line defines a new area named MY_ASM_PROG1 which has CODE and READONLY attributes:

```
AREA  MY_ASM_PROG1, CODE, READONLY
```

Among commonly used attributes are CODE, DATA, READONLY, READWRITE, and ALIGN. The following paragraphs describe them in more details.

READONLY is an attribute given to an area of memory which can only be read from. Since it is READONLY section of the program it is by default for CODE. In Arm assembly language we use this area to write our instructions for machine code execution. All the READONLY sections of the same program are put next to each other in the flash memory by the linker.

READWRITE is an attribute given to an area of memory which can be read from and written to. Since it is READWRITE section of the program it is by default for DATA. In Arm assembly language we use this area to set aside SRAM memory for variables and stack. The linker puts all the READWRITE sections of the same program next to each other in the SRAM memory.

Note

In Keil, the memory space of **READONLY** and **READWRITE** are defined in the *Target* tabs of the **Project-Options**. Keil project wizard sets the default values according to the memory map of the chosen device.

CODE is an attribute given to an area of memory used for executable machine instructions. Since it is used for code section of the program it is by default READONLY memory. In Arm assembly language we use this area to write our instructions. The following line defines a new area for writing programs:

```
AREA  OUR_ASM_PROG, CODE, READONLY
```

DATA is an attribute given to an area of memory used for data and no instructions (machine instructions) can be placed in this area. Since it is used for data section of the program it is by default a READWRITE memory. In Arm assembly language we use this area to set aside SRAM memory for variables and stack. The following line defines a new area for defining variables:

```
AREA  OUR_VARIABLES, DATA, READWRITE
```

To define constant values in the flash memory we write the following:

```
AREA  OUR_CONSTS, DATA, READONLY
```

ALIGN is another attribute given to an area of memory to indicate how memory should be allocated according to the addresses. The ALIGN attribute of AREA has a number after like ALIGN=3 which indicates the information should be placed in memory with addresses of 2^3, that is 0x50000, 0x50008, 0x50010, 0x50018, and so on. The usage and importance of ALIGN attribute is discussed in Chapter 6.

EXPORT and IMPORT

To inform the assembler that a name or symbol will be referenced by other modules (in other files), it is marked by the **EXPORT** directive. If a module is referencing a name outside itself, that name must be declared as **IMPORT**. Correspondingly, in the module where the variable is defined, that variable must be declared as **EXPORT** in order to allow it to be referenced by other modules. The following example shows how the IMPORT and EXPORT directives are used:

```
; File1.s
; from the main program:
IMPORT MY_FUNC
...
BL      MY_FUNC        ;call MY_FUNC function
...


; File2.s (a different file)
AREA   OUR_EXAMPLE,CODE,READONLY
EXPORT MY_FUNC
IMPORT DATA1
MY_FUNC
LDR     R1,=DATA1
...
...
```

Notice that the *IMPORT* directive is used in the main procedure to show that *MY_FUNC* is defined in another module. This is needed because *MY_FUNC* is not defined in that module. Correspondingly, *MY_FUNC* is defined as *EXPORT* in the module where it is defined. *IMPORT* is used in the *MY_FUNC* module to declare that operand DATA1 has been defined in another module. Correspondingly, DATA1 is declared as EXPORT in the calling module.

END

Another pseudocode is the END directive. This indicates to the assembler the end of the source file. The END directive is the last line of the Arm assembly program, meaning that anything after the END directive in the source file is ignored by the assembler. Program 2-1 shows how the AREA and END directives are used.

Program 2-1

```
; Arm Assembly language program to add some data and store the SUM in R3.

      EXPORT  __main
      AREA    PROG_2_1, CODE, READONLY
__main
      MOV   R1, #0x25   ; R1 = 0x25
      MOV   R2, #0x34   ; R2 = 0x34
      ADD   R3, R2, R1  ; R3 = R2 + R1
HERE  B     HERE        ; stay here forever
      END
```

EQU (equate)

This is used to define a constant value or a fixed address by a name to make the program easier to read. The EQU directive does not set aside storage for a data item in the program, it merely associates an identifier with the constant value. The following code uses EQU for the counter constant, and then the constant is used to load the R2 register:

```
COUNT        EQU    0x25
 . . .        . . .    . . . .
    MOV    R2, #COUNT   ; R2 = 0x25
```

The assembler remembers the association between the word "COUNT" and the value 0x25 when it encounters the line with EQU. When it assembles the line with #COUNT, it replaces COUNT by the value 0x25. So the instruction "MOV R2, #COUNT" is converted to "MOV R2, #0x25". When executing the above instruction "MOV R2, #COUNT", the register R2 will be loaded with the value 0x25.

What are the advantages of using EQU? First, it enhances the readability. The meaning is more obvious in the word "COUNT" than the value "0x25." Furthermore, if a constant is used multiple times throughout the program, and the programmer wants to change its value everywhere. By the use of EQU, the programmer can change it once and the assembler will change all of its occurrences in the program. This allows the programmer to avoid searching the entire program trying to find and change every occurrence which is tedious and error prong.

Using EQU for special register address assignment

EQU is used to assign special function register (including peripheral registers) addresses to more readable names. This is so widely used, many manufacturers supply files with all the registers defined for the devices they make.

Examine the following code:

```
FIO2SET0     EQU    0x3FFFC058   ; PORT2 output set register 0 address
    MOV    R6, #0x01            ; R6 = 0x01
    LDR    R2, =FIO2SET0        ; R2 = 0x3FFFC058
    STRB   R6, [R2]             ; Write 0x01 to FIO2SET0
```

Each identifier may only be used by EQU once. If you try to use EQU to assign a name with a new value, an assembler error occurs.

Review Questions

1. Give an example of hex data representation in the Arm assembler.

2. Show how to represent decimal 20 in formats of (a) hex, (b) decimal, and (c) binary in the Arm assembler.

3. What is the advantage in using the EQU directive to define a constant value?

4. Show the hexadecimal value of the numbers used by the following directives:

   (a) ASC_DATA EQU '4'   (b) MY_DATA EQU 2_00011111

5. Give the value in R2 after the execution of the following instruction:

```
MYCOUNT       EQU    15
              MOV    R2, #MYCOUNT
```

6. Give the value in memory location 0x200000 after the execution of the following instructions:
```
MYCOUNT       EQU    0x95
MYMEM         EQU    0x200000
              MOV    R0, #MYCOUNT
              LDR    R2, =MYMEM
              STRB   R0, [R2]
```

7. Give the value in data memory 0x630000 after the execution of the following instructions:
```
MYDATA        EQU    12
MYMEM         EQU    0x00630000
FACTOR        EQU    0x10
              MOV    R1, #MYDATA
              MOV    R2, #FACTOR
              LDR    R3, =MYMEM
              ADD    R1 R2, R1
              STRB   R1, [R3]
```

8. Write an Arm assembly program that loads R2 and R3 with 22 and 33, respectively. Then, adds R2 to R3.

Section 2-6: Assembler data allocation directives

In most assembly languages there are some directives to allocate memory and initialize its value. In Arm assembly language DCB, DCD, and DCW allocate memory and initialize them. The SPACE directive allocates memory without initializing it.

DCB directive (define constant byte)

The DCB directive allocates a byte size memory and initializes their values.

```
MYVALUE       DCB    5                  ; MYVALUE = 5
MYMSAGE       DCB    "HELLO WORLD"       ; ASCII string
```

Each alphanumeric letter in a string is converted to its ASCII encoding value.

DCW directive (define constant half-word)

The DCW directive allocates a half-word size memory and initializes the values.

```
MYDATA        DCW    0x20, 0xF230, 5000, 0x9CD7
```

DCD directive (define constant word)

The DCD directive allocates a word size memory and initializes the values.

```
MYDATA        DCD    0x200000, 0x30F5, 5000000, 0xFFFF9CD7
```

See Tables 2-8 and 2-9.

Directive	Description
DCB	Allocates one or more bytes of memory, and defines the initial runtime contents of the memory
DCW	Allocates one or more halfwords of memory, aligned on two-byte boundaries, and defines the initial runtime contents of the memory.
DCWU	Allocates one or more halfwords of memory, and defines the initial runtime contents of the memory. The data is not aligned.
DCD	Allocates one or more words of memory, aligned on four-byte boundaries, and defines the initial runtime contents of the memory.
DCDU	Allocates one or more words of memory and defines the initial runtime contents of the memory. The data is not aligned.

Table 2-8: Some Widely Used Arm Memory Allocation Directives

Data Size	Bits	Decimal	Hexadecimal	Directive	Instruction
Byte	8	0 – 255	0 - 0xFF	DCB	STRB/LDRB
Half-word	16	0 – 65535	0 - 0xFFFF	DCW	STRH/LDRH
Word	32	$0 - 2^{32}-1$	0 - 0xFFFFFFFF	DCD	STR/LDR

Table 2-9: Unsigned Data Range in Arm and associated Instructions

In Program 2-2 you see an example of storing constant values in the program memory using the directives. Figure 2-12 shows how the data is stored in memory. In the example, the program goes from location 0x00 to 0x0F. The DCB directive stores data in addresses 0x10–0x17. As you see one byte is allocated for each data. The DCD allocates 4 bytes for each data. As a result, the lowest byte of 0x23222120 (which is 0x20) is stored in location 0x18 and the next bytes are stored in the next locations. In this order, the least significant byte of the word is stored at the lowest address and the most significant byte of the word is stored at the highest address. The ordering of bytes in a word is called "endian" and we will discuss it in more details Chapter 6.

Program 2-2: Sample of Storing Fixed Data in Program Memory

```
        EXPORT  __main
        AREA   PROG2_2, CODE, READONLY
__main
        LDR    R2, =OUR_FIXED_DATA      ; point to OUR_FIXED_DATA
        LDRB   R0, [R2]     ; load R0 with the contents
                            ; of memory pointed to by R2
                            ; Now, R0 contains 0x55
        ADD    R1, R1, R0   ; add R0 to R1
HERE    B      HERE         ; stay here forever
        AREA   LOOKUP_EXAMPLE, DATA, READONLY
OUR_FIXED_DATA
        DCB    0x55, 0x33, 1, 2, 3, 4, 5, 6
        DCD    0x23222120, 0x30
        DCW    0x4540, 0x50
        END
```

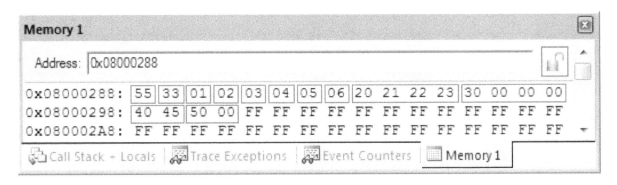

Figure 2-12: Memory Dump for Program 2-2

The DCW directive allocates 2 bytes for each data. For example, the low byte of 0x4540 is located in address 0x20 and the high byte of it goes to address 0x21. Similarly, the low byte of 0x50 is located in address 0x22 and the high byte of it in address 0x23.

In the program, to access the data, first the R2 register is loaded with the address of OUR_FIXED_DATA. In this example, OUR_FIXED_DATA has address 0x10. So, R2 is loaded with 0x10. Then, the contents of location 0x10 is loaded into register R0, using the LDRB instruction.

Notice that the ADR pseudo-instruction can also be used to load addresses into registers. For example, in Program 2-2 we can load R2 with the address of OUR_FIXED_DATA using the following pseudo-instruction:

```
ADR   R2, OUR_FIXED_DATA        ;point to OUR_FIXED_DATA
```

Strings

You can use DCB together with double quotations to store strings. The following snippet of code, stores "Hello World!" in the flash memory:

```
       AREA MY_STRINGS, CODE, READONLY
MY_MSG DCB  "Hello World!"
```

SPACE directive

Using the SPACE directive, we can allocate memory for variables without initial values. The following lines allocate 4 and 2 bytes of memory and name them as LONG_VAR and OUR_ALFA:

```
LONG_VAR    SPACE      4     ; Allocate 4 bytes
OUR_ALFA    SPACE      2     ; Allocate 2 bytes
```

The following snippet of code allocates 400 bytes of memory. Since the AREA is defined as READWRITE, the space is located in SRAM:

```
       AREA MYSTACK, DATA, READWRITE
       SPACE 400
```

In the following program, three variables are defined: A, B, and C. A and B are initialized with values 5 and 4, respectively. In the next step A and B are added together and the result is stored in C:

Program 2-3

```
        EXPORT  __main
        AREA OUR_PROG, CODE, READONLY
__main
        ; A = 5
        LDR   R0, =A        ; R0 = Addr. of A
        MOV   R1, #5        ; R1 = 5
        STR   R1, [R0]      ; init. A with 5

        ; B = 4
        LDR   R0, =B        ; R0 = Addr. of B
        MOV   R1, #4        ; R1 = 4
        STR   R1, [R0]      ; init. B with 4

        ; R1 = A
        LDR   R0, =A        ; R0 = Addr. of A
        LDR   R1, [R0]      ; R1 = value of A

        ; R2 = B
        LDR   R0, =B        ; R0 = Addr. of A
        LDR   R2, [R0]      ; R2 = value of A

        ; C = R1 + R2 (C = A + B)
        ADD   R3, R1, R2   ; R3 = A + B
        LDR   R0, =C       ; R0 = Addr. of C
        STR   R3, [R0]     ; C = R3

        loop  B    loop

        AREA  OUR_DATA, DATA, READWRITE
        ; Allocates the followings in SRAM memory
A       SPACE 4
B       SPACE 4
C       SPACE 4
        END
```

ALIGN

This is used to make sure data is aligned on the 32-bit word or 16-bit half word address boundary. The following uses ALIGN to make the data 32-bit word aligned:

```
ALIGN 4       ; the next instruction is word (4 bytes) aligned
...
ALIGN 2       ; the next instruction is half-word (2 bytes) aligned
...
```

Example 2-7 shows the result of using the ALIGN directive.

Example 2-7

Compare the result of using ALIGN in the following programs:

a)

```
        AREA   E2_7A, READONLY, CODE
__main
        ADR    R2, DTA
        LDRB   R0, [R2]
        ADD    R1, R1, R0
H1      B      H1

DTA     DCB    0x55
        DCB    0x22
        END
```

b)

```
        AREA   E2_7B, READONLY, CODE
__main
        ADR    R2, DTA
        LDRB   R0, [R2]
        ADD    R1, R1, R0
H1      B      H1

DTA     DCB    0x55
        ALIGN 2
        DCB    0x22
        END
```

c)

```
        AREA   E2_7C, READONLY, CODE
__main
        ADR    R2, DTA
        LDRB   R0, [R2]
        ADD    R1, R1, R0
H1      B      H1

DTA     DCB    0x55
        ALIGN 4
        DCB    0x22
        END
```

Solution:

a) When there is no ALIGN directive the DCB directive allocates the first empty location for its data. In this example, address 0x10 is allocated for 0x55. So 0x22 goes to address 0x11.

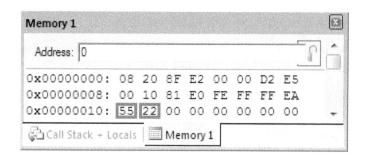

b) In the example the ALIGN is set to 2 which means the data should be put in a location with even address. The 0x55 goes to the first empty location which is 0x10. The next empty location is 0x11 which is not a multiple of 2. So, it is filled with 0 and the next data goes to location 0x12.

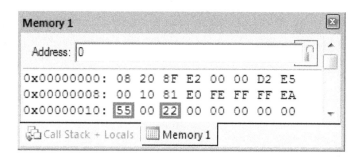

c) In the example the ALIGN is set to 4 which means the data should go to locations whose address is multiple of 4. The 0x55 goes to the first empty location which is 0x10. The next empty locations are 0x11, 0x12, and 0x13 which are not a multiple of 4. So, they are filled with 0s and the next data goes to location 0x14.

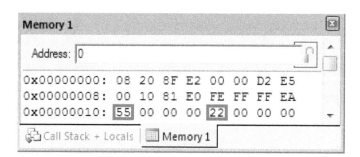

Rules for labels in assembly language

By choosing label names that are meaningful, a programmer can make a program much easier to read and maintain. There are several rules that label names must follow. First, each label name must be unique in the file. The names used for labels in assembly language programming consist of alphabetic letters in both uppercase and lowercase, the digits 0 through 9, and the special character underscore '_'. The first character of the label must be an alphabetical letter or underscore and cannot be a numeral. Every assembler has some reserved words that must not be used as labels in the program. Foremost among the reserved words are the mnemonics for the instruction opcodes and the directives. For example, "MOV" and "ADD" are reserved because they are instruction mnemonics. Check your assembler manual for the list of reserved words.

Review Questions
1. True or false. DCB allocates a byte of memory.
2. True or false. DCB cannot initialize memory.
3. True or false. SPACE allocates memory.

Section 2.7: Introduction to Arm Assembly Programming

In this section we discuss assembly language format and define some widely used terminology associated with assembly language programming.

While the CPU can work only in binary, it can do so at a very high speed. It is quite tedious and slow for humans, however, to deal with 0s and 1s in order to program the computer. A program that consists of 0s and 1s is called machine language. In the early days of the computer, programmers coded programs in machine language. Although the octal or hexadecimal system was used as a more efficient way to represent binary numbers, the process of working in machine code was still cumbersome for humans. Eventually, assembly languages were developed, which provided mnemonics for the machine code instructions, plus other features that made programming easier and less prone to error. The term mnemonic is frequently used in computer science and engineering literature to refer to codes and abbreviations that are relatively easy to remember. Assembly language programs must be translated into machine code by a program called assembler. Assembly language is referred to as a low-level language because it deals directly with the internal structure of the CPU. To program in assembly language, the programmer must know all the registers of the CPU and the size of each, as well as other details.

Today, one can use many different programming languages, such as C, C++, Java, Python, and numerous others. These languages are called *high-level* languages because the programmer does not have to be concerned with the internal details of the CPU. Whereas an assembler is used to translate an assembly language program into machine code, high-level languages are translated into machine code by a program called a compiler. For instance, to write a program in C, one must use a C compiler to translate the program into machine language.

Next we look at the Arm assembly language format.

Structure of assembly language

An assembly language program consists of, among other things, a series of lines of assembly language instructions. An assembly language instruction consists of a mnemonic of opcode, optionally followed by one, two or three operands. The operands are the data items being manipulated, and the opcodes are the commands to the CPU, telling it what to do with the operands. See Program 2-4.

Program 2-4: Sample of an Arm Assembly Language Program

```
; Arm Assembly language program to add some data and store the SUM in R3.
```

```
        EXPORT  __main
        AREA    PROG_2_4, CODE, READONLY
__main
        MOV     R1, #0x25    ; R1 = 0x25
        MOV     R2, #0x34    ; R2 = 0x34
        ADD     R3, R2, R1   ; R3 = R2 + R1
HERE    B       HERE         ; stay here forever
        END
```

In addition to the instructions, an assembly language program contains directives. While instructions tell the CPU what to do, directives give directions to the assembler. For example, in Program 2-4, the MOV and ADD instructions are commands to the CPU, AREA and END are directives to the assembler.

An assembly language instruction consists of four fields:

[label]   opcode   [operands]   [; comment]

Brackets indicate that a field is optional and not all lines have them. Brackets should not be typed in. Regarding the above format, the following points should be noted:

1. The label field allows the program to refer to the address of a line of code by name.

2. The assembly language opcode and operand(s) fields together perform the real work of the program and accomplish the tasks for which the program was written for. In assembly language statements such as

        MOV   R3, #0x55
        MOV   R2, #0x67
        ADD   R2, R2, R3   ; R2 = R2 + R3

    ADD and MOV are the mnemonics of the opcodes; the "0x55" and "0x67" are the operands.

3. Instead of instructions, the program may contain directives. The following line is an assembly directive that tells the assembler that the following lines are for program instructions.

        AREA   PROG_2_4, CODE, READONLY

4. The comment field begins with a semicolon comment indicator "; ". Comments may be at the end of a line or on a line by themselves. The assembler ignores comments, but they are indispensable to programmers. Although comments are optional, it is recommended that they be used to describe the program in a way that makes it easier for someone else to read and understand.

5. Notice the label "HERE" in the label field in Program 2-4. In the B (Branch) statement the Arm is told to stay in this loop indefinitely.

> **Note!**
>
> The first column of each line is always considered as label. Thus, be careful to press a Tab at the beginning of each line that does not have a label; otherwise, your instruction is considered as a label and an error message will appear when compiling.

Review Questions

1. What is the purpose of assembler directives?
2. _____ are translated by the assembler into machine code, whereas _____ are not.
3. True or false. Assembly language is a high-level language.
4. Which of the following instructions produces machine code? List all that do.
   (a) MOV R6, #0x25 (b) ADD R2, R1, R3 (c) END (d) HERE B HERE
5. True or false. Assembler directives are not used by the CPU itself. They are simply a guide to the assembler.
6. In Question 4, which one is an assembler directive?

Section 2.8: Creating an Arm Assembly Program

Now that the basic form of an Assembly language program has been given, the next question is: How it is created, assembled, and made ready to run? The steps to create an executable assembly language program (Figure 2-13) are outlined as follows:

1. First we use a text editor to type in a program similar to Program 2-4. In the case of Arm, we can use the Keil IDE, which has a text editor, assembler, simulator, debugger, and much more all in one software package. It is an excellent development software that supports all the Arm chips. A free version with 32k byte limit is available at www.keil.com. Many editors or word processors also can be used to create or edit the program. A widely used editor is the Notepad in Windows, which comes with all Microsoft operating systems. Notice that the editor must be able to produce an ASCII file. For assemblers, the file names follow the usual DOS conventions, but the source file should have the extension ".s", ".a" or ".asm". The ".asm" extension for the source file is used by an assembler in the next step.

2. The ".asm" source file containing the program code created in step 1 is fed to the Arm assembler. The assembler produces an object file, and a listing file. The object file has the extension ".o", and the listing file has ".lst" extension.

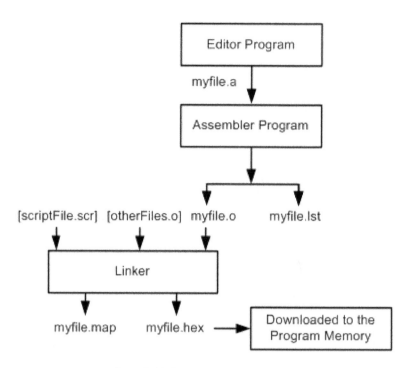

Figure 2-13: Steps to Create a Program

3. The object file plus a linker script file are used by the linker to produce the map file and the memory image files. The map file has the extension ".map"

4. The memory image file contains the binary code that can be downloaded to the target device or the simulator for execution. By default, Keil IDE produces a ".axf" file that contains the binary code and the symbols for debugger. Optionally, an Intel hex format file may be produced. The hex file has ".hex" extension. The linker script file (or scatter file in Keil) is optional and can be replaced by some command line options. After a successful link, the hex file is ready to be burned into the Arm processor's program FLASH memory and is downloaded into the Arm chip.

More about asm and object files

The assembler converts the assembler source file's assembly language instructions into machine code and provides the ".o" (object) file. The object file, as mentioned earlier, has a ".o" as its extension. The object file is used as input to the linker.

Before we can assemble a program to create a ready-to-run program, we must make sure that it is free of syntax errors. The Keil uVision IDE provides us error messages and we examine them to see the location and nature of the syntax error. The assembler will not assemble the program until all the syntax errors are fixed. A sample of an error message is shown in Figure 2-14.

```
Build target 'Target 1'
assembling a1.asm...a1.asm(7): error: A1163E: Unknown opcode MOVE, expecting opcode or Macro
Target not created
```

Figure 2-14: Sample of an Error Message

"lst" and "map" files

The map file shows the labels defined in the program together with their values. Examine Figure 2-15. It shows the Map file of Program 2-4.

The lst (listing) file shows the source code and the machine code; it also shows which instructions are used in the source code, and the amount of memory the program uses. See Figure 2-16.

These files can be accessed by a text editor such as Notepad and displayed on the monitor, or sent to the printer to get a hard copy. The programmer uses the listing and map files to help debugging the program.

There are many different Arm assemblers available for evaluation nowadays. If you use the Windows operating system, IAR IDE and Keil uVision can be used. They have great features and nice environments. GNU assembler is popular because it if free and open source. It is also generic and can be used for most of the processors in the market. The drawbacks of GNU assembler are that its syntax is slightly different from the other Arm assembler and it takes some effort to set up the toolchain.

```
Memory Map of the image

  Image Entry point : 0x080000ed

  Load Region LR_1 (Base: 0x08000000, Size: 0x0000028c, Max: 0xffffffff, ABSOLUTE)

    Execution Region ER_RO (Exec base: 0x08000000, Load base: 0x08000000, Size: 0x0000028c,
Max: 0xffffffff, ABSOLUTE)

    Exec Addr     Load Addr     Size          Type    Attr    Idx   E Section Name      Object

    0x08000000    0x08000000    0x000000ec    Data    RO       8     RESET     startup_stm32f10x_md.o
    0x080000ec    0x080000ec    0x00000040    Code    RO       9   * .text     startup_stm32f10x_md.o
    0x08000134    0x08000134    0x0000000e    Code    RO       1     PROG_2_4            main.o

    Execution Region ER_RW (Exec base: 0x20000000, Load base: 0x0800028c, Size: 0x00000000,
Max: 0xffffffff, ABSOLUTE)

    **** No section assigned to this execution region ****


    Execution Region ER_ZI (Exec base: 0x20000000, Load base: 0x0800028c, Size: 0x00000600,
Max: 0xffffffff, ABSOLUTE)

    Exec Addr     Load Addr     Size        Type    Attr    Idx   E Section Name      Object

    0x20000000       -          0x00000200  Zero    RW       7     HEAP      startup_stm32f10x_md.o
    0x20000200       -          0x00000400  Zero    RW       6     STACK     startup_stm32f10x_md.o


===========================================================================================
```

Figure 2-15: Sample of a Map File

```
Arm Macro Assembler        Page 1
 1  00000000                      ; Arm Assembly language program to add some data and store the SUM in
                                  R3.
 2  00000000
 3  00000000                      EXPORT  __main
 4  00000000                      AREA    PROG_2_4, CODE, READONLY
 5  00000000            __main
 6  00000000  F04F 0125           MOV     R1, #0x25       ; R1 = 0x25
 7  00000004  F04F 0234           MOV     R2, #0x34       ; R2 = 0x34
 8  00000008  EB02 0301           ADD     R3, R2,R1       ; R3 = R2 + R1
 9  0000000C  E7FE       HERE     B       HERE            ; stay here forever
10  0000000E                      END
```

Figure 2-16: Sample of a List File for Arm

Review Questions

1. True or false. The editor of Keil IDE and Windows Notepad text editor both produce an ASCII file.

2. True or false. The extension for the assembly program source file may be ".a".

3. Which of the following files is usually produced by a text editor?

    (a) myprog.asm (b) myprog.obj (c) myprog.hex (d) myprog.lst

4. Which of the following files is produced by an assembler?

    (a) myprog.asm (b) myprog.obj (c) myprog.hex (d) myprog.lst

Section 2.9: The Program Counter and Program Memory Space in the Arm

In this section we discuss the role of the program counter (PC) in executing a program and show how the code is fetched from ROM and executed. We will also discuss the program (code) memory space for various Arm family members.

Program counter in the Arm

The most important register in Arm is the PC (program counter). As we mentioned earlier, register R15 is the program counter in Arm CPU. The program counter is used by the CPU to point to the address of the next instruction to be executed. As the CPU fetches the opcode from the program memory, the program counter is incremented automatically to point to the next instruction.

The program counter in the Arm family is 32 bits wide. This means that the Arm family can access addresses 00000000 to 0xFFFFFFFF, a total of 4 gigabytes of memory space locations.

Power up location for Arm

One question that we must ask about any microcontroller (or microprocessor) is: "at what address does the CPU wake up to when power is applied or when the CPU is reset?" Each microprocessor is different. In the case of the Cortex-M microcontrollers, when the CPU is powered up or reset, the PC (program counter) is loaded with the contents of memory location 0x00000004. For this reason, the address of the first instruction must be burned into memory location 0x00000004 of program ROM. As, we will see in Chapter 4, in the startup file, the location 4 is loaded with the address of a subroutine (named Reset_Handler) which initializes the Arm CPU and then branches to __main. Next, we discuss the step-by-step action of the program counter in fetching and executing a sample program.

Placing code in program ROM

To get a better understanding of the role of the program counter in fetching and executing a program, we examine the action of the program counter as each instruction is fetched and executed. First, we examine once more the listing file of the sample program and show how the code is placed into the Flash ROM of the Arm chip. As we can see in Figure 2-16, the machine code for each instruction is listed in the third column of the listing file and the address offset of each instruction is in the second column. The address offsets are given based on the __main label. For example, the listing shows that address offset 0x00000000 contains "F04F 0125", which is the machine code for moving an immediate value (in this case 0x25) into a register (in this case R1). Therefore, the instruction "MOV R1, #0x25" has a machine code of "F04F 0125". See Figure 2-16. Similarly, the machine code "F04F 0234" is located in location 0x00000004 and represents the opcode and the operands for the instruction "MOV R2, #0x34". In the same way, machine code "EB02 0301" is located in address offset 0x00000008 and represents the opcode and the operand for the instruction "ADD R3, R1, R2". The opcode for "B HERE" and its target address offset are located in location 0x0000000C.

Executing a program instruction by instruction

Assuming that the above program is burned into the ROM of an Arm chip, the following is a step-by-step description of the action of the Arm upon applying power to it:

1. When the Arm is powered up, the CPU reads the contents of locations 0x000000004 to 0x00007 and loads to the PC (program counter). So, the PC is loaded with the address of the subroutine which is in the startup file and the CPU executes the subroutine and then there is a branch to __main and the PC is loaded with the address of __main and the CPU starts to fetch the first instruction from location __main of the program ROM. In the case of the above program, the first machine code in the __main routine is "F04F 0125", which is the code for moving operand 0x25 to R1. Upon executing the code, the CPU places the value of 25 in R1. Now one instruction is finished. The program counter is already incremented by 0x00000004, which contains code "F04F 0234", the machine code for the instruction "MOV R2, #0x34".

2. Upon executing the machine code "F04F 0234", the value 0x34 is loaded to R2. The program counter is incremented by 0x00000004.

3. The next location has the machine code for instruction "ADD R3, R2, R1". This instruction is executed and the PC is incremented by 4.

4. PC points to the next instruction, which is "HERE B HERE". After the execution of this instruction, although the PC is incremented by 2 (the branch instruction is 2-byte long), the execution of the instruction loads PC with the address of the B instruction and the B instruction is executed infinitely. This keeps the program in an infinite loop. More on branch instructions will be in Chapter 4.

The actual steps of running a code in Arm is slightly different from what mentioned above because of the use of pipeline in Arm architecture. We will examine pipelines later in Section 2-11.

Review Questions

1. In the Arm, the program counter is _____ bits wide.
2. True or false. Every member of the Arm family wakes up at memory 0x00000008 when it is powered up.
3. At what ROM location do we store the address of the first opcode of a Cortex-M program?

Section 2.10: Some Arm Addressing Modes

The various ways operands are specified in the instruction are called addressing modes. In the narrower definition, it is the way CPU generates address from instruction to read/write the operands in the memory. But the term addressing mode is used to cover a broader definition including the operands that are not in the memory. With the RISC architecture, the destinations of all Arm instructions are always a register except the "store" instructions, which is a mirror of "load."

Using advanced addressing modes to access different data types and data structures (e.g. arrays, pointers) are discussed in Chapter 6. Some of the simple Arm addressing modes are:

1. register
2. immediate
3. register indirect (indexed addressing mode)

Register addressing mode

The register addressing mode involves the use of registers to hold the data to be manipulated. Memory is not accessed when this addressing mode is executed; therefore, it is relatively fast. See Figure 2-17.

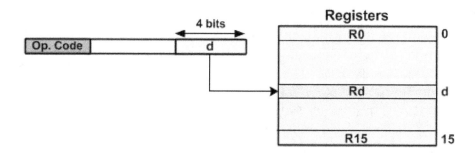

Figure 2-17: Register Addressing Mode

Examples of register addressing mode are as follow:

```
MOV    R6, R2        ; copy the contents of R2 into R6
ADD    R1, R1, R3    ; add the contents of R3 to contents of R1
SUB    R7, R7, R2    ; subtract R2 from R7
```

Immediate addressing mode

In the immediate addressing mode, the source operand is a literal constant. In immediate addressing mode, as the name implies, when the instruction is assembled, the operand comes immediately after the opcode in the encoding of the instruction. For this reason, this addressing mode executes quickly. See Figure 2-18. Examples:

```
MOV   R9, #0x25        ; move 0x25 into R9
MOV   R3, #62          ; load the decimal value 62 into R3
ADD   R6, R6, #0x40    ; add 0x40 to R6
```

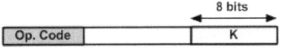

Figure 2-18: Immediate Addressing Mode

In the first two addressing modes, the operands are either inside the CPU or tagged along with the instruction, which is fetched into the CPU before the instruction is executed. In most programs, the data to be processed are originally in some memory location outside the CPU. There are many ways of accessing the data in the memory space. The following describes one of the methods. We will discuss more ways of accessing data memory in Chapter 6.

Register Indirect Addressing Mode (Indexed addressing mode)

In the register indirect addressing mode, the address of the memory location where the operand resides is held by a register. See Figure 2-19. For example.

```
STR   R5, [R6]      ; write the content of R5 into the memory location
                    ; pointed to by R6
LDR   R10, [R3]     ; load into R10 the content of the
                    ; memory location pointed to by R3.
```

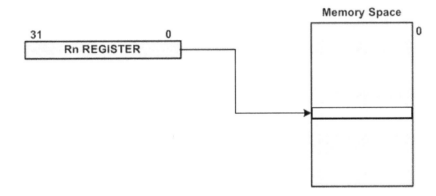

Figure 2-19: Register Indirect Addressing Mode

Using register indirect addressing mode, we can implement the different pointers. Since the registers are 32-bit they can address the entire memory space. Here you see a simple code in C and its equivalent in Assembly:

C Language:

```
char *ourPointer;
ourPointer = (char*) 0x12456; //Point to location 12456
*ourPointer = 25;  //store 25 in location 0x12456
ourPointer ++;    //point to next location
```

Assembly Language:
```
LDR     R2, =0x12456  ; point to location 0x12456
MOV     R0, #25    ; R0 = 25
STRB    R0, [R2]   ; store R0 in location 0x12456
ADD     R2, R2, #1  ; increment R2 to point to next location
```

Depending on the data type that the pointer points to, STR/LDR, STRH/LDRH, or STRB/LDRB might be used. In the above example, since it points to char (which is 8-bit) STRB is used.

Review Questions

1. Can the Arm programmer make up new addressing modes?
2. Which registers can be used for the register indirect addressing mode?
3. Where is the data located in immediate addressing mode?

Section 2.11: Pipelining and Harvard Architecture in Arm

There are three ways available to microprocessor designers to increase the processing power of the CPU:

1. Increase the clock frequency of the chip: Some drawbacks of this method are that the higher the frequency, the more power consumption and more heat dissipation. Power consumption is especially a problem for portable devices.

2. Use Harvard architecture by increasing the number of buses to bring more information (code and data) into the CPU to be processed concurrently. As we will see in this section, the new Arm chips, including the Cortex series, have Harvard architecture.

3. Change the internal architecture of the CPU and use Pipelining and the RISC architecture.

Arm has used all three methods to increase the processing power of the Arm microcontrollers.

Pipelining

In early microprocessors such as the 8085 or 6800, the CPU could either fetch or execute at a given time. In other words, the CPU had to fetch an instruction from memory, decode, and then execute it, and then fetch the next instruction, decode and execute it, and so on as shown in Figure 2-20. All steps of running a program occur serially. The idea of pipelining in its simplest form is to allow the CPU to fetch and execute at the same time. That is an instruction is being fetched while the previous instruction is being executed.

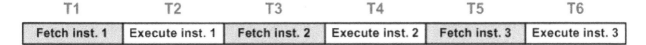

Figure 2-20: Non-pipeline execution

We can use a pipeline to speed up execution of instructions. In pipelining, the process of executing instructions is split into small steps that are executed in parallel. In this way, the executions of many instructions are overlapped. One limitation of pipelining is that the speed of execution is limited to the slowest stage of the pipeline. Compare this to making pizza. You can split the process of making pizza into many stages, such as flattening the dough, putting on the toppings, and baking, but the process is limited to the slowest stage, baking, no matter how fast the rest of the stages are performed.

Arm multistage execution pipeline

As shown in Figure 2-21, in the Arm Cortex-M (except Cortex-M0), each instruction is executed in 3 stages: Fetch, Decode, and Execute.

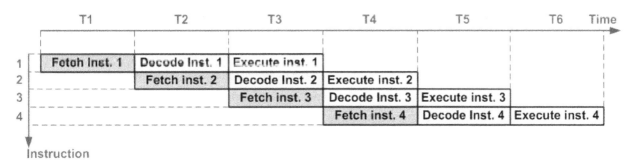

Figure 2-21: Pipeline in Arm

In step 1, the opcode is fetched. In step 2, the opcode is decoded. In step 3, the instruction is executed and result is written into the destination register.

Harvard Architecture

In Chapter 0, we discussed Harvard and Von Neumann architecture. Cortex-M (except Cortex-M0) uses Harvard architecture, which means that there are separate buses for the code and the data memory. See Figure 2-22. The Harvard architecture feeds the CPU with both code and data at the same time via two sets of buses, one for code and one for data. This lets the CPU fetch the next instruction while accessing the memory using the LDR/STR instructions.

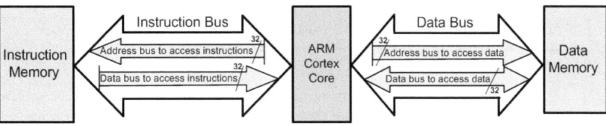

Figure 2-22: Harvard Architecture

In Sections 2-2 and 2-3, we learned about data memory space and how to use the STR and LDR instructions. When the CPU wants to execute the "LDR Rd, [Rx]" instruction, it puts the value of Rx on the address bus of the data bus, and receives data through the data bus. For example, to execute "LDR R2, [R5]", assuming that R5 = 0x20000200, the CPU puts 0x20000200 on the address bus. The location 0x20000200 is in the SRAM (See Figure 2-4). Thus, SRAM puts the contents of location 0x20000200 on the data bus. The CPU gets the contents of location 0x20000200 through the data bus and puts it in R2.

Review Questions
1. True or false. Cortex-M series use a 3-stage pipeline.
2. True or false. Cortex-M uses von Neumann architecture.
3. True or false. Cortex-M series (except Cortex-M0) use Harvard architecture.
4. True or false. Harvard architecture uses the same address and data buses to fetch both code and data.

Section 2.12: RISC Architecture in Arm

In the early 1980s, a controversy broke out in the computer design community, but unlike most controversies, it did not go away. Since the 1960s, in all mainframes and minicomputers, designers put as many instructions as they could think of into the CPU. Some of these instructions performed complex tasks like string operations. Naturally, microprocessor designers followed the lead of minicomputer and mainframe designers. Because these processors used such a large number of instructions, many of which performed highly complex activities, they came to be known as CISC (complex instruction set computer) processors. According to several studies in the 1970s, many of these complex instructions etched into CPUs were never used by programmers and compilers. The huge cost of implementing a large number of instructions (some of them complex) into the microprocessor, plus the fact that a good portion of the transistors on the chip are used by the instruction decoder, made some designers think of simplifying and reducing the number of instructions. As this concept developed, the resulting processors came to be known as RISC (reduced instruction set computer).

Features of RISC
The following are some of the features of RISC as implemented by the Arm microcontroller.

Feature 1
RISC processors have a fixed instruction size. In a CISC microprocessors such as the x86, instructions can be 1, 2, 3, or even 5 bytes. For example, look at the following instructions in the x86:

```
CLR    C                      ; clear Carry flag, a 1-byte instruction
ADD    Accumulator, #mybyte   ; a 2-byte instruction
LJMP   target_address         ; a 5-byte instruction
```

This variable instruction size makes the task of the instruction decoder very difficult because the size of the incoming instruction is never known. In a RISC architecture, the size of all instructions is fixed. Therefore, the CPU can decode the instructions quickly. This is like a bricklayer working with bricks of the same size as opposed to using bricks of variable sizes.

But fixed-size instruction has one drawback as well; the program memory usage is not optimized. In a 32-bit RISC CPU all the instructions are 32-bit and if not all the 32 bits are needed to form the instruction it fills with zeros. In the original Arm all the instructions were 32-bit. To decrease the memory size, Arm introduced the Thumb 16-bit instruction set. Thumb programs use less memory but since the Thumb instruction set is very limited compared to the Arm instruction set, the Thumb programs have lower performance. In the next step, Arm introduced Thumb2. In Thumb2, instructions can be 16-bit and 32-bit and it covers most of the features of the original Arm. So, the performance is almost the same as the original Arm while the program space is optimized.

Feature 2

One of the major characteristics of RISC architecture is a large number of registers. All RISC architectures have at least 8 or 16 registers. Of these 16 registers, only a few are assigned to dedicated functions. One advantage of a large number of registers is that it avoids the need for a large stack to store temporary data. Accessing data on the stack is a memory read/write and is much slower than CPU register access. Although a stack is implemented on a RISC processor, it is not as essential as in CISC because so many registers are available. In Arm the use of a large number of general purpose registers satisfies this RISC feature. The stack for the Arm is covered in Chapter 4.

Feature 3

RISC processors have a smaller instruction set. RISC processors have only basic instructions such as ADD, SUB, MUL, LOAD, STORE, AND, ORR, EOR, CALL, B, and so on. The limited number of instructions is one of the criticisms leveled at the RISC processor because it makes the task of assembly language programmers much more tedious and difficult compared to CISC assembly language programming. It is interesting to note that some defenders of CISC have called it "complete instruction set computer" instead of "complex instruction set computer" because it has a complete set of every kind of instruction. How many of these instructions are used and how often is another matter. In the recent years, almost all the new programs are written in high level languages such as C or Java. The advantage of CISC in this regard is no longer valid. The limited number of instructions in RISC leads to programs that are larger. Although these programs can use more memory, this is not a problem because memory is cheaper. Before the advent of semiconductor memory in the 1960s, however, CISC designers had to pack as much action as possible into a single instruction to get the maximum bang for their buck. In the Arm we have around 50 instructions. We will examine more of the instruction set for the Arm in future chapters.

Feature 4

At this point, one might ask, with all the difficulties associated with RISC programming, what is the gain? The most important characteristic of the RISC processor is that more than 99% of instructions are executed with only one clock cycle because the instructions are much simpler, in contrast to CISC instructions which take various number of clock cycles to execute. Even some of the 1% of the RISC instructions that are executed with two clock cycles can be executed with one clock cycle by juggling instructions around (code scheduling).

Feature 5

Because CISC has such a large number of instructions, each with so many different addressing modes, microinstructions (microcode) are used to implement them. The implementation of microinstructions inside the CPU employs more than 40–60% of transistors in many CISC processors. RISC instructions, however, due to the small set of instructions, are implemented using the hardwire method. Hardwiring of RISC instructions takes no more than 10% of the transistors. With much smaller circuit, the RISC processor consumes much less power. This is a major reason Arm processor is used in majority of the portable devices like cellphone or tablet.

Feature 6

RISC uses load/store architecture. In CISC microprocessors, data can be manipulated while it is still in memory. For example, in instructions such as "ADD Reg, Memory", the microprocessor must bring the contents of the external memory location into the CPU, add it to the contents of the register, then move the result back to the external memory location. The problem is there might be a delay in accessing the data from external memory then the whole process would be stalled, preventing other instructions from proceeding in the pipeline. In RISC, designers did away with these kinds of instructions. In RISC, instructions can only load from external memory into registers or store registers into external memory locations. There is no direct way of doing arithmetic and logic operations between a register and the contents of external memory locations. All these instructions must be performed by first bringing both operands into the registers inside the CPU, then performing the arithmetic or logic operation, and then sending the result back to memory. This idea was first implemented by the Cray 1 supercomputer in 1976 and is commonly referred to as load/store architecture. In the last section, we saw that the arithmetic and logic operations are between the GPRs registers, but none involves a memory location. For example, there is no "ADD R1, RAM-Loc" instruction in Arm. Operating only on the CPU registers guarantees that the memory bus contention will not slow down the instruction execution.

In concluding this discussion of RISC processors, it is interesting to note that RISC technology was explored by the scientists at IBM in the mid-1970s, but it was David Patterson of the University of California at Berkeley who in 1980 brought the merits of RISC concepts to the attention of computer scientists. It must also be noted that in recent years CISC processors such as the Pentium have used some RISC features in their design. This was the only way they could enhance the processing power of the x86 processors and stay competitive. Of course, they had to add circuits in the CPU to translate the x86 instructions into an internal RISC instruction set, because they had to deal with all the CISC instructions of the x86 processors and the legacy software of DOS/Windows.

Review Questions
1. What do RISC and CISC stand for?
2. True or false. The CISC architecture executes the vast majority of its instructions in 2, 3, or more clock cycles, while RISC executes them in one clock.
3. RISC processors normally have a _____ (large, small) number of general-purpose registers.
4. True or false. Instructions such as "ADD R16, ROMmemory" do not exist in RISC microprocessors such as the Arm.

5. True or false. While CISC instructions are of variable sizes, RISC instructions are all the same size.
6. Which of the following operations do not exist for the ADD instruction in RISC?
   (a) register to register (b) immediate to register (c) memory to memory

Problems

Section 2.1: The General-Purpose Registers in the Arm

1. Arm is a(n) _____-bit microprocessor.
2. The general-purpose registers are _____ bits wide.
3. The value in MOV R2, #value is _____ bits wide.
4. The largest number that an Arm GPR register can have is _____ in hex.
5. What is the result of the following code and where is it kept?

          MOV   R2, #0x15
          MOV   R1, #0x13
          ADD   R2, R1, R2
6. Which of the followings is (are) illegal?
   (a) MOV R1, #0x52   (b) MOV R2, #0x50   (c) MOV R1, #0x00
   (d) MOV R1, 255   (e) MOV R17, #25   (f) MOV R23, #0xF5
   (g) MOV 123, 0x50   (h) MOV R1,#0x1234
7. Which of the following is (are) illegal?
   (a) ADD R2, #20, R1   (b) ADD R1, R1, R2   (c) ADD R5, R16, R3
8. What is the result of the following code and where is it kept?

          MOV R9, #0x25
          ADD R8, R9, #0x1F
9. What is the result of the following code and where is it kept?

          MOV R1, #0x15
          ADD R6, R1, #0xEA
10. True or false. We have 32 general purpose registers in the Arm.
11. True or false. R13 and R14 are special function registers.
12. Show the lowest and highest values (in hex) that the Arm program counter can take.

Section 2.2: The Arm Memory Map
13. True or false. The peripheral registers are mapped to memory space.
14. True or false. The On-chip Flash is the same size in all members of Arm.
15. True or false. The On-chip data SRAM is the same size in all members of Arm.
16. What is the maximum number of bytes that the Arm can access?
17. Find the address of the last location of on-chip Flash for each of the following, assuming the first location is 0:
    (a) Arm with 32 KB   (b) Arm with 8 KB
    (c) Arm with 64 KB   (d) Arm with 16 KB
    (e) Arm with 128 KB   (f) Arm with 256 KB

18. A given Arm has 0x7FFF as the address of the last location of its on-chip ROM. What is the size of on-chip Flash for this Arm?
19. Repeat Question 18 for 0x3FFF.
20. Find the on-chip program memory size in K for the Arm chip with the following address ranges:
    (a) 0x0000–0x1FFF   (b) 0x0000–0x3FFF
    (c) 0x0000–0x7FFF   (d) 0x0000–0xFFFF
    (e) 0x0000–0x1FFFF   (f) 0x00000–0x3FFFF
21. Find the on-chip program memory size in K for the Arm chips with the following address ranges:
    (a) 0x00000–0xFFFFFF   (b) 0x00000–0x7FFFF
    (c) 0x00000–0x7FFFFF   (d) 0x00000–0xFFFFF
    (e) 0x00000–0x1FFFFF   (f) 0x00000–0x3FFFFF

Section 2.3: Load and Store Instructions in Arm

22. Show a simple code to store values 0x30 and 0x97 into locations 0x20000015 and 0x20000016, respectively.
23. Show a simple code to load the value 0x55 into locations 0x20000030–0x20000038.
24. True or false. We cannot load immediate values into the data SRAM directly.
25. Show a simple code to load the value 0x11 into locations 0x20000010–0x20000015.
26. Show a simple code to load the value 0x19 into locations 0x20000034–0x2000003C.
27. Show the contents of the memory locations after the execution of each instruction.

    (a) LDR R2, =0x129F   (b) LDR R4, =0x8C63
        LDR R1, =0x20001450   LDR R1, =0x20002400
        LDR R2, [R1]   LDRH R4, [R1]

        0x20001450 = (........)   0x20002400 = (........)
        0x20001451 = (........)   0x20002401 = (........)

Section 2.4: Arm CPSR (Current Program Status Register)

28. The status register is a(n) _____ -bit register.
29. Which bits of the status register are used for the C and Z flag bits, respectively?
30. Which bits of the status register are used for the V and N flag bits, respectively?
31. In the ADD instruction, when is C raised?
32. In the ADD instruction, when is Z raised?
33. What is the status of the C and Z flags after the following code?
            LDR   R0, =0xFFFFFFFF
            LDR   R1, =0xFFFFFFF1
            ADDS   R1, R0, R1
34. Find the C flag value after each of the following codes:

    (a) LDR R0, =0xFFFFFF54   (b) MOV R3, #0   (c) LDR R3, =0xFFFFFFFF
        LDR R5, =0xFFFFFFC4   LDR R6, =0xFFFFFFFF   LDR R8, =0xFFFFFF05
        ADDS R2, R5, R0   ADDS R3, R3, R6   ADDS R2, R3, R8

35. Write a simple program in which the value 0x55 is added 5 times.

Section 2.5: Arm Data Format and Directives

36. State the value (in hex) used for each of the following data:

        MYDAT_1 EQU 55
        MYDAT_2 EQU 98
        MYDAT_3 EQU 'G'
        MYDAT_4 EQU 0x50
        MYDAT_5 EQU 200
        MYDAT_6 EQU 'A'
        MYDAT_7 EQU 0xAA
        MYDAT_8 EQU 255
        MYDAT_9 EQU 2_10010000
        MYDAT_10 EQU 2_01111110
        MYDAT_11 EQU 10
        MYDAT_12 EQU 15

37. State the value (in hex) for each of the following data:

        DAT_1 EQU 22
        DAT_2 EQU 0x56
        DAT_3 EQU 2_10011001
        DAT_4 EQU 32
        DAT_5 EQU 0xF6
        DAT_6 EQU 2_11111011

38. Show a simple code to load the value 0x10102265 into locations 0x20000030–0x2000003F.

39. Show a simple code to (a) load the value 0x23456789 into locations 0x20000060–0x2000006F, and (b) add them together and place the result in R9 as the values are added. Use EQU to assign the names TEMP0–TEMP3 to locations 0x20000060–0x2000006F.

Section 2.6: Assembler data allocation directives

40. Allocate 4 bytes of memory and initialize them with 1, 2, 3, and 4.
41. Using DCB, store your first name in the flash memory.
42. Allocate 60 bytes of SRAM.
43. Allocate 4 bytes of RAM and name it as Temp.

Sections 2.7 and 2.8: Introduction to Arm Assembly Programming and Assembling an Arm Program

44. Assembly language is a _____ (low, high)-level language while C is a _____ (low, high)-level language.
45. Of C and assembly language, which is more efficient in terms of code generation (i.e., the amount of program memory space it uses)?
46. Which program produces the .obj file?
47. True or false. The assembly source file may have the extension ".asm".

48. True or false. The source code file can be a non-ASCII file.
49. True or false. Every source file must have EQU directive.
50. Do the EQU and END directives produce opcodes?
51. The file with the _____ extension is downloaded into Arm Flash ROM.
52. Give three file extensions produced by Arm Keil.

Section 2.9: The Program Counter and Program ROM Space in the Arm

53. Every Cortex-M loads PC with _____ when it is powered up.
54. Write a program to add each of your 5-digit ID to a register and place the result into memory location 0x2000100. Use the program listing to show the Flash memory addresses and their contents.

Section 2.10: Some Arm Addressing Modes

55. Give the addressing mode for each of the following:
    (a) MOV R5, R3   (b) MOV R0, #56

    (c) LDR R5, [R3]   (d) ADD R9, R1, R2

    (e) LDR R7, [R2]   (f) LDRB R1, [R4]

Section 2.11: Pipelining and Harvard Architecture in Arm
56. The Arm Cortex-M3 uses a pipeline of _____ stages.
57. Give the names of the pipeline stages in the Cortex.

Section 2.12: RISC Architecture in Arm
58. What do RISC and CISC stand for?
59. In _____ (RISC, CISC) architecture we can have 1-, 2-, 3-, or 4-byte instructions.
60. In _____ (RISC, CISC) architecture instructions are fixed in size.
61. In _____ (RISC, CISC) architecture instructions are mostly executed in one or two cycles.
62. In _____ (RISC, CISC) architecture we can have an instruction to ADD a register to external memory.
63. True or false. Most instructions in CISC are executed in one or two cycles.

Answers to Review Questions

Section 2.1
```
1. MOV    R2, #0x34
2.

   MOV    R1, #0x16
   MOV    R2, #0xCD
   ADD    R1, R1, R2

   or
```

```
MOV    R1, #0x16
ADD    R1, R1, #0xCD
```

3. False
4. 32
1. True
2. general-purpose registers
3. 32
4. Special function registers (SFRs)
5. 32

Section 2.2

1. True
2. The flash ROM is used to store the program code. The memory can also be used for storage of static data such as text strings and look-up tables

Section 2.3

1. True
2.

```
MOV    R1, #0x20
MOV    R2, #0x95
STRB   R2, [R1]
```

3. `STR    R2, [R8]`
4.

```
MOV    R1, #0x20
LDR    R4, [R1]
```

5. 0xFF in hex or 255 in decimal
6. R6
7. It copies the lower 8 bits of R1 into location pointed to by R2.
8. 0xFFFFFFFF in hex or 4,294,967,295 in decimal ($2^{32}-1$)

Section 2.4

1. CPSR (current program status register)
2. 32 bits
3.

Hex	Binary
FFFFFF9F	1111 1111 1111 1111 1111 1111 1001 1111
+00000061	+ 0000 0000 0000 0000 0000 0000 0110 0001
1 00000000	1 0000 0000 0000 0000 0000 0000 0000 0000

This leads to C = 1 and Z = 1.

4.

Hex	Binary
00000022	0000 0000 0000 0000 0000 0000 0010 0010
+00000022	+ 0000 0000 0000 0000 0000 0000 0010 0010
0 00000000	0000 0000 0000 0000 0000 0000 0100 0100

This leads to Z = 0.

5.

Hex	Binary
0000 0067	0000 0000 0000 0000 0000 0000 0110 0111
+ 0000 0099	+ 0000 0000 0000 0000 0000 0000 1001 1001
0000 0100	0000 0000 0000 0000 0000 0001 0000 0000

This leads to C = 0 and Z = 0.

Section 2.5

1. MOV R1, #0x20
2. (a) MOV R2, #0x14   (b) MOV R2, #20   (c) MOV R2, #2_00010100
3. If the value is to be changed later, it can be done once in one place instead of at every occurrence in the file and the code becomes more readable, as well.
4. (a) 0x34   (b) 0x1F
5. 15 in decimal (0x0F in hex)
6. Value of location 0x00000200 = 0x95
7. 0x0C + 0x10 = 0x1C will be in data memory location 0x00000630.

Section 2.6

1. True
2. False
3. True

Section 2.7

1. Assembly directives direct the assembler in doing its job.
2. The instructions, assembler directives
3. False
4. All except (c)
5. True
6. (c)

Section 2.8

1. True
2. True

3. (a)
4. (b), (c) and (d)

Section 2.9
1. 32
2. False
3. 0x00000004

Section 2.10
1. No
2. The general purpose registers (R0 to R15)
3. It is a part of the instruction

Section 2.11
1. True
2. False
3. True
4. False

Section 2.12
1. RISC is Reduced Instruction Set Computer; CISC stands for Complex Instruction Set Computer.
2. True
3. Large
4. True
5. True
6. (c)

Chapter 3: Arithmetic and Logic Instructions and Programs

In this chapter, most of the arithmetic and logic instructions are discussed and program examples are given to illustrate the application of these instructions. Unsigned numbers are used in this discussion of arithmetic and logic instructions. In Section 3.1 we examine the arithmetic instructions for unsigned numbers. The logic instructions and programs are covered in Section 3.2. In Section 3.3 we discuss the Arm instructions for rotate and shift. In Section 3.4 we perform the shift and rotate operations as part of the other data processing instructions. Section 3.5 is dedicated to BCD and ASCII data conversion.

Section 3.1: Arithmetic Instructions

Unsigned numbers are numbers that represent only zero or positive numbers. All the bits are used to represent data and no bits are set aside for the positive or negative sign. This means that the operand can be between 00 and 0xFF (0 to 255 decimal) for 8-bit data and between 0x0000 and 0xFFFF (0 to 65535 decimal) for 16-bit data. For the 32-bit operand it can be between 0 and 0xFFFFFFFF (0 to 2^{32} -1). See Table 3-1. This section covers the ADD, SUB, multiply and divide instructions for unsigned number.

Data Size	Bits	Decimal	Hexadecimal	Load instruction used
Byte	8	0 – 255	0 – 0xFF	STRB
Half-word	16	0 – 65535	0 – 0xFFFF	STRH
Word	32	0 – 2^{32}-1	0 – 0xFFFFFFFF	STR

Table 3-1: Unsigned Data Range Summary in Arm

Affecting flags in Arm instructions

A unique feature of the execution of Arm arithmetic instructions is that it does not affect (updates) the flags in the CPSR register unless we explicitly request it. This is different from most of other microprocessors and microcontrollers. In other processors the arithmetic/logic instructions (and sometimes other instructions) automatically change the N, Z, C, and V flags according to the result of the operation. To update the flags in CPSR register in Arm CPU by the data processing instructions, the S flag in the instruction must be set. This is done by appending the 'S' suffix to the opcode of the instruction. With the S suffix, the Arm assembler will set the S flag in the instruction. For example, we use SUBS instead of SUB if we want the instruction to update the flags in CPSR. The SUBS means subtract and set the flags, while the SUB simply subtracts without having any effect on the flags. See Table 3-2.

Instruction (Flags unchanged)		Instruction (Flags updated)	
ADD	Add	ADDS	Add and set flags
ADC	Add with carry	ADCS	Add with carry and set flags
SUB	SUBS	SUBS	Subtract and set flags
SBC	Subtract with carry	SBCS	Subtract with carry and set flags
RSB	Reverse subtract	RSBS	Reverse subtract and set flags
RSC	Reverse subtract with carry	RSCS	Reverse subtract with carry and set flags
Note: The above instructions affect all the N, Z, C, and V flag bits of CPSR (current program status register) but the N and V flags are for signed data and are discussed in Chapter 5.			

Table 3-2: Arithmetic Instructions and Flag Bits for Unsigned Data

Addition of unsigned numbers

The form of the ADD instruction is

```
ADD    Rd, Rn, Op2          ; Rd = Rn + Op2
```

The instructions ADD and ADC are used to add two operands. The destination operand must be a register. The Op2 (or operand 2) can be a register or an immediate value. Remember that memory-to-register or memory-to-memory arithmetic and logic operations are never allowed in Arm processor since it is a RISC processor. The instruction could change any of the N, Z, C, or V bits of the program status register, as long as we use the ADDS instead of ADD. The effects of the ADDS instruction on the V (overflow) and N (negative) flags are discussed in Chapter 5 since they are used in signed number operations. Look at Examples 3-1 and 3-2 for the effect of ADDS instruction on Z and C flags.

Example 3-1

Show the flag bits of status register for the following cases:

```
a)    LDR    R2, =0xFFFFFFF5   ; R2 = 0xFFFFFFF5 (notice the = sign)
      MOV    R3, #0x0B
      ADDS   R1, R2, R3        ; R1 = R2 + R3 and update the flags


b)    LDR    R2, =0xFFFFFFFF
      ADDS   R1, R2, #0x95     ; R1 = R2 + 95 and update the flags
```

Solution:

a)

```
        0xFFFFFFF5        1111 1111 1111 1111 1111 1111 1111 0101
     +  0x0000000B     +  0000 0000 0000 0000 0000 0000 0000 1011
        0x100000000     1 0000 0000 0000 0000 0000 0000 0000 0000
```

First, notice how the "LDR R2, =0xFFFFFFF5" pseudo-instruction loads the 32-bit value into R2 register. Also notice the use of ADDS instruction instead of ADD since the ADD instruction does not update the flags. Now, after the addition, the R1 register (destination) contains 0 and the flags are as follows:
C = 1, since there is a carry out from D31
7 = 1, the result of the action is zero (for all 32 bits)

b)

```
        0xFFFFFFFF        1111 1111 1111 1111 1111 1111 1111 1111
     +  0x00000095     +  0000 0000 0000 0000 0000 0000 1001 0101
        0x100000094     1 0000 0000 0000 0000 0000 0000 1001 0100
```

After the addition, the R1 register (destination) contains 0x94 and the flags are as follows:
C = 1, since there is a carry out from D31
Z = 0, the result of the action is not zero (for the 32 bits)

Example 3-2

Show the flag bits of status register for the following case:

```
LDR    R2, =0xFFFFFFF1    ; R2 = 0xFFFFFFF1
MOV    R3, #0x0F
ADDS   R3, R3, R2         ; R3 = R3 + R2 and update the flags
ADD    R3, R3, #0x7       ; R3 = R3 + 0x7 and flags unchanged
MOV    R1, R3
```

Solution:

```
    0xFFFFFFF1       1111 1111 1111 1111 1111 1111 1111 0001
+   0x0000000F     + 0000 0000 0000 0000 0000 0000 0000 1111
   ----------      -----------------------------------------
    0x100000000    1 0000 0000 0000 0000 0000 0000 0000 0000
```

After the ADDS addition, the R3 register (destination) contains 0 and the flags are as follows:

C = 1, since there is a carry out from D31

Z = 1, the result of the action is zero (for the 32 bits)

After the "ADD R3, R3, #0x7" addition, the R3 register (destination) contains 0x7 (0 + 07 = 07) and the flags are unchanged from previous instruction since we used ADD instead of ADDS. Therefore, the Z = 1 and C = 1. If we used "ADDS R3, R3, #0x7" instruction instead of "ADD R3, R3, #0x7", we would have Z = 0 and C = 0. Use the Keil Arm simulator to verify this.

Comment

Microsoft Windows comes with a calculator. Use the Programmer mode to verify the calculations in this and future chapters. The calculator supports data size of up to 64-bit

---

ADC (add with carry)

This instruction is used for adding multiword (data larger than 32-bit) numbers. The form of the ADC instruction is

```
ADC    Rd, Rn, Op2 ; Rd = Rn + Op2 + Carry
```

In discussing addition, the following two cases will be examined:

- Addition of single word data
- Addition of multiword data

CASE 1: Addition of single word data

The result of adding two 32-bit registers can be more than 32-bit. So, whenever some big 32-bit values are added, after each addition, the carry flag should be considered. See Example 3-3.

Example 3-3: Write a program to calculate the total sum of five words of data. Each data value represents the mass of a planet in integer. The decimal data are as follow: 1000000000, 2000000000, 3000000000, 4000000000, and 4100000000. The results should be in R9:R8.

Solution:

```
        EXPORT  __main
        AREA        EXAMPLE3_3, CODE, READONLY
__main
        LDR     R1, =1000000000
        LDR     R2, =2000000000
        LDR     R3, =3000000000
        LDR     R4, =4000000000
        LDR     R5, =4100000000
        MOV     R8, #0      ; R8 = 0 for saving the lower word
        MOV     R9, #0      ; R9 = 0 for accumulating the carries

        ADDS    R8, R8, R1  ; R8 = R8 + R1
        ADC     R9, R9, #0  ; R9 = R9 + 0 + Carry
                            ; (increment R9 if there is carry)
        ADDS    R8, R8, R2  ; R8 = R8 + R2
        ADC     R9, R9, #0  ; R9 = R9 + 0 + Carry
        ADDS    R8, R8, R3  ; R8 = R8 + R3
        ADC     R9, R9, #0  ; R9 = R9 + 0 + Carry
        ADDS    R8, R8, R4  ; R8 = R8 + R4
        ADC     R9, R9, #0  ; R9 = R9 + 0 + Carry
        ADDS    R8, R8, R5  ; R8 = R8 + R5
        ADC     R9, R9, #0  ; R9 = R9 + 0 + Carry
HERE    B       HERE
        END
```

CASE 2: Addition of multi-word numbers

Assume a program is needed that will add the total U.S. budget for the last 100 years or the mass of all the planets in the solar system. In cases like this, the numbers being added could be up to 8 bytes wide or more. Since Arm registers are only 32 bits wide (4 bytes), it is the job of the programmer to write the code to break down these large numbers into smaller chunks to be processed by the CPU. If a 32-bit register is used and the operand is 8 bytes wide, that would take a total of two iterations. See Example 3-4. However, if a 16-bit register is used, the same operands would require four iterations. This obviously takes more time for the CPU, one reason to have wide registers in the design of the CPU.

Example 3-4

Analyze the following program which adds 0x35F62562FA to 0x21F412963B:

```
        LDR     R0, =0xF62562FA     ; R0 = 0xF62562FA
        LDR     R1, =0xF412963B     ; R1 = 0xF412963B
        MOV     R2, #0x35           ; R2 = 0x35
```

```
MOV    R3, #0x21        ; R3 = 0x21
ADDS   R5, R1, R0       ; R5 = 0xF62562FA + 0xF412963B
                        ; now C = 1
ADC    R6, R2, R3       ; R6 = R2 + R3 + C
                        ;     = 0x35 + 21 + 1 = 0x57
```

Solution:

After the R5 = R0 + R1 the carry flag is one. Since C = 1, when ADC is executed, R6 = R2 + R3 + C = 0x35 + 0x21 + 1 = 0x57.

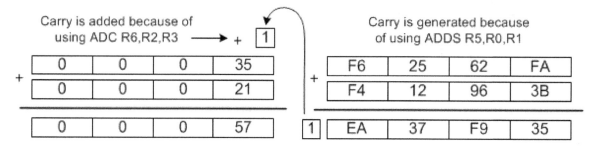

Microsoft Windows calculator support data size of up 64-bit (double word). Use it to verify the above calculations.

Subtraction of unsigned numbers

```
SUB    Rd, Rn, Op2  ; Rd = Rn - Op2
```

In subtraction, the Arm microprocessors (and almost all modern CPUs) use the 2's complement method. All CPUs contain adder circuitry. It would be redundant to design a separate subtractor circuitry if subtraction can be performed with adder. Assuming that the Arm is executing simple subtract instructions, one can summarize the steps of the hardware of the CPU in executing the SUB instruction for unsigned numbers as follows:

1. Take the 2's complement of the subtrahend (Operand 2).
2. Add it to the minuend (Rn operand).
3. Place the result in destination Rd.
4. Update the flags in CPSR if the S flag is set in the instruction.

These four steps are performed for every SUBS instruction by the internal hardware of the Arm CPU. It is after these four steps that the result is obtained and the flags are set. Examples 3-5 through 3-7 illustrates the four steps.

Example 3-5

Show the steps involved for the following cases:

a)
```
MOV    R2, #0x4F        ; R2 = 0x4F
MOV    R3, #0x39        ; R3 = 0x39
SUBS   R4, R2, R3       ; R4 = R2 - R3
```

80

b)

```
        MOV   R2, #0x4F          ; R2 = 0x4F
        SUBS  R4, R2, #0x05      ; R4 = R2 - 0x05
```

Solution:

a)

```
        0x4F        0000004F
      - 0x39      + FFFFFFC7  2's complement of 0x39
        0x16      1 00000016  (C = 1 step 4)
```

The flags would be set as follows: C = 1, and Z = 0.

b)

```
        0x4F          0000004F
      - 0x05        + FFFFFFFB  2's complement of 0x05
        0x4A        1 0000004A  (C=1 step 4)
```

Example 3-6

Analyze the following instructions:

```
        LDR   R2, =0x88888888   ; R2 = 0x88888888
        LDR   R3, =0x33333333   ; R3 = 0x33333333
        SUBS  R4, R2, R3        ; R4 = R2 - R3
```

Solution:

Following are the steps for "SUB R4, R2, R3":

```
        88888888      88888888
      - 33333333    + CCCCCCCD  (2's complement of 0x33333333)
        55555555    1 55555555  (C = 1 step 4)
```

After the execution of SUBS, if C=1, there was no borrow; if C = 0, borrow occurred at the most significant bit. Since we are only dealing with unsigned numbers in this chapter, the result is incorrect with a borrow.

Example 3-7

Analyze the following instructions:

```
MOV   R1, #0x4C   ; R1 = 0x4C
MOV   R2, #0x6E   ; R2 = 0x6E
SUBS  R0, R1, R2  ; R0 = R1 - R2
```

Solution:

Following are the steps for "SUB R0, R1, R2":

```
    4C     0000004C
  - 6E   + FFFFFF92   (2's complement of 0x6E)
  - 22   0 FFFFFFDE   (C = 0 step 4) result is incorrect
```

SBC (subtract with borrow)

```
SBC   Rd, Rn, Op2 ; Rd = Rn - Op2 - 1 + C
```

This instruction is used for subtraction of multiword (data larger than 32-bit) numbers. Notice that in some other architectures, the CPU inverts the C flag after subtraction so the content of carry flag is the borrow bit of subtract operation. But in Arm the carry flag is not inverted after subtraction and the carry flag after the subtraction is the invert of the borrow. This difference does not affect the use of SBC instruction because in those architectures the subtract with borrow is implemented as "Rd = Rn − Op2 − C" but in Arm, it is implemented as "Rd = Rn − Op2 − 1 + C". So the polarity of the carry bin in subtraction is compensated by SBC instruction. See Example 3-8.

Example 3-8

Analyze the following program which subtracts 0x21F62562FA from 0x35F412963B:

```
LDR   R0, =0xF62562FA   ; R0 = 0xF62562FA,
                        ; notice the syntax for LDR
LDR   R1, =0xF412963B   ; R1 = 0xF412963B
MOV   R2, #0x21         ; R2 = 0x21
MOV   R3, #0x35         ; R3 = 0x35
SUBS  R5, R1, R0        ; R5 = R1 - R0
                        ;    = 0xF412963B - 0xF62562FA, and C = 0
SBC   R6, R3, R2        ; R6 = R3 - R2 - 1 + C
                        ;    = 0x35 - 0x21 - 1 + 0 = 0x13
```

Solution:

After the R5 = R1 − R0 there is a borrow so the carry flag is cleared. Since C = 0, when SBC is executed, R6 = R3 − R2 − 1 + C = 0x35 − 0x21 − 1 + 0 = 0x35 − 0x21 − 1= 0x13.

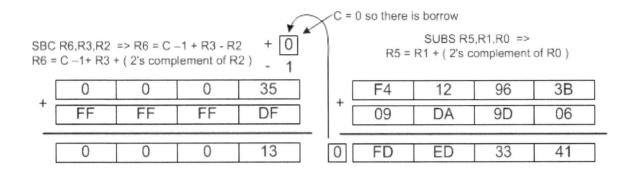

Multiplication and division of unsigned numbers

Because multiplication and division circuits are complex, not all processors have instructions for multiplication and division. All the Arm processors have multiplication instructions but not all have the division. Some family members such as Arm Cortex-A, Cortex-M3, and M4 have both the division and multiplication instructions. In this section we examine the multiplication of unsigned numbers. Signed numbers multiplication is treated in Chapter 5.

Multiplication of unsigned numbers in Arm

The Arm gives you two choices of unsigned multiplication: regular multiply and long multiply. The regular multiply instruction (MUL) is used when the result is less than 32-bit, while the long multiply (MULL) is used when the result is greater than 32-bit. See Table 3-3. In this section we examine both of them.

Instruction	Source 1	Source 2	Destination	Result
MUL	Rn	Op2	Rd (32 bits)	Rd=Rn×Op2
UMULL	Rn	Op2	RdLo, RdHi (64 bits)	RdLo:RdHi=Rn×Op2
Note 1: Using MUL for word × word multiplication preserves only the lower 32-bit result in Rd and the rest are dropped. If the result is greater than 0xFFFFFFFF, then we must use UMULL (unsigned Multiply Long) instruction. Note 2: In some CPUs the C flag is used to indicate the result is greater than 32-bit but this is not the case with Arm MUL instruction.				

Table 3-3: Unsigned Multiplication (UMUL Rd, Rn, Op2) Summary

MUL (multiply)

```
MUL    Rd, Rn, Op2 ; Rd = Rn × Op2
```

In multiplication, all the operands must be in register. Immediate value is not allowed as an operand. After the multiplication, the destination register will contain the result. See the following example:

```
MOV   R1, #0x25   ; R1=0x25
MOV   R2, #0x65   ; R2=0x65
MUL   R3, R1, R2  ; R3 = R1 × R2 = 0x65 × 0x25
```

Note that in the case of half-word times half-word or smaller sources since the destination register is 32-bit there is no problem in keeping the result of 65,535 × 65,535, the highest possible unsigned 16-bit data. That is not the case in word times word multiplication because 32-bit × 32-bit can produce a result greater than 32-bit. If the MUL instruction is used, the destination register will only hold the lower word (32-bit) and the portion beyond 32-bit is dropped. So, it is not safe to use MUL for multiplication of numbers greater than 65,536. See the following example:

```
LDR    R1, =100000 ; R1=100,000
LDR    R2, =150000 ; R2=150,000
MUL    R3, R2, R1  ; R3 is not 15,000,000,000 because
                   ; it cannot fit in 32 bits.
```

For this reason, we must use UMULL (unsigned multiply long) instruction if the result is going to be greater than 0xFFFFFFFF.

UMULL (unsigned multiply long)
```
UMULL  RdLo, RdHi, Rn, Op2     ; RdHi:RdLoRd = Rn × Op2
```

In unsigned long multiplication, all the operands must be in register and no immediate value is allowed. After the multiplication, the destination registers will contain the result. Notice that the left most register in the instruction, RdLo in our case, will hold the lower word and the higher portion beyond 32-bit is saved in the second register, RdHi. See the following example:

```
LDR    R1, =0x54000000   ; R1 = 0x54000000
LDR    R2, =0x10000002   ; R2 = 0x10000002
UMULL  R3, R4, R2, R1    ; 0x54000000 × 0x10000002 = 0x054000000A8000000
                         ; R3 = 0xA8000000, the lower 32 bits
                         ; R4 = 0x05400000, the higher 32 bits
```

Notice that it is the job of programmer to choose the best type of multiplication depending on the size of operands and the result. See Example 3-9.

Example 3-9

Write a short program to multiply 0xFFFFFFFF by itself.
Solution:

```
MOV R1, #0xFFFFFFFF      ; R1 = 0xFFFFFFFF
UMULL R3, R4, R1, R1
```

Since 0xFFFFFFFF × 0xFFFFFFFF = 0xFFFFFFFE00000001, then R4=0xFFFFFFFE and R3=0x00000001. If we had used MUL instruction, then the 0xFFFFFFFF would have been dropped and only 0x00000001 would have been kept by the destination register.

Notice that in Cortex-M which use Thumb-2 the multiply instructions do not affect the flags.

Division of unsigned numbers in Arm

To divide unsigned numbers UDIV can be used:

```
UDIV   Rd, Rn, Op2  ; Rd = Rn / Op2
```

The following example divides 8 by 3 and stores the result in R5:

```
MOV   R1, #8        ; R1 = 8
MOV   R2, #3        ; R2 = 3
UDIV  R5, R1, R2    ; R5 = 8/3 = 2
```

Other Arithmetic Instructions (Case study)

RSB (reverse subtract)

The format for the RSB instruction is

```
RSB   Rd, Rn, Op2          ; Rd = Op2 - Rn
```

Notice the difference between the RSB and SUB instruction. They are essentially the same except the way the source operands are subtracted is reversed. See Example 3-10.

Example 3-10

Find the result of R0 for the following:

```
MOV   R1, #0x1   ; R1 = 1
RSB   R0, R1, #0  ; R0 = 0 - R1 = 0 - 1
```

Solution:

Following are the steps for "RSB R0, R1, #0":

```
        0     00000000
    -  6E   + FFFFFF92   (2's complement)
    -  6E     FFFFFF92   (C = 0) result is negative
```

Multiply and Accumulate Instructions in Arm

In some applications such as digital signal processing (DSP) we need to multiply two variables and add the result to another variable. The Arm has an instruction to do both in a single instruction. The format of MLA (multiply and accumulate) instruction is as follows:

```
MLA   Rd, Rm, Rs, Rn    ; Rd = Rm × Rs + Rn
```

In multiplication and accumulate, all the operands must be in register. After the multiplication and add, the destination register will contain the result. See the following example:

```
MOV   R1, #100          ; R1 = 100
MOV   R2, #5            ; R2 = 5
```

```
MOV    R3, #40            ; R3 = 40
MLA    R4, R1, R2, R3     ; R4 = R1 × R2 + R3 = 100 × 5 + 40 = 540
```

To accumulate the products of the multiplication, just use the same register for Rd and Rn:

```
MLA    R3, R1, R2, R3     ; R3 = R1 × R2 + R3 or R3 += R1 × R2
```

Notice that multiply and accumulate can produce a result greater than 32-bit, if the MLA instruction is used, the destination register will only hold the lower word (32 bits) of the sum and the portion beyond 32-bit is dropped. For this reason, we must use UMLAL (unsigned multiply and accumulate long) instruction if the result is going to be greater than 0xFFFFFFFF. The format of UMLAL instruction is as follows:

```
UMLAL RdLo, RdHi, Rn, Op2    ; RdHi:RdLo = Rn × Op2 + RdHi:RdLo
```

In UMLAL instruction, all the operands must be in register. Notice that the addend and the destination use the same registers, the two left most registers in the instruction. It means that the contents of the registers which have the addend will be changed after execution of UMLAL instruction. See the following example:

```
LDR    R1, =0x34000000   ; R1 = 0x34000000
LDR    R2, =0x2000000    ; R2 = 0x2000000
MOV    R3, #0            ; R3 = 0x00
LDR    R4, =0x00000BBB   ; R4 = 0x00000BBB
UMLAL R4, R3, R2, R1     ; 0x34000000×0x2000000+0xBBB
                         ;   = 0x068000000000000BBB
```

Review Questions

1. Explain the difference between ADDS and ADD instructions.
2. The ADC instruction that has the syntax "ADC Rd, Rn, Op2" means _____.
3. Explain why the Z=0 for the following:

```
MOV        R2, #0x4F
MOV        R4, #0xB1
ADDS       R2, R4, R2
```

4. Explain why the Z=1 for the following:

```
MOV        R2, #0x4F
LDR        R4, =0xFFFFFFB1
ADDS       R2, R4, R2
```

5. Show how the CPU would subtract 0x05 from 0x43.
6. If C = 1, R2 = 0x95, and R3 = 0x4F prior to the execution of "SBC R2, R2, R3", what will be the contents of R2 after the subtraction?
7. In unsigned multiplication of "MUL R2, R3, R4", the product will be placed in register _____.
8. In unsigned multiplication of "MUL R1, R2, R4", the R2 can be maximum of _____ if R4 = 0xFFFFFFFF so that there are no bits lost by the operation.

Section 3.2: Logic Instructions

In this section we discuss the logic instructions AND, OR, and Ex-OR in the context of many examples. Just like arithmetic instruction, we must use the S suffix in the instruction if we want to update the flags. If the S suffix is used the Z flag will be set if and only if the result is all zeros, and the N flag will be set to the logical value of bit 31 of the result. The V flag in the CPSR will be unaffected, and the C flag will be updated according to the calculation of the Operand 2. See Table 3-4.

Instruction (Flags Unchanged)	Action	Instruction (Flags Changed)	Hexadecimal
AND	ANDing	ANDS	Anding and set flags
ORR	ORRing	ORS	Oring and set flags
EOR	Exclusive-ORing	EORS	Exclusive ORing and set flags
BIC	Bit Clearing	BICS	Bit clearing and set flags

Table 3-4: Logic Instructions and Flag Bits

AND

```
AND    Rd, Rn, Op2        ; Rd = Rn ANDed Op2
```

Inputs		Output	Symbol
X	Y	X AND Y	
0	0	0	
0	1	0	
1	0	0	
1	1	1	

This instruction will perform a bitwise logical AND on the operands and place the result in the destination. The destination and the first source operand are registers. The second source operand can be a register or an immediate value of less than 0xFF with even bits of rotate.

If we use ANDS instead of AND it will change the N and Z flags according to the result (and C flag during the calculation of operand 2). As seen in Example 3-11, AND can be used to mask certain bits of the operand.

Example 3-11

Show the results of the following cases

a)
```
        MOV    R1, #0x35
        AND    R2, R1, #0x0F      ; R2 = R1 ANDed with 0x0F
```

b)
```
        MOV    R0, #0x97
        MOV    R1, #0xF0
        AND    R2, R0, R1         ; R2 = R0 ANDed with R1
```

Solution:

a)

```
         0x35    0 0 1 1 0 1 0 1
AND      0x0F    0 0 0 0 1 1 1 1
         0x05    0 0 0 0 0 1 0 1
```

b)

```
         0x97    1 0 0 1 0 1 1 1
AND      0xF0    1 1 1 1 0 0 0 0
         0x90    1 0 0 1 0 0 0 0
```

ORR

```
ORR    Rd, Rn, Op2 ; Rd = Rn ORed Op2
```

Inputs		Output	Symbol
X	Y	X OR Y	
0	0	0	
0	1	1	
1	0	1	
1	1	1	

The operands are ORed and the result is placed in the destination. ORR can be used to set certain bits of an operand to one. The destination and the first source operand are registers. The second source operand can be either a register or an immediate value of less than 0xFF with even bits of rotate.

If we use ORRS instead of ORR, the flags will be updated, just the same as for the ANDS instruction. See Example 3-12.

Example 3-12

Show the results of the following cases:

a)

```
MOV    R1, #0x04        ; R1 = 0x04
ORRS   R2, R1, #0x68    ; R2= R1 ORed 0x68
```

b)

```
MOV    R0, #0x97
MOV    R1, #0xF0
ORR    R2, R0, R1       ; R2= R0 ORed with R1
```

Solution:

a)

```
         0x04    0000 0100
OR       0x68    0110 1000        Flag will be: Z = 0
         0x6C    0110 1100
```

88

b)

```
        0x97   1001 0111
OR      0xF0   1111 0000      Flag will be unchanged
        0xF7   1111 0111
```

The ORR instruction can also be used to test for a zero operand. For example, "ORRS R2, R2, #0" will OR the register R2 with zero and make Z = 1 if R2 is zero.

EOR

```
EOR   Rd, Rn, Op2 ; Rd = Rn Ex-ORed with Op2
```

Inputs		Output	Symbol
X	Y	X EOR Y	
0	0	0	
0	1	1	
1	0	1	
1	1	0	

The EOR instruction will perform an Exclusive-OR of the two operands and place the result in the destination register. EOR sets the result bits to 1 if the corresponding source bits are not equal; otherwise, they are clear to 0. The flags are updated if we use EORS instead of EOR. The rules for the operands are the same as in the AND and OR instructions. See Examples 3-13 and 3-14.

Example 3-13

Show the results of the following:

```
MOV   R1, #0x54
EOR   R2, R1, #0x78      ; R2 = R1 ExOred with 0x78
```

Solution:

```
        0x54   0 1 0 1 0 1 0 0
EOR     0x78   0 1 1 1 1 0 0 0
        0x2C   0 0 1 0 1 1 0 0
```

Example 3-14

The EOR instruction can be used to clear the contents of a register by Ex-ORing it with itself.
Show how "EOR R1, R1, R1" clears R1, assuming that R1 = 0x45.

Solution:

```
          0x45   0 1 0 0 0 1 0 1
EOR       0x45   0 1 0 0 0 1 0 1
          0x00   0 0 0 0 0 0 0 0
```

Another application of EOR is to toggle bits of an operand. For example, to toggle bit 2 of register R2:

```
EOR   R2, R2, #0x04       ; EOR R2 with 0000 0100
```

This would cause bit 2 of R2 to change to the complement value; all other bits would remain unchanged.

BIC (bit clear)

```
BIC   Rd, Rn, Op2       ; clear certain bits of Rn specified by
                        ; the Op2 and place the result in Rd
```

Inputs		Output
X	Y	X AND (NOT Y)
0	0	0
0	1	0
1	0	1
1	1	0

The BIC (bit clear) instruction is used to clear the selected bits of the Rn register. The selected bits are held by Op2. The bits that are HIGH in Op2 will be cleared and bits with LOW will be left unchanged. For example, assuming that R3 = 0000000000001000 binary, the instruction "BIC R2, R2, R3" will clear bit 3 of R2 and leaves the rest of the bits unchanged. In reality, the BIC instruction performs AND operation on Rn register with the complement of Op2 and places the result in destination register. Look at the following example:

```
MOV   R2, #0xAA
BIC   R3, R2, #0x0F       ; now R3 = 0xAA AND 0xF0 = 0xA0
```

We can use the AND operation with complement to achieve the same result:

```
MOV R2, #0xAA
AND R3, R2, #~0x0F       ; AND R2 with the complement of #0x0F
                        ; and store the result in R3
```

If we want the flags to be updated, then we must use BICS instead of BIC.

MVN (move not)

```
MVN   Rd, Rn           ; move the complement of Rn to Rd
```

The MVN (move not) instruction is used to generate one's complement of an operand. For example, the instruction "MVN R2, #0" will make R2=0xFFFFFFFF. Look at the following example:

```
LDR     R2, =0xAAAAAAAA          ; R2 = 0xAAAAAAAA
MVN     R2, R2                   ; R2 = 0x55555555
```

We can also use Ex-OR instruction to generate one's complement of an operand. Ex-ORing an operand with 0xFFFFFFFF will generate the 1's complement. See the following code:

```
LDR     R2, =0xAAAAAAAA          ; R2 = 0xAAAAAAAA
MVN     R0, #0                   ; R0 = 0xFFFFFFFF
EOR     R2, R2, R0               ; R2 = R2 ExORed with 0xFFFFFFFF
                                 ;     = 0x55555555
```

Review Questions

1. Use operands 0x4FCA and 0xC237 to perform:

    (a) AND   (b) OR   (c) XOR

2. ANDing a word operand with 0xFFFFFFFF will result in what value for the word operand? To set all bits of an operand to 0, it should be ANDed with _____.

3. To set all bits of an operand to 1, it could be ORed with _____.

4. XORing an operand with itself results in what value for the operand?

5. Write an instruction that sets bit 4 of R7.

6. Write an instruction that clears bit 3 of R5.

Section 3.3: Shift and Rotate Instructions

In this section we explore the shift and rotate instructions.

LSL (Logical Shift Left) instruction

```
LSL     Rd, Rm, Rn
```

Shift left is a logical shift. It is the reverse of LSR. After every shift, the LSB is filled with 0 and the MSB goes to C flag in CPSR if the 'S' suffix is used in the 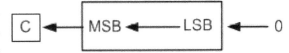 instruction. One can use an immediate value or a register to hold the number of times it is to be shifted left. See Example 3-15. One can use the LSL to multiply a number by 2. See Example 3-16.

Example 3-15

Show the effects of LSL in the following:

```
LDR     R1, =0x0F000006
LSLS    R2, R1, #8
```

Solution:

```
       00001111 00000000 00000000 00000110
C=0    00011110 00000000 00000000 00001100 (shifted left once)
C=0    00111100 00000000 00000000 00011000
```

```
C=0    01111000 00000000 00000000 00110000
C=0    11110000 00000000 00000000 01100000
C=1    11100000 00000000 00000000 11000000
C=1    11000000 00000000 00000001 10000000
C=1    10000000 00000000 00000011 00000000
C=1    00000000 00000000 00000110 00000000 (shifted eight times)
```

After eight shifts left, the R2 register has 0x00000600 and C = 1. The eight MSBs are lost through the carry, one by one, and 0s fill the eight LSBs. Another way to write the above code is:

```
LDR    R1, =0x0F000006
MOV    R0, #0x08
LSLS   R2, R1, R0
```

Notice that the LSL instruction multiplies the content of the register by power of 2 as long as there is no carry out. For example, when a number is shifted left 3 times, it is multiplied by 2^3. See the following example.

Example 3-16

Show the results of LSL in the following:

```
TIMES EQU    0x5
      LDR    R1, #0x7          ; R1=0x7
      MOV    R2, #TIMES        ; R2=0x05
      LSL    R1, R1, R2        ; shift R1 left R2 number of times
                               ; and place the result in R1
```

Solution:
After the five shifts, the R1 will contain 0x000000E0. 0xE0 is 224 in decimal. Notice that $7 \times 2^5 = 7 \times 32 = 224 = 0xE0$. So, it multiplies number by 32 (2 to the power of 5 is 32).

LSR (Logical Shift Right) Instruction

```
LSR    Rd, Rm, Rn
```

The operand is shifted right bit by bit, and for every shift the LSB (least significant bit) will go to the carry flag if the 'S' suffix is used in the instruction and the MSB (most significant bit) is filled with 0. At the end of the execution of the instruction, the carry flag will hold the last bit shifted out if the 'S' suffix is used in the instruction. One can use an immediate value or a register to hold the number of times it is to be shifted. Example 3-17 should help to clarify LSR.

Example 3-17

Show the result of the MOVS instruction with LSR in the following:

```
MOV    R0, #0x9A        ; R0 = 0x9A
LSRS   R1, R0, #3       ; shift R0 to right 3 times
                        ; then store the result in R1
```

Solution:

0x9A = 00000000 00000000 0000000 00000000 10011010

first shift: 00000000 00000000 0000000 00000000 01001101 C = 0

second shift: 00000000 00000000 0000000 00000000 00100110 C = 1

third shift: 00000000 00000000 0000000 00000000 00010011 C = 0

After shifting right three times, R1 = 0x00000013 and C = 0. Another way to write the above code is:

```
MOV    R0, #0x9A
MOV    R2, #0x03
LSRS   R1, R0, R2   ; shift R0 to right R2 times
                    ; and move the result to R1
```

One can use the LSR to divide a number by 2. See Example 3-18.

Example 3-18

Show the results of LSR in the following:

```
LDR    R0, =0x88        ; R0=0x88
MOVS   R1, R0, LSR #3   ; shift R0 right three times (R1 = 0x11)
```

Solution:

After the three shifts, the R1 will contain 0x11. This divides the number by 8 since 2 to the power of 3 is 8.

Table 3-5 lists the logical shift operations in Arm.

Operation	Destination	Source	Number of shifts
LSR (Shift Right)	Rd	Rn	Immediate value
LSR (Shift Right)	Rd	Rn	register Rm
LSL (Shift Left)	Rd	Rn	Immediate value
LSL (Shift Left)	Rd	Rn	register Rm
Note: Number of shifts cannot be more than 32.			

Table 3-5: Logic Shift operations for unsigned numbers in Arm

ROR (Rotate Right) instruction

```
ROR    Rd, Rm, Rn   ; Rd=rotate Rm right Rn bit positions
```

As each bit of Rm register is shifted from left to right, they exit from the end (LSB) and entered from left end (MSB). The number of bits to be rotated right is given by Rn and the result is placed in Rd register. To update the flags, use RORS instruction.

Example 1:
```
        LDR    R2, =0x00000010
        ROR    R0, R2, #8   ; R0=R2 is rotated right 8 times
                            ; now, R0 = 0x10000000, C=0
```

Example 2:
```
        LDR    R0, =0x00000018
        MOV    R1, #12
        ROR    R2, R0, R1   ; R2=R0 is rotated right R1 number of times
                            ; now, R2 = 0x01800000, C=0
```

Example 3:
```
        LDR    R0, =0x0000FF18
        MOV    R1, #16
        ROR    R2, R0, R1   ; R2=R0 is rotated right R1 number of times
                            ; now, R2 = 0xFF180000, C=0
```

RRX (Rotate Right with extend) instruction

```
        RRXS   Rd, Rm        ; Rd=rotate Rm right 1 bit through C flag
```

Each bit of Rm register is rotated from left to right one bit through C flag when RRXS instruction is used. If the 'S' suffix is not used, the LSB is lost and the current C flag is shifted into MSB.

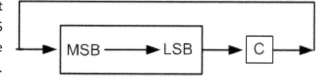

Example:
```
        LDR    R2, =0x00000002
        RRX    R0, R2        ; R0=R2 is shifted right one bit
                            ; now, R0=0x00000001
```

Table 3-6 lists the rotate instructions of the Arm.

Operation	Destination	Source	Number of Rotates
ROR (Rotate Right)	Rd	Rn	Immediate value
ROR (Rotate Right)	Rd	Rn	register Rm
RRX (Rotate Right Through Carry)	Rd	Rn	1 bit

Table 3-6: Rotate operations for unsigned numbers in Arm

Review Questions
  1. Find the contents of R2 after executing the following code:

```
MOV    R1, #0x08
ROR    R2, R1, #2
```

2. Find the contents of R4 after executing the following code:

```
MOV    R3, #0x3
LSL    R4, R3, #2
```

Section 3.4: Rotate and Shift in Data Processing Instructions (Case Study)

In previous sections, we discussed that as the second operand of data process instructions we can use register or immediate values. We can also perform the shift and rotate operations as part of the other data processing instructions (arithmetic and logic instructions) such as MOV, ADD, or SUB. Examples 3-19 through 3-22 should help to clarify.

Example 3-19

Show the result of the MOVS instruction with LSR in the following:

```
MOV    R0, #0x9A        ; R0 = 0x9A
MOVS   R1, R0, LSR #3   ; shift R0 to right 3 times
                        ; then store the result in R1
```

Solution:

0x9A =   00000000 00000000 0000000 00000000 10011010

first shift: 00000000 00000000 0000000 00000000 01001101 C = 0

second shift: 00000000 00000000 0000000 00000000 00100110 C = 1

third shift: 00000000 00000000 0000000 00000000 00010011 C = 0

After shifting right three times, R1 = 0x00000013 and C = 0. Another way to write the above code is:

```
MOV    R0, #0x9A
LSR    R1, R0, #0x03    ; shift R0 to right 3 times
```

Example 3-20

Show the results of the ADDS with LSR in the following:

```
LDR    R1, =0x777       ; R1=0x777
LDR    R2, =0xA6D       ; R2=0xA6D
ADDS   R3, R1, R2, LSR #4   ; shift R2 right 4 times then add it to
                            ; R1 and place the result in R3
                            ; R3 = 0x777 + 0xA6 = 0x81D
```

Solution:

After four shifts, the R2 will contain 0xA6. The four LSBs are lost through the carry, one by one, and 0s fill the four MSBs. 0xA6 is added to 0x777 in R1 and the sum 0x81D is placed in R3. Unlike MOVS operation, which does not affect the C flag itself, the ADDS operation generates C flag depending on the carry out of the MSB by the ADD, which will overwrite the C flag generated from the shift of operand 2. In this example, "R2, LSR #4" generates C = 1 but the add results in C = 0. So at the end of the ADDS instruction, C = 0.

Example 3-21

Show the effects of LSL in the following:

```
LDR     R1, =0x0F000006
MOVS    R2, R1, LSL #5
```

Solution:

	00001111 00000000 00000000 00000110
C=0	00011110 00000000 00000000 00001100 (shifted left once)
C=0	00111100 00000000 00000000 00011000
C=0	01111000 00000000 00000000 00110000
C=0	11110000 00000000 00000000 01100000
C=1	11100000 00000000 00000000 11000000 (shifted five times)

After five shifts left, the R2 register has 0x000000C0 and C = 1. The five MSBs are lost through the carry, one by one, and 0s fill the five LSBs. Another way to write the above code is:

```
LSL     R2, R1, #0x05
```

Example 3-22

Show the results of the ADDS with ROR in the following:

```
LDR     R1, =0x777          ; R1=0x777
LDR     R2, =0x444          ; R2=0x444
ADDS    R3, R1, R2, ROR #2  ; rotate R2 right 2 times then add it to
                            ; R1 and place the result in R3
                            ; R3 = 0x777 + 0x111 = 0x888
```

Solution:

After two shifts, 0x444 becomes 0x10000011. The six LSBs are rotated to the MSB bits, one by one. 0x10000011 is added to 0x777 in R1 and the sum 0x10000788 is placed in R3.

Table 3-7 lists the available shift and rotate instructions in Cortex-M3.

Operation	Source	Number of Rotates
ASR (Arithmetic Shift Right)	Rn	Immediate value
LSL (Logic Shift Left)	Rn	Immediate value
LSR (Logic Shift Right)	Rn	Immediate value
ROR (Rotate Right)	Rn	Immediate value
RRX (Rotate Right Through Carry)	Rn	1 bit

Table 3-7: Rotate operations for unsigned numbers in Arm

Review Questions

1. Find the contents of R3 after executing the following code:

```
MOV    R0, #0x04
MOV    R3, R0, LSR #2
```

2. Find the contents of R4 after executing the following code:

```
LDR    R1, =0xA0F2
MOV    R2, #0x3
MOV    R4, R1, LSR R2
```

3. Find the contents of R3 after executing the following code:

```
LDR    R1, =0xA0F2
MOV    R2, #0x3
MOV    R3, R1, LSL R2
```

4. Find the contents of R5 after executing the following code:

```
SUBS   R0, R0, R0
MOV    R0, #0xAA
MOV    R5, R0, ROR #4
```

5. Find the contents of R0 after executing the following code:

```
LDR    R2, =0xA0F2
MOV    R1, #0x1
MOV    R0, R2, ROR R1
```

6. Give the result in R1 for the following:

```
MVN    R1, #0x01, #2
```

7. Give the result in R2 for the following:

```
MVN    R2, #0x02, #28
```

Section 3.5: BCD and ASCII Conversion

This section covers binary, BCD, and ASCII conversions with some examples.

BCD number system

BCD stands for Binary Coded Decimal. Most of the computers these days perform arithmetic in binary because binary arithmetic is easier and faster to implement in electronic circuit. But most of the numbers used in real lift are decimal, so it requires to convert the decimal numbers to binary before the computations can be done. Earlier computers did arithmetic in decimal because it does not require the decimal to binary and binary to decimal conversions.

To perform arithmetic in decimal, data need to be encoded in decimal format but using binary system of the computer so binary coded decimal (BCD) is often used. See Table 3-8. In the modern computing, you may still encounter the usage of BCD in some applications. For example, BCD is used in many real-time clock (RTC) of the embedded systems.

Digit	BCD
0	0000
1	0001
2	0010
3	0011
4	0100
5	0101
6	0110
7	0111
8	1000
9	1001

Table 3-8: BCD Codes

There are two formats for BCD numbers: (1) unpacked BCD, and (2) packed BCD.

Unpacked BCD

In unpacked BCD, each decimal digit is represented by a byte (8-bit). The lower 4 bits of the byte represent the BCD number and the rest of the bits are 0. For example, "0000 1001" and "0000 0101" are unpacked BCD for 9 and 5, respectively.

Packed BCD

In the case of packed BCD, two decimal digits are packed in one byte, one in the lower 4 bits and one in the upper 4 bits. For example, "0101 1001" is packed BCD for 59. Obviously, packed BCD is more efficient in memory usage but to perform arithmetic with packed BCD, the circuit has to be able to detect the decimal carry from the lower digit to the upper digit in the same byte. Arm CPU does not do that.

ASCII encoding

The American Standard Code for Information Interchange was established in the early 1960's as a character encoding standard for telegraph in United States. It was adopted by computer developers to be used as the encoding for transmitting text between computer and peripherals, text file storage, and communication between computers. The ASCII code has the advantage over other encode of the earlier time as the code within the three groups of code (numerals, uppercase alphabets, and alphabets) are all in consecutive order. That makes conversion between ASCII and BCD or between uppercase and lowercase easier.

ASCII codes are 7-bit long. The ASCII encoding of numerals starts from "011 0000" (0x30) for "0". Since all the numeral codes are consecutive, "1" is encoded as "011 0001" (0x31) "2" is encoded as "011 0010" (0x32) and so on.

For example, in an ASCII keyboard when a key is pressed, the ASCII encoding of that key is transmitted to the computer. So, when key "0" is pressed, "011 0000" (0x30) is sent to the computer. In the same way, when key "5" is pressed, "011 0101" (0x35) is sent. The ASCII codes of numerals are shown in the following table together with the corresponding BCD code:

Key	ASCII	Binary(hex)	BCD (unpacked)
0	30	011 0000	0000 0000
1	31	011 0001	0000 0001
2	32	011 0010	0000 0010
3	33	011 0011	0000 0011
4	34	011 0100	0000 0100
5	45	011 0101	0000 0101
6	36	011 0110	0000 0110
7	37	011 0111	0000 0111
8	38	011 1000	0000 1000
9	39	011 1001	0000 1001

Though we mentioned earlier that processing decimal data in BCD does not require to convert the data to binary. But often the input/output data and the data stored in the files are in ASCII code for ease of human reading. Input/output devices like keyboard and LCD display are usually using ASCII encoding. These ASCII data need to be converted to BCD before performing decimal data processing and be converted back to ASCII afterward. These are the subjects covered next.

ASCII to unpacked BCD conversion

The lower nibble (least significant four bits) of the ASCII codes for numeral contain the binary value of that digit. To convert ASCII data to unpacked BCD, the programmer must get rid of the "011" in the upper 3 bits of the 7-bit ASCII. To do that, each ASCII number is ANDed with "0000 1111" (0x0F).

ASCII to packed BCD conversion

To convert ASCII numbers to packed BCD, they are first converted to unpacked BCD (remove the upper 3 bits) and then combined every two digits to make a packed BCD. For example, if the user typed digit 2 and 7 on an ASCII keyboard, the keyboard transmits ASCII codes 0x32 and 0x37 to the computer. The goal is to produce 0x27 or "0010 0111", which is called packed BCD. This process is illustrated in detail in the program snippet below.

Key	ASCII	Unpacked BCD	Packed BCD
2	32	00000010	
7	37	00000111	00100111 (0x27)

```
MOV   R1, #0x37          ; R1 = 0x37
MOV   R2, #0x32          ; R2 = 0x32
AND   R1, R1, #0x0F      ; mask 3 to get unpacked BCD
AND   R2, R2, #0x0F      ; mask 3 to get unpacked BCD
ORR   R3, R1, R2, LSL #4 ; shift R2 4 bits to the left and combine
                         ; with R1 to get packed BCD in R3 = 0x27
```

Packed BCD to ASCII conversion

For data to be displayed or printed on a device that accepts only ASCII format, they need to be converted to ASCII first. Conversion from packed BCD to ASCII is discussed next. To convert packed BCD to ASCII, it must first be unpacked and then tagged with 011 0000 (0x30) to encode in ASCII. The following code snippet shows the process of converting from packed BCD to ASCII.

Packed BCD	Unpacked BCD	ASCII
0x29	0x02 & 0x09	0x32 & 0x39
0010 1001	0000 0010 & 0000 1001	011 0010 & 011 1001

```
MOV    R0,  #0x29
AND    R1,  R0,  #0x0F      ; mask upper four bits
ORR    R1,  R1,  #0x30      ; combine with 30 to get ASCII
MOV    R2,  R0,  LSR #04    ; shift right 4 bits to get unpacked BCD
ORR    R2,  R2,  #0x30      ; combine with 30 to get ASCII
```

Review Questions

1. For the following decimal numbers, give the packed BCD and unpacked BCD representations in binary

   (a) 15   (b) 99

2. For the following packed BCD numbers, give the decimal and unpacked BCD representations.

   (a) 0x41   (b) 0x09

3. Repeat question 2 for ASCII.

Problems

Section 3.1: Arithmetic Instructions

1. Find C and Z flags for each of the following. Also indicate the result of the addition and where the result is saved.

   (a)

   ```
   MOV    R1,  #0x3F
   MOV    R2,  #0x45
   ADDS   R3,  R1,  R2
   ```

   (b)

   ```
   LDR    R0,  =0x95999999
   LDR    R1,  =0x94FFFF58
   ADDS   R1,  R1,  R0
   ```

   (c)

   ```
   LDR    R0,  =0xFFFFFFFF
   ADDS   R0,  R0,  #1
   ```

   (d)

   ```
   LDR    R2,  =0x00000001
   LDR    R1,  =0xFFFFFFFF
   ADDS   R0,  R1,  R2
   ADCS   R0,  R0,  #0
   ```

   (e)

```
LDR    R0,  =0xFFFFFFFE
ADDS   R0,  R0,  #2
ADC    R1,  R0,  #0x0
```

2. State the three steps involved in a SUB and show the steps for the following data.

    (a) 0x23 – 0x12   (b) 0x43 – 0x51   (c) 0x99 – 0x99

Section 3.2: Logic Instructions

3. Assume that the following registers contain these hex contents: R0 = 0xF000, R1 = 0x3456, and R2 = 0xE390. Perform the following operations. Indicate the result and the register where it is stored.

    Note: the operations are independent of each other.

    (a) AND R3, R2, R0   (b) ORR R3, R2, R1

    (c) EOR R0, R0, #0x76   (d) AND R3, R2, R2

    (e) EOR R0, R0, R0   (f) ORR R3, R0, R2

    (g) AND R3, R0, #0xFF   (h) ORR R3, R0, #0x99

    (i) EOR R3, R1, R0   (j) EOR R3, R1, R1

4. Give the value in R2 after the following code is executed:

```
MOV    R0,  #0xF0
MOV    R1,  #0x55
BIC    R2,  R1,  R0
```

5. Give the value in R2 after the following code is executed:

```
LDR    R1,  =0x55555555
MVN    R0,  #0
EOR    R2,  R1,  R0
```

Section 3.3: Shift and Rotate Instructions

6. Assuming C = 0, what is the value of R1 after the following?

```
MOV    R1,  #0x25
RORS   R1,  R1,  #4
```

7. Assuming C = 0, what are the values of R0 and C after the following?

```
LDR    R0,  =0x3FA2
MOV    R2,  #8
RORS   R0,  R0,  R2
```

8. Assuming C = 0 what is the value of R2 and C after the following?

```
MOV    R2, #0x55
RRX    R2, R2
```

9. Assuming C = 0 what is the value of R1 after the following?

```
MOV    R1, #0xFF
MOV    R3, #5
RORS   R1, R1, R3
```

Section 3.4: Rotate and Shift in Data Processing Instructions

10. Find the contents of registers and C flag after executing each of the following codes:

a)

```
MOV  R0, #0x04
MOVS R1, R0, LSR #2
```

b)

```
LDR  R1, =0xA0F2
MOVS R3, R1, LSL #3
```

c)

```
LDR  R1, =0xB085
MOVS R4, R1, LSR #3
```

11. Find the contents of registers and C flag after executing each of the following codes:

a)

```
MOV  R0, #0xAA
MOVS R1, R0, ROR #4
```

b)

```
MOV  R2, #0xAA
MOVS R1, R2, ROR #1
```

c)

```
LDR  R0, =0x1234
MOVS R1, R0, ROR #4
```

d)

```
MOV  R0, #0xAA
MOVS R1, R0, RRX
```

Section 3.5: BCD and ASCII Conversion

12. Write a program to convert 0x76 from packed BCD number to ASCII. Place the ASCII codes into R1 and R2.

13. For "3" and "2" the keyboard gives 0x33 and 0x32, respectively. Write a program to convert 0x33 and 0x32 to packed BCD and store the result in R2.

Answers to Review Questions

Section 3.1: Arithmetic Instructions

1. The ADDS instruction updates the flag bits in CPSR register while ADD does not do that.
2. Rd = Rn + Op2 + C
3. 0x4F + 0xB1 = 0x100, since the result is less than 32-bit the C = 0 and Z = 0.
4. 0x4F + 0xFFFFFFB1 = 0x00000000, since the result is greater than 32-bit, there is a carry out from the MSB and the remaining 32 bits are all 0, the C = 1 and Z = 1.

5.

0x43	0100 0011		00000000000000000000000001000011
−0x05	0000 0101	2's complement =	+ 11111111111111111111111111111011
0x3E			1 00000000000000000000000000111110

C = 1; therefore, the result is positive

6. R2 = R2 − R3 − C + 1 = 0x95 − 0x4F − 1 + 1 = 0x46
7. R2
8. R2 = 1

Section 3.2: Logic Instructions

1. (a) 0x4202   (b) 0xCFFF   (c) 0x8DFD
2. The operand will remain unchanged; all zeros
3. All ones
4. All zeros
5. ORR R7, R7, #0x10   ; R7 = R7 ORed 0001 0000
6. BIC R5, R5, #0x8   ; R5 = R5 ANDed 1111 1111 1111 0111

Section 3.3: Shift and Rotate Instructions

1. 0x02
2. 0x0C

Section 3.4: Rotate and Barrel Shifter Operation

1. R3 = 1
2. R4 = 0x0000141E
3. R3 = 0x00050790
4. R5 = 0xA000000A
5. R0 = 0x00005079
6. 0xBFFFFFFF
7. 0xFFFFFFDF

Section 3.5: BCD and ASCII Conversion

1. (a) 15 = 0001 0101 packed BCD = 0000 0001 0000 0101 unpacked BCD

   (b) 99 = 1001 1001 packed BCD = 0000 1001 0000 1001 unpacked BCD

2. (a) 0x41 = 0000 0100 0000 0001 unpacked BCD = 41 in decimal

   (b) 0x09 = 0000 0000 0000 1001 unpacked BCD = 9 in decimal

3. (a) 0x34, 0x31

   (b) 0x30, 0x39

Chapter 4: Branch, Call, and Looping in Arm

In the sequence of instructions to be executed, it is often necessary to transfer program control to a different location (e.g. when a function is called, execution of a loop is repeated, or an instruction executes conditionally). There are many instructions in Arm to achieve this. This chapter covers the control transfer instructions available in Arm assembly language. In Section 4.1, we discuss instructions used for looping, as well as instructions for conditional and unconditional branches (jumps). In Section 4.2, we examine the instructions associated with calling subroutine. In Section 4.3, instruction timing and time delay subroutines are discussed. The Stack of Arm is discussed in Section 4.4. The Startup file is explored in Section 4.5.

Section 4.1: Looping and Branch Instructions

In this section we first discuss how to perform a looping action in Arm and then the branch (jump) instructions, both conditional and unconditional.

Looping in Arm

Repeating a sequence of instructions or an operation for a certain number of times is called a *loop*. The loop is one of the most widely used programming techniques. In the Arm, there are several ways to repeat an operation many times. One way is to repeat the operation over and over until it is finished, as shown below:

```
MOV    R0, #0         ; R0 = 0
MOV    R1, #9         ; R1 = 9
ADD    R0, R0, R1     ; R0 = R0 + R1, add 9 to R0  (Now R0 is 0x09)
ADD    R0, R0, R1     ; R0 = R0 + R1, add 9 to R0  (Now R0 is 0x12)
ADD    R0, R0, R1     ; R0 = R0 + R1, add 9 to R0  (Now R0 is 0x1B)
ADD    R0, R0, R1     ; R0 = R0 + R1, add 9 to R0  (Now R0 is 0x24)
ADD    R0, R0, R1     ; R0 = R0 + R1, add 9 to R0  (Now R0 is 0x2D)
ADD    R0, R0, R1     ; R0 = R0 + R1, add 9 to R0  (Now R0 is 0x36)
```

In the above program, we add 9 to R0 six times. That makes 6 × 9 = 54 = 0x36. One problem with the above technique is that too much code space would be needed for a large number of repetitions like 50 or 1000. A much better way is to use a loop. Next, we describe the method to do a loop in Arm.

Using instruction BNE for looping

The BNE (branch if not equal) instruction uses the zero flag in the status register (CPSR). The BNE instruction is used as follows:

```
BACK   .........          ; start of the loop
       .........          ; body of the loop
       .........          ; body of the loop
       SUBS  Rn, Rn, #1   ; Rn = Rn - 1, set the flag Z = 1 if Rn = 0
       BNE   BACK         ; branch if Z = 0
```

In the last two instructions, the Rn (e.g. R2 or R3) is decremented; if it is not zero, it branches (jumps) back to the target address referred to by the label. Prior to the start of the loop, the Rn is loaded with the counter value for the number of repetitions (loop count). Notice that the BNE instruction refers to the Z flag of the status register affected by the previous instruction, SUBS. This is shown in Example 4-1.

Example 4-1

Write a program to (a) clear R0, (b) add 9 to R0 a thousand times, then (c) place the sum in R4. Use the zero flag and BNE instruction.

Solution:

```
        ; --- this program adds value 9 to the R0 1000 times ---
        EXPORT  __main
        AREA    EXAMPLE4_1, CODE, READONLY
__main
        LDR     R2, =1000    ; R2 = 1000 (decimal) for counter
        MOV     R0, #0       ; R0 = 0 (sum)
AGAIN   ADD     R0, R0, #9   ; R0 = R0 + 9 (add 09 to R1, R1 = sum)
        SUBS    R2, R2, #1   ; Decrement counter and set the flags.
        BNE     AGAIN        ; repeat until COUNT = 0 (when Z = 1)
        MOV     R4, R0       ; store the sum in R4
HERE    B       HERE         ; stay here
        END
```

INSTRUCTIONS

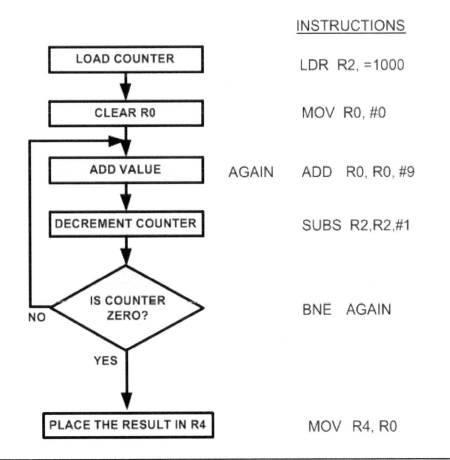

LDR R2, =1000

MOV R0, #0

AGAIN   ADD   R0, R0, #9

SUBS R2,R2,#1

BNE   AGAIN

MOV R4, R0

105

In the program in Example 4-1, register R2 is used as a counter. The counter is first set to 1000. In each iteration, the SUBS instruction decrements the R2 and sets the flag bits accordingly. If R2 is not zero (Z = 0), it jumps to the target address associated with the label "AGAIN". This looping action continues until R2 becomes zero. After R2 becomes zero (Z = 1), it falls through the loop and executes the instruction immediately below it, in this case "MOV R4, R0".

It must be emphasized again that we must use SUBS instead of SUB since the SUB instruction will not change (update) the flags in CPSR. As we mentioned in Chapter 3, many of the Arm instructions have the option of affecting the flags. In these instructions the default is not to affect the flags. Therefore, to update the flag we must add 'S' suffix to the instruction. That means SUBS and ADDS instructions are different from SUB and ADD, as far as the flags are concerned. As another example see Example 4-2.

Example 4-2

Write a program to place value 0x55 into 100 consecutive bytes of RAM locations.

Solution:

```
        RAM_ADDR EQU 0x20000000

        EXPORT  __main
        AREA    EXAMPLE4_2, CODE, READONLY
__main
        MOV     R2, #25          ; counter (25 x 4 = 100 byte block size)
        LDR     R1, =RAM_ADDR    ; R1 = RAM Address
        LDR     R0, =0x55555555  ; R0 = 0x55555555

OVER    STR     R0, [R1]         ; send it to RAM
        ADD     R1, R1, #4       ; R1 = R1 + 4 to increment pointer
        SUBS    R2, R2, #1       ; R2 = R2 - 1 for decrement counter
        BNE     OVER             ; keep doing it

HERE    B       HERE
        END
```

Looping a trillion times with loop inside a loop

As shown in Example 4-3, the maximum count is 2^{32}-1. What happens if we want to repeat an action more times than that? To do that, we use a loop inside a loop, which is called a nested loop. In a nested loop, we use two registers to hold the loop counts. See Example 4-3.

Example 4-3

Explain what is the maximum number of times that the loop in Example 4-1 can be repeated. Then, write a program to load the R0 register with the value 0x55, and complement it 16,000,000,000 (16 billion) times.

Solution:

Because Arm registers are 32-bit long, they can hold a maximum of 0xFFFFFFFF ($2^{32} - 1$ decimal); therefore, the loop can be repeated a maximum of $2^{32} - 1$ times. This example shows how to create a nesting loop to go beyond 4 billion times. Because 16,000,000,000 is larger than 0xFFFFFFFF (the maximum capacity of any R0–R12 registers), we use two registers to hold the counts. The following code shows how to use R2 and R1 as a register for counters in a nesting loop.

```
        EXPORT  __main
        AREA    EXAMPLE4_3, CODE, READONLY
__main
        MOV     R0, #0x55           ; R0 = 0x55
        MOV     R2, #16             ; load 16 into R2 (outer loop count)
L1      LDR     R1, =1000000000     ; R1 = 1,000,000,000 (inner loop count)
L2      EOR     R0, R0, #0xFF       ; complement R0 (R0 = R0 Ex-OR 0xFF)
        SUBS    R1, R1, #1          ; R1 = R1 - 1, decrement R1 (inner loop)
        BNE     L2                  ; repeat it until R1 = 0
        SUBS    R2, R2, #1          ; R2 = R2 - 1, decrement R2 (outer loop)
        BNE     L1                  ; repeat it until R2 = 0
HERE    B       HERE                ; stay here
        END
```

In this program, R1 is used to keep the inner loop count. In the instruction "BNE L2", whenever R1 becomes 0 it falls through and "SUBS R2, R2, #1" is executed. The next instructions force the CPU to load the inner count with 1,000,000,000 if R2 is not zero, and the inner loop starts again. This process will continue until R2 becomes zero and the outer loop is finished. If you use the Keil IDE to verify the operation of the above program use smaller values for counter to go through the iterations. See Figure 4-1.

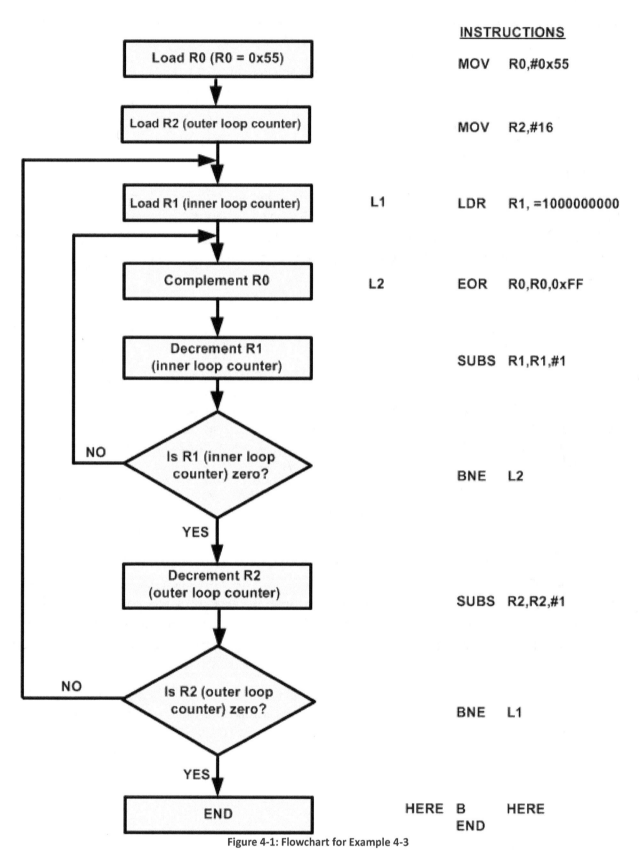

Figure 4-1: Flowchart for Example 4-3

Other conditional Branches

As we mentioned in Chapter 3, C and Z flags reflect the result of calculation on unsigned numbers. Table 4-1 lists available conditional branches for unsigned numbers that use C and Z flags. More details of each instruction are provided in Appendix A. In Table 4-1 notice that the instructions, such as BEQ (Branch if Z = 1) and BCS (Branch if carry set, C = 1), jump only if a certain condition is met. Next, we examine some conditional branch instructions with examples. The other conditional branch instructions associated with the signed numbers are discussed in Chapter 5 when arithmetic operations for signed numbers are discussed.

Instruction		Action
BCS/BHS	branch if carry set/branch if higher or same	Branch if C = 1
BCC/BLO	branch if carry clear/branch if lower	Branch if C = 0
BEQ	branch if equal	Branch if Z = 1
BNE	branch if not equal	Branch if Z = 0
BLS	branch if lower or same	Branch if Z = 1 or C = 0
BHI	branch if higher	Branch if Z = 0 and C = 1

Table 4-1: Arm Conditional Branch Instructions for Unsigned Data

BCC (branch if carry is clear, branch if C = 0)

In this instruction, the carry flag bit in program status registers (CPSR) is used to make the decision whether to branch or not. In executing "BCC label", the processor looks at the carry flag to see if it is cleared (C = 0). If it is, the CPU starts to fetch and execute instructions from the address of the label. If C = 1, it will not jump but will execute the next instruction below BCC. See Example 4-4.

Example 4-4

Examine the following code and give the result in registers R0, R1, and R2.

```
        MOV     R1, #0              ; clear high word (R1 = 0)
        MOV     R0, #0              ; clear low word (R0 = 0)
        LDR     R2, =0x99999999     ; R2 = 0x99999999
        ADDS    R0, R0, R2          ; R0 = R0 + R2 and set the flags
        BCC     L1                  ; if C = 0, jump to L1 and add next number
        ADDS    R1, R1, #1          ; ELSE, increment (R1 = R1 + 1)
L1      ADDS    R0, R0, R2          ; R0 = R0 + R2 and set the flags
        BCC     L2                  ; if C = 0, add next number
        ADDS    R1, R1, #1          ; if C = 1, increment
L2      ADDS    R0, R2              ; R0 = R0 + R2 and set the flags
        BCC     L3                  ; if C = 0, add next number
        ADDS    R1, R1, #1          ; C = 1, increment
L3      ADDS    R0, R2              ; R0 = R0 + R2 and set the flags
        BCC     L4                  ; if C = 0, add next number
        ADDS    R1, R1, #1          ; if C = 1, and set the flags
L4
```

Solution:

This program adds 0x99999999 together four times.

	R1 (high word)	R0 (low word)
At first	0	0
Just before L1	0	0x99999999
Just before L2	1	0x33333332
Just before L3	1	0xCCCCCCCB
Just before L4	2	0x66666664

Here is the loop version of the above program that runs 10 times.

```
        EXPORT  __main
        AREA    EXAMPLE4_4, CODE, READONLY
__main
        MOV     R1, #0              ; clear high word (R1 = 0)
        MOV     R0, #0              ; clear low word (R0 = 0)
        LDR     R2, =0x99999999     ; R2 = 0x99999999
        MOV     R3, #10             ; counter
L1      ADDS    R0, R2              ; R0 = R0 + R2 and set the flags
        BCC     NEXT                ; if C = 0, add next number
        ADD     R1, R1, #1          ; if C = 1, increment the upper word
NEXT    SUBS    R3, R3, #1          ; R3 = R3 - 1 and set the flags
                                    ; (Decrement counter)
        BNE     L1                  ; next round if Z = 0
HERE    B       HERE                ; stay here
        END
```

Note that there is also a "BCS label" instruction. In the BCS instruction, if C = 1 it jumps to the target address. We will give more examples of these instructions in the context of applications.

Comparison of unsigned numbers

```
        CMP     Rn, Op2     ; compare Rn with Op2 and set the flags
```

The CMP instruction compares two operands and set or clear the flags according to the result of the comparison. The operands themselves remain unchanged. There is no destination register and the second source operands can be a register or an immediate value (an 8-bit value with even number or rotate). It must be emphasized that "CMP Rn, Op2" instruction is really a subtract operation (SUBS) without a destination. Op2 is subtracted from Rn (Rn – Op2), the result is discarded and flags are set accordingly. Although all the C, S, Z, and V flags reflect the result of the comparison, only C and Z are used for unsigned numbers, as outlined in Table 4-2.

Instruction	C	Z
Rn > Op2	1	0
Rn = Op2	1	1
Rn < Op2	0	0

Table 4-2: Flag Settings for Compare (CMP Rn, Op2) of Unsigned Data

110

Look at the following case:

```
        LDR    R1, =0x35F    ; R1 = 0x35F
        LDR    R2, =0xCCC    ; R2 = 0xCCC
        CMP    R1, R2        ; compare 0x35F with 0xCCC
        BCC    OVER          ; branch if C = 0
        MOV    R1, #0        ; if C = 1, then clear R1
OVER    ADD    R2, R2, #1    ; R2 = R2 + 1 = 0xCCC + 1 = 0xCCD
```

Figure 4-2 shows the diagram and the C language version of the code.

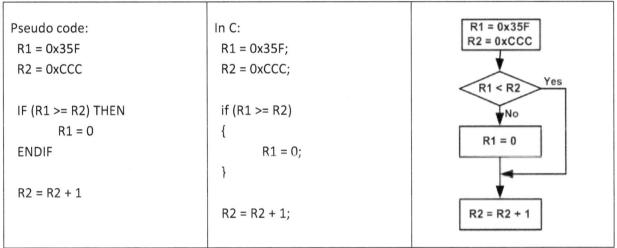

Figure 4-2: Flowchart of if Instruction

In the above program, R1 is less than the R2 (0x35F < 0xCCC); therefore, C = 0 and BCC (branch if carry clear) will go to target OVER. In contrast, look at the following:

```
        LDR    R1, =0xFFF
        LDR    R2, =0x888
        CMP    R1, R2              ; compare 0xFFF with 0x888
        BCC    NEXT
        ADD    R1, R1, #0x40
NEXT    ADD    R1, R1, #0x25
```

In the above, R1 is greater than R2 (0xFFF > 0x888), which sets C = 1, the branch, "BCC NEXT," is not taken and the execution falls through so that "ADD R1, R1, 0x40" is executed.

Again, it must be emphasized that in CMP instructions, the operands are unaffected regardless of the result of the comparison. Only the flags are affected. It also may be noted that, unlike other arithmetic and logic instructions, there is no need to put the 'S' suffix in the CMP instruction to update the flags. In other words, the CMP instruction always updates the flags.

Program 4-1 uses the CMP instruction to search for the highest byte in a series of 5 data bytes. To search for the highest value, the instruction "CMP R1, R3" works as follows where R1 is the contents of the memory location brought into R1 register by the [R2] pointer.

111

a) If R1 < R3, then C = 0 and R3 becomes the basis of the new comparison.

b) If R1 ≥ R3, then C = 1 and R1 is the larger of the two values and remains the basis of comparison.

Program 4-1

Assume that there is a class of five people with the following grades:

69, 87, 96, 45, and 75. Find the highest grade.

```
; searching for highest value in a list
COUNT       RN   R0              ; COUNT is the new name of R0
MAX         RN   R1              ; MAX is the new name of R1
                                 ; (MAX has the highest value)
POINTER     RN   R2              ; POINTER is the new name of R2
NEXT        RN   R3              ; NEXT is the new name of R3

            AREA  PROG_4_1D, DATA, READONLY
MYDATA      DCD   69, 87, 96, 45, 75

            EXPORT __main
            AREA  PROG_4_1, CODE, READONLY
__main
            MOV   COUNT, #5         ; COUNT = 5
            MOV   MAX, #0           ; MAX = 0
            LDR   POINTER, =MYDATA  ; POINTER has the address of first data
AGAIN       LDR   NEXT, [POINTER]   ; load NEXT with contents at address
                                    ; in POINTER
            CMP   MAX, NEXT         ; compare MAX and NEXT
            BHS   CTNU              ; if MAX > NEXT branch to CTNU
            MOV   MAX, NEXT         ; MAX = NEXT
CTNU        ADD   POINTER, POINTER, #4   ; increment POINTER for next word
            SUBS  COUNT, COUNT, #1  ; decrement counter
            BNE   AGAIN             ; branch AGAIN if counter is not zero

HERE        B     HERE
            END
```

Program 4-1 searches through five data items to find the highest value. The program has a variable called "MAX" that holds the highest grade found so far. One by one, the grades are brought into the register and compared to MAX. If any of them is higher, that value is placed in MAX. This continues until all data items are checked. A REPEAT-UNTIL structure was chosen in the program design. Figure 4-3 shows the flowchart for Program 4-1. This design could be used to code the program in many different languages.

Program 4-1 also demonstrates using aliasing for registers. This practice improves the readability of the small programs.

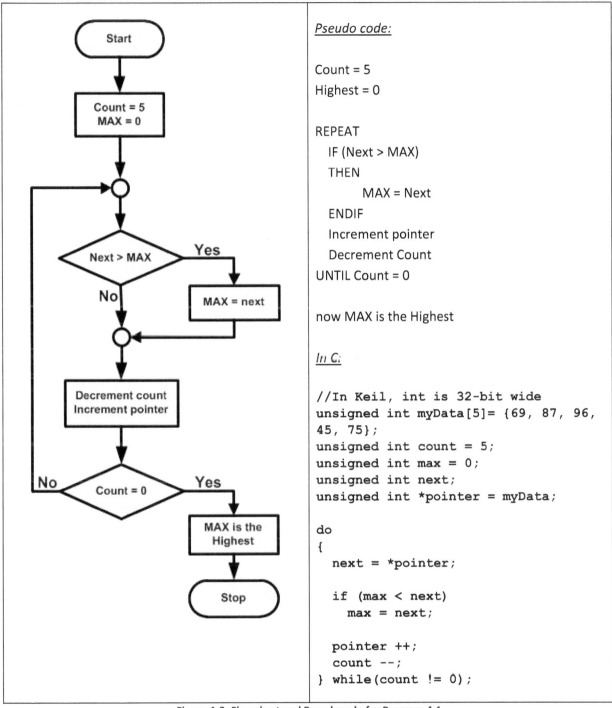

Pseudo code:

Count = 5
Highest = 0

REPEAT
  IF (Next > MAX)
  THEN
        MAX = Next
  ENDIF
  Increment pointer
  Decrement Count
UNTIL Count = 0

now MAX is the Highest

In C:

```c
//In Keil, int is 32-bit wide
unsigned int myData[5]= {69, 87, 96,
45, 75};
unsigned int count = 5;
unsigned int max = 0;
unsigned int next;
unsigned int *pointer = myData;

do
{
  next = *pointer;

  if (max < next)
    max = next;

  pointer ++;
  count --;
} while(count != 0);
```

Figure 4-3: Flowchart and Pseudocode for Program 4-1

Using CMP instruction followed by conditional branches we can make comparison on numbers, as shown in Table 4-3. Although BCS (branch carry set) and BCC (branch carry clear) check the carry flag and can be used after a compare instruction, it is recommended that BHS (branch higher or same) and BLO (branch lower) be used because "branch higher" and "branch lower" are easier to understand than "branch carry set" and "branch carry clear, " since it is more immediately apparent that one number is larger than another than whether a carry would be generated if the two numbers were subtracted.

Instruction		Action
BCS/BHS	branch if carry set/branch if higher or same	Branch if Rn ≥ Op2
BCC/BLO	branch if carry clear/branch lower	Branch if Rn < Op2
BEQ	branch if equal	Branch if Rn = Op2
BNE	branch if not equal	Branch if Rn ≠ Op2
BLS	branch if less or same	Branch if Rn ≤ Op2
BHI	branch if higher	Branch if Rn > Op2

Table 4-3: Arm Conditional Branch Instructions for Unsigned Data

Division of unsigned numbers in Arm

Some of the older Arm family members do not have instructions for division since it took too many gates to implement it. In ARMs with no divide instructions we can use SUB instruction to perform the division. Program 4-2 shows an example of an unsigned division using simple subtract operation. In the program the numerator is placed in a register and the denominator is subtracted from it repeatedly. The quotient is the number of times we subtracted and the remainder is in the register upon completion. This program is to demonstrate the used of conditional branch in a loop. The program is not efficient in calculating the quotient. There are much more efficient algorithms to perform division but they are beyond the scope here. See Figure 4-4 for the flowchart of the simple division program.

Program 4-2: Division by Repeated Subtractions

```
        EXPORT  __main
        AREA    PROG_4_2, CODE, READONLY      ; Division by subtractions
__main
        LDR   R0, =2012    ; R0 = 2012 (numerator)
                           ; it will contain remainder
        MOV   R1, #10      ; R1 = 10 (denominator)
        MOV   R2, #0       ; R2 = 0 (quotient)
L1      CMP   R0, R1       ; Compare R0 with R1 to see if less than 10
        BLO   FINISH       ; if R0 < R1 jump to finish
        SUB   R0, R0, R1   ; R0 = R0 - R1 (division by subtraction)
        ADD   R2, R2, #1   ; R2 = R2 + 1 (quotient is incremented)
        B     L1           ; go to L1 (B is discussed in the next section)
FINISH B     FINISH
```

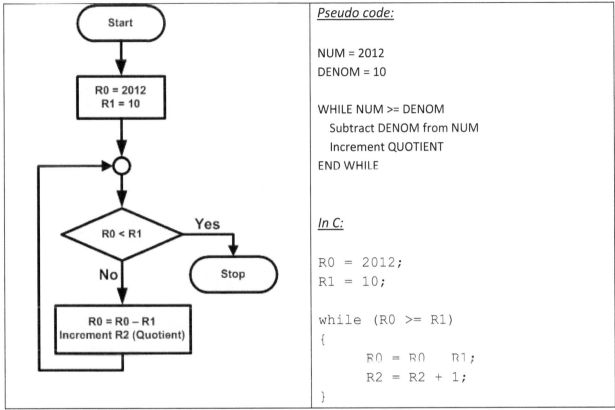

Figure 4-4: Flowchart and Pseudo-code for Program 4-2

TST (Test)

```
        TST    Rn, Op2              ; Rn AND with Op2 and flag bits are updated
```

The TST instruction is used to test the contents of register to see if one or multiple bits are HIGH. Similar to CMP instruction, TST is an ANDS instruction without a destination. After the operands are ANDed together the flags are updated. If the result is zero, then Z flag is raised and one can use BEQ (branch equal) to make decision. In the following example below, the program execution stays in the loop between OVER and BEQ OVER until bit 2 (0x04) of the content at "myport" becomes high.

```
        MOV    R0, #0x04            ; R0=00000100 in binary
        LDR    R1, =myport          ; port address
OVER    LDRB   R2, [R1]             ; load R2 from myport
        TST    R2, R0               ; is bit 2 HIGH?
        BEQ    OVER                 ; keep checking
```

In TST, like other data processing instructions, the Op2 can be an immediate value (an 8-bit value with even number of rotate). Look at the following example, which does the same as the program snippet above.

```
        LDR    R1, =myport          ; port address
OVER    LDRB   R2, [R1]             ; load R2 from myport
        TST    R2, #0x04            ; is bit 2 HIGH?
        BEQ    OVER                 ; keep checking
```

See Example 4-5.

Example 4-5

Assume address location 0x40010808 is assigned to an input port address and connected to 8 DIP switches. Write a short program to check the input port and whenever both pins 4 or 6 are LOW, R4 register is incremented.

Solution:

```
MYPORT EQU 0x40010808
       MOV  R0, #2_01010000   ; R0=0x50 (01010000 in binary)
       LDR  R1, =MYPORT       ; R1 = port address
OVER   LDRB R2, [R1]          ; get a byte from PORT and place it in R2
       TST  R2, R0            ; are bits 4 and 6 LOW?
       BNE  OVER              ; keep checking
       ADD  R4, R4, #1
```

TEQ (test equal)
```
       TEQ  Rn, Op2           ; Rn EX-ORed with Op2 and flag bits are set
```

The TEQ instruction is used to test to see if the contents of two registers or one register and the immediate value are equal. Like CMP and TST, TEQ is an EORS instruction without a destination. After the source operands are Ex-ORed together the flag bits are set according to the result. If result is 0, then Z flag is raised and one can use BEQ (branch zero) to make decision. Recall that if we Exclusive-OR a value with itself, the result is zero. Look at the following example for checking to see whether the temperature on a given port is equal to 100 or not:

```
TEMP   EQU  100
       MOV  R0, #TEMP         ; R0 = Temp
       LDR  R1, =myport       ; port address
OVER   LDRB R2, [R1]          ; load R2 from myport
       TEQ  R2, R0            ; is it 100?
       BNE  OVER              ; keep checking
```

Unconditional branch (jump) instruction

The unconditional branch is a jump in which control is transferred unconditionally to the target location. In the Arm there are two unconditional branches: B (branch) and BX (branch and exchange). This is discussed next.

B (Branch)

B (branch) is an unconditional jump that can go to any memory location within the ±32M byte address range.

B has different usages like implementing if/else, while, and for instructions. In the following code you see an example of implementing the if/else instruction:

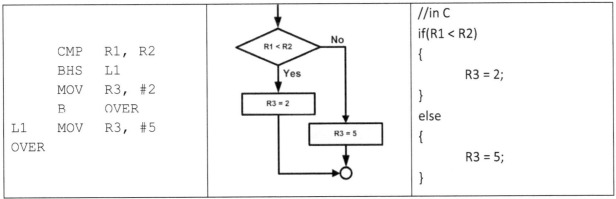

In the above code, R3 is initialized with 2 when R1 is lower than R2. Otherwise, it is initialized with 5.

As an example of implementing the while instruction see the following program. It calculates the sum of numbers between 1 and 5:

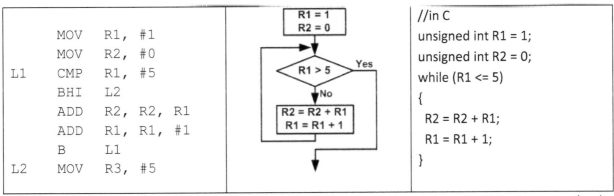

The *for* instruction can be implemented the same way as the *while* instruction. For example, the above assembly program can be considered as a *for* loop.

In cases where there is no operating system or monitor program, we use the "branch to itself" in order to keep the program from running away. In a stand-alone program, if we allow it to continue beyond the end of program, there is no telling what it is going to happen. A simple way of keeping the program from running away is shown below:

```
HERE    B       HERE    ; stay here
```

Calculating the branch address

The branch instruction is made of two parts: the opcode and the relative address. See Figure 4-5. The target address is relative to the value in the program counter. If the relative address is positive, the jump is forward. If the relative address is negative, then the jump is backward. Because the Thumb-2 instructions are either 2-byte or 4-byte long, the lowest bit of the addresses for instructions are always 0.

There is no need to keep the lowest bit in the relative address and the offset in the instruction does not hold the lowest bit. When the instruction is decoded, the offset is shifted left for one bit to form the offset.

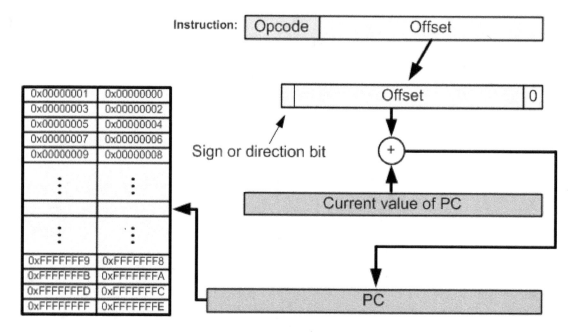

Figure 4-5: B (Branch) Instruction

Notice that in Cortex-M series (except Cortex-M0) pipeline has 3 stages: fetch, decode, and execute. So, when the branch instruction is executed, the next instruction is already fetched into the pipeline and the program counter is pointing to 4 bytes below. So, we add the relative address to the address of 4 bytes below the current instruction. See Example 4-6.

Example 4-6

In Cortex-M, the program counter points to 4 bytes below the current executing instruction. Using the following list file verify the jump forward address calculation.

LINE	ADDRESS	Machine		Mnemonic	Operand
1	00000000			EXPORT	__main
2	00000000			AREA	EXAMPLE_4_6, CODE, READONLY
3	00000000	F04F 0115	__main	MOV	R1, #0x15 ; R1 = 0x15
4	00000004	E005		B	THERE
5	00000006	F04F 0125		MOV	R1, #0x25 ; R1 = 0x25
6	0000000A	F04F 0235		MOV	R2, #0x35 ; R2 = 0x35
7	0000000E	F04F 0345		MOV	R3, #0x45 ; R3 = 0x45
8	**00000012**	F04F 0455	THERE	MOV	R4, #0x55 ; R4 = 0x55
9	00000016	E7FE	H	B	H
10	00000018	END			

Solution:

First notice that the B instruction in line 4 jumps forward. To calculate the target address, the relative address (offset) is shifted left and added to the PC. The value of PC is 4 bytes below the current

118

instruction and the current instruction has address 00000004. So the current value of PC is 0000004 + 4 = 0000008.

In line 4 the instruction "B THERE" has the machine code of E005. If we compare it with the B instruction format, we see that the operand is 000005. Recall that to calculate the target address, the relative address (offset) is shifted left and added to the current value of the PC (Program Counter). Shifting the offset (000005) left results in 00000A and then adding it to the PC (00000008) we have 000008 + 0000000A = 00000012 which is exactly the address of THERE label.

It must also be noted that for the backward branch the relative value is negative (2's complement). That is shown in Example 4-7.

Example 4-7

Verify the calculation of backward jumps for the listing of Example 4-1, shown below.

LINE	ADDRESS	Machine	Mnemonic		Operand
1	00000000			EXPORT	__main
2	00000000			AREA	EXAMPLE_4_7, CODE, READONLY
3	00000000	F44F 727A	__main	LDR	R2, =1000 ; R2 = 1000
4	00000004	F04F 0000		MOV	R0, #0 ; R0 = 0, sum
5	00000008	F100 0009	AGAIN	ADD	R0, R0, #9 ; R0 = R0 + 9
6	0000000C	1E52		SUBS	R2, R2, #1 ; R2 = R2 - 1
7	0000000E	D1FB		BNE	AGAIN ; repeat
8	00000010	4604		MOV	R4, R0 ; store the sum in R4
9	00000012	E7FE	H	B	H
10	00000014			END	

Solution:

In the program list, "BNE AGAIN" in line 7 has machine code D1FB. To separate the operand and opcode, we compare the instruction with the branch instruction format. The operand (relative offset address) is FB. The FB gives us −5, which means the displacement is (−5 × 2 = −10 = **−0x0A**).

The branch is located in address 0x000E. The current value of PC is 0x000E + 4 = **0x0012**.

When the relative address of −0x0A is added to 00000012, we have −0x000A + 0x0012 = 0x08

Notice that 00000008 is the address of the label AGAIN.

FC is a negative number and that means it will branch backward. For further discussion of the addition of negative numbers, see Chapter 5.

119

Branching beyond the limits

Although branch instruction does not cover the whole 4 GB memory space of Arm, it is more than adequate for most of the applications. In rare cases that there is need to branch to whole 4 GB, we use BX (branch and exchange) instruction. The "BX Rn" instruction uses register Rn to hold target address. Since Rn can be any of the R0–R14 registers and they are 32-bit registers, the "BX Rn" instruction can land anywhere in the 4G bytes address space of the Arm. In the instruction "BX R2" the content of R2 is loaded into the program counter (R15) and CPU starts to fetch instructions from the target address pointed to by the program counter. See Figure 4-6.

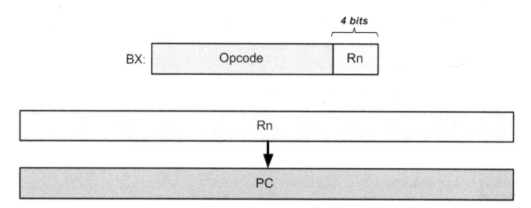

Figure 4-6: BX (Branch and exchange) Instruction Target Address

Since the instructions are either 16-bit or 32-bit long, the lowest bit of the Rn are 0s. The BX instruction is also used to switch between Arm and THUMB modes using bit 0 of the register operand. If bit 0 is set, the mode changes to Thumb mode; otherwise, the mode will be changed to Arm mode.

Review Questions

1. The mnemonic BNE stands for _____.
2. True or false. "BNE BACK" makes its decision based on the last instruction affecting the Z flag.
3. "BNE HERE" is a ____ -byte instruction.
4. In "BEQ NEXT", which flag bit is checked to see if it is high?
5. B(ranch) is a(n) ____ -byte instruction.
6. Compare B and BX instructions.

Section 4.2: Calling Subroutine with BL

Another control transfer instruction is the BL (branch with link) instruction, which is used to call a subroutine. Subroutines are often used to perform tasks that need to be performed frequently. This makes a program more structured in addition to saving memory space.

BL (Branch and Link) instruction and calling subroutine

The BL instruction is made of two parts: the opcode and the offset. The offset is used to address the target subroutine, as shown in Figure 4-7.

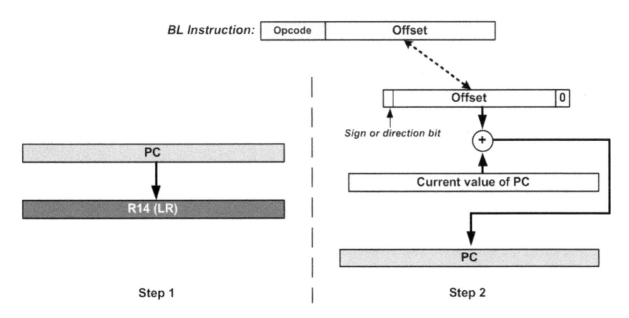

BL Instruction:

| Opcode | Offset |

Offset | 0

Sign or direction bit

(+)

Current value of PC

PC

PC

R14 (LR)

Step 1 | Step 2

Figure 4-7: BL (Branch and Link) Instruction

The link register and returning from subroutine

To make sure that the Arm knows where to return to after execution of the called subroutine, the BL instruction automatically saves the address of the instruction immediately below the BL (the return address) in the link register (LR), the R14. After finishing the subroutine, the program execution should return to where the caller is left off. This is done by putting the return address into the program counter. To return, we may use "BX LR" instruction, which copies the content of LR to PC, to transfer control back to the caller.

To further understand the role of the R14 register in BL instruction and the return, examine the Examples 4-8. The following points should be noted for the Example 4-8:

1. Notice the DELAY subroutine. Upon executing the first "BL DELAY", the address of the instruction right below it, "MOV R0, #0xAA", is saved onto the R14 register, and the CPU starts to execute instructions at DELAY subroutine.

2. In the DELAY subroutine, first the counter R3 is set to 5 (R3 = 5); therefore, the inner loop is repeated 5 times. When R3 becomes 0, control falls to the "BX LR" instruction, which restores the address into the program counter and returns to main program to resume executing the instructions after the BL.

Example 4-8

Write a program to toggle all the bits of address 0x40000000 by sending to it the values 0x55 and 0xAA continuously. Put a time delay between each issuing of data to address 0x40000000 location.

Solution:

```
        EXPORT       __main
        AREA    EXAMPLE4_8, CODE, READONLY
RAM_ADDR     EQU   0x40000000  ; change the address for your Arm
__main
        LDR    R1, =RAM_ADDR      ; R1 = RAM address
AGAIN MOV    R0, #0x55    ; R0 = 0x55
        STRB   R0, [R1]     ; send it to RAM
        BL     DELAY        ; call delay (R14 = PC of next instruction)
        MOV    R0, #0xAA    ; R0 = 0xAA
        STRB   R0, [R1]     ; send it to RAM
        BL     DELAY        ; call delay
        B      AGAIN        ; keep doing it

        ; -------------------DELAY SUBROUTINE
DELAY LDR    R3, =5       ; R3=5, modify this value for different delay
L1      SUBS   R3, R3, #1  ; R3 = R3 - 1
        BNE    L1
        BX     LR           ; return to caller
        ; -------------------end of DELAY subroutine
        END           ; notice the place for END directive
```

Use Keil IDE simulator for Arm to simulate the above program and examine the registers and memory location 0x40000000. You might have to change the address 0x40000000 to some other value depending on the RAM address of the Arm chip you use.

In above program, in place of "BX LR" for return, we could have used "BX R14", "MOV R15, R14", or "MOV PC, LR" instructions. All of them will copy the content of LR to PC; but it is recommended to use the "BX LR" instruction.

The amount of time delay in Example 4-8 depends on the frequency of the Arm chip. The time calculation will be explained in the next section of this chapter.

PROC and ENDP directives

The PROC (procedure) and ENDP (end procedure) assembler directives are used to mark the beginning and the end of subroutines in Keil. If the directives are not used, Keil does not go into the subroutine if it is called from another file, while debugging. For example, the DELAY subroutine should be defined as follows in the Keil IDE:

```
DELAY PROC
   ...
   BX LR
   ENDP
```

Main Program and Calling Subroutines

In real world projects we divide the programs into small subroutines (also called functions) and the subroutines are called from the main program. Figure 4-8 shows the format.

```
       ; MAIN program calling subroutines
       AREA   ProgramName, CODE, READONLY


__main      PROC
       BL     SUBR_1              ; Call Subroutine 1
       BL     SUBR_2              ; Call Subroutine 1
       BL     SUBR_3              ; Call Subroutine 1
HERE   B      HERE         ; stay here
       ENDP   ; ----end of MAIN


       ; -------------------SUBROUTINE 1
SUBR_1      PROC

       ....
       ....
       BX     LR      ; return to main
       ENDP   ; --- end of subroutine 1


       ; -------------------SUBROUTINE 2
SUBR_2      PROC

       ....
       ....
       BX     LR      ; return to main
       ENDP   ; ------    end of subroutine 2


       ; -------------------SUBROUTINE 3
SUBR_3      PROC

       ....
       ....
       BX     LR      ; return to main
       ENDP   ; ------    end of subroutine 3
       END             ; notice the END of file
```

Figure 4-8: Arm Assembly Main Program That Calls Subroutines

Program 4-3 shows an example of the main program calling subroutine.

Program 4-3

```
       ; This program fills a block of memory with a fixed value and
       ; then transfers (copies) the block to new area of memory
RAM1_ADDR   EQU   0x40000000   ; Change the address for your Arm
RAM2_ADDR   EQU   0x40000100   ; Change the address for your Arm
       EXPORT __main
       AREA   PROGRAM4_3, CODE, READONLY
__main      PROC
       BL     FILL                ; call block fill subroutine
       BL     COPY                ; call block transfer subroutine
```

```
HERE    B       HERE                ; Brach here
        ENDP
        ; ----------------BLOCK FILL SUBROUTINE
FILL    PROC
        LDR     R1, =RAM1_ADDR      ; R1 = RAM Address pointer
        MOV     R0, #10             ; counter
        LDR     R2, =0x55555555
L1      STR     R2, [R1]            ; send it to RAM
        ADD     R1, R1, #4          ; R1 = R1 + 4 to increment pointer
        SUBS    R0, R0, #1          ; R0 = R0 - 1 to decrement counter
        BNE     L1                  ; keep doing it until R0 is 0
        BX      LR                  ; return to caller
        ENDP
        ; ----------------BLOCK COPY SUBROUTINE
COPY    PROC
        LDR     R1, =RAM1_ADDR      ; R1 = RAM Address pointer (source)
        LDR     R2, =RAM2_ADDR      ; R2 = RAM Address pointer (destination)
        MOV     R0, #10             ; counter
L2      LDR     R3, [R1]            ; get from RAM1
        STR     R3, [R2]            ; send it to RAM2
        ADD     R1, R1, #4          ; R1 = R1 + 4 to increment pointer for RAM1
        ADD     R2, R2, #4          ; R2 = R2 + 4 to increment pointer for RAM2
        SUBS    R0, R0, #1          ; R0 = R0 - 1 for decrementing counter
        BNE     L2                  ; keep doing it
        BX      LR                  ; return to caller
        ENDP    ; ----------
        END
```

Register usage in a subroutine

The registers can be used to pass data between the caller and the subroutine. When calling a subroutine, the parameters may be stored in the registers. When exiting the subroutine, the return value may also be left in the register. To improve the compatibility of the reusable software modules, Arm published Arm Architecture Procedure Call Standard (AAPCS), which defines the register usages among other things. According to AAPCS, the first four registers (R0-R3) are used to pass data between caller and subroutine. Caller should not expect the data in these four registers to be preserved. The rest of the registers except PC (program counter, R15) and SP (stack pointer, R13) should be preserved across a subroutine call.

Review Questions

1. The mnemonic BL stands for _____.
2. True or false. "BL DELAY" saves the address of the instruction below BL in LR register.
3. "BL DELAY" is a ____ -byte instruction.
4. LR is an ____ -bit register.
5. LR is the same as _____register.
6. Explain the difference between B and BL instructions.

Section 4.3: Time Delay

In this section we discuss how to generate various time delays and calculate time delays for the Arm. We will also discuss the impact of pipelining on execution time.

Delay calculation for the Arm

In creating a time delay using assembly language instructions, one must be mindful of two factors that can affect the accuracy of the delay:

1. **The core clock frequency:** The frequency of the core clock connected to the CPU is one factor in the time delay calculation. The duration of the clock period for the instruction cycle is a function of this core clock frequency.

2. **The Arm design:** Since the 1970s, both the field of IC technology and the architectural design of microprocessors have seen great advancements. Due to the limitations of IC technology and limited CPU design experience for many years, the instruction cycle duration was longer. Advances in both IC technology and CPU design in the 1980s and 1990s have made the single instruction cycle a common feature of many microprocessors. Indeed, one way to increase performance without losing code compatibility with the older generation of a given family is to reduce the number of instruction cycles it takes to execute an instruction. One might wonder how microprocessors such as Arm are able to execute an instruction in one cycle. There are three ways to do that: (a) Use Harvard architecture to get the maximum amount of code and data into the CPU, (b) use RISC architecture features, and finally (c) use pipelining to overlap fetching and execution of instructions.

Instruction cycle time for the Arm

It takes certain amount of time for the CPU to execute an instruction. The unit of time is referred to as *machine cycles*. Thanks to the RISC architecture, Arm executes most instructions in one machine cycle. The length of the machine cycle depends on the frequency of the oscillator connected to the core clock of the CPU. The oscillator circuitry with external crystal or on-chip clock reference, provides the clock source for the Arm CPU. To calculate the machine cycle for the CPU, we take the inverse of the oscillator frequency, as shown in Example 4-9.

Example 4-9

The following shows the oscillator frequency for four different Arm-based systems. Find the period of the instruction cycle in each case.
(a) 8 MHz   (b) 72 MHz   (c) 100 MHz   (d) 50 MHz

Solution:

(a) instruction cycle is 1/8 MHz = 0.125 ms (microsecond) = 125 ns (nanosecond)
(b) instruction cycle = 1/72 MHz = 0.01389 ms = 13.89 ns
(c) instruction cycle = 1/100 MHz = 0.01 ms = 10 ns

(d) instruction cycle = 1/50 MHz = 0.02 ms = 20 ns

Branch penalty

The overlapping of fetch and execution of the instruction is widely used in today's microprocessors such as Arm. For the concept of pipelining to work, we need a buffer or queue in which instructions are pre-fetched and ready to be executed. In some circumstances, the CPU must flush out the queue. For example, when a branch is taken and the CPU starts to fetch code from a new memory location, the code in the queue that was previously fetched becomes useless. In this case, the execution unit must wait until the new instruction is fetched. This is called a branch penalty. The penalty is an extra instruction cycle time to fetch the instruction from the new location instead of executing the instruction already in the queue. This means that while the vast majority of Arm instructions take only one machine cycle, some instructions take three machine cycles. These are Branch, BL (call), and all the conditional branch instructions such as BNE, BLO, and so on. The conditional branch instruction can take only one machine cycle if the condition is not met and the branch is not taken. For example, the BNE will jump if Z = 0 and that takes three machine cycles. If Z = 1, then it falls through and it takes only one machine cycle. See Examples 4-10 and 4-11.

Example 4-10

For an Arm system with the core clock running at 100 MHz, find how long it takes to execute each of the following instructions:

(a) MOV   (b) SUB   (c) B

(d) ADD   (e) NOP   (f) BHI

(g) BLO   (h) BNE   (i) EQU

Solution:

The machine cycle for a system of 100 MHz clock is 10 ns, as shown in Example 4-9. Therefore, we have:

Instruction	Instruction cycles	Time to execute
(a) MOV	1	1 × 10 ns = 10 ns
(b) SUB	1	1 × 10 ns = 10 ns
(c) B	3	3 × 10 ns = 30 ns
(d) ADD	1	1 × 10 ns = 10 ns
(e) NOP	1	1 × 10 ns = 10 ns

For the following, due to branch penalty, 3 clock cycles if taken and 1 if it falls through:

(f) BHI	3/1	3 × 10 ns = 30 ns
(g) BLO	3/1	3 × 10 ns = 30 ns
(h) BNE	3/1	3 × 10 ns = 30 ns
(i) EQU	0	(directives do not produce machine instructions)

Delay calculation for Arm

A delay subroutine consists of two parts: (1) setting a counter, and (2) a loop. Most of the time delay is performed by the body of the loop, as shown in Example 4-11.

Example 4-11

Find the size of the delay of the code snippet below if the system clock frequency is 100 MHz:

```
DELAY MOV    R0, #255
AGAIN PROC
      NOP
      NOP
      SUBS  R0, R0, #1
      BNE   AGAIN
      BX    LR          ; return
      ENDP
```

Solution:

We have the following machine cycles for each instruction of the DELAY subroutine:

	Instruction			Machine Cycle
DELAY	PROC			
	MOV	R0, #255	;	1
AGAIN	NOP		;	1
	NOP		;	1
	SUBS	R0, R0, #1	;	1
	BNE	AGAIN	;	3/1
	BX	LR	;	3
	ENDP			

Therefore, we have a time delay of [1 + ((1+1+1+3) × 255) + 3] × 10 ns = 15, 340 ns.
Notice that BNE takes three instruction cycles if it jumps back, and takes only one cycle when falling through the loop. That means the above number should be 15,320 ns. Because the last time, when R0 is zero, the BNE takes only one cycle because it falls through the loop

---

Often, we calculate the time delay based on the instructions inside the loop and ignore the clock cycles associated with the instructions outside the loop.

In Example 4-11, the largest value the R0 register can take is 2^{32} = 4G. One way to increase the delay is to use many NOP instructions within the loop. NOP, which stands for "no operation," simply wastes time, but takes 2 bytes of program memory and that is too heavy a price to pay for just one instruction cycle. A better way is to use a nested loop.

Loop inside a loop delay

Another way to get a large delay is to use a loop inside a loop, which is also called a *nested loop*. See Example 4-12.

Example 4-12

In a given Arm trainer an I/O port is connected to 8 LEDs. The following program toggles the LEDs by sending to it 0x55 and 0xAA values continuously. Calculate the time delay for toggling of LEDs. Assume the system clock frequency of 100 MHz.

Solution:

```
        AREA   Example4_12, CODE, READONLY
        EXPORT __main
PORT_ADDR   EQU   0x4001080C  ; change the address for your Arm and PORT
__main      PROC
        LDR    R1, =PORT_ADDR   ; R1 = port address
AGAIN MOV    R0, #0x55   ; R0 = 0x55
        STRB   R0, [R1]    ; send it to LEDs
        BL     DELAY       ; call delay
        MOV    R0, #0xAA   ; R0 = 0xAA
        STRB   R0, [R1]    ; send it to LEDs
        BL     DELAY       ; call delay
        B      AGAIN       ; keep doing it forever
        ENDP


        ; -------------------DELAY SUBROUTINE
DELAY PROC
        MOV    R3, #100   ; R3 = 100, modify this value for different size delay
L1      LDR    R4, =250000 ; R4 = 250, 000 (inner loop count)
L2      SUBS   R4, R4, #1  ; 1 clock
        BNE    L2          ; 3 clock
        SUBS   R3, R3, #1  ; R3 = R3 - 1
        BNE    L1
        BX     LR    ; return to caller
        ENDP
        END
```

Ignoring the delay associated with the outer loop, we have the following time delay:
$[(1 + 3) \times 250, 000 \times 100] \times 10$ ns = 1 second since 1/100 MHz = 10 ns.

Examine the working frequency for your Arm trainer, change the above address 0x4001080C to your Arm trainer port address and verify the time delay using oscilloscope.

From these discussions we conclude that the use of instructions in generating time delay is not the most reliable method. To complicate the matter, newer performance enhancements of the CPU hardware or the compiler software may affect the loop timing.

To get more accurate time delay Timers are used. All Arm microcontrollers come with on-chip Timers. We can use Keil uVision's simulator to verify delay time and number of cycles used. Meanwhile, to get an accurate time delay for a given Arm microcontroller, we must use an oscilloscope to verify the exact time delay.

Review Questions

1. True or false. In the Arm, the machine cycle lasts 1 clock period of the core clock frequency.
2. The minimum number of machine cycles needed to execute an Arm instruction is _____.
3. Find the machine cycle for a core clock frequency of 66 MHz.
4. Assuming a core clock frequency of 100 MHz, find the time delay associated with the loop section of the following DELAY subroutine:

```
DELAY       PROC
            LDR     R2,  =50000000
HERE        NOP
            NOP
            NOP
            NOP
            NOP
            SUBS    R2,  R2,  #1
            BNE     HERE
            MOV     PC,  LR
            ENDP
```

5. Find the machine cycle for an Arm if the core clock frequency is 50 MHz.
6. True or false. In the Arm, the instruction fetching and execution are done at the same time.
7. True or false. B and BL will always take 2 machine cycles.
8. True or false. The BNE instruction will always take 3 machine cycles.

Section 4.4: Stack in Arm Cortex

Stack is like a stack of pancakes that you usually put a new one on the top or take the one on the top off. Similarly, stack is a part of RAM memory. You store data on top of the stack and you get the top most data. Storing data on to the stack is called push and getting data from the stack is called pop.

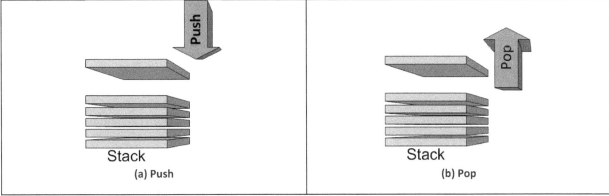

(a) Push

(b) Pop

Figure 4-9: Push and Pop

Since stack is used extensively in handling subroutine call and interrupt, most of the processors have hardware support to facilitate the creation of stacks in assembly language programming.

Each stack has a stack pointer (SP) that points to the top-of-stack where data is pushed onto or popped off. In Arm Cortex, register R13 is used as stack pointer. The PUSH instruction stores the contents of registers onto the stack and POP pops off the data from top of stack to registers.

Pushing onto the stack

To push registers onto stack the PUSH instruction is used:

```
PUSH {register list}    ;push the contents of the registers on to the stack
```

The register list can contain any of the CPU registers (R0-R15). For example, to store the value of R10 we can write the following instruction:

```
PUSH {R10}    ;store R10 onto the stack
```

The register list can be more than one register, as well. See the following examples:

```
PUSH {R4,R9,R11}   ;store R4, R9, and R11
PUSH {R5-R9}       ;store R5, R6, R7, R8, and R9 onto the stack
```

When we push data onto the stack, the SP is decremented by 4 to point to the word above the top of stack and then the content of a register is copied into that space. Assuming that SP is pointing to location 0x194, and R4 contains 20, the following figure shows pushing R4 onto the stack.

Popping from the stack

Popping the contents of the stack back into a given register is the opposite process of pushing. When the POP instruction is executed, the data in the top location of the stack is copied the registers and the SP is incremented by 4. The POP instruction is used to pop off data from top of stack:

```
POP {register list}     ;pop from the top of stack to the register list
```

For example, the following instruction pops from the top of stack and copies to R7:

```
POP {R7}
```

See Example 4-13.

Example 4-13

The following program places some data into registers and pushes them onto the stack. Assuming that SP is initialized with 0x20000600, examine the stack, stack pointer, and the registers after the execution of each instruction.

```
      EXPORT  __main
      AREA    STACK_SAMPLE1, CODE, READONLY
__main      PROC
      LDR   R0, =0x123        ; R0 = 0x123
```

130

```
        LDR     R1, =0x455          ; R1 = 0x455
        LDR     R2, =0x6677         ; R2 = 0x6677
        PUSH    {R0}
        PUSH    {R1}
        PUSH    {R2}
        MOV     R0,#0
        MOV     R1,#0
        MOV     R2,#0
        POP     {R2}
        POP     {R1}
        POP {R0}
HERE    B       HERE                ; stay here
        ENDP
```

After the execution of	Contents of some the registers (in Hex)				Stack
	R0	R1	R2	SP (R13)	
LDR R0, =0x123 LDR R1, =0x455 LDR R2, =0x6677	123	455	6677	20000600	MSB / LSB; 200005F4, 200005F8, 200005FC, 20000600 ← SP
PUSH {R0}	123	455	6677	200005FC	200005F4; 200005F8; 200005FC 00 00 01 23 ← SP; 20000600
PUSH {R1}	123	455	6677	200005F8	200005F4; 200005F8 00 00 04 55 ← SP; 200005FC 00 00 01 23; 20000600
PUSH {R2}	123	455	6677	200005F4	200005F4 00 00 66 77 ← SP; 200005F8 00 00 04 55; 200005FC 00 00 01 23; 20000600

131

Instruction	R0	R1	R2	SP	Memory				

MOV R0,#0
 	
MOV R1,#0
 	
MOV R2,#0
	0	0	0	200005F4	200005F4	00	00	66	77 ◄SP

					200005F8	00	00	04	55
					200005FC	00	00	01	23
					20000600				
POP {R2}	0	0	6677	200005F8	200005F4				
					200005F8	00	00	04	55 ◄SP
					200005FC	00	00	01	23
					20000600				
POP {R1}	0	455	6677	200005FC	200005F4				
					200005F8				
					200005FC	00	00	01	23 ◄SP
					20000600				
POP {R0}	123	455	6677	20000600	200005F4				
					200005F8				
					200005FC				
					20000600				◄SP

Initializing the stack pointer in Arm Cortex

When the Cortex is powered up, the R13 (SP) register is loaded with the contents of memory location 0. Therefore, the address of stack must be stored in locations 0x00000000 to 0x00000003 of memory so that SP points to somewhere in the internal SRAM. In Arm, when we push onto the stack, the SP is decremented and then data is stored. So, the stack pointer should be initialized with the address of the location next to the last location of stack memory. For example, if we want to use locations 0x20000000 to 0x200000FF for stack, we should store 0x20000100 in location 0x00000000. In the case, when the Arm is powered up, the SP is loaded with 0x20000100 and when the first PUSH is executed, the SP is decremented by 4 and the data is stored in the last word of stack (0x200000FC to 0x200000FF).

Some Stack Usages

Nested calls

In Arm, the stack is used for subroutine calls and interrupt handling. We must remember that upon calling a subroutine from the main program using the BL instruction, R14, the linker register, keeps track of where the CPU should return to after completing the subroutine. Now, if we have another call inside the subroutine using the BL instruction, then it is our job to store the original R14 on the stack.

Failure to do that will end up crashing the program. In the Program 4-4, FUNC2 is called in FUNC1. The execution of BL changes the link register (LR). So, before calling FUNC2, the LR register is stored onto the stack and it is restored after returning.

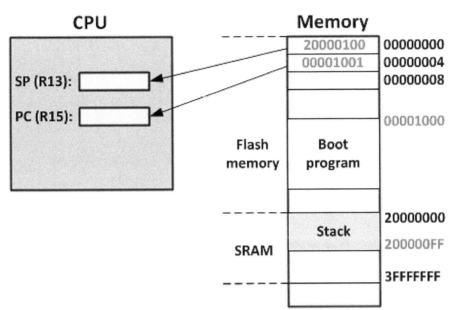

Figure 4-10: Power-up Initialization in Cortex

Program 4-4

```
      EXPORT  __main
      AREA  MY_CODE, CODE, READONLY
__main
      BL    FUNC1
HERE  B     HERE


      ; subroutine FUNC1
FUNC1
      PUSH  {LR}  ;store the LR on the stack
      BL    FUNC2 ;the func. call changes the LR
      POP   {LR}  ;restore the LR value from the stack
      BX    LR


      ; subroutine FUNC2
FUNC2
      ;do something
      BX LR ; return
      END
```

Preserving registers values

Each subroutine may use some registers to hold data temporarily. The subroutine has no knowledge whether the caller left any data in the registers that need to be preserved. So each subroutine should preserve any register it is going to use and restore them before return. In Program 4-5, R7 is used in the delay subroutine. So, its value is preserved using the stack.

Program 4-5

```
        EXPORT  __main
        AREA  MY_CODE, CODE, READONLY
__main
        BL    DELAY
HERE    B     HERE
DELAY
        PUSH  {R7}  ;store R7 onto the stack
        LDR   R7,=120000
D_1     SUBS  R7,R7,#1
        BNE   D_1
        POP   {R7}  ;restore R7
        END
```

Review Questions

1. The _____ register is the default stack pointer.
2. Write an instruction that pushes R5, R6, R7, and R8 into the stack.
3. Write an instruction that pops R5, R6, R7, and R8 from the stack.
4. On power-up, the SP is loaded with the contents of location _____ .

Section 4-5: Exploring the Startup File

In your real projects, you will need to change the Startup file. So, in this section we will explore the contents of the startup file. Figure 4-11 is a generic startup file which shows the important parts of startup files. The file has 4 important parts: stack, heap, vector table, and reset handler.

Figure 4-11: A Generalized startup.s file

```
1                      ;---- Stack -----------------------------------------
2      Stack_Size      EQU     0x00000400 ;stack size in bytes
3                      AREA    STACK, NOINIT, READWRITE, ALIGN=3
4      Stack_Mem       SPACE   Stack_Size ;allocating memory for stack
5      __initial_sp
6
7                      ;---- Heap  -----------------------------------------
8      Heap_Size       EQU     0x00000200 ;Heap size in bytes
9
10                     AREA    HEAP, NOINIT, READWRITE, ALIGN=3
11     __heap_base
12     Heap_Mem        SPACE   Heap_Size
13     __heap_limit
14
15
16                     ;---- Vector Table Mapped to Address 0 at Reset ----
```

```
17                    AREA      RESET, DATA, READONLY
18                    EXPORT    __Vectors
19                    EXPORT    __Vectors_End
20                    EXPORT    __Vectors_Size
21
22      __Vectors     DCD       __initial_sp           ; Top of Stack
23                    DCD       Reset_Handler          ; Reset Handler
24                    DCD       NMI_Handler            ; NMI Handler
25                    DCD       HardFault_Handler      ; Hard Fault Handler
26                    DCD       MemManage_Handler      ; MPU Fault Handler
27                    DCD       BusFault_Handler       ; Bus Fault Handler
28                    DCD       UsageFault_Handler     ; Usage Fault Handle
29                    DCD       0                      ; Reserved
30                    DCD       0                      ; Reserved
31                    DCD       0                      ; Reserved
32                    DCD       0                      ; Reserved
33                    DCD       SVC_Handler            ; SVCall Handler
34                    DCD       DebugMon_Handler       ; Debug Monitor Handler
35                    DCD       0                      ; Reserved
36                    DCD       PendSV_Handler         ; PendSV Handler
37                    DCD       SysTick_Handler        ; SysTick Handler
38
39
40            ;---- Reset handler -------------------------------
41                    AREA      |.text|, CODE, READONLY
42      Reset_Handler PROC
43                    EXPORT    Reset_Handler              [WEAK]
44                    IMPORT    __main
45                    IMPORT    SystemInit
46                    LDR       R0, =SystemInit
47                    BLX       R0
48                    LDR       R0, =__main
49                    BX        R0
50                    ENDP
```

Stack

Lines 1 to 5 are related to stack memory allocation. The AREA is defined as READWRITE, since the stack memory must be allocated from the RAM memory. In line 4, the space is allocated for the stack. Since Stack_Size is equal to 0x400, 0x400 bytes (1024 bytes) are allocated for the stack. If you like to change the stack size, you can change the value of Stack_Size in line 1. As we saw in the stack section, the stack pointer should be initialized with the address of the next byte to the last location of the stack space. The next byte to the stack space is labeled as __initial_sp and it is used in the vector table to initialize the stack pointer.

Heap

The heap memory is used in C language for dynamic variables. Lines 7 to 13, allocate a part of RAM for heap. If you like to change the heap size, you can change the value of Heap_Size.

Vector Table

Lines 16 to 38 are for vector table. The AREA is defined as READONLY in line 17. So, the vector table will be placed in the flash memory. Since it is labeled as __vector in line 18, it will be placed in first locations of memory starting from location 0x0000. As we saw in this chapter and Chapter 2, when the Arm is powered up, the contents of memory locations 0000 to 0x0003 are loaded to the stack pointer, and the PC is loaded with contents of locations 0x0004 to 0x0007. In line 18, the value of __initial_sp is stored in location 0 (the first location of vector table). DCD allocates 4 bytes of memory. So, __initial_sp is stored in locations 0x0000 to 0x0003. For example, if the stack space ends in location 0x200005FF, __initial_sp has the address of 0x20000600; and 0x20000600 will be stored in locations 0 to 3. Similarly, the address of Reset_Handler is stored in locations 0x0004 to 0x0007. See Figure 4-12.

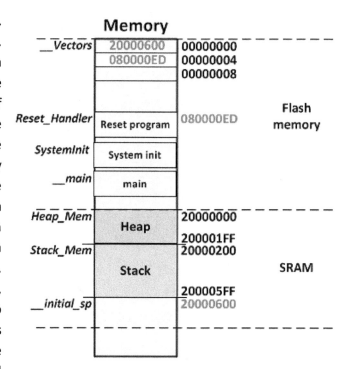

Figure 4-12: Memory Allocation for the Program

The Vector table is discussed in more detail in the interrupt chapter.

Reset Handler

Lines 40 to 50 are for Reset Handler. The address of Reset_Handler is stored in locations 0x0004 to 0x0007. So, when the CPU is powered up, the program counter is loaded with the address of Reset_Handler. Reset Handler is a piece of code which is executed first.

In line 46, R0 is loaded with the address of the SystemInit subroutine and then SystemInit is called in line 47. SystemInit sets the CPU clock frequency and initializes the system. SystemInit is in the system_xxx.c file.

In lines 48 and 49, R0 is loaded with the address of __main and then branches to the __main subroutine. That is why we label our first line of code as __main. In C programs, the programs begin from the main function, as well.

Problems

Section 4.1: Looping and Branch Instructions

1. In the Arm, looping action using a single register is limited to _____ iterations.
2. If a conditional branch is not taken, what is the next instruction to be executed?

3. In calculating the target address for a branch, a displacement is added to the contents of register _____.
4. The mnemonic BNE stands for _____.
5. What is the advantage of using BX over B?
6. True or false. The target of a BNE can be anywhere in the 4G word address space.
7. True or false. All Arm branch instructions can branch to anywhere in the 4G byte address space.
8. Dissect the B instruction and indicate how it can branch.
9. True or false. The program counter has the address of the instruction which the CPU is decoding.
10. Show code for a nested loop to perform an action 10, 000, 000, 000 times.
11. Show code for a nested loop to perform an action 200, 000, 000, 000 times.
12. Find the number of times the following loop is performed:

```
        MOV   R0, #0x55
        MOV   R2, #40
L1      LDR   R1, =10000000
L2      EOR   R0, R0, #0xFF
        SUB   R1, R1, #1
        BNE   L2
        SUB   R2, R2, #1
        BNE   L1
```

13. Indicate the status of Z and C after CMP is executed in each of the following cases.

(a)
```
MOV   R0, #50
MOV   R1, #40
CMP   R0, R1
```

(b)
```
MOV   R1, #0xFF
MOV   R2, #0x6F
CMP   R1, R2
```

(c)
```
MOV   R2, #34
MOV   R3, #88
CMP   R2, R3
```

(d)
```
SUB   R1, R1, R1
MOV   R2, #0
CMP   R1, R2
```

(e)
```
EOR   R2, R2, R2
MOV   R3, #0xFF
CMP   R2, R3
```

(f)
```
EOR   R0, R0, R0
EOR   R1, R1, R1
CMP   R0, R1
```

(g)
```
MOV   R4, #0x78
MOV   R2, #0x40
CMP   R4, R2
```

(h)
```
MOV   R0, #0xAA
AND   R0, R0, #0x55
CMP   R0, #0
```

14. Rewrite Program 4-1 to find the lowest grade in that class.
15. The target address of a BNE is backward if the relative address portion of opcode is _____ (negative, positive).
16. The target address of a BNE is forward if the relative address portion of opcode is _____ (negative, positive).

Section 4.2: Calling Subroutine with BL

17. BL is used to _____.
18. In Arm, which register is the linker register?
19. True or false. The BL target address can be anywhere in the 4G byte address space.

20. Describe how we can return from a subroutine in Arm.

21. In Arm, which address is saved when BL instruction is executed.

Section 4.3: Arm Time Delay and Instruction Pipeline

22. Find the core clock frequency if the machine cycle = 1.25 ns.

23. Find the machine cycle if the core clock frequency is 200 MHz.

24. Find the machine cycle if the core clock frequency is 100 MHz.

25. Find the machine cycle if the core clock frequency is 160 MHz.

26. Find the time delay for the delay subroutine shown below if the system has an Arm with a core clock frequency of 80 MHz:

```
           MOV    R8, #200
BACK       LDR    R1, =400000000
HERE       NOP
           SUBS   R1, R1, #1
           BNE    HERE
           SUBS   R8, R8, #1
           BNE    BACK
```

27. Find the time delay for the delay subroutine shown below if the system has an Arm with a core clock frequency of 50 MHz:

```
           MOV    R2, #100
BACK       LDR    R0, =50000000
HERE       NOP
           NOP
           SUBS   R0, R0, #1
           BNE    HERE
           SUBS   R2, R2, #1
           BNE    BACK
```

28. Find the time delay for the delay subroutine shown below if the system has an Arm with a core clock frequency of 40 MHz:

```
           MOV    R1, #200
BACK       LDR    R0, #20000000
HERE       NOP
           NOP
           NOP
           SUBS   R0, R0, #1
           BNE    HERE
           SUBS   R1, R1, #1
           BNE    BACK
```

29. Find the time delay for the delay subroutine shown below if the system has an Arm with a core clock frequency of 100 MHz:

```
           MOV    R8, #500
```

```
BACK        LDR    R1, =20000
HERE        NOP
            NOP
            NOP
            SUBS   R1, R1, #1
            BNE    HERE
            SUBS   R8, R8, #1
            BNE    BACK
```

Section 4.4: Stack and Stack Usage in Arm

30. True or false. In Arm the R13 is designated as stack pointer.
31. In Arm, stack pointer is a_____ bit register.

Answers to Review Questions

Section 4.1: Looping and Branch Instructions

1. Branch if not Equal
2. True
3. 4
4. Z flag of CPSR (status register)
5. 4
6. The B uses immediate value for offset and can only branch to an address location within ±32 MB address space, while the BX uses register operand to hold the branch target address and can go anywhere in the 4 GB address space of Arm.

Section 4.2: Calling Subroutine with BL

1. Branch and Link
2. True
3. 4
4. 32
5. R14
6. In both of them the target address is relative to the value of the program counter and the relative address can cover memory space of ±32MB from current location of program counter. The BL instruction saves the address of the next instruction in the LR register before jumping, while the B instruction just jumps without saving anything.

Section 4.3: Arm Time Delay and Instruction Pipeline

1. True
2. 1
3. MC = 1/66 MHz = 0.015 ms = 15 ns
4. [50, 000, 000 × (1 + 1 + 1 + 1 + 1 + 1 + 3)] × 10 ns = 4.5 seconds
5. Machine Cycle = 1 / 50 MHz = 0.02 ms = 20 ns
6. True
7. False
8. False. It takes 3 cycles, only if it branches to the target address.

Section 4.4: Stack and Stack Usage in Arm
1. R13
2. PUSH {R5-R8}
3. POP {R5-R8}
4. 0x00000000 to 0x00000003

Chapter 5: Signed Integer Numbers Arithmetic

This chapter deals with signed integer number instructions and operations. In Section 5.1, we focus on the concept of signed numbers in software engineering. Signed number arithmetic operations and instructions are explained along with examples in Section 5.2. Signed number comparison is discussed in Section 5-3 and Sign extension is covered in Section 5-4.

Section 5.1: Signed Numbers Concept

All data items used so far have been unsigned integer numbers, meaning that the entire 8-bit, 16-bit or 32-bit operand was used for the magnitude and the numbers represented are all positive or zero. Many applications require the use of negative numbers or signed data. In this section the concept of signed integer numbers is discussed.

Concept of signed numbers in computers

In everyday life, numbers are used that could be positive or negative or zero. For example, a temperature of 5 degrees below zero can be represented as -5, and 20 degrees above zero as +20. Computers must be able to accommodate such numbers. To do that, computer scientists have devised the following arrangement for the representation of signed positive and negative numbers: The most significant bit (MSB) is set aside for the sign (+ or -) and the rest of the bits are used for the magnitude. The sign is represented by 0 for positive (+) numbers and 1 for negative (-) numbers. Signed byte and word representations are discussed below.

Sign-magnitude format

In sign-magnitude format the sign and magnitude of the number are represented independently. For a byte, D7 (MSB) is the sign and D0 to D6 are set aside for the magnitude of the number. If D7 = 0, the number is positive, and if D7 = 1, it is negative.

The range of magnitude that can be represented by the above format is 2^7 (0 to 127). And the range of the number that can be represented is -127 to +127.

Dec.	Binary
0	0000 0000
+1	0000 0001
...	....
+5	0000 0101
...	....
+127	0111 1111

Dec.	Binary
-0	1000 0000
-1	1000 0001
...	....
-5	1000 0101
...	....
-127	1111 1111

The sign-magnitude format is easier for human to understand but more complex for computer to process.

Negative numbers using 2's complement

To simplify ALU circuitry, we often use 2's complement for the negative number representation. Adding a negative number in 2's complement has the same result as a subtraction. Using 2's complement representation, the adder in the ALU will be able to perform both add and subtract. With 2's complement, the most significant bit is still 0 for positive numbers and 1 for negative numbers like sign magnitude format.

In writing negative numbers in program source file, we usually use sign-magnitude format (e.g.: -75 for negative 75) and the assembler/compiler will convert the negative number to 2's complement. We will demonstrate the following process used to convert a positive number to a negative number represented in 2's complement to help you understand the 2's complement representation. Follow these steps:

1. Write the positive number in binary.
2. Invert bit.
3. Add 1 to it.

Unlike sign-magnitude format that has a positive zero and a negative zero, with 2's complement representation, there is only one zero. And there is one more negative number than positive.

Dec.	Binary
0	0000 0000
+1	0000 0001
...	....
+5	0000 0101
...	....
+127	0111 1111

Dec.	Binary
0	0000 0000
-1	1111 1111
...	....
-5	1111 1011
...	....
-127	1000 0001
-128	1000 0000

Examples 5-1, 5-2, and 5-3 demonstrate these three steps.

Example 5-1

Show how the computer would represent -5 in 8-bit 2's complement.

Solution:

1. 0000 0101   5 in 8-bit binary
2. 1111 1010   invert each bit
3. 1111 1011   add 1 (0xFB)

This is the signed number representation in 2's complement for -5.

Example 5-2

Show -34 hex as it is represented in 2's complement.

Solution:

1. 0011 0100   (0x34)
2. 1100 1011
3. 1100 1100   (which is 0xCC)

Example 5-3

Show the 2's complement representation for -128_{10}.

Solution:

1. 1000 0000   (128_{10})
2. 0111 1111
3. 1000 0000   Notice that this is not negative zero (−0).

From the examples above it is clear that the range of byte-sized negative numbers is -1 to -128. The following lists byte-sized signed number ranges:

Decimal	Binary	Hex
-128	1000 0000	80
-127	1000 0001	81
-126	1000 0010	82
...	....	..
-2	1111 1110	FE
-1	1111 1111	FF
0	0000 0000	00
+1	0000 0001	01
+2	0000 0010	02
...	....	..
+127	0111 1111	7F

Halfword-sized signed numbers

In Arm CPU a half-word is 16 bits in length. Using 2's complement representation, the MSB (D15) is used for the sign leaving a total of 15 bits (D14–D0) for the magnitude. This gives a range of –32,768 (–2^{15}) to +32,767 (2^{15}–1).

D15	D14	...	D0
sign	magnitude		

The following table shows the range of signed half-word numbers. To convert a half-word positive number to a negative half-word number in 2's complement representation, the same steps discussed above for byte size number are used.

Decimal	Binary	Hex
-32,768	1000 0000 0000 0000	8000
-32,767	1000 0000 0000 0001	8001
-32,766	1000 0000 0000 0010	8002
...	....	..
-2	1111 1111 1111 1110	FFFE
-1	1111 1111 1111 1111	FFFF
0	0000 0000 0000 0000	0000
+1	0000 0000 0000 0001	0001
+2	0000 0000 0000 0010	0002
...	....	...
+32,766	0111 1111 1111 1110	7FFE
+32,767	0111 1111 1111 1111	7FFF

Using Microsoft Windows calculator for signed numbers

All Microsoft Windows operating systems come with a handy calculator. Use it in programmer's mode to verify the signed number operations in this section.

Word-sized signed numbers

In Arm CPU a word is 32 bits in length. Using 2's complement representation, the MSB (D31) is used for the sign leaving a total of 31 bits (D30–D0) for the magnitude. This gives a range of -(2^{31}) to +(2^{31}-1).

D31	D30	...	D0
sign	magnitude		

To convert a word-size positive number to a negative word-size number in 2's complement representation, the same steps discussed above for byte size number are used. See Example 5-4.

Example 5-4

Show how the computer would represent -5 in 2's complement for (a) 8-bit, (b) 16-bit, and (c) 32-bit data sizes.

Solution:

(a) 8-bit

      1. 0000 0101   5 in 8-bit binary

      2. 1111 1010   invert each bit

      3. 1111 1011   add 1   (0xFB)

(b) 16-bit

      1. 0000 0000 0000 0101   5 in 16-bit binary

      2. 1111 1111 1111 1010   invert each bit

      3. 1111 1111 1111 1011   add 1 (0xFFFB)

(c) 32-bit

      1. 0000 0000 0000 0000 0000 0000 0000 0101 5 in 32-bit binary

      2. 1111 1111 1111 1111 1111 1111 1111 1010 invert each bit

      3. 1111 1111 1111 1111 1111 1111 1111 1011 add 1   (0xFFFFFFFB)

Use the Windows calculator to verify these examples.

If a number is larger than 32-bit, it must be treated as a 64-bit double-word number and be processed word by word the same way as unsigned numbers. The following shows the range of signed word-size numbers.

Decimal	Binary	Hex
-2,147,483,648	10000000000000000000000000000000	80000000
-2,147,483,647	10000000000000000000000000000001	80000001
-2,147,483,646	10000000000000000000000000000010	80000002
...	...	...
-2	11111111111111111111111111111110	FFFFFFFE
-1	11111111111111111111111111111111	FFFFFFFF
0	00000000000000000000000000000000	00000000
+1	00000000000000000000000000000001	00000001
+2	00000000000000000000000000000010	00000002
...	...	...
+2,147,483,646	01111111111111111111111111111110	7FFFFFFE
+2,147,483,647	01111111111111111111111111111111	7FFFFFFF

Table 5-1 shows a summary of signed data ranges.

Data Size	Bits	2^n	Decimal	Hexadecimal
Byte	8	-2^7 to $+2^7-1$	-128 to +127	0x80–0x7F
Half-word	16	-2^{15} to $+2^{15}-1$	-32,768 to +32,767	0x8000–0x7FFF
Word	32	-2^{31} to $+2^{31}-1$	-2,147,483,648 to +2,147,483,647	0x80000000–0x7FFFFFFF

Table 5-1: Signed Data Range Summary

Review Questions

1. In an 8-bit number, bit _____ is used for the sign bit, whereas in a 16-bit number, bit _____ is used for the sign bit. Repeat for 32-bit signed data.
2. Compute the byte-sized 2's complement of 0x16.
3. The range of byte-sized signed operands is -_____ to +_____. The range of half word-sized signed numbers is -_____ to +_____.
4. The range of word-sized signed numbers is -_____ to +_____.
5. Compute the 2's complement of 0x00500000.

Section 5.2: Signed Number Instructions and Operations

In this section we examine issues associated with signed number arithmetic operations. We will also discuss the Arm instructions for signed numbers and how to use them.

N flag

In Arm the N flag bit in CSPR is the sign bit. For positive results, N becomes 0 and for negative results it becomes 1. See Example 5-5.

Example 5-5

Write the following program in Keil and monitor the N flag:

```
        EXPORT  __main
        AREA MY_PROG,  CODE,  READONLY
__main
        MOV     R1,#6
        SUBS  R0,R1,#8
        ADDS  R3,R1,#3
H       B       H
```

Solution:

Following are the steps for "SUB R4, R2, R3":

After the execution of	Result	N
SUBS R0,R1,#8	6 − 8 = -2	1 (negative)
ADDS R3,R1,#3	-2 + 3 = +1	0 (positive)

Overflow problem in signed number operations

When using signed numbers, a serious problem arises that must be dealt with. This is the overflow problem. The CPU indicates the existence of the problem by raising the V (oVerflow) flag, but it is up to the programmer to take care of it. Now what is an overflow? If the result of an operation on signed numbers is too large for the register and resulted in an error, an overflow occurs and the programmer is notified. Look at Example 5-6.

Example 5-6

Look at the following case for 8-bit data size:

$$
\begin{array}{rll}
+\,96 & 0110\ 0000 & +96 \\
+\,70 & +0100\ 0110 & +\ +70 \\
\hline
+166 & 1010\ 0110 & -90
\end{array}
$$

We are adding two positive numbers together and the result is -90 in 2's complement notation, which is wrong.

In the example above, +96 is added to +70 and the result according to 2's complement notation is -90. Why? The reason is that the result was more than 8 bits could handle. The largest positive number an 8-bit registers can hold is +127. The sum of +96 and +70 is +166, which is more than +127. The designers of the CPU created the overflow flag specifically for the purpose of informing the programmer that the result of the signed number operation is erroneous.

Overflow flag in Arm

In a 32-bit operation, V is set to 1 in either of two cases:

1. There is a carry from D30 to D31 but no carry out of D31 (C = 0).

2. There is a carry from D31 out (C = 1) but no carry from D30 to D31.

The overflow flag is set to 1 when there is a carry into the most significant bit (D31) or out of the most significant bit, but not both. See Examples 5-7 and 5-8.

Example 5-7

Observe the result of executing the following program:

```
        EXPORT  __main
        AREA MY_PROG, CODE, READONLY
__main
        LDR R1,=0x6E2F356F
        LDR R2,=0x13D49530
        ADDS R3,R1,R2
H       B       H
        END
```

Solution:

```
  +6E2F356F    0110 1110 0010 1111 0011 0101 0110 1111
+ +13D49530   +0001 0011 1101 0100 1001 0101 0011 0000
  +8203CA9F    1000 0010 0000 0011 1100 1010 1001 1111   = −0x7DFC3561
```

The result is incorrect and the V flag will be set to show the over flow. (V = 1, C = 0, N = 1)

Example 5-8

Observe the result of executing the following instructions:

```
LDR R1,=0x6E2F356F
LDR R2,=0x13D49530
ADDS R3,R1,R2
```

Solution:

```
  +542F356F    0101 0100 0010 1111 0011 0101 0110 1111
+ +12E09530    0001 0010 1110 0000 1001 0101 0011 0000
  +670FCA9F    0110 0111 0000 1111 1100 1010 1001 1111   = +670FCA9F
```

The result is correct; V = 0, C = 0, N = 0

Avoiding Overflow

To avoid the overflow problems associated with signed number operations, the variables that we use in our programs should be big enough to be able to store the results.

Signed number multiplication

Signed number multiplication is similar in its operation to the unsigned multiplication described in Chapter 3. The only difference between them is that the operands including the result in signed number operations are treated as 2's complement representation of positive or negative numbers. In Arm we have SMULL (signed multiply long) that multiplies two 32-bit signed numbers and resulted in a 64-bit signed number. Arm also have 16-bit × 16-bit and 32-bit × 16-bit signed multiplication instructions but they are not available in all versions of the processors. Table 5-2 shows the 32-bit × 32-bit signed multiplication; it is similar to Table 3-3 in Chapter 3. See Examples 5-9 and 5-10.

Multiplication	Operand 1	Operand 2	Result
word×word	Rm	Rs	RdHi= upper 32-bit,RdLo=lower 32-bit
Note: Using SMULL (signed multiply long) for word × words multiplication provides the 64-bit result in RdLo and RdHi register. This is used for 32-bit × 32-bit numbers in which result can go beyond 0xFFFFFFFF.			

Table 5-2: Signed Multiplication (SMULL RdLo, RdHi, Rm, Rs) Summary

Example 5-9

Observe the results of the following multiplication of signed numbers:

```
LDR   R1, =-3500   ; R1 = -3500 (0xFFFFF254)
LDR   R0, =-100    ; R0 = -100 (0xFFFFFF9C)
SMULL R2, R3, R0, R1
```

Solution:

-3500 × -100 = 350,000 = 0x55730 in hex. After executing the above program R2 and R3 will contain 0x55730 and 00000000, respectively.

Example 5-10

The following program is similar to Example 5-9. But, instead of SMULL, the UMULL instruction is used. Observe the results of the following multiplication:

```
LDR   R1, =-3500   ; R1 = -3500 (0xFFFFF254)
MOV   R0, #-100    ; R0 = -100 (0xFFFFFF9C)
UMULL R2, R3, R0, R1
```

Solution:

0xFFFFF254 × 0xFFFFFF9C = 0xFFFFF1F000055730. Thus, R2 and R3 will contain 0x00055730 and 0xFFFFF1F0, respectively. As you can see, the results of the programs are completely different. In the previous program the SMULL instruction considers the operands signed numbers and the product of two negative numbers becomes positive. As a result, the sign bit becomes zero, but in this example the operands are considered as unsigned numbers.

Arithmetic shift

There are two types of shifts: logical and arithmetic. Logical shift, which is used for unsigned numbers, was discussed in Chapter 3. The arithmetic shift is used for signed numbers. It is basically the same as the logical shift, except that the old sign bit is copied to the new sign bit so that the sign of the number does not change.

ASR (arithmetic shift right)
```
ASR   Rn, Op2, count
```

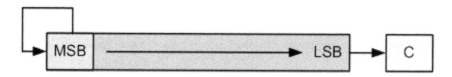

The number of bits to shift can be a register or an immediate value. As the bits of the source are shifted to the right into C, the empty bits are filled with the original sign bit. One can use the ASR instruction to divide a signed number by 2, as shown below:

```
MOV   R0, #-10    ; R0 = -10 = 0xFFFFFFF6
ASR   R3, R0, #1  ; R0 is arithmetic shifted right once
                  ; R3 = 0xFFFFFFFB = -5
```

Review Questions
1. Explain the difference between an overflow and a carry.
2. The instruction for signed long multiplication is _____.

Section 5.3: Signed Number Comparison

In Chapter 4 we saw that the CMP instruction affects the Z and C flags; using the flags we compared unsigned numbers. This instruction affects the N and V flags as well. We can use flags Z, V, and N to compare signed numbers. The Z flag shows if the numbers are equal or not. When the numbers are equal the Z flag is set to one. N and V flags show if the left operand is bigger than the right operand or not. When N and V have the same value, the first operand has a greater value.

In summary, after executing the instruction *CMP Rn, Op2* the flags are changed as follows:

$$Op2 > Rn \qquad V = N$$
$$Op2 = Rn \qquad Z = 1$$
$$Op2 < Rn \qquad N \neq V$$

Instruction		Action
BEQ	Branch equal	Branch if Z = 1
BNE	Branch not equal	Branch if Z = 0
BMI	Branch minus (branch negative)	Branch if N = 1
BPL	Branch plus (branch positive)	Branch if N = 0
BVS	Branch if V set (branch overflow)	Branch if V = 1
BVC	Branch if V clear (branch if no overflow)	Branch if V = 0
BGE	Branch greater than or equal	Branch if N = V
BLT	Branch less than	Branch if N ≠ V
BGT	Branch greater than	Branch if Z = 0 and N = V
BLE	Branch less than or equal	Branch if Z = 1 or N ≠ V

Table 5-3: Arm Conditional Branch (Jump) Instructions for Signed Data

Table 5-3 lists the branch instructions which check the Z, V, and N flags. The instructions can be used together with the CMP instruction to compare signed numbers.

Program 5-1 finds the lowest number among a list of numbers. The lowest number known so far is kept in R2. The numbers are brought into R1 and compared to R2. If a smaller one is found, it replaces the one in R2. The program starts by putting the first number in R2 since it is the lowest number known so far.

Program 5-1

```
        ; Finding the lowest of signed numbers
        AREA    PROG5_1, CODE, READONLY
__main  LDR R0, =SIGN_DAT
        MOV R3, #9
        LDRSB   R2, [R0]     ; bring first number into R2 and sign extend it
LOOP    ADD     R0, R0, #1   ; point to next
        SUBS    R3, R3, #1   ; decrement counter
        BEQ     DONE         ; if R3 is zero, done
        LDRSB   R1, [R0]     ; bring next number into R1 and sign extend it
        CMP     R1, R2       ; compare R1 and R2
        MOVLT   R2, R1       ; if R1 is smaller, keep it in R2
        B       LOOP
DONE    LDR     R0, =LOWEST  ; R0 = address of LOWEST
        STR     R2, [R0]     ; store R2 in location SUM
HERE    B       HERE

SIGN_DAT DCB    +13, -10, +19, +14, -18, -9, +12, -19, +16
        ALIGN
        AREA    VARIABLES, DATA, READWRITE
LOWEST      DCD     0
        END
```

CMN instruction

```
CMN    Rn, Op2
```

In Arm we have two compare instructions: CMP and CMN. While the CMP instruction sets the flags by subtracting operand2 from operand1, the CMN sets the flags by adding operand2 from operand1. As the result CMN compares the destination operand with the negative of the source operand:

destination > (-1 × source)	V = N
destination = (-1 × source)	Z = 1
destination < (-1 × source)	N ≠ V

When the source operand is an immediate value, the instructions can be used interchangeably. Example 5-11 is an example of using the CMN instruction.

Example 5-11

Assuming R5 has a positive value, write a program that finds its negative match in an array of data (OUR_DATA).

Solution:

```
            EXPORT __main
            AREA   EXAMPLE5_11, CODE, READONLY
__main      MOV    R5, #13
            LDR    R0, =OUR_DATA
            MOV    R3, #9
BEGIN
            LDRSB R1, [R0]      ; R1 = contents of loc. pointed to by R0
                                ; (sign extended)
            CMN   R1, R5        ; compare R1 and negative of R5
            BEQ   FOUND         ; branch if R1 is equal to negative of R5

            ADDS  R0, R0, #1    ; increment pointer
            SUBS  R3, R3, #1    ; decrement counter
            BNE   BEGIN         ; if R3 is not zero branch BEGIN

NOT_FOUND   B     NOT_FOUND
FOUND       B     FOUND

OUR_DATA    DCB   +13, -10, -13, +14, -18, -9, +12, -19, +16
            END
```

In the above program R5 is initialized with 13. Therefore, it finishes searching when it gets to -13.

Review Questions
1. For each of the following instructions, indicate the flag condition necessary for each branch to occur: (a) BLE (b) BGT

Section 5-4: Sign Extension

The Arm arithmetic instructions work on the 32-bit general purpose registers. But we might have 8-bit or 16-bit signed variables in the memory. If we load them using LDRB or LDRH, the most significant bits, including the sign bit, will be filled with zeros and the result of calculations might be incorrect. See Example 5-12 for instance.

Example 5-12

Run the following program in Keil and watch the value of R3.

```
      EXPORT __main
      AREA  NotExtendingExample, CODE, READONLY
__main
      LDR   R0,=AA
      LDRB  R1,[R0]        ;read AA

      LDR   R0,=BB
      LDRB  R2,[R0]        ;read BB

      SDIV  R3,R1,R2       ;R3 = R1 / R2
```

152

```
HERE    B   HERE

AA      DCB    -25
BB      DCB    -2
```

Solution:

The program initializes AA and BB with -25 and -2. So, AA and BB are loaded with 0xE7 and 0xFE, respectively. Then, R1 and R2 are loaded with AA and BB. So, they will contain 0xE7 and 0xFE. Then, they are divided using the SDIV instruction. Considering 0x000000E7 and 0x000000FE as 32-bit numbers, they are in fact 231 and 254. So, the result becomes 231/254 = 0 and R3 will be loaded with 0.

To solve the problem, we need to sign extend the 8-bit and 16-bit signed variables before doing calculations. The LDRSB (load register signed byte) instruction and the LDRSH (load register signed half-word) do just that. They work as follows:

LDRSB loads into the destination register a byte from memory and sign extends (copy D7, the sign bit) to the remaining 24 bits. This is illustrated in Figure 5-1.

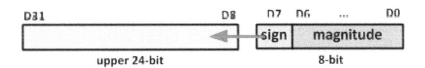

Figure 5-1: Sign Extending a Byte

Look at the following example:

```
; assume memory location 0x80000 has +96 = 0110 0000 and R1=0x80000
LDRSB R0, [R1]      ; now R0 =   00000000000000000000000001100000
; assume memory location 0x80000 contains -2 = 1111 1110 and R2=0x80000
LDRSB R4, [R2]   ; now R4 = 11111111111111111111111111111110
```

As can be seen in the above examples, LDRSB does not alter the lower 8 bits. The sign bit of the 8-bit data is copied to the rest of the 32-bit register.

LDRSH loads the destination register with a 16-bit signed number and sign-extends to the rest of the 16 bits of the 32-bit register. This is used for signed half-word operand and is illustrated in Figure 5-2.

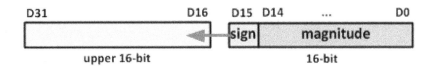

Figure 5-2: Sign Extending a Half-word

Look at the following example:

```
; assume 0x80000 contains +260 = 0000 0001 0000 0100 and R1=0x80000
LDRSH R0, [R1]   ; R0=0000 0000 0000 0000 0000 0001 0000 0100
```

Another example:

```
; assume location 0x20000 has -327660=0x8002 and R2=0x20000
LDRSH R1, [R2]            ; R1=FFFF8002
```

As we see in the above examples, LDRSH does not alter the lower 16 bits. The sign bit of the 16-bit data is copied to the rest of the 32-bit register. In Example 5-13, we correct the program of Example 5-13 using sign extending.

Example 5-13

Rewrite the program of Example 5-12 and correct the problem using sign extending.

Solution:

```
      EXPORT  __main
      AREA  ExtendingExample, CODE, READONLY
__main
      LDR    R0,=AA
      LDRSB R1,[R0]      ;read AA

      LDR    R0,=BB
      LDRSB R2,[R0]      ;read BB

      SDIV R3,R1,R2      ; R3 = R1 / R2
HERE  B HERE

AA    DCB -25
BB    DCB    -2
```

In the above program, LDRSB sign extends the values of AA and BB. So, R1 and R2 are loaded with 0xFFFFFFE7 and 0xFFFFFFFE, respectively. Now, the result is -25/-2 = 12 and R3 will be loaded with 12.

Sign extending can be used to avoid overflow errors, as well. In Example 5-14, the calculation of Example 5-6 is done using 32-bit variables and the overflow problem is avoided.

Example 5-14

Write a program for Example 5-5 to handle the overflow problem.

Solution:

```
      EXPORT  __main
      AREA  EXAMPLE5_14, CODE, READONLY
__main
      LDR    R1, =DATA1
      LDR    R2, =DATA2
      LDR    R3, =RESULT
```

```
        LDRSB R4, [R1]     ; R4 = +96
        LDRSB R5, [R2]     ; R5 = +70
        ADD   R4, R4, R5   ; R4 = R4 + R5 = 96 + 70 = +166
        STR   R4, [R3]     ; Store +166 in location RESULT
HL      B     HL

DATA1         DCB   +96
DATA2         DCB   +70
        ALIGN
        AREA  VARIABLES, DATA, READWRITE    ; The following is stored in RAM
RESULT        DCW   0
        END
```

The following is an analysis of the values in Example 5-14. Each is sign-extended and then added as follows:

Sign	Binary numbers	Decimal
0	000 0000 0000 0000 0000 0000 0110 0000	+96 after sign ext.
0	000 0000 0000 0000 0000 0000 0100 0110	+70 after sign ext.
0	000 0000 0000 0000 0000 0000 1010 0110	+166

As a rule, if the possibility of overflow exists, all byte-sized signed numbers should be sign-extended into a word, and similarly, all halfword-sized signed operands should be sign-extended to a word before they are processed. This is shown in Program 5-2. Program 5-2 finds total sum of a group of signed number data.

Program 5-2

```
    ; This program calculates the sum of signed numbers

    AREA   PROG5_2, CODE, READONLY

    LDR    R0, =SIGN_DAT
    MOV    R3, #9
    MOV    R2, #0
LOOP LDRSB          R1, [R0]
    ; Load into R1 and sign extend it.
    ADD    R2, R2, R1   ; R2 = R2 + R1
    ADD    R0, R0, #1   ; point to next
    SUBS   R3, R3, #1   ; decrement counter
    BNE    LOOP
    LDR    R0, =SUM
    STR    R2, [R0]     ; Store R2 in location SUM
HERE B      HERE

SIGN_DAT DCB      +13, -10, +19, +14, -18, -9, +12, -19, +16
```

```
        AREA    VARIABLES, DATA, READWRITE
SUM     DCD     0
        END
```

Review Questions
1. Explain the purpose of the LDRSB and LDRSH instructions.
2. Demonstrate the effect of LDRSB on R0 = 0xF6.
3. Demonstrate the effect of LDRSH on R1 = 0x124C.

Problems

Section 5.1: Signed Numbers Concept
1. Show how the 32-bit computers would represent the following numbers in 2's complement notation and verify each with a calculator.

(a) -23	(b) +12	(c) -0x28
(d) +0x6F	(e) -128	(f) +127
(g) +365	(h) -32,767	

2. Show how the 32-bit computers would represent the following numbers in 2's complement and verify each with a calculator.

(a) -230	(b) +1200	(c) - 0x28F
(d) +0x6FF		

Section 5.2: Signed Number Instructions and Operations
3. In a program, A and B are defined as 8-bit signed integer variables. They are loaded with -65 and -98 respectively. Then B is added to A (A = A + B;). Check if the result is valid or overflow occurs.
4. In a program, A and B are defined as 8-bit signed integer variables. They are loaded with 15 and -95 respectively. Then B is subtracted from A (A = A − B;). Check if the result is valid or overflow occurs.

Section 5.3: Signed Number Comparison
5. Modify Program 5-2 to find the highest number. Verify your program.

Section 5.4: Sign Extension

6. Find the overflow flag for each case and verify the result using an Arm IDE. Do byte-sized calculation on them.

(a) (+15) + (-12)	(b) (-123) + (-127)	(c) (+0x25) + (+34)
(d) (-127) + (+127)	(e) (+100) + (-100)	

7. Sign-extend the following values into 32 bits using Arm instructions in the Keil IDE.

(a) -122	(b) -0x999	(c) +0x17
(d) +127	(e) -129	

Answers to Review Questions

Section 5.1

1. D7, D15, and D31 for 32-bit signed data.
2. 0x16 = 0001 0110; its 2's complement is: 1110 1001 + 1 = 1110 1010
3. −128 to +127; −32,768 to +32,767 (decimal)
4. -2,147,483,648 to +2,147,483,647
5. 0x500000 = 0000 0000 0101 0000 0000 0000 0000 0000;

   Its 2's complement is: 1111 1111 1010 1111 1111 1111 1111 1111 + 1 =

   1111 1111 1011 0000 0000 0000 0000 0000 = 0xFFB00000

Section 5.2

1. C flag is raised when there is a carry out from the operation, but V flag is raised when there is a carry into the sign bit and no carry out of the sign bit or when there is no carry into the sign bit and there is a carry out of the sign bit. C flag is used to indicate overflow in unsigned arithmetic operations while V flag is involved in signed operations.
2. SMULL

Section 5.3

1.

   (a) BLE will jump if V is different from N, or if Z = 1.

   (b) BGT will jump if V equals N, and if Z = 0.

Section 5.4

1. The LDRSB instruction sign extends the sign bit of a byte into a word; the LDRSH instruction sign extends the sign bit of a half-word into a word.
2. In 0xF6 the sign bit is 1; thus, it is sign-extended into 0xFFFFFFF6
3. 0x124C sign-extended into R1 would be 0x0000124C.

Chapter 6: Arm Addressing Modes

This chapter discusses the issue of memory access and the stack. Section 6.1 is dedicated to Arm memory map and memory access. We will also explain the concepts of align, non-align, little endian, and big endian data access. Advanced indexed addressing mode is explained in Section 6.2. In Section 6.3, we discuss the bit-addressable (bit-banding) SRAM and peripherals. In Section 6.4, we describe the PC relative addressing mode and its use in implementing ADR and LDR.

Section 6.1: Arm Memory Access

The Arm CPU uses 32-bit addresses to access memory and peripherals. This gives us a maximum of 4 GB (gigabytes) of memory space. This 4GB of directly accessible memory space has addresses 0x00000000 to 0xFFFFFFFF, meaning each byte is assigned a unique address (Arm is a byte-addressable CPU). See Figure 6-1.

D31	D24 D23	D16 D15	D8 D7	D0
0x00000003	0x00000002	0x00000001	0x00000000	0x00000000
0x00000007	0x00000006	0x00000005	0x00000004	0x00000004
0x0000000B	0x0000000A	0x00000009	0x00000008	0x00000008
0x0000000F	0x0000000E	0x0000000D	0x0000000C	0x0000000C
⋮	⋮	⋮	⋮	
0xFFFFFFF3	0xFFFFFFF2	0xFFFFFFF1	0xFFFFFFF0	0xFFFFFFF0
0xFFFFFFF7	0xFFFFFFF6	0xFFFFFFF5	0xFFFFFFF4	0xFFFFFFF4
0xFFFFFFFB	0xFFFFFFFA	0xFFFFFFF9	0xFFFFFFF8	0xFFFFFFF8
0xFFFFFFFF	0xFFFFFFFE	0xFFFFFFFD	0xFFFFFFFC	0xFFFFFFFC

Figure 6-1: Memory Byte Addressing in Arm

The Arm buses and memory access

D31–D0 Data bus

See Figure 6-2. The 32-bit data bus of the Arm provides the 32-bit data path to the on-chip and off-chip memory and peripherals. They are grouped into 8-bit data bytes, D0–D7, D8–D15, D16–D23, and D24–D31.

A31–A0

These signals provide the 32-bit address path to the on-chip and off-chip memory and peripherals. Since the Arm supports data access of byte (8 bits), half word (16 bits), and word (32 bits), the buses must be able to access any of the 4 banks of memory connected to the 32-bit data bus. The A0 and A1 are used to select one of the 4 bytes of the D31-D0 data bus. See Figure 6-3.

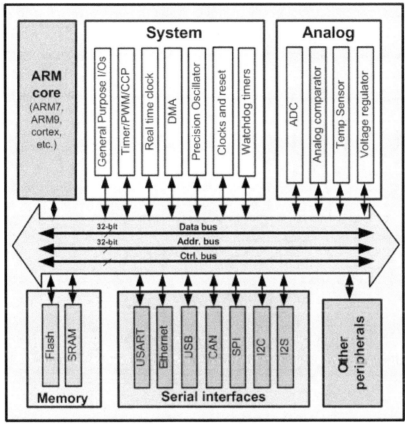

Figure 6-2: Memory Connection Block Diagram in Arm

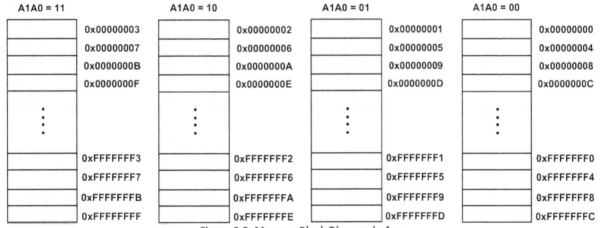

Figure 6-3: Memory Block Diagram in Arm

AHB and APB buses

The Arm CPU is connected to the on-chip memory via an AHB (advanced high-performance bus). The AHB is used not only for connection to on-chip ROM and RAM, it is also used for connection to some of the high speed I/Os (input/output) such as GPIO (general purpose I/O). Arm chip also has the APB (advanced peripherals bus) bus dedicated for communication with the on-chip peripherals such as timers, ADC, UART, SPI, I2C, and other peripheral ports.

While we need the 32-bit data bus between CPU and the memory (RAM and ROM), many slower peripherals have no need for fast data bus pathway. For this reason, Arm uses the AHB-to-APB bridge to access the slower on-chip devices such as peripherals. Also since peripherals do not need a high speed bus, a bridge between AHB and APB allows going from the higher speed bus of AHB to lower speed bus of peripherals. The AHB bus allows a single-cycle access. See Figure 6-4 for AHB-to-APB bridge.

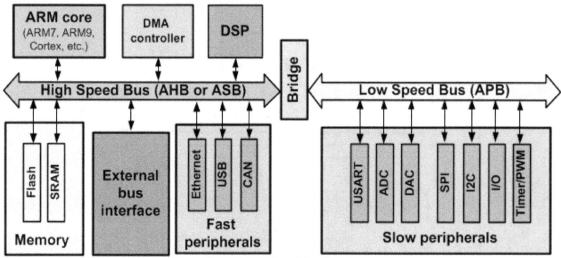

Figure 6-4: AHB and APB in Arm

Bus cycle time

To access a device such as memory or I/O, the CPU provides a fixed amount of time called a bus cycle time. During this bus cycle time, the read or write operation of memory or I/O must be completed. The bus cycle time used for accessing memory is often referred to as MC (memory cycle) time. The time from when the CPU provides the addresses at its address pins to when the data is expected at its data pins is called memory read cycle time. While for on-chip memory the cycle time can be 1 clock, in the off-chip memory the cycle time is often 2 clocks or more. If memory is slow and its access time does not match the MC time of the CPU, extra time can be requested from the CPU to extend the read cycle time. This extra time is called a wait state (WS). In the 1980s, the clock speed for memory cycle time was the same as the CPU's clock speed. For example, in the 20 MHz processors, the buses were working at the same speed of 20 MHz. This resulted in 2 × 50 ns = 100 ns for the memory cycle time (1/20 MHz = 50 ns). See Example 6-1.

Example 6-1

Calculate the memory cycle time of a 50-MHz bus system with
(a) 0 WS,
(b) 1 WS, and
(c) 2 WS.
Assume that the bus cycle time for off-chip memory access is 2 clocks.

Solution:

1/50 MHz = 20 ns is the bus clock period. Since the bus cycle time of zero wait states is 2 clocks, we have:

(a) Memory cycle time with 0 WS   2 × 20 = 40 ns
(b) Memory cycle time with 1 WS   40 + 20 = 60 ns
(c) Memory cycle time with 2 WS   40 + 2 × 20 = 80 ns

It is preferred that all bus activities be completed with 0 WS. However, if the read and write operations cannot be completed with 0 WS, we request an extension of the bus cycle time. This extension is in the form of an integer number of WS. That is, we can have 1, 2, 3, and so on WS, but not 1.25 WS.

When the CPU's speed was under 100 MHz, the bus speed was comparable to the CPU speed. In the 1990s the CPU speed exploded to 1 GHz (gigahertz) while the bus speed maxed out at around 200 MHz. The gap between the CPU speed and the bus speed is one of the biggest challenges in the design of high-performance systems. To avoid the use of too many wait states in interfacing memory to CPU, cache memory and other high-speed DRAMs are used.

Bus bandwidth

The rate of data transfer is generally called bus bandwidth. In other words, bus bandwidth is a measure of how fast buses transfer information between the CPU and memory or peripherals. The wider the data bus, the higher the bus bandwidth. However, the advantage of the wider external data bus comes at the cost of increasing the die size for system on-chip (SOC) or the printed circuit board size for off-chip memory. Now you might ask why we should care how fast buses transfer information between the CPU and outside, as long as the CPU is working as fast as it can. The problem is that the CPU cannot process information that it does not have. This is like driving a Porsche or Ferrari in first gear; it is a terrible under usage of CPU power. Bus bandwidth is measured in MB (megabytes) per second and is calculated as follows:

bus bandwidth = (1/bus cycle time) × bus width in bytes

In the above formula, bus cycle time can be for both memory and I/O since the Arm uses the memory mapped I/O. Example 6-2 clarifies the concept of bus bandwidth. As can be seen from Example 6-2, there are two ways to increase the bus bandwidth: Either use a wider data bus or shorten the bus cycle time (or do both). That is exactly what many processors have done. Again, it must be noted that although the processor's speed can go to 1 GHz or higher, the bus speed for off-chip memory is limited to around 200 MHz. The reason for this is that the signals become too noisy for the circuit board if they are above 200 MHz.

Example 6-2

Calculate memory bus bandwidth for the following CPU if the bus speed is 100 MHz.

(a) Arm Thumb with 0 WS and 1 WS (16-bit data bus)

(b) Arm with 0 WS and 1 WS (32-bit data bus)

Assume that the bus cycle time for off-chip memory access is 2 clocks.

Solution:

The memory cycle time for both is 2 clocks, with zero wait states. With the 100 MHz bus speed we have a bus clock of 1/100 MHz = 10 ns.

(a)   Bus bandwidth = $(1/(2 \times 10$ ns$)) \times 2$ bytes = 100M bytes/second (MB/s)

With 1 wait state, the memory cycle becomes 3 clock cycles

$3 \times 10 = 30$ ns and the memory bus bandwidth is = $(1/30$ ns$) \times 2$ bytes = 66.6 MB/s

(b)   Bus bandwidth = $(1/(2 \times 10$ ns$)) \times 4$ bytes = 200 MB/s

With 1 wait state, the memory cycle becomes 3 clock cycles

$3 \times 10 = 30$ ns and the memory bus bandwidth is = $(1/30$ ns$) \times 4$ bytes = 126.6 MB/s

From the above it can be seen that the two factors influencing bus bandwidth are:

1. The read/write cycle time of the CPU

2. The width of the data bus

Code memory region

The 4 GB of Arm memory space is organized as 1G × 32 bits since the Arm instructions are 32-bit. The internal data bus of the Arm is 32-bit, allowing the transfer of one instruction into the CPU every clock cycle. This is one of the benefits of the RISC fixed instruction size. The fetching of an instruction in every clock cycle can work only if the code is word aligned, meaning each instruction is placed at an address location ending with 0, 4, 8, or C. Example 6-3 shows the placement of code in Arm memory. Notice that the code addresses go up by 2 or 4 since the Thumb-2 Arm instructions are 2 bytes or 4 bytes. While compilers ensure that codes are word aligned, it is job of the programmer to make sure the data in SRAM is word aligned too. We will examine this important topic soon.

Example 6-3

Compile and debug the following code in Keil and see the placement of instructions in memory locations.

```
        EXPORT  __main
        AREA   ARMex, CODE, READONLY
__main PROC
```

```
        MOV    R2, #0x00    ; R2=0x00
        MOV    R3, #0x35    ; R3=0x35
        ADD    R4, R3, R2
HERE    B      HERE
        ENDP
        END                 ; Mark end of code
```

Solution

As you can see in the figure, the first MOV instruction starts from location 0x08000134, the second MOV instruction starts from location 0x08000138 and the ADD instruction starts from location 0x0800013C.

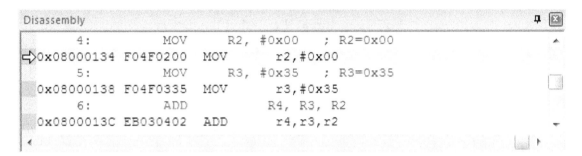

The following image displays the first locations of memory. The code of the first MOV instruction is in location 0x08000134 of memory which is word aligned. The same rule applies for the other instructions. Note that the code of MOV R2, #0x00 is F0 4F 02 00 but 4F F0 00 02 is stored in the memory. We will discuss the reason in this chapter when we focus on the concept of big endian and little endian.

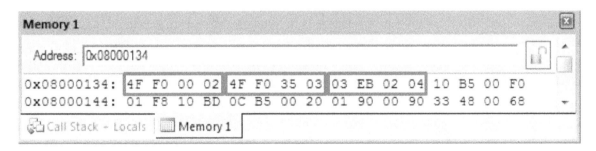

SRAM memory region

A section of the memory space is used by SRAM. The SRAM can be on-chip or off-chip (external). For small embedded systems, this on-chip SRAM is used by the CPU for scratch pad to store parameters and is also used by the CPU for the purpose of the stack. We will examine the stack usage by the Arm in the next section. For larger systems, the operating system and application programs may be run in Flash ROM or copied into the SRAM and run everything in the SRAM just like your laptop computer.

In using the SRAM memory for storing parameters, we must be careful when loading or storing data in the SRAM lest we use unaligned data access. Next, we discuss this important issue.

Data misalignment in SRAM

The case of misaligned data has a major effect on the Arm bus performance. If the data is aligned, for every memory read cycle, the Arm brings in 4 bytes of information (data or code) using the D31–D0 data bus. Such data alignment is referred to as word alignment. To make data word aligned, the least significant digits of the hex addresses must be 0, 4, 8, or C (in hex).

While the compilers make sure that program codes (instructions) are always aligned (Example 6-3), it is the placement of data in SRAM by the programmer that can be nonaligned and therefore subject to memory access penalty. In other words, the single cycle access of memory is also used by Arm to bring into registers 4 bytes of data every clock cycle assuming that the data is aligned. To make sure that data are also aligned we use the ALIGN directive. The use of ALIGN directive for RAM data makes sure that each word is located at an address location ending with address of 0, 4, 8, or C. If our data is word size (using DCDU directive) then the use of ALIGN directive at the start of the data section guaranties all the data placements will be word aligned. When a word size data is defined using the DCD directive, the assembler aligns it to be word aligned.

Accessing non-aligned data

As we have stated many times before, Arm defines 32-bit data as a word. The address of a word can start at any address location. For example, in the instruction "LDR R1, [R0]" if R0 = 0x20000004, the address of the word being fetched into R1 starts at an aligned address. In the case of "LDR R1, [R0]" if R0 = 0x20000001 the address starts at a non-aligned address. In systems with a 32-bit data bus, accessing a word from a non-aligned addressed location can be slower. This issue is important and applies to all 32-bit processors.

In the 8-bit system, accessing a word (4 bytes) is treated like accessing four consecutive bytes regardless of the address location. Since accessing a byte takes one memory cycle, accessing 4 bytes will take 4 memory cycles. In the 32-bit system, accessing a word with an aligned address takes one memory cycle. That is because each byte is carried on its own data path of D0–D7, D8–D15, D16–D23, and D24–D31 in the same memory cycle. However, accessing a word with a non-aligned address requires two memory cycles. For example, see how accessing the word in the instruction "LDR R1, [R0]" works as shown in Figure 6-5. As a case of aligned data, assume that R0 = 0x80000000. In this instruction, the contents of 4 bytes of memory (locations 0x80000000 through 0x80000003) are being fetched in one cycle. In only one cycle, the Arm CPU accesses locations 0x80000000 through 0x80000003 and puts them in R1.

Now assuming that R0 = 0x80000001 in this instruction, the contents of 8 bytes of memory (locations 0x80000000 through 0x80000007) are being fetched in two consecutive cycles but only 4 bytes of it are used. In the first cycle, the Arm CPU accesses locations 0x80000000 through 0x80000003 and puts them in R1 only the desired three bytes of locations 0x800000001 through 0x80000003. In the second cycle, the contents of memory locations 0x8000004 through 0x80000007 are accessed and only the desired byte of 0x80000004 is put into R1. See Example 6-4.

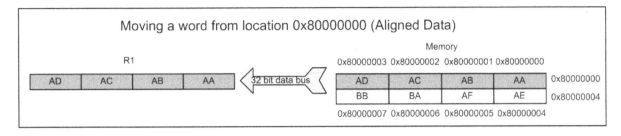

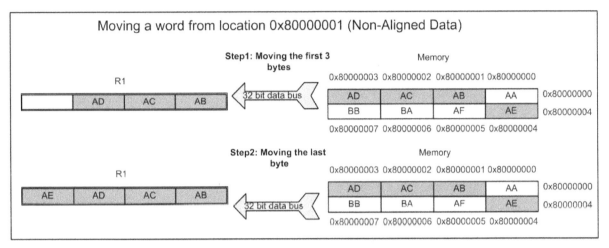

Figure 6-5: Memory Access for Aligned and Non-aligned Data

Example 6-4

Show the data transfer of the following cases and indicate the number of memory cycle times it takes for data transfer. Assume that R2 = 0x4598F31E.

```
LDR   R1, =0x40000000    ; R1=0x40000000
LDR   R2, =0x4598F31E    ; R2=0x4598F31E
STR   R2, [R1]           ; Store R2 to location 0x40000000
ADD   R1, R1, #1         ; R1 = R1 + 1 = 0x40000001
STR   R2, [R1]           ; Store R2 to location 0x40000001
ADD   R1, R1, #1         ; R1 = R1 + 1 = 0x40000002
STR   R2, [R1]           ; Store R2 to location 0x40000002
ADD   R1, R1, #1         ; R1 = R1 + 1 = 0x40000003
STR   R2, [R1]           ; Store R2 to location 0x40000003
```

Solution:

For the first STR R2, [R1] instruction, the entire 32 bits of R2 is stored into locations with addresses of 0x40000000, 0x40000001, 0x40000002, and 0x40000003. The 4-byte content of register R2 is stored into memory locations with starting address of 0x40000000 via the 32-bit data bus of D31–D0. This address is word aligned since address of the least significant digit is 0. Therefore, it takes only one memory cycle to transfer the 32-bit data.

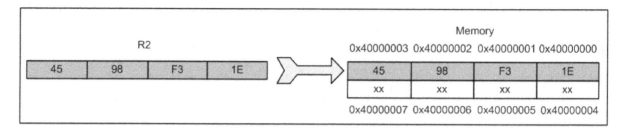

For the second STR R2, [R1] instruction, in the first memory cycle, the lower 24 bits of R2 is stored into locations 0x40000001, 0x40000002, and 0x40000003. In the second memory cycle, the upper 8 bits of R2 is stored into the 0x40000004 location.

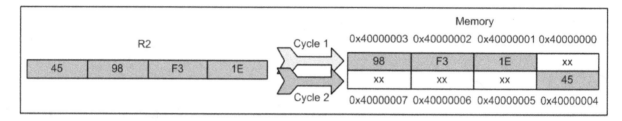

For the third STR R2, [R1] instruction, in the first memory cycle, the lower 16 bits of R2 is stored into locations 0x40000002 and 0x40000003. In the second memory cycle, the upper 16 bits of R2 is stored into locations 0x40000004 and 0x40000005.

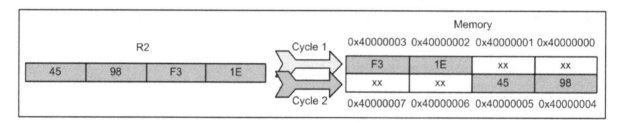

For the fourth STR R2, [R1] instruction, in the first memory cycle, the lower 8 bits of R2 is stored into locations 0x40000003. In the second memory cycle, the upper 24 bits of R2 is stored into the locations 0x40000004, 0x40000005, and 0x40000006.

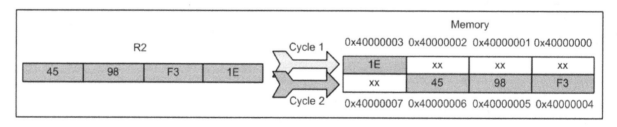

The lesson to be learned from this is to try not to put any words on a non-aligned address location in a 32-bit system. Indeed, this is so important that directive ALIGN is specifically designed for this purpose. Next, we discuss the issue of aligned data.

Using LDR instruction with DCD and ALIGN directives

The DCD and DCDU directives are used for 32-bit (word) data. The DCD directive ensures 32-bit data types are aligned, in contrast to DCDU which does not. DCD is used as follows:

```
VALUE1      DCD     0x99775533
```

This ensures that VALUE1, a word-sized operand, is located in a word aligned address location. Therefore, an instruction accessing it will take only a single memory cycle. Since performance of the CPU depends on how fast it can fetch the data we must ensure that any memory access reading 32-bit data is done in a single clock cycle. This means we must make sure all 32-bit data are word aligned. This is so important that Arm has an interrupt (exception) dedicated to misaligned data, meaning any time it accesses misaligned data, it lets us know that there is a problem. The one-time use of ALIGN directive at the beginning of data area using DCDU makes the data aligned for that group of data.

Different versions of Arm handle unaligned access differently. Some may allow unaligned access without generating interrupt. In this case, the programmer must be careful to allocate data on the word aligned boundary so that the system will perform at its optimal bus bandwidth.

Using LDRH with DCW and ALIGN directives

The problem of misaligned data is also an issue when the data size is in half-words (16-bit). In many cases using DCWU, we must use the ALIGN directive multiple times in the data area of a given program to ensure they are aligned. This is in contrast to the DCW directive which ensures data type to be half-word aligned. This is especially the case when we use the LDRH instruction. See Example 6-5. Aligned data is also an issue for the Thumb version of the Arm.

Example 6-5

Show the data transfer of the following LDRH instructions and indicate the number of memory cycle times it takes for data transfer.

```
LDR    R1, =0x80000000    ; R1=0x80000000
LDR    R3, =0xF31E4598    ; R3=0xF31E4598
LDR    R4, =0x1A2B3D4F    ; R4=0x1A2B3D4F
STR    R3, [R1]           ; (STR R3, [R1]) stores R3 to location 0x80000000
STR    R4, [R1, #4]       ; (STR R4, [R1+4]) stores R4 to location 0x80000004
LDRH   R2, [R1]           ; loads two bytes from location 0x80000000 to R2
LDRH   R2, [R1, #1]       ; loads two bytes from location 0x80000001 to R2
LDRH   R2, [R1, #2]       ; loads two bytes from location 0x80000002 to R2
LDRH   R2, [R1, #3]       ; loads two bytes from location 0x80000003 to R2
```

Solution:
In the LDRH R2, [R1] instruction, locations with addresses of 0x80000000, 0x80000001, 0x80000002, and 0x80000003 are accessed but only 0x80000000 and 0x80000001 are used to get the 16 bits to R2. This address is halfword aligned since the least significant digit is 0. Therefore, it takes only one memory cycle to transfer the data. Now, R2=0x00004598

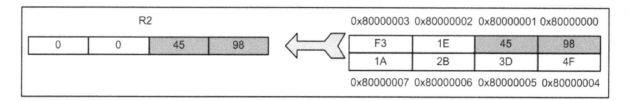

For the LDRH R2, [R1, #1], instruction, locations with addresses of 0x80000000, 0x80000001, 0x80000002, and 0x80000003 are accessed, but only 0x80000001 and 0x80000002 are used to get the 16 bits to R2. Therefore, it takes only one memory cycle to transfer the data. Now, R2=0x00001E45.

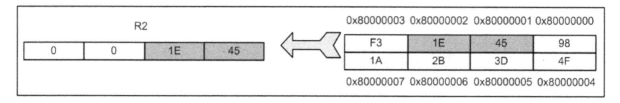

For the LDRH R2, [R1, #2], instruction, locations with addresses of 0x80000000, 0x80000001, 0x80000002, and 0x80000003 are accessed, but only 0x80000002 and 0x80000003 are used to get the 16 bits to R2. Therefore, it takes only one memory cycle to transfer the data. Now, R2=0x0000F31E.

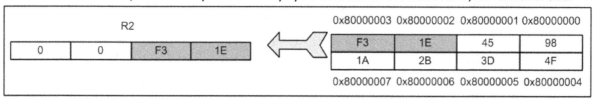

For the LDRH R2, [R1, #3] instruction, in the first memory cycle, locations with addresses of 0x80000000, 0x80000001, 0x80000002, and 0x80000003 are accessed, but only 0x80000003 is used to get the lower 8 bits to R2. In the second memory cycle, the address locations 0x80000004, 0x80000005, 0x80000006, and 0x80000007 are accessed where only the 0x80000004 location is used to get the upper 8 bits to R2. Now, R2=0x00004FF3.

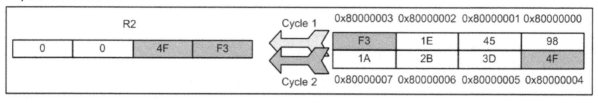

Using LDRB with DCB and ALIGN directives

The problem of misaligned data does not exist when the data size is bytes. A single byte of data will never straddle across a word boundary. In cases such as using the string of ASCII characters with the DCB directive, accessing a byte takes the same amount of time (one memory cycle) as an aligned word (4 bytes), regardless of the address location of the data. See Example 6-6.

Example 6-6

Show the data transfer of the following LDRB instructions and indicate the number of memory cycle times it takes for data transfer.

```
LDR    R1, =0x80000000   ; R1=0x80000000
LDR    R3, =0xF31E4598   ; R3=0xF31E4598
LDR    R4, =0x1A2B3D4F   ; R4=0x1A2B3D4F
STR    R3, [R1]     ; Store R3 to location 0x80000000
STR    R4, [R1, #4] ; (STR R4, [R1+4]) Store R4 to location 0x80000004
LDRB   R2, [R1]     ; load one byte from location 0x80000000 to R2
LDRB   R2, [R1, #1] ; (LDRB R2, [R1+1]) load one byte from location 0x80000001
LDRB   R2, [R1, #2] ; (LDRB R2, [R1+2]) load one byte from location 0x80000002
LDRB   R2, [R1, #3] ; (LDRB R2, [R1+3]) load one byte from location 0x80000003
```

Solution:

In the LDRB R2, [R1] instruction, locations with addresses of 0x80000000, 0x80000001, 0x80000002, and 0x80000003 are accessed but only 0x80000000 is used to get the 8 bits to R2. Therefore, it takes only one memory cycle to transfer the data. Now, R2=0x00000098.

In the LDRB R2, [R1, #1] instruction, locations with addresses of 0x80000000, 0x80000001, 0x80000002, and 0x80000003 are accessed but only 0x80000001 is used to get the 8 bits to R2. Therefore, it takes only one memory cycle to transfer the data. Now, R2=0x00000045.

In the LDRB R2, [R1, #2] instruction, locations with addresses of 0x80000000, 0x80000001, 0x80000002, and 0x80000003 are accessed but only 0x80000002 is used to get the 8 bits to R2. Therefore, it takes only one memory cycle to transfer the data. Now, R2=0x0000001E.

In the LDRB R2, [R1, #3] instruction, locations with addresses of 0x80000000, 0x80000001, 0x80000002, and 0x80000003 are accessed but only 0x80000003 is used to get the 8 bits to R2. Therefore, it takes only one memory cycle to transfer the data. Now, R2=0x000000F3.

Little Endian vs. Big Endian war

In storing data in memory, there are two major byte orderings used. The little endian places the least significant byte (little end of the data) in the low address and the big endian places the most significant byte in the low address. The origin of the terms *big endian* and *little endian* was from a Gulliver's Travels story about how an egg should be opened: from the big end or the little end. Arm supports both little and big endian. In most of the Arm devices little endian is the default. Some Arm chip manufacturers provide an option for changing the endian by software. See Example 6-7 to understand little endian and big endian data storage.

Example 6-7

Show how data is placed after execution of the following code using
a) little endian and

b) big endian.

```
LDR    R2,  =0x7698E39F    ;  R2=0x7698E39F
LDR    R1,  =0x80000000
STR    R2,  [R1]
```

Solution:

a) For little endian we have:

Location 80000000 = (9F)

Location 80000001 = (E3)

Location 80000002 = (98)

Location 80000003 = (76)

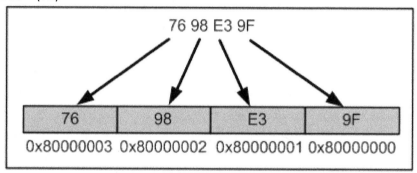

b) For big endian we have:

Location 80000000 = (76)

Location 80000001 = (98)

Location 80000002 = (E3)

Location 80000003 = (9F)

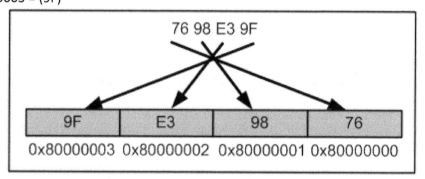

In Example 6-7, notice how the least significant byte (the little end of the data) 0x9F goes to the low address 0x80000000, and the most significant byte of the data 0x76 goes to the high address 0x80000003. This means that the little end of the data goes in first, hence the name little endian. In the Arm with big endian option enabled, data is stored the opposite way: The big end (most significant byte) goes into the low address first, and for this reason it is called big endian. Many of recent RISC processors allow selection of mode in software, big endian or little endian.

Review Questions
1. Who makes sure that instructions are aligned on word boundary?
2. In most of the Arm devices, the _____ endian is the default.
3. A 66 MHz system has a memory cycle time of _____ ns if it is used with a zero wait state.
4. To interface a 100 MHz processor to a 50 ns access time ROM, how many wait states are needed?
5. True or false. Arm uses big endian format when is powered up.

Section 6.2: Advanced Indexed Addressing Modes

In previous chapters we discussed the use of STR and LDR instructions in the form of "LDR Rd, [Rx]" and "STR Rx, [Rd]", where the registers within the brackets hold the pointer (the address where the data resides). These registers within the brackets are referred as the index register or base register. Arm provides three advanced indexed addressing mode that allow the modification of the value in the index register. We will discuss them in this section.

Base plus offset addressing modes

The Arm provides three advanced indexed addressing modes called base plus offset addressing modes. In addition to the base register specified within the bracket, an offset can be added to the value of the base register. These modes are: pre-index, pre-index with writeback, and post-index modes. Table 6-1 summarizes these modes. Each of these addressing modes can be used with offset of fixed value. The pre-index addressing mode can be used with offset of fixed value, register, or a shifted register. See Table 6-2. In this section we will discuss each mode in detail.

Addressing Mode	Syntax	Effective Address of Memory	Rm Value After Execution
Pre-index	LDR Rd, [Rm, #k]	Rm + #k	Rm
Pre-index with WB*	LDR Rd, [Rm, #k]!	Rm + #k	Rm + #k
Post-index	LDR Rd, [Rm], #k	Rm	Rm ı #k
*WB means Writeback ** Rd and Rm are any of registers and #k is a signed immediate value			

Table 6-1: Indexed Addressing in Arm

Offset	Syntax	Pointing Location
Fixed value	LDR Rd, [Rm, #k]	Rm + #k
Register	LDR Rd, [Rm, Rn]	Rm + Rn
Shifted register	LDR Rd, [Rm, Rn, <shift>]	Rm + (Rn shifted <shift>)
* Rn and Rm are any registers and #k is a signed immediate value ** <shift> is any of the shift operations studied in Chapter3 like LSL #2		

Table 6-2: Offset of Fixed Value vs. Offset of Shifted Register

Pre-indexed addressing mode with fixed offset

In this addressing mode, a register and a positive or negative immediate value are used as a pointer to the data's memory location. The value of register does not change after instruction is executed.

See Figure 6-6. This addressing mode can be used with STR, STRB, STRH, LDR, LDRB, and LDRH. See Example 6-8.

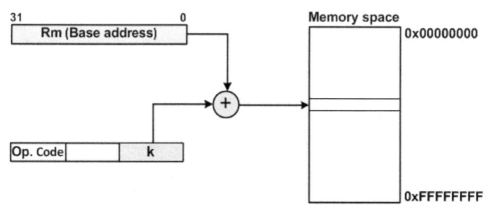

Figure 6-6: Pre-indexed Addressing Mode with Fixed Offset

Example 6-8

Write a program to store contents of R5 to the SRAM location 0x20000000 to 0x20000000F using pre-indexed addressing mode with fixed offset.

Solution:

```
LDR    R5, =0x55667788
LDR    R1, =0x20000000   ; load the address of first location
STR    R5, [R1]          ; store R5 to location 0x10000000
STR    R5, [R1, #4] ; store R5 to location 0x20000000 + 4 (0x20000004)
STR    R5, [R1, #8] ; store R5 to location 0x20000000 + 8 (0x20000008)
STR    R5, [R1, #0x0C]
               ; store R5 to location 0x20000000 + 0x0C (0x2000000C)
```

Notice that after running this code the content of R1 is still 0x20000000

It is a common practice to use a register to point to the first location of the memory space and access the different locations using proper offsets. For example, see the following program:

```
          ADR    R0, OUR_DATA       ; point to OUR_DATA
          LDRB   R2, [R0, #1]       ; load R2 with offset of BETA
          . . .
OUR_DATA
ALFA      DCB    0x30
BETA      DCB    0x21
```

Pre-indexed addressing mode with writeback and fixed offset

This addressing mode is like pre-indexed addressing mode with fixed offset except that the calculated pointer is written back to the pointing register. We put '!' after the instruction to tell the assembler to enable writeback in the instruction. See Figure 6-7 and Example 6-9.

172

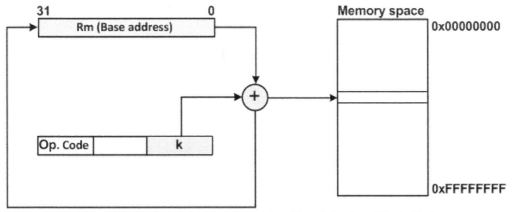

Figure 6-7: Pre-indexed addressing mode with writeback and fixed offset

---

Example 6-9

Rewrite Example 6-8 using pre-indexed addressing mode with writeback and fixed offset.

Solution:

```
LDR    R1, =0x20000000    ; load the address of first location
STR    R5, [R1]           ; store R5 to location 0x20000000
STR    R5, [R1, #4]!
                ; store R5 to location 0x20000000 + 4 (0x20000004)
                ; writeback makes R1 = 0x20000004
STR    R5, [R1, #4]!
                ; store R5 to location 0x20000004 + 4 (0x20000008)
                ; writeback makes R1 = 0x20000008
STR    R5, [R1, #4]!
                ; store R5 to location 0x20000008 + 4 (0x2000000C)
                ; writeback makes R1 = 0x2000000C
```

Notice that after running this code the content of R1 is 0x2000000C

Post-indexed addressing mode with fixed offset

This addressing mode is like pre-indexed addressing mode with fixed offset and writeback except that the instruction is executed on the location that Rn is pointing to regardless of offset value. See Figure 6-8. The new value of the pointer is calculated after the load/store operation and written back to the index register. Examine the following instructions:

```
STR    R1, [R2], #4       ; store R1 into memory pointed to by
                          ; R2 and then write back R2 + 4 to R2
LDRB   R5, [R3], #1       ; load a byte from memory pointed to
                          ; by R3 and then write back R3 + 1 to R3
```

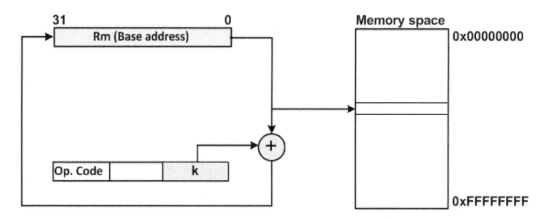

Figure 6-8: Post-indexed addressing mode

Notice that writeback is by default enabled in post-indexed addressing and there is no need to put '!' after instructions because in post-indexing without writeback the offset is neither used in the load/store operation nor written back to the index register. See Example 6-10.

Example 6-10

Rewrite Example 6-9 using post-indexed addressing mode with fixed offset.

Solution:

```
LDR    R1, =0x20000000    ; load the address of first location
STR    R5, [R1], #4        ; store R5 to location 0x20000000 and writeback
                           ; 0x20000000 + 4 (0x20000004) to R1
STR    R5, [R1], #4        ; store R5 to location 0x20000004 and writeback
                           ; 0x20000004 + 4 (0x20000008) to R1
STR    R5, [R1], #4        ; store R5 to location 0x20000008 and writeback
                           ; 0x20000008 + 4 (0x2000000C) to R1
STR    R5, [R1], #4        ; store R5 to location 0x2000000C and writeback
                           ; 0x2000000C + 4 (0x20000010) to R1
```

Notice that after running this code the content of R1 is 0x20000010.

Pre-indexed address mode with offset of a shifted register

See Figure 6-9. This advanced addressing mode is an important feature in the Arm. We start describing this mode from simple cases with no shift and then we will move on to more complex formats.

Simple format of pre-indexed address mode with offset register

The following is the simple syntax for LDR and STR.

```
LDR    Rd, [Rm, Rn]    ; Rd is loaded from location Rm + Rn of memory
STR    Rs, [Rm, Rn]    ; Rs is stored to location Rm + Rn of memory
```

This addressing mode is often used in implementing array access. Rm holds the base address of the array (the address of the first element of the array) and Rn holds the array index. Example 6-11 shows how we use this addressing mode in accessing different locations of an array with byte size elements in memory.

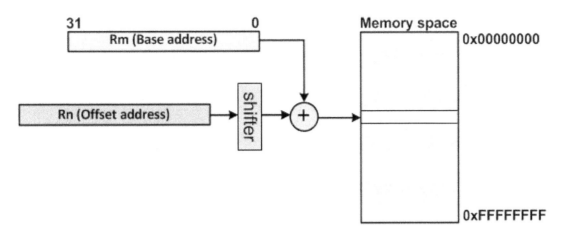

Figure 6-9: Pre-indexed address mode with offset of a shifted register

Example 6-11

Examine the value of R5 and R6 after the execution of the following program.

```
INDEX        RN    R2
ARRAY1       RN    R1
      EXPORT __main
      AREA  EXAMPLE_6_11, CODE, READONLY
__main
      LDR   ARRAY1, =MYDATA   ; use array address as base address
      LDRB  R4, [ARRAY1]
            ; load R4 with first element of ARRAY1 (R4=0x45)

      MOV   INDEX, #1   ; INDEX = 1 to point to location 1 of array
      LDRB  R5, [ARRAY1, INDEX]
                  ; Load R5 with second element of ARRAY1 (R5=0x24)
      MOV   INDEX, #2   ; INDEX = 2 to point to location 2 of array
      LDRB  R6, [ARRAY1, INDEX]
                  ; Load R5 with third element of ARRAY1 (R6=0x18)
HERE  B     HERE
MYDATA       DCB   0x45, 0x24, 0x18, 0x63
      END
```

Solution:

After running the LDRB R4, [ARRAY1] instruction, first element with offset 0 of MYDATA is loaded into R4. Now R4=0x45.

Next, after running the LDRB R5, [ARRAY1, INDEX] instruction, second element with offset 1 of MYDATA is loaded into R5. Now R5 = 0x24.

Next, after running the LDRB R6, [ARRAY1, INDEX] instruction, third element with offset 2 of MYDATA is loaded into R6. So the content of R6 = 0x18.

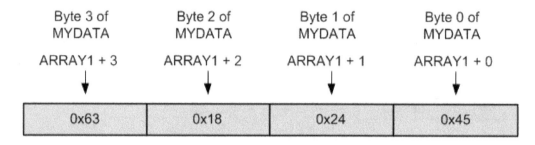

Notice that in Example 6-11, the array MYDATA contains byte size element so DCB and LDRB were used. It will not work if the data size of the array elements is different from a byte. In the next example, the array elements are word size (4 bytes each). We define the array using DCD, then we will not be able to use LDRB R5, [ARRAY1, INDEX] to load the INDEX location of ARRAY1. See Example 6-12 for clarification.

Example 6-12

In Example 6-11, change MYDATA DCB 0x45, 0x24, 0x18, 0x63 to MYDATA DCD 0x45, 0x2489ACF5 and examine the value of R5 and R6 after the execution of the following program.

```
INDEX        RN    R2
ARRAY1       RN    R1
      EXPORT __main
      AREA   EXAMPLE_6_12, CODE, READONLY
__main
      LDR    ARRAY1, =MYDATA    ; use array address as base address
      LDRB   R4, [ARRAY1]       ; load R4 with first element of ARRAY1
                                ; (R4=0x45)
      MOV    INDEX, #1          ; INDEX = 1 to point to location 1 of array
      LDRB   R5, [ARRAY1, INDEX]
                                ; Load INDEX location of ARRAY1 to R5
      MOV    INDEX, #2          ; INDEX = 2 to point to location 2 of array
      LDRB   R6, [ARRAY1, INDEX]
                                ; load INDEX location of ARRAY1 to R6
HERE  B      HERE
MYDATA       DCD    0x45, 0x2489ACF5
      END
```

Solution:
After running the LDRB R4, [ARRAY1] instruction, location 0 of MYDATA is loaded to R4. Now R4=0x45.
Next, after running the LDRB R5, [ARRAY1, INDEX] instruction, location 1 of MYDATA is loaded to R5. Now R5 = 0x00.
Next, after running the LDRB R6, [ARRAY1, INDEX] instruction, location 2 of MYDATA is loaded to R6. So the content of R6 = 0x00.

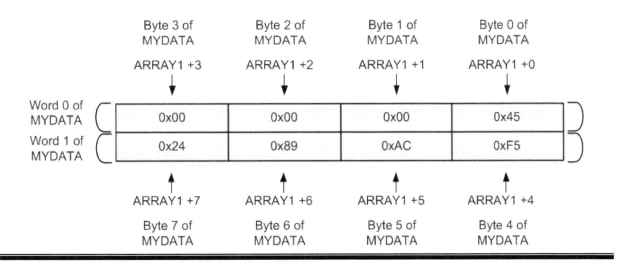

To access locations of a word size array we have to multiply the array index by four to yield the offset. Similarly, to access locations of a half-word size array we have to multiply the array index by two to get the offset. For example, we can correct program of Example 6-12 by replacing instruction

```
MOV    INDEX, #2    ; INDEX = 2 to point to location 2 of array
```

with following instructions:

```
MOV    INDEX, #2   ; INDEX = 2
MOV    INDEX, INDEX, LSL #2    ; INDEX is shifted left two bits (×4)
                               ; to point to word 2 of the array
```

Notice that by shifting left a value by two bits, we multiply it by four. Next, we will see how we can use indexed addressing with shifted registers to combine multiplication with the LDR and STR instructions.

General format of pre-indexed address mode with offset register

The general format of indexed addressing with shifted register for LDR and STR is as follows:

```
LDR Rd, [Rm, Rn, <shift>]  ; (Shifted Rn) + Rm is used as the address
STR Rd, [Rm, Rn, <shift>]  ; (Shifted Rn) + Rm is used as the address
```

In the above instructions <shift> can be any of shift instructions studied in Chapter 3 such as LSL, LSR, ASR and ROR. But for array indexing, LSL is most often used because it is the equivalent of signed multiply by power of two. Examine the following instructions:

```
LDR    R1, [R2, R3, LSL #2]    ; R2 + (R3 × 4) is used as the address
                  ; content at location R2 + (R3 × 4) is loaded into R1
STR    R1, [R2, R3, LSL #1]    ; R2 + (R3 × 2) is used as the address
                  ; R1 is stored at location R2 + (R3 × 2)
STRB   R1, [R2, R3, LSL #2]    ; R2 + (R3 × 4) is used as the address
      ; least significant byte of R1 is stored at location R2 + (R3 × 4)
LDR    R1, [R2, R3, LSR #2]    ; R2 + (R3 / 4) is used as the address
                  ; content at location R2 + (R3 / 4) is loaded into R1
```

From the above code we can see that indexed addressing with shifted register is used to multiply the offset by a power of two and that is why it is also called indexed addressing with scaled register. Examine Example 6-13 to see how we can use scaled register indexing to access an array of words.

Example 6-13

Examine the value of R5 and R6 after the execution of the following program.

```
INDEX        RN    R2
ARRAY1       RN    R1
      EXPORT   __main
      AREA    EXAMPLE_6_13, CODE, READONLY
__main
      LDR    ARRAY1, = MYDATA
      MOV    INDEX, #0                 ; INDEX = 0
      LDR    R4, [ARRAY1, INDEX, LSL #2]    ;

      MOV    INDEX, #1                 ; INDEX = 1
      LDR    R5, [ARRAY1, INDEX, LSL #2]    ;

      MOV    INDEX, #2                 ; INDEX = 2
      LDR    R6, [ARRAY1, INDEX, LSL #2]    ;

HERE   B     HERE
MYDATA       DCD    0x45, 0x2489ACF5, 0x2489AC23
      END
```

Solution:

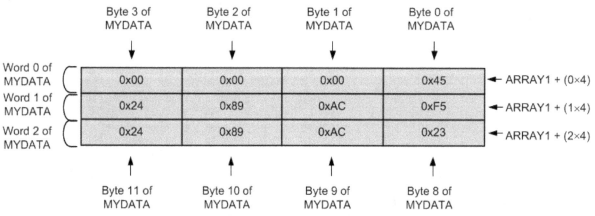

After running the LDR R4, [ARRAY1, INDEX, LSL #2]] instruction, element 0 of array MYDATA is loaded to R4. Now R4=0x45.

Next, after running the LDR R5, [ARRAY1, INDEX, LSL #2] instruction, element 1 of array MYDATA is loaded to R5. Now R5 = 0x2489ACF5.

Next, after running the LDR R6, [ARRAY1, INDEX, LSL #2] instruction, element 2 of array of MYDATA is loaded to R6. So the content of R6 = 0x2489AC23.

Look-up table

One application of indexed addressing mode is for implementing look-up tables. The look-up table is an array of pre-calculated constants. It allows obtaining frequently used values with no complex arithmetic operations during run-time at the cost of the memory space for the table. This technique is often used in embedded systems with lower computing power and stringent real time demand. The constant data in the look-up table may be calculated when the program is written or they may be calculated during the initialization of the program. In the Examples 6-14 through 6-16, the look-up tables are stored in program memory space and accessed as an array using indexed addressing mode.

Example 6-14

Write a program to use the x value in R9 and leave the value of $x^2 + 2x + 3$ in R10. Assume R9 has the x value range of 0–9. Use a look-up table instead of a multiply instructions.

Solution:

```
        EXPORT __main
        AREA LOOKUP_EXAMP6_14, READONLY, CODE
__main      PROC
        ADR   R2, LOOKUP         ; point to LOOKUP
        LDRB  R10, [R2, R9]      ; R10 = entry of lookup table index by R9
HERE    B     HERE               ; stay here forever
        ENDP

LOOKUP DCB  3, 6, 11, 18, 27, 38, 51, 66, 83, 102
        END
```

Example 6-15

Write a program to use the x value in R9 and get the factorial of x in R10. Assume R9 has the x value range of 0–10. Use a look-up table instead of a multiply instruction.

Solution:

```
        EXPORT __main
        AREA LOOKUP_EXAMP6_15, READONLY, CODE
__main PROC
        MOV   R9, #5
        ADR   R2, LOOKUP              ; point to LOOKUP
        LDR   R10, [R2, R9, LSL #2]   ; R10 = entry of lookup table index by R9
HERE    B     HERE                    ; stay here forever
        ENDP
LOOKUP DCD  1, 1, 2, 6, 24, 120, 720, 5040, 40320, 362880, 3628800
        END
```

Example 6-16

Write a program that calculates 10 to the power of R2 and stores the result in R3. Assume R2 has the x value range of 0–6. Use a look-up table instead of a multiply instruction.

Solution:

```
        EXPORT  __main
        AREA LOOKUP_EXAMP_6_16, READONLY, CODE
__main      PROC
        ADR  R1, LOOKUP              ; point to LOOKUP
        LDR  R3, [R1, R2, LSL #2]    ; R3 = entry of lookup table index by R2
HERE  B    HERE                      ; stay here forever
        ENDP
LOOKUP DCD  1, 10, 100, 1000, 10000, 100000, 1000000
        END
```

Review Questions
1. Indexed addressing mode in Arm uses (register, memory) as pointer to data location.
2. List the three types of Indexed addressing mode in Arm
3. True or false. In the preindexed addressing mode the value of register does not change after the instruction is executed.
4. What is the difference between the preindexed and preindexed with write back?
5. What symbol do we use to indicate the preindexed with write back?

Section 6.3: ADR, LDR, and PC Relative Addressing

In indexed addressing modes, any registers including the PC (R15) register can be used as the pointer register. For example, the following instruction reads the contents of memory location PC+4:

```
LDR   R0, [PC, #4]
```

In this way, the data which has a known distance from the current executing line can be accessed. As discussed in Chapter 4, the PC register points 4 bytes ahead of executing instruction. As a result, "LDR R0, [PC, #4]" accesses a memory location whose address is 4+4 bytes ahead of the current instruction. Generally speaking, the address of the memory location which is being accessed using "LDR R0, [PC, offset]" can be found using this formula: the address of current instruction + 4 + offset. For instance, if "LDR R0, [PC, #4]" is located in address 0x10 the effective address is: 0x10 + 4 + 4 = 0x18.

Calculating the offset from current PC is a tedious job that needs be done every time new instructions are inserted or deleted. There are two pseudo-instructions using PC relative addressing mode, ADR and LDR with "=" to make programming easier.

The ADR Pseudo-instruction

The ADR pseudo-instruction uses the PC relative addressing mode to load a register with an address. It has the syntax of

180

```
ADR    Rn, Label
```

The assembler calculates the offset from the current PC value to the line where Label is and translates the pseudo-instruction into:

```
ADD    Rn, PC, #offset
```

For example, see the following program:

```
       AREA   LOOKUP_EXAMPLE, READONLY, CODE
__main
       ADR    R2, OUR_FIXED_DATA      ; R2 points to OUR_FIXED_DATA
       LDRB   R0, [R2]                ; load R0 with the contents
                                      ; of memory pointed to by R2
       ADD    R1, R1, R0              ; add R0 to R1
HERE   B      HERE                    ; stay here forever
OUR_FIXED_DATA
       DCB    0x55, 0x33, 1, 2, 3, 4, 5, 6
       END
```

See Figure 6-10. At compile time, the ADR is replaced with A201 which is the machine code for "ADD R2, PC, #0x04". Since the instruction is at address 0x0800012C, the instruction accesses location 0x0800012C + 4 + 0x04 = 0x08000134. As shown in the Figure, where 0x08000134 is the address of OUR_FIXED_DATA.

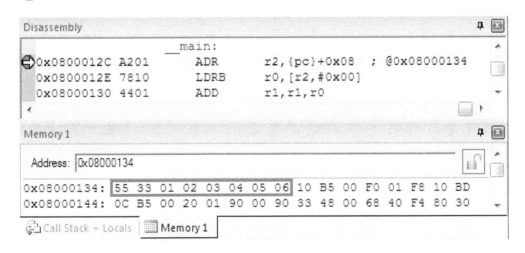

Figure 6 10: Memory Dump for ADR Instruction

Implementing the LDR Pseudo-instruction

Arm instructions are 12-bit or 32-bit long. It is impossible to incorporate a 32-bit immediate data in a 32-bit instruction. To load a register with a 32-bit immediate data, the Arm assembler stores the value as a constant data in program memory and accesses it using the LDR instruction and the PC relative addressing mode. Figure 6-11 shows the implementation of Program 6-1. For example, 0x12345678 is stored in memory locations 0x08000140–0x08000143, and the LDR directive is replaced with LDR R0, [PC, #8]. The LDR R0,=0x1234567 is located at address 0x08000134. Now we have 0x08000134+4+8= 0x08000140.

```
        EXPORT  __main
        AREA    PROG6_1, READONLY, CODE
__main
        LDR     R0, =0x12345678
        LDR     R1, =0x86427531
        ADD     R2, R0, R1
H1      B       H1
        END
```

Figure 6-11: Memory Dump for LDR Instruction

The memory region that the assembler reserved to store the constant data for LDR pseudo-instruction is called the "Literal pool." Literal pool is normally located at the end of the current section. (Section is terminated either by END or by another AREA directive.) The LDR instruction with immediate offset allows a range of -4095 to 4095. It is a good programming practice not to write a program section longer than 1000 instructions. Just in case you have a need to have a long section that the distance from the LDR instruction to the end of the section is beyond the range, Arm assembler allows you to designate a location within the section for an additional literal pool using LTORG directive. Remember, you should not designate a location for literal pool where the program execution may encounter and attempt to execute the constant data as instructions.

Review Questions
1. Which register is used as the pointer in PC relative addressing mode?
2. Which directive is more optimized ADR or LDR? Why?

Section 6.4: Arm Bit-Addressable Memory Region

Arm memory/peripheral is byte-addressable, that means the smallest size of memory access by the CPU is a single byte (8-bit). This presents an issue. In order to modify a single bit in the memory or peripheral, the whole byte where the bit resides is read, the bit is modified using AND, OR, or exclusive-OR operation, then the whole byte is written back. This is a procedure commonly referred to as the Read-Modify-Write (RMW) operation. One problem with RMW is that it incurs three steps and generally three separate instructions. In a multi-tasking environment, two RMW operations to the same byte may occur simultaneously, one interrupts the other. In that case, the second write will wipe out the modification of the first write.

There are software techniques to mitigate the issue of RMW but Arm introduced the "bit-banding" option for Cortex-M in hardware. It is generally available in M3 and M4 controllers though the manufacturers have the option to decide whether they implement it or not in the controllers.

Bit-banding is a feature that creates an alias word address for each bit of the selected memory and peripheral regions. Writing to the bit-banded alias address will change only the corresponding bit in the memory or peripheral without affecting all other bits of the same byte. Reading from the bit-banded alias address will return only the value of the addressed bit. Although internally writing to bit-banded memory or peripheral is still implemented in a read-modify-write in hardware, it is done with a single write instruction and therefore an "atomic" operation that will not be interrupted by the other instruction.

Since bit-banded alias addresses are word-size addresses, they are read and written in word size. Writing a word to a bit-banded alias address will transfer Bit n of the word to the selected bit. All other bits in the word have no effect. Reading from a bit-banded alias address will return the bit value in Bit n of the word. All other bits will be 0.

Bit-addressable (bit-banded) memory region

Since bit-banding assigns each bit with a word address and each word has 32 bits, bit-banding requires 32 times the address space than the regular memory addressing. Of the 4GB memory space of the Arm, only few small regions of memory or peripheral are bit-addressable, or the bit-banded regions. Recall bit-banding is created to mitigate the issues of Read-Modify-Write, so none of the read-only memory is bit-banded. Only the RAM and peripheral regions might be bit-banded.

The bit-banded RAM and peripheral locations vary among the family members and manufacturers. The Arm Cortex-M generic manual defines the address regions of bit-banding as 0x20000000 to 0x200FFFFF for SRAM and 0x40000000 to 0x400FFFFF for peripherals. We will use these address regions for the discussion in the rest of this section. Notice they are located at the lowest 1 MB address space of SRAM and peripherals. It must be also noted that the bit-banded (bit-addressable) regions are the only region that can be accessed in both bit and byte/halfword/word formats while the other area of memory must be accessed in byte/halfword/word size.

For the Arm Cortex-M, the bit-banded SRAM has addresses of 0x20000000 to 0x200FFFFF. This 1M bytes bit-addressable region is given 32M-byte bit-banded alias addresses of 0x22000000 to 0x23FFFFFF. The higher addresses are called alias address because they are addressing the exact same memory as the lower direct addressed SRAM, only each word address represents a single bit in the lower address. Because each bit occupies a word (4-byte) address in the bit-banded alias address region, to find out the alias address of a bit, the following formula is used:

Bit alias address = Bit alias base address + Byte offset $\times$ 32 + Bit number $\times$ 4

For example, to calculate the bit alias address of bit 3 of 0x20000004

Bit alias base address = 0x22000000
Byte offset = 0x20000004 − 0x20000000 = 4
Bit number = 3

Bit alias address = 0x22000000 + 4 * 32 + 3 * 4 = 0x22000000 + 128 + 12 = 0x22000000 + 0x8C
= 0x2200008C

Therefore, the bit addresses 0x22000000 to 0x2200001F are for the first byte of SRAM location 0x20000000, and 0x22000020 to 0x2200003F are the bit addresses of the second byte of SRAM location 0x20000001, and so on. See Figure 6-12.

SRAM Byte addresses	SRAM Bit addresses (We use these addresses to access the individual bits)							
	D7	D6	D5	D4	D3	D2	D1	D0
200FFFFF	23FFFFFC	F8	F4	F0	EC	E8	E4	23FFFFE0
200FFFFE	23FFFFDC	D8	D4	D0	CC	C8	C4	23FFFFC0
200FFFFD	23FFFFBC	B8	B4	B0	AC	A8	A4	23FFFFA0
200FFFFC	23FFFF9C	98	94	90	8C	88	84	23FFFF80
200FFFFB	23FFFF7C	78	74	70	6C	68	64	23FFFF60
200FFFFA	23FFFF5C	x8	x4	x0	xC	x8	x4	23FFFF40
200XXXXX	2XXXXXXC	X8	X4	X0	XC	X8	X4	2XXXXXX0
20000008	2200011C	118	114		...		104	22000100
20000007	220000FC	F8	F4	F0	EC	E8	E4	220000E0
20000006	220000DC	D8	D4	D0	CC	C8	C4	220000C0
20000005	220000BC	B8	B4	B0	AC	A8	A4	220000A0
20000004	2200009C	98	94	90	8C	88	84	22000080
20000003	2200007C	78	74	70	6C	68	64	22000060
20000002	2200005C	58	54	50	4C	48	44	20000040
20000001	2200003C	38	34	30	2C	28	24	22000020
20000000	2200001C	18	14	10	0C	08	04	22000000

Figure 6-12: SRAM bit-addressable region and their alias addresses

Since each byte of SRAM has 8 bits, we need an address for each bit. This means we need at least 8M address locations to access 8M bits, one address for each bit. However, to make the addresses word-aligned the Arm provides 4-byte alias address for each bit. For example, 0x22000000 to 0x2200001F is assigned to a single byte location of 0x20000000. That means we have 0x22000000 to 0x23FFFFFF (total of 32M locations, as alias addresses) for 1M bytes of address.

Bit map for SRAM

From Figure 6-12 once again notice the following facts:

1. The bit address 0x22000000 is assigned to D0 of SRAM location of 0x20000000.
2. The bit address 0x22000004 is assigned to D1 of SRAM location of 0x20000000.
3. The bit address 0x22000008 is assigned to D2 of SRAM location of 0x20000000.
4. The bit address 0x2200000C is assigned to D3 of SRAM location of 0x20000000.
5. The bit address 0x22000010 is assigned to D4 of SRAM location of 0x20000000.
6. The bit address 0x22000014 is assigned to D5 of SRAM location of 0x20000000.
7. The bit address 0x22000018 is assigned to D6 of SRAM location of 0x20000000.
8. The bit address 0x2200001C is assigned to D7 of SRAM location of 0x20000000.

Notice that SRAM locations 0x20000000 – 0x200FFFFF are both byte-addressable and bit-addressable. The only difference is when we access it in byte (or halfword or word) we use direct addresses 0x20000000 to 0x200FFFFF, but when they are accessed in bit, they are accessed via their alias addresses of 0x22000000 to 0x23FFFFFF. The reason they are called aliases is because it is the same physical memory but accessed by two different addresses. It is like a same person but different names (aliases) See Examples 6-17 through 6-19.

Example 6-17

The generic Arm chip has the following address assignments. Calculate the space and the amount of memory given to each region.
(a)   Address range of 0x20000000–200FFFFF for SRAM bit-addressable region
(b)   Address range of 0x22000000–23FFFFFF for alias addresses of bit-addressable SRAM

Solution:

(a)   200FFFFF – 20000000 = FFFFF bytes. Converting FFFFF to decimal, we get 1,048,575 + 1 = 1,048,576, which is equal to 1M bytes
(b)   23FFFFFF – 22000000 = 1FFFFFF bytes. Converting 1FFFFFF to decimal, we get 33,554,431 + 1 = 33,554,432, which is equal to 32M bytes.

Example 6-18

Write a program to set HIGH the D6 of the SRAM location 0x20000001 using a) byte address and b) the bit alias address.

Solution:
a)

```
        LDR    R1, =0x20000001    ; load the address of the byte
        LDRB   R2, [R1]           ; get the byte
        ORR    R2, R2, #2_01000000    ; make D6 bit high
                          ; (binary representation in Keil for 0b01000000)
        STRB   R2, [R1]           ; write it back
```

b)   From Figure 6-12 we have address 0x22000038 as the bit address of D6 of SRAM location 0x20000001.

```
        LDR    R1, =0x22000038    ; load the alias address of the bit
        MOV    R2, #1             ; R2 = 1
        STR    R2, [R1]           ; Write one to D6
```

Example 6-19

Write a program to set LOW the D0 bit of the SRAM location 0x20000005 using a) byte address and b) the bit alias address.

Solution:

a)

```
LDR   R1, =0x20000005    ; load the address of byte
LDRB  R2, [R1]           ; get the byte
AND   R2, R2, #2_11111110    ; make D0 bit low
STRB  R2, [R1]           ; write it back
```

b)   From Figure 6-12 we have address 0x220000A0 as the bit address of D0 of SRAM location 0x20000005.

```
LDR R2, =0x220000A0     ; load the alias address of the bit
MOV R0, 0               ; R0 = 0
STR R0, [R2]            ; write zero to D0
```

Peripheral I/O port bit-addressable region

The general purpose I/O (GPIO) and peripherals such as ADC, DAC, RTC, and serial COM port are widely used in the embedded system design. In many Arm-based trainer boards we see the connection of LEDs, switches, and LCD to the GPIO pins of the Arm chip. In such trainers the vendor provides the details of I/O port and peripheral connections to the Arm chip in addition to their address map. As we discussed earlier, the Arm Cortex-M3 and M4 have set aside 1M bytes of address space to be used bit-banding (bit-addressable) I/O and peripherals. The address space assigned to bit-banded peripherals and GPIO is 0x40000000 to 0x400FFFFF with bit-banded alias addresses of 0x42000000 to 0x43FFFFFF. Examine your trainer board data sheet for the bit-banded addresses implemented on the Arm chip.

Bit map for I/O peripherals

From Figure 6-13 once again notice the following facts.

1. The bit address 0x42000000 is assigned to D0 of peripherals location of 0x40000000.
2. The bit address 0x42000004 is assigned to D1 of peripherals location of 0x40000000.
3. The bit address 0x42000008 is assigned to D2 of peripherals location of 0x40000000.
4. The bit address 0x4200000C is assigned to D3 of peripherals location of 0x40000000.
5. The bit address 0x42000010 is assigned to D4 of peripherals location of 0x40000000.
6. The bit address 0x42000014 is assigned to D5 of peripherals location of 0x40000000.
7. The bit address 0x42000018 is assigned to D6 of peripherals location of 0x40000000.
8. The bit address 0x4200001C is assigned to D7 of peripherals location of 0x40000000.

When accessing a peripheral port in a single-bit manner, we must use the address aliases of 0x42000000 – 0x43FFFFFF.

Peripherals Byte addresses	Peripherals Bit addresses (We use these addresses to access the individual bits)							
	D7	D6	D5	D4	D3	D2	D1	D0
400FFFFF	43FFFFFC	F8	F4	F0	EC	E8	E4	43FFFFE0
400FFFFE	43FFFFDC	D8	D4	D0	CC	C8	C4	43FFFFC0
400FFFFD	43FFFFBC	B8	B4	B0	AC	A8	A4	43FFFFA0
400FFFFC	43FFFF9C	98	94	90	8C	88	84	43FFFF80
400FFFFB	43FFFF7C	78	74	70	6C	68	64	43FFFF60
400FFFFA	43FFFF5C	x8	x4	x0	xC	x8	x4	43FFFF40
400XXXXX	4XXXXXXC	X8	X4	X0	XC	X8	X4	4XXXXXX0
40000008	4200011C	118	114		...		104	42000100
40000007	420000FC	F8	F4	F0	EC	E8	E4	420000E0
40000006	420000DC	D8	D4	D0	CC	C8	C4	420000C0
40000005	420000BC	B8	B4	B0	AC	A8	A4	420000A0
40000004	4200009C	98	94	90	8C	88	84	42000080
40000003	4200007C	78	74	70	6C	68	64	42000060
40000002	4200005C	58	54	50	4C	48	44	42000040
40000001	4200003C	38	34	30	2C	28	24	42000020
40000000	4200001C	18	14	10	0C	08	04	42000000

Figure 6-13. Peripherals bit-addressable region and their alloc addresses.

Review Questions

1. True or false. All bytes of SRAM in Arm are bit-addressable.
2. True or false. All bits of the I/O peripherals in Arm are bit-addressable.
3. True or false. All ROM locations of the Arm are bit-addressable.
4. Of the 4G bytes of memory in the Arm, how many bytes are bit-addressable? List them.
5. How would you check to see whether bit D0 of location 0x20000002 is high or low?
6. Find out to which byte each of the following bits belongs. Give the address of the RAM byte in hex.

   (a) 0x23000030   (b) 0x23000040   (c) 0x23000048
   (d) 0x4200003C   (e) 0x43FFFFFC

Problems

Section 6.1: Arm Memory Map and Memory Access

1. What is the bus bandwidth unit?
2. Give the variables that affect the bus bandwidth.
3. True or false. One way to increase the bus bandwidth is to widen the data bus.
4. True or false. An increase in the number of address bus pins results in a higher bus bandwidth for the system.
5. Calculate the memory bus bandwidth for the following systems.
   (a) Arm of 100 MHz bus speed and 0 WS
   (b) Arm of 80 MHz bus speed and 1 WS

6. Indicate which of the following addresses is word aligned.

   (a) 0x1200004A   (b) 0x52000068   (c) 0x66000082
   (d) 0x23FFFF86   (e) 0x23FFFFF0   (f) 0x4200004F
   (g) 0x18000014   (h) 0x43FFFFF3   (i) 0x44FFFF05

7. Show how data is placed after execution of the following code using (a) little endian and (b) big endian.

   ```
   LDR    R2, =0xFA98E322
   LDR    R1, =0x20000100
   STR    [R1], R2
   ```

8. True or false. In Arm, instructions are always word aligned.
9. True or false. In a word aligned address the lower digit of the address is 0, 4, 8, or C.
10. Show how many memory cycles does it take to fetch the following data into register

    ```
    LDR    R1, =0x20000004
    LDR    R2, [R1]
    ```

11. Show how many memory cycles does it take to fetch the following data into register

    ```
    LDR    R1, =0x20000102
    LDR    R2, [R1]
    ```

12. Show how many memory cycles does it take to fetch the following data into register

    ```
    LDR    R1, =0x20000103
    LDR    R2, [R1]
    ```

13. Show how many memory cycles does it take to fetch the following data into register

    ```
    LDR     R1, =0x20000006
    LDRH    R2, [R1]
    ```

14. Show how many memory cycles does it take to fetch the following data into register

    ```
    LDR     R1, =0x20000C10
    LDRB    R2, [R1]
    ```

Section 6.2: Advanced Indexed Addressing Mode

15. True or false. Writeback is by default enabled in pre-indexed addressing mode.
16. Indicate the addressing mode in each of the following instructions
    (a) LDR R1, [R5], R2, LSL #2   (b) STR R2, [R1, R0]
    (c) STR R2, [R1, R0, LSL #2]!   (d) STR R9, [R1], R0
17. Which addressing mode uses the register as pointer to data location?
18. True or false. In the preindexed addressing mode with write back the value of register does not change after the instruction is executed.
19. How many Indexed addressing modes do we have in Arm? Name them.

20. In which indexed addressing mode the value of register does not change after the instruction is executed.
21. True or false. In the preindexed addressing mode only a fixed value can be used as offset.
22. True or false. In the preindexed addressing mode both fixed value and a register can be used as offset.

Section 6.3: ADR, LDR, and PC Relative Addressing
23. Assuming that the instruction "LDR R2, [PC, #8] is located in address 0x300, calculate the address of the memory location which is accessed.
24. Using PC relative addressing mode, write an LDR instruction that accesses a memory location which is 0x20 bytes ahead of itself.
25. Why ADR is called pseudo-instruction?

Section 6.4: Arm Bit-Addressable Memory Region
26. Give the bit-addressable SRAM region address for generic Arm.
27. What bit addresses are assigned to byte address of 0x20000004?
28. What bit addresses are assigned to byte address of 0x20000010?
29. What bit addresses are assigned to byte address of 0x200FFFFF?
30. What bit addresses are assigned to byte address of 0x20000020?
31. What bit addresses are assigned to byte address of 0x40000008?
32. What bit addresses are assigned to byte address of 0x4000000C?
33. What bit addresses are assigned to byte address of 0x40000020?
34. The following are bit addresses. Indicate where each one belongs.

    (a) 0x2200004C   (b) 0x22000068   (c) 0x22000080
    (d) 0x23FFFF80   (e) 0x23FFFF00   (f) 0x4200004C
    (g) 0x42000014   (h) 0x43FFFFF0   (i) 0x43FFFF00

35. Of the 4G bytes of memory locations in the Arm, how many of them are also assigned a bit address as well? Indicate which bytes those are.
36. True or false. The bit-addressable region cannot be accessed in byte.
37. True or false. The bit-addressable region cannot be accessed in word.
38. Write a program to see whether the D7 bit of RAM location 0x20000020 is high. If so, send a 1 to D1 of RAM location 0x20000000.
39. Write a program to see whether the D7 bit of I/O location 0x40000000 is low. If so, send a 0 to the D0 of location 0x400FFFFF.

Answers to Review Questions

Section 6.1
1. Compilers ensure that codes are word aligned.
2. little endian
3. 1/66 MHz = 15.15 ns is the bus clock period. Since the bus cycle time of zero wait states is 2 clocks, we have 2 × 15.15 = 30.3 ns

4. 1/100 MHz = 10 ns is the bus clock period. 50 ns - 10 ns = 40 ns. The Number of WS is 40 ns / 10 ns = 4.
5. False, most of the Arm devices use little endian as default.

Section 6.2
1. Register
2. Preinseded, postindexed and preindexed with write back.
3. True
4. In the preindexed write back the calculated value is written back to the pointing register. That is not the case for preindexed mode.
5. !

Section 6.3
1. PC (R15)
2. ADR, To implement the LDR directive the value is stored in memory; as a result, it uses more memory while the ADR uses no memory.

Section 6.4
1. False
2. False
3. False
4. 2MBytes; locations 0x20000000 to 0x200FFFFF of SRAM and 0x40000000 to 0x400FFFFF of GPIO

5.

```
LDR    R0, =0x22000040
LDR    R1, [R0]
CMP    R1, #0
BNE    L1
. . .
L1
```

6.

(a)   0x23000030 - 0x22000000 = 0x1000030; 0x1000030 / 0x20 = 0x80001; thus, it is in location 0x20000000 + 0x80001 = 0x20080001
       (0x1000030 % 32) / 4 = (48 % 32) / 4 = 16 / 4 = 4; it is D4 of 0x20080001.
(b)   0x1000040 / 0x20 = 0x80002; it is in location 0x20080002
       (0x1000040 % 0x20) / 4 = 0; it is D0 of location 0x20080002
(c)   0x1000048 / 0x20 = 80002; it is in location 0x20080002
       (0x1000048 % 0x20) / 4 = 2; it is D2 of location 0x20080002
(d)   0x4200003C - 0x42000000 = 0x03C; 0x03C / 0x20 = 0x01
       (0x3C % 0x20) / 4 = 0x1C / 4 = 7; it is D7 of 0x40000001
(e)   0x43FFFFFC − 0x42000000 = 0x1FFFFFC; 0x1FFFFFC / 0x20 = 0xFFFFF
       (0x1FFFFFC % 0x20) / 4 = 0x1C / 4 = 7; D7 of 0x400FFFFF

Chapter 7: C for Embedded Systems

In reading this book we assume you already have some understanding of how to program in C language. In this chapter, we will examine some important concepts widely used in embedded system design that you may not be familiar with due to the fact that many generic C programming books do not cover them. In section 7.1, we examine the C data types for 32-bit systems. The bit-wise operators are covered in section 7.2.

Section 7.1: C Data types for Embedded systems

In general C programming textbooks, we see *char*, *short*, *int*, *long*, *float*, and *double* data types. We need to examine the size of C data types in the light of 32-bit processors such as ARM. The C standards do not specify the size of data types. The compiler designers are free to decide the size for each data type. The *float* and *double* data types are standardized by the IEEE754. The sizes of char and short are often set at 1 byte and 2 bytes. The size of int is often depending on the data size of the CPU but rarely go below 16 or above 32. The sizes of long and long long are implemented the same way everywhere.

For now, we will discuss the data types defined by Keil MDK-ARM.

char

The *char* data type is a byte size data whose bits are designated as D7-D0. It can be *signed* or *unsigned*. In the signed format the D7 bit is used for the + or - sign and takes values between -128 to +127. In the *unsigned char* we have values between 0x00 to 0xFF in hex or 0 to 255 in decimal since there is no sign and the entire 8 bits are used for the magnitude. (For more information about signed numbers see Chapter 5.)

short int

The *short int* (or usually referring as *short*) data type is a 2-byte size data whose bits are designated as D15-D0. It can be *signed* or *unsigned*. In the signed format, the D15 bit is used for the + or - sign and takes values between -32,768 to +32,767. In the *unsigned short int* we have values between 0x0000 to 0xFFFF in hex or 0 to 65,535 in decimal since there is no sign and the entire 16 bits are used for the magnitude.

A 32-bit processor such as the ARM architecture reads the memory with a minimum of 32 bits on the 4-byte boundary (address ending in 0, 4, 8, and C in hex). If a short int variable is allocated straddling the 4-byte boundary, access to that variable is called an *unaligned access*. Not all the ARM processor support unaligned access. Those devices supporting unaligned access pay a performance penalty by having to read/write the memory twice to gain access to one variable (see Example 7-1). Unaligned access can be avoided by either padding the variables with unused bytes (Keil) or rearranging the sequence of the variables in allocation. By default, the compilers usually generate aligned variable allocation.

Example 7-1

Show how memory is assigned to the following variables in aligned and unaligned allocation. Begin from memory location 0x20000000.

```
unsigned char a;
unsigned short int b;
unsigned short int c;
```

Solution:

Unaligned allocation of variable c

a	b	b	c
20000000	20000001	20000002	20000003
c			
20000004	20000005	20000006	20000007

Aligned allocation of variables by padding one byte between variable a and b

a		b	b
20000000	20000001	20000002	20000003
c	c		
20000004	20000005	20000006	20000007

Aligned allocation of variables by rearranging the variable sequence

b	b	c	c
20000000	20000001	20000002	20000003
a			
20000004	20000005	20000006	20000007

int

The *int* data type usually represents for the native data size of the processor. For example, it is a 2-byte size data for a 16-bit processor and a 4-byte size data for a 32-bit processor. This may cause confusion and portability issue. The C99 standard addressed the issue by creating a new set of integer variable types that will be discussed later. For now, we will stick to the conventional data types.

The *int* data type of the ARM processors is 4-byte size and identical to *long int* data type described below.

long int

The long int (or *long*) data type is a 4-byte size data whose bits are designated as D31-D0. It can be signed or unsigned. In the signed format the D31 bit is used for the + or - sign and takes values between -2^{31} to $+2^{31}-1$. In the unsigned long we have values between 0x00000000 to 0xFFFFFFFF in hex. In the 32-bit microcontroller when we declare a long variable, the compiler sets aside 4 bytes of storage in SRAM. But it also makes sure they are aligned, meaning it places the data in locations with addresses ending with 0,4,8 and C in hex. This avoids unaligned data access performance penalty. The unsigned long is widely used in ARM for defining addresses since ARM address size is 32-bit long.

Example 7-2

Show how memory is assigned to the following variables in aligned and unaligned allocation. Begin from memory location 0x20000000.

```
unsigned char a;
unsigned short int b;
unsigned short int c;
unsigned int d;
```

Solution:

Unaligned allocation of variable c

a	b	b	c
20000000	20000001	20000002	20000003
c	d	d	d
20000004	20000005	20000006	20000007
d			
20000008	20000009	2000000A	2000000B
2000000C	2000000D	2000000E	2000000F

Aligned allocation of variables by padding byte(s) between variable a and b

a		b	b
20000000	20000001	20000002	20000003
c	c		
20000004	20000005	20000006	20000007
d	d	d	d
20000008	20000009	2000000A	2000000B
2000000C	2000000D	2000000E	2000000F

Aligned allocation of variables by rearranging the variable sequence

d 20000000	d 20000001	d 20000002	d 20000003
b 20000004	b 20000005	c 20000006	c 20000007
a 20000008	20000009	2000000A	2000000B

long long

The *long long* data type is an 8-byte size data whose bits are designated as D63-D0. It can be signed or unsigned. In the signed format the D63 bit is used for the + or - sign and takes values between -2^{63} to $+2^{63}-1$. In the *unsigned long long* we have values between 0x0000000000000000 to 0xFFFFFFFFFFFFFFFF in hex. In the 32-bit microcontroller, when we declare a long long variable, the compiler sets aside 8 bytes of storage in SRAM and it makes sure they are aligned, meaning it places the data in locations with addresses ending with 0 and 8. This avoids unaligned data access performance penalty.

Data type	Size	Range
char	1 byte	-128 to 127
unsigned char	1 byte	0 to 255
short int	2 bytes	-32,768 to 32,767
unsigned int	2 bytes	0 to 65,535
long	4 bytes	-2,147,483,648 to 2,147,483,647
unsigned long	4 bytes	0 to 4,294,967,295
long long	8 bytes	-9,223,372,036,854,775,808 to 9,223,372,036,854,775,807
unsigned long long	8 bytes	0 to 18,446,744,073,709,551,615

Table 7-1: ANSI C (ISO C89) integer data types and their ranges

Why should I care about which data type to use?

There are three major reasons why a programmer should care about data type, performance, overflow, and coercion.

Performance

It must be noted that while in the 8-bit microcontrollers we need to use the proper data type for the variables to improve the performance, this is less of problem in 32-bit CPUs such as ARM. For example, for the number of days working in a month (or number of hours in a day) we use unsigned char since it is less than 255. Using unsigned char in 8-bit microcontroller is important since it saves RAM space, memory access time, and computation clock cycles. If we use int instead, the compiler allocates 2 bytes in RAM and that is a waste of RAM resource. The CPU will have to access the additional byte and perform

194

additional arithmetic instructions with it even if the byte contains zero and has no effect on the result. This is a problem that we should avoid since an 8-bit microcontroller usually has few RAM bytes with slower clock speed for bus and CPU. In the case of 32-bit systems such as ARM, 1, 2, or 4 bytes of data will result in the same memory access time and computation time. Most of the 32-bit systems also have more generous amount of RAM to alleviate the concern of memory usage and allow padding for aligned access.

Overflow

Unlike assembly language programming, high level program languages do not provide indications when an overflow occurs and the program just fails silently. For example, if you use a short int to hold the number of seconds of a day, 9 hours 6 minutes and 7 seconds into the day, the second count will overflow from 32,767 to -32,768. Even if your program handles negative second count, the time will jump back to the day before.

With 32-bit int in a 32-bit ARM processor, overflow is a much less frequent problem because a 32-bit int will hold a number up to 2,147,483,647 but it does not eliminate the potential of the problem. One of the critical overflow problem waiting to happen is the Unix Millennium Bug. Unix keeps track of time using a 32-bit int for the number of seconds since January 1st 1970. This variable is going to overflow comes January 19, 2038. Because of the popularity of Unix, not only Unix systems are extensively used, many other systems use the same or similar format to keep track of time. So far, there is no universal solution to mitigate this problem yet.

Coercion

In C language, the data types of the operands must be identical for binary operations (the operator with two operands such as A + B). If you write a statement with different operand data types for a binary operation, the compiler will convert the smaller data type to the bigger data type. If it is an assignment operator (A = B), the right-hand side operand is converted to the left-hand side data type before the assignment. This implicit data type conversion is called *coercion*. The compiler may or may not give you warning when coercion occurs.

In two conditions, coercion may result in undesirable result. If the variable is signed and the data sized is increased, the new bits are filled with the sign bit (most significant bit) of the original value. For example, if an 8-bit number 0b10010010 is coerced into a 16-bit signed number, the result will be 0b1111111110010010. This may work just fine in most cases, but there are few occasions that will become an issue.

The other problem is when you assign a larger data type to a smaller data type variable, the higher order bits will be truncated. For example, in the statement "A = B;" if A is 8-bit and B is 16-bit, the upper 8 bits of B is discarded before the assignment.

There is not a simple solution for the data type size issues. As a programmer, you have to be cognizant about them all the time.

Notes

1. By default variables are considered as *signed* unless the *unsigned* keyword is used. As a result, *signed long* is the same as *long*; the *long long* is the same as *signed long long*, and so on with the exception of *char*. Whether *char* is signed or unsigned by default varies from compiler to compiler. In some compilers, including Keil, there is an option to choose if the char variable should be considered as *signed char* or *unsigned char* by default. (To choose this in Keil, go to *Project* menu and select *Options*. Then, in the *C/C++* tab, check or uncheck the choice *Plain char is signed*, as you desire.) It is a good practice to write out the *signed* keyword explicitly, when you want to define a variable as *signed char*.

2. In some compilers (including Keil and IAR) the *int* type is considered as long int while in some other compilers (including AVR and PIC compilers) it is considered as *short int*. In other words, the *int* type is commonly defined so that the processor can handle it easily. As we will see next, we can use int16t and int32t instead of short and long in order to prevent any kind of ambiguity and make the code portable between different processors and compilers.

Data types in ISO C99 standard

While every C programmer has used ANSI C (ISO C89) data types, many C programmers are not familiar with the ISO C99 standard. In C standards, the sizes of integer data types were not defined and are up to the compilers to decide. By conventions, char is one byte and short is two-byte size. But int and long varies greatly among the compilers.

In ISO C99 standard, a set of data types were defined with number of bits and sign clearly defined in the data type names. (See Table 7-2.) The C ISO C99 standard is used extensively by embedded system programmer for RTOS (real time operating system) and system design. It is also supported by many C compilers. Notice the range is the same as ANSI C standard except it uses more descriptive syntax.

These integer data types are defined in a header file called *stdint.h*. You need to include this header file in order to use these data types.

Data type	Size	Range
int8_t	1 byte	-128 to 127
uint8_t	1 byte	0 to 255
int16_t	2 bytes	-32,768 to 32,767
uint16_t	2 bytes	0 to 65,535
int32_t	4 bytes	-2,147,483,648 to 2,147,483,647
uint32_t	4 bytes	0 to 4,294,967,295
int64_t	8 bytes	-9,223,372,036,854,775,808 to 9,223,372,036,854,775,807
uint64_t	8 bytes	0 to 18,446,744,073,709,551,615

Table 7-2: ISO C99 integer data types and their ranges

Review questions

1. In an 8-bit system we use (char, unsigned char) for the number of months in a year.
2. For a system with 16-bit address, bus we use (int, unsigned int) for address definition.
3. For an ARM system the address is _____bit wide and we use _____data type for it.

4. True or false. In C programming of ARM, compiler makes sure data are aligned.

Section 7.2: Bit-wise Operations in C

One of the most important and powerful features of the C language is its ability to perform bit manipulation. Because many books on C do not cover this important topic, it is appropriate to discuss it in this section. This section describes the action of bit-wise logic operators and provides some examples of how they are used.

Bit-wise operators in C

While every C programmer is familiar with the logical operators AND (&&), OR (||), and NOT (!), many C programmers are less familiar with the bitwise operators AND (&), OR (|), EX-OR (^), inverter (~), shift right (>>), and shift left (<<). These bit-wise operators are widely used in software engineering for embedded systems and control; consequently, their understanding and mastery are critical in microprocessor-based system design and interfacing. See Table 7-3.

A	B	AND (A & B)	OR (A | B)	EX-OR (A^B)	Invert ~B
0	0	0	0	0	1
0	1	0	1	1	0
1	0	0	1	1	1
1	1	1	1	0	0

Table 7-3: Bit-wise Logic Operators for C

The following shows some examples using the C bit-wise operators:

```
0x35 & 0x0F results in0x05        /* ANDing */
0x04 | 0x68 results in 0x6C       /* ORing:    */
0x54 ^ 0x78 results in 0x2C       /* XORing */
~0x55 results in 0xAA             /* Inverting 0x55 */
```

Examples 7-3 and 7-4 show how the bit-wise operators are used in C. Run the following programs on your simulator and examine the results.

Example 7-3

Run the following program on your simulator and examine the results.

```
int main(void)
{
    volatile unsigned char temp;    /* declare volatile otherwise
the optimizer will remove it. */
    temp = 0x35 & 0x0F;       /* ANDing      : 0x35 & 0x0F = 0x05 */
    temp = 0x04 | 0x68;       /* ORing : 0x04 | 0x68 = 0x6C */
    temp = 0x54 ^ 0x78;       /* XORing      : 0x54 | 0x78 = 0x2C */
    temp = ~0x55;             /* Inverting   : ~0x55 = 0xAA */
    while (1);
}
```

Setting and Clearing (masking) bits

As discussed in Chapter 3, OR can be used to set a bit, and AND can be used to clear a bit. If you examine Table 7-3 closely, you will see that:

- Anything ORed with a 1 results in a 1; anything ORed with a 0 results in no change.

- Anything ANDed with a 1 results in no change; anything ANDed with a 0 results in a zero.

- Anything EX-ORed with a 1 results in the complement; anything EX-ORed with a 0 results in no change.

See Example 7-4.

Example 7-4

The following program toggles only bit 4 of var1 continuously without disturbing the rest of the bits.

```
...
int main(void)
{
    unsigned char var1;
    while(1)
    {
        var1 = var1 | 0x10;      /* Set bit 4 (5th bit) of var1 */
        var1 = var1 & 0xEF;      /* Clear bit 4 (5th bit) of var1 */
    }

    return 0;
}
...
```

Notice that we can also toggle the bit using XOR as shown below:

```
var1 = var1 ^ 0x10;
```

Testing bit with bit-wise operators in C

In many cases of system programming and hardware interfacing, it is necessary to test a given bit to see if it is high or low. For example, many devices send a high signal to state that they are ready for an action or to indicate that they have data. How can the bit (or bits) be tested? In such cases the unused bits are masked and then the remaining data is tested. See Example 7-5.

Example 7-5

Write a C program to monitor bit 5 of var1. If it is HIGH, change value of var2 to 0x55; otherwise, change value of var2 to 0xAA.

Solution:

198

```
...
    while(1)
    {
        if (var1 & 0x20)    /* check bit 5 (6th bit) of var1 */
            var2 = 0x55;    /* this statement is executed if bit 5 is a 1 */
        else
            var2 = 0xAA;    /* this statement is executed if bit 5 is a 0 */
    }
...
```

Bit-wise shift operation in C

There are two bit-wise shift operators in C. See Table 7-4.

Operation	Symbol	Format of Shift Operation
Shift Right	>>	data >> number of bit-positions to be shifted right
Shift Left	<<	data << number of bit-positions to be shifted left

Table 7-4: Bit-wise Shift Operators for C

The following shows some examples of shift operators in C:

1. 0b00010000 >> 3   /* it equals 00000010. Shifting right 3 times */
2. 0b00010000 << 3   /* it equals 10000000. Shifting left 3 times */
3. 1 << 3   /* it equals 00001000. Shifting left 3 times */

Compound Operators

In C language, whenever the left-hand-side of the assignment operator (=) and the first operand on the right-hand-side are identical we can avoid repeating the operand by using the compound operands. As shown in Table 7-5, in compound operators, the operators are mentioned just on the left-hand-side of the equal sign and the first operand is omitted.

Instruction	Its equivalent using compound operators
a = a + 6;	a += 6;
a = a − 23;	a −= 23;
y = y * z;	y *= z;
z = z / 25;	z /= 25;
w = w | 0x20;	w |= 0x20;
v = v & mask;	v &= mask;
m = m ^ togBits;	m ^= togBits;

Table 7-5: Some Compound Operator Examples

Bit-wise operations using compound operators

The majority of hardware access level code involves setting a bit or bits in a register, clearing a bit or bits in a register, toggling a bit or bits in a register, and monitoring the status bits. For the first three cases, the operations read the content of the register, modify a bit of bits then write it back to the same register. The compound operators are very suitable for these operations.

199

To set bit(s) in a register,

```
register |= MASK;
```

where MASK is a number that has '1' at the bit(s) to be set.

```
register |= 0x08;
```

The number 0x08 has a '1' at bit 3, therefore the statement sets bit 3 of the register.

```
register |= 0x42;
```

The number 0x42 has a '1' at bit 6 and bit 1, therefore the statement sets bit 6 and bit 1 of the register.

To clear bit(s) in a register,

```
register &= ~MASK;
```

where MASK is a number that has '1' at the bit(s) to be cleared.

```
register &= ~0x20;
```

The number 0x20 has a '1' at bit 5, therefore the statement clears bit 5 of the register.

```
register &= ~0x12;
```

The number 0x12 has a '1' at bit 4 and bit 1, therefore the statement clears bit 4 and bit 1 of the register. Notice the mask for clearing the bits is the same as the mask for setting the bits, where the bits to be modified are '1' and the rest of the bits are '0' except that in clearing the bits, the mask is complemented in the statements.

To toggle bit(s) in a register,

```
register ^= MASK;
```

The examples are similar to setting bits so we will skip them here.

Using shift operator to generate mask

With the statements above, one challenge may be to generate the mask with the correct bit(s) set to 1 depending on how proficient you are with converting binary numbers to hexadecimal. Some compilers allow you to write a literal binary number in the format of 0b00000000 but since it is not in the C standards, many compilers do not accept this notation.

One way to ease the generation of the mask is to use the left shift operator. To generate a mask with bit n set to 1, use the expression:

```
1 << n
```

If more bits are to be set in the mask, they can be "or" together. To generate a mask with bit n and bit m set to 1, use the expression:

```
(1 << n) | (1 << m)
```

Now to set bit 3 of the register, we can rewrite the statement

```
register |= 0x08;
```

as

```
register |= 1 << 3;
```

And to set bit 6 and bit 1 of the register, we can rewrite the statement

```
register |= 0x42;
```

as

```
register |= (1 << 6) | (1 << 1);
```

The same goes for clearing bit 5 of the register, we can use the statement

```
register &= ~(1 << 5);
```

And to clear bit 4 and bit 1 of the register

```
register &= ~((1 << 4) | (1 << 1));
```

Notice that regardless of setting or clearing bits, the mask always has 1s at the bit locations for the bits to be modified and when multiple bits are used in the mask, they are always ORed together. We will leave the toggling of the bits for the readers.

Setting the value in a multi-bit field

Some of the bits in a register form a field with meaningful values. For example, if register bits 30-28 determine the divisor value to divide the clock and we would like to set the divisor to 5. One way of doing so is to set or clear the bits one by one.

```
register |= 1 << 30;
register &= ~(1 << 29);
register |= 1 << 28;
```

Although this method will achieve the desired result, the divisor value 5 is not apparent from reading the code. An alternative way is to clear the field first then set the value.

```
register &= ~(7 << 28);
register |= 5 << 28;
```

The first statement clears bit 30-28 and the second statement set the value 5 in the field. With this method, the divisor 5 is visible in the second statement.

These two statements may be combined into a single statement:

```
register = (register & ~(7 << 28)) | (5 << 28);
```

Review Questions
1. What is result of 0x2F &0x27?
2. What is result of 0x2F | 0x27?
3. What is result of 0x2F ^ 0x27?
4. What is result of 0x2F >> 3?
5. What is result of 0x27 << 4?
6. In Example 7-5 what is stored in var2 if the value of var1 is 0x03?

Problems

Section 7.1: C Data types for Embedded systems
1. Indicate what data type you would use for the following variables:
   a. weather temperature
   b. the number of days in a week
   c. the number of days in a year
   d. the number of months in a year
   e. a counter to track the number of people getting on a bus
   f. a counter to track the number of people going to a class
   g. an address of 64K RAM space
   h. the age of a person
   i. a string for a message to welcome people to a building

Section 7.2: Bitwise Operations in C
2. Assuming that c is uint8_t, indicate the data loaded into it in each of the following cases: (*Note*: The operations are independent to each other.)
   (a) c=0xF0&0x45;   (b) c=0xF0&0x56;
   (c) c=0xF0^0x76;   (d) c=0xF0&0x90;
   (e) c=0xF0^0x90;   (f) c=0xF0|0x90;
   (g) c=0xF0&0xFF;   (h) c=0xF0|0x99;
   (i) c=0xF0^0xEE;   (j) c=0xF0^0xAA;
3. Find the contents of b after each of the following operations. b is uint8_t:
   (a) b=0x65&0x76;   (b) b=0x70|0x6B;
   (c) b=0x95^0xAA;   (d) b=0x5D&0x78;
   (e) b=0xC5|0x12;   (f) b=0x6A^0x6E;
   (g) b=0x37|0x26;
4. Find the port value after each of the following is executed:
   (a) b=0x65>>2;   (b) b=0x39<<2;
   (c) b=0xD4>>3;   (d) b=0xA7<<2;
5. Show the C code to swap 0x95 to make it 0x59.

Answer to Review Questions

Section 7.1: C Data types for Embedded systems
1. unsigned char
2. unsigned int
3. 32 – unsigned long (or uint32_t)
4. True

Section 7.2: Bitwise Operations in C
3. 0x27
4. 0x2F
5. 0x08
6. 0x05
7. 0x70
8. 0xAA

Chapter 8: STM32F103 I/O Programming

We use the general-purpose input output (GPIO) pins of microcontrollers to interface with LED, switch (SW), LCD, Keypad, and so on. This chapter covers the programming of GPIO using LED, switches, and seven segment LEDs as examples. Because some of the core materials covered in this chapter are widely used in subsequent chapters, we urge you to study this chapter thoroughly before moving on to other chapters. Section 8.1 shows how to access the special function registers associated with the GPIO of STM32F103. In section 8.1, we also use simple LEDs and SWs to show the programming of GPIO. Section 8.2 examines the 7-segment LED connection to the microcontroller and how to programming it.

Section 8.1: I/O Port Programming in STM32F1xx

In microcontrollers we have two types of I/O ports. They are:

a. **General Purpose I/O (GPIO):** The GPIO ports are used for interfacing devices such as LEDs, switches, LCD, keypad, and so on.
b. **Special purpose I/O:** These I/O ports have designated function such as ADC (Analog-to-Digital), Timer, UART (universal asynchronous receiver transmitter), and so on.

We have dedicated many chapters to these special purpose I/O ports. In this chapter, we examine the GPIO and its interfacing to LEDs, switches, and 7-segment LEDs and show how to access them.

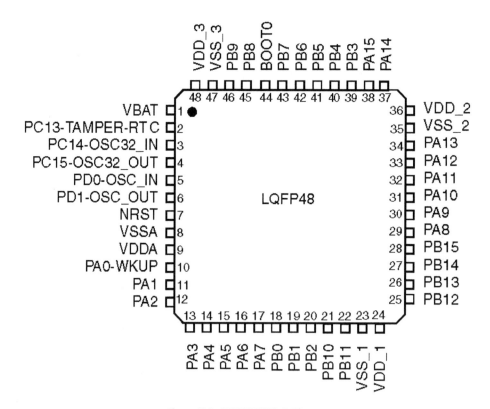

Figure 8-1: STM32F103xB Pinout

204

GPIO Ports

The STM32F103x8 has ports A, B, C, and D. The pins are designated as PA0-PA15, PB0-PB15, PC0-PC15, and PD0-PD15. See Figure 8-1. The rest of the pins are designated as VSS, VDD, RESET, VREF, and so on. They are discussed in Section 8.3.

The number of ports in the STM32 family varies depending on the number of pins on the chip. The 20-pin chips have ports A and B, while the 144-pin chips have ports A through G. To use any of these ports as an input or output port, it must be programmed, as we will explain throughout this section. Not all ports have 16 pins. For example, in the STM32F103x8, Port C has 3 pins. Each port has seven I/O registers associated with it, as shown in Figure 8-2. They are designated as CRL (Configuration Register Low), CRH (Configuration Register High), IDR (Input Data Register), ODR (Output Data Register), BSSR (Bit Set Reset Register), BRR (Bit Reset Register), and LCKR (Lock Register). Next, we describe how to access the I/O registers associated with the ports.

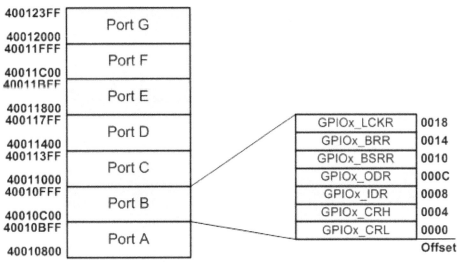

Figure 8-2: GPIO Memory Map

CRH and CRL Configuration Registers

Before using each pin, you should configure it. The CRL and CRH registers are used to configure the pins of a port. See Figure 8-3. STM32F1xx has a rich I/O configuration register and there are 4 bits to configure each pin of a port. This means that each pin can be configured into 16 different modes. The MODEn bits choose if the pin is input or output. The pin is input if the MODE bits are 00. In other cases, the pin is output. We do not go into details right now and we consider that to make a pin output, we should set the CNFn:MODEn bits to 0x3 (0011 in binary). To make a pin input, we should set the CNFn:MODEn bits to 0x4 (0100 in binary). CRL has the configuration bits for pins 0 to 7 and CRH configures pins 8 to 15. For example, the following line of code makes pins 0 and 1 of port B, output and makes pins 2 to 7 input:

```
GPIOB->CRL = 0x44444433;      //PB2-PB7 as inputs, PB0 and PB1 as outputs
```

The following instruction configures PB8 and PB10 to PB15 as inputs and configures the PB9 as output:

```
GPIOB->CRH = 0x44444434;  //PB10-PB15 as inputs, PB9: output, PB8: input
```

205

Notice that upon reset, all ports have the value 0x4 in their CRH:CRL registers. This means that all ports are configured as input.

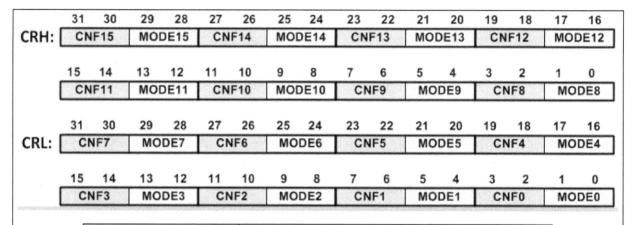

MODEx bits	Direction	Max speed
00	Input	
01		10 MHz
10	Output	2 MHz
11		50 MHz

CNFx when the MODEx is set to input (MODEx = 00):

CNFx bits	Configuration	Description
00	Analog mode	Select this mode when you use a pin as an ADC input.
01	Floating input	In this mode, the pin is high-impedance.
10	Input with pull-up/pull-down	The value of ODR chooses if the pull-up or pull-down resistor is enabled. (1: pull-up, 0:pull-down)
11	reserved	

CNFx when the MODEx is set to output (MODEx > 00):

CNFx bits	Configuration
00	General purpose output push-pull
01	General purpose output Open-drain
10	Alternate function output Push-pull
11	Alternate function output Open-drain

Figure 8-3: CRH and CRL (Configuration Registers)

ODR (Output Data Register) and IDR (Input Data Register)

Each port has an ODR (output data register) and an IDR (Input Data Register). To send out data, we write into the ODR register. IDR gives the current values of the pins. See Figures 8-4 and 8-5. ODR and IDR are 16 bits wide, and each port has a maximum of 16 pins; therefore, each bit of the I/O registers is related to one pin of the port. For example, bit 3 of ODR is related to pin 3 of the port and pin 5 of the port is related to bit 5 of IDR.

ODR: | ODR15 | ODR14 | ODR13 | ODR12 | ODR11 | ODR10 | ODR9 | ODR8 | ODR7 | ODR6 | ODR5 | ODR4 | ODR3 | ODR2 | ODR1 | ODR0 |

(GPIOB->ODR)

Figure 8-4: ODR (Output Data Register)

IDR: | IDR15 | IDR14 | IDR13 | IDR12 | IDR11 | IDR10 | IDR9 | IDR8 | IDR7 | IDR6 | IDR5 | IDR4 | IDR3 | IDR2 | IDR1 | IDR0 |

(GPIOx->IDR)

Figure 8-5: IDR (Input Data Register)

Figure 8-6 shows the simplified view of a GPIO pin and the relation between ODR, CRH:CRL, and IDR. There is a tri-state buffer between the ODR register and the pin. If the direction is set to output, the tri-state buffer becomes enabled and ODR is connected to the pin. Otherwise, the buffer is disabled.

To get data from the GPIO pins we read the IDR register. If you want to use a port as input, you must configure the pin as input, otherwise, the value of ODR conflicts with the input signal. So, in our program first we configure the pins. Then we use the ODR register to write to the pin and IDR to read data from the pin.

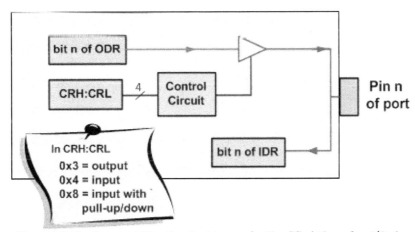

Figure 8-6: The Data and Direction Registers and a Simplified View of an I/O pin

The Clock enable registers

APB1ENR, APB2ENR, and AHBENR are used to enable the clock source for the peripherals. If a peripheral is not used, the clock source to it can be disabled in order to save power. See Figure 8-7 through 8-9 and Table 8-1. On reset all the bits, except SRAMEN (SRAM clock enable) and FLITFEN, are disabled. So, when we want to use a peripheral, we should enable the clock of it. To enable the clock, the bit should be set to 1. The enable bits for I/O ports (IOPA, IOPB, etc) are located in the APB2ENR register. For example, the following instruction enables the clock for GPIO port B:

```
RCC->APB2ENR |= (1<<3); //set the IOPBEN bit to 1 (IOPBEN is bit 3 of APB2ENR)
```
The following line of code enables the clock for all GPIO ports:

```
RCC->APB2ENR |= 0x1FC; //set IOPAEN,IOPBEN,..., and IOPGEN (1FC = 0000 0001 1111 1100)
```

207

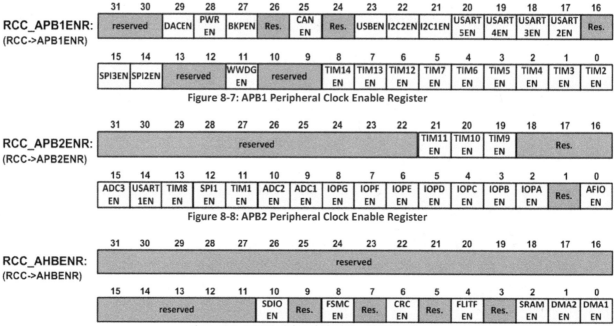

Figure 8-7: APB1 Peripheral Clock Enable Register

Figure 8-8: APB2 Peripheral Clock Enable Register

Figure 8-9: AHB Peripheral Clock Enable Register

Label	Description	Label	Description
IOPx	I/O port x clock enable	ADCnEN	ADCn clock enable
USARTnEN	USARTn clock enable	DACnEN	DACn clock enable
USBEN	USB clock enable	TIMnEN	TIMn timer clock enable
CANEN	CAN clock enable	SPInEN	SPI n clock enable
PWREN	Power interface clock enable	BKPEN	Backup interface clock enable
WWDG	Window watchdog clock enable	SDIOEN	SDIO clock enable
DMAnEN	DMAn clock enable	FSMCEN	FSMC clock enable
CRCEN	CRC clock enable	I2CnEN	I2Cn clock enable
Note: (0: clock disabled, 1: clock enabled)			

Table 8-1: The bits of AHBENR, APB1ENR, and APB2ENR Registers

Notice that without clock the port will not work. Any access to the registers associated with the port before the clock is enabled will result in a hard fault and the program crashes. Also notice that the available enable bits vary depending on how many I/O ports that chip supports. For example, in STM32F103C8, the I/O ports are from A to D. In this case, IOPEEN to IOPGEN are not used.

Program 8-1 toggles all the pins of port A.

Program 8-1: Toggling Port A

```
#include "stm32f10x.h"

void delay_ms(uint16_t t);

int main()
{
    RCC->APB2ENR |= 0xFC; //Enable the clocks for GPIO ports
```

```
   GPIOA->CRL = 0x33333333; //PA0 to PA7 as outputs
   GPIOA->CRH = 0x33333333; //PA8 to PA15 as outputs

   while(1)
   {
      GPIOA->ODR = 0x0000; //make all the pins of Port A low
      delay_ms(1000);        //wait 1000ms
      GPIOA->ODR = 0xFFFF; //make all the pins of Port A high
      delay_ms(1000);        //wait 1000ms
   }
}
/* The following delay is tested with Keil and 72MHz */
void delay_ms(uint16_t t)
{
   volatile unsigned long l = 0;
   for(uint16_t i = 0; i < t; i++)
      for(l = 0; l < 6000; l++)
      {
      }
}
```

Note: *If you want to write programs for I/O ports in Assembly language, read Section 8-4. Then, continue reading the section.*

Program 8-2 toggles PA2. It is similar to Program 8-1. But we used Ex-OR to toggle. As we saw in Chapter 3 and Chapter 7, if we Ex-OR a bit with 1, it toggles. In the program we want to toggle PA2. So, we shift left 1 twice and then Ex-OR it with the ODR.

Program 8-2: Toggling PA2

```
#include <stm32f10x.h>

void delay_ms(uint16_t t);

int main()
{
   RCC->APB2ENR |= 0xFC; /* Enable clocks for GPIO ports */
   GPIOA->CRL = 0x44444344; /* PA2 as output */
   while(1)
   {
      GPIOA->ODR ^= (1<<2); /* toggle PA2 */
      delay_ms(1000);
   }
}
 //copy delay_ms from Program 8-1 to here
```

Measuring time delay in a C program loop

One simple way of creating a time delay is using a **for** loop in C language. The size of time delay loop for a given system is function of two factors: a) the CPU frequency and b) the compiler. It must be noted that a time delay C loop measured using a given compiler (e.g. Keil) may not give the same result if a different compiler such as IAR is used. Regardless of clock source to CPU and the C compiler used, always use oscilloscope to measure the size of time delay loop for a given system with a given compiler and compiler option setting.

Program 8-3 toggles PC13. If you have a Blue pill board an LED is connected to the PC13. Otherwise, you can connect an LED together with a resistor as shown in Figure 8-10.

Program 8-3: Toggle PC13

```c
#include <stm32f10x.h>

void delay_ms(uint16_t t);

int main()
{
   RCC->APB2ENR |= 0xFC; //Enable GPIO ports clocks

   GPIOC->CRH = 0x44344444; //PC13 as output

   while(1)
   {
      GPIOC->ODR ^= (1<<13); //toggle PC13
      delay_ms(1000);
   }
}
//copy delay_ms from Program 8-1 to here
```

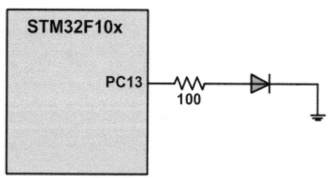

Figure 8-10: connecting an LED

See Example 8-1.

Example 8-1

Connect 8 LEDs to pins PA0 to PA7. Then write a program that makes the LEDs one after another ON. Then, make the LEDs in the reverse order Off.

Solution:

```c
#include <stm32f10x.h>

void delay_ms(uint16_t t);
```

210

```
int main()
{
    RCC->APB2ENR |= 0xFC;          //ENABLE clocks for GPIOs
    GPIOA->CRL = 0x33333333;       //PA0-PA7 as outputs

    while(1)
    {
        int8_t n;
        for(n = 0; n <= 7; n++)     //from 0 to 7
        {
            GPIOA->ODR |= (1<<n);    //make PAn high
            delay_ms(500);
        }

        for(n = 7; n >= 0; n --)    //from 7 to 0
        {
            GPIOA->ODR &= ~(1<<n);   //make PAn low
            delay_ms(500);
        }
    }
}
//copy delay_ms from Program 8-1 to here
```

Configuring a pin and the other pins configurations remain unchanged (For experts)

You might like to configure the PA5 as output, while the configurations of other pins remain unchanged. To do so, you can write the following code:

```
uint32_t t = GPIOA->CRL; //get the contents of CRL
t = t & 0xFF0FFFFF;      //clear the configuration bits for PA5
t = t | 0x00300000;      //configure the bits for PA5 as output
GPIOA->CRL = t;          //load CRL with the new value
```

The above code, gets the current value of CRL and clears the bits for PA5. Then, sets the bits for PA5 to 0x3 to configure the PA5 as output and then writes the value to CRL. We can also write the above code in a single line as shown below:

```
GPIOA->CRL = (GPIOA->CRL&0xFF0FFFFF)|(0x300000); //PA5 as output
```

You can also write the above instruction using shifts:

```
GPIOA->CRL = (GPIOA->CRL&(~(0xF<<(5*4))))|(0x3<<(5*4)); //PA5 as output
```

In the following program, we implement the ioPinConfig function for configuring single pins. In the program, we define the value of 0x3 as OUTPUT by writing "#define OUTPUT 0x3" and we use it for configuration. Then, we write "ioPinConfig(GPIOC,13,0x3);" to configure the PC13 as output.

Program 8-4: Writing a function for configuring pins

```
#include <stm32f10x.h>

#define OUTPUT 0x03
#define INPUT 0x04

void delay_ms(uint16_t t);
```

```
void ioPinConfig(GPIO_TypeDef *portName, uint8_t pinNum, uint8_t configValue);

int main()
{
    RCC->APB2ENR |= 0xFC;  //Enable GPIO ports clocks

    ioPinConfig(GPIOC,13,OUTPUT);
    while(1)
    {
        GPIOC->ODR ^= (1<<13);
        delay_ms(1000);
    }
}

void ioPinConfig(GPIO_TypeDef *portName, uint8_t pinNum, uint8_t configValue)
{
    if(pinNum < 8)    /* for pins 0 to 7 */
        portName->CRL = (portName->CRL&(~(0xF<<(4*pinNum))))|(configValue<<(4*pinNum));
    else /* for pins 8 to 15 */
        portName->CRH = (portName->CRH&(~(0xF<<(4*(pinNum-8)))))
            |(configValue<<(4*(pinNum-8)));
}
//copy delay_ms from Program 8-1 to here
```

BSRR (Bit set/reset register) and BRR (Bit reset register)

In the previous examples we used OR and AND to make a pin HIGH and LOW. There are also BSRR and BRR specifically for making a single port HIGH and LOW.

See Figure 8-11. For each pin there is a bit in the BRR register. If you write 1 to each bit, the related pin becomes low while the other pins remain unchanged. For example, the following line of code makes PB5 low while the other pins remain unchanged:

```
GPIOB->BRR = 1<<5; //make PB5 low
```

As another example, the following instruction makes PA3 and PA5 low:

```
GPIOA->BRR = (1<<3)|(1<<5); //make PA3 and PA5 low
```

	15	14	13	12	11	10	9	8	7	6	5	4	3	2	1	0
BRR:	BR15	BR14	BR13	BR12	BR11	BR10	BR9	BR8	BR7	BR6	BR5	BR4	BR3	BR2	BR1	BR0

(GPIOx->BRR)

Figure 8-11: BRR (Bit Reset Register)

See Figure 8-12. Bits 0 to 15 of the BSRR register are named as BSn (Bit Set n) and bits 16 to 31 are named as BRn (Bit Reset n). Writing 1 to bits 0 to 15 makes the related pin High. If we write 1 to bits 16 to 31, the related pin becomes low. The following instruction makes PC5 high while the other pins remain unchanged:

```
GPIOC->BSRR = (1<<5);
```

As another example, the following line of code, makes PB5 high and makes PB3 low:

```
GPIOB->BSRR = (1<<5)|(1<<19);
```

212

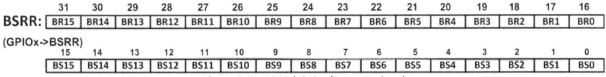

BSRR:	31	30	29	28	27	26	25	24	23	22	21	20	19	18	17	16
	BR15	BR14	BR13	BR12	BR11	BR10	BR9	BR8	BR7	BR6	BR5	BR4	BR3	BR2	BR1	BR0

(GPIOx->BSRR)

	15	14	13	12	11	10	9	8	7	6	5	4	3	2	1	0
	BS15	BS14	BS13	BS12	BS11	BS10	BS9	BS8	BS7	BS6	BS5	BS4	BS3	BS2	BS1	BS0

Figure 8-12: BSRR (Bit Set/Reset Register)

The following program toggles PC13 using BSRR and BRR. If you use a Blue Pill, a green LED is connected to PC13. Otherwise, you can connect an LED to the pin to blink.

Program 8-5: Toggling PC13 using BSRR and BRR

```c
#include <stm32f10x.h>

void delay_ms(uint16_t t);

int main()
{
   RCC->APB2ENR |= 0xFC;           /* Enable clocks for GPIOs */
   GPIOC->CRH = 0x44344444;        /* PC13 as output */
   while(1)
   {
      GPIOC->BRR = 1<<13;          /* make the pin low */
      delay_ms(500);               /* wait 0.5 sec. */
      GPIOC->BSRR = 1<<13;         /* make the pin high */
      delay_ms(500);               /* wait 0.5 sec. */
   }
}
// copy delay_ms from Program 8-1
```

Configuring a port as input

The following program, configures port B as input, and port A as output. Then, gets the data present at the pins of port B and sends it to port A, continuously.

Program 8-6: Reading from port B and writing to port A

```c
#include <stm32f10x.h>

int main()
{
   RCC->APB2ENR |= 0xFC; /* Enable GPIO ports clocks */

   GPIOA->CRL = 0x33333333; /* PA0-PA7 as outputs */
   GPIOA->CRH = 0x33333333; /* PA8-PA15 as outputs */
   GPIOB->CRL = 0x44444444; /* PB0-PB7 as inputs */
   GPIOB->CRH = 0x44444444; /* PB8-PB15 as inputs */

   while(1)
   {
      GPIOA->ODR = GPIOB->IDR; /* read from port B and write to port A */
   }
}
```

See Program 8-7. It monitors the PB10 pin. When PB10 is high, it makes PC13 high. Otherwise, it makes PC13 low.

Notice that in Blue Pill, the anode pin of the green LED is connected to VCC and the cathode pin is connected to PC13. So, the green LED lights up when PC13 is low.

Program 8-7: Monitoring PB10

```
#include <stm32f10x.h>

int main()
{
    RCC->APB2ENR |= 0xFC;  /* Enable GPIO ports clocks */

    GPIOB->CRH = 0x44444444;  /* PB8-PB15 as inputs */
    GPIOC->CRH = 0x44344444;  /* PC13 as output */

    while(1)
    {
        if((GPIOB->IDR & (1<<10)) != 0)  /* is PB10 high */
            GPIOC->ODR |= (1 << 13);  /* make PC13 high */
        else
            GPIOC->ODR &= ~(1 << 13);  /* make PC13 low */
    }
}
```

Connecting a switch to a pin

You can connect a key to the GPIO pins as shown in Figure 8-13. When the key is pressed, the key connects the pin to ground and makes the pin low. When the key is released, the resistor pulls the pin up and the pin becomes high.

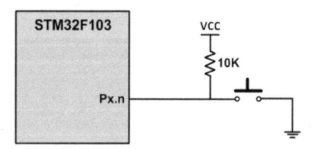

Figure 8-13: Connecting a Switch

The GPIO internal pull-up/pull-down resistor

According to Figure 8-3, in input mode (MODEn=00), if CNFn =10, a pull-up/pull-down resistor is enabled which connects the pin through a resistor to the ODR. See Figure 8-14.

Whenever nothing is connected to the pin or the connected devices are high impedance, the resistor pulls up/down the pin, depending on the value of ODR. If the bit of ODR is 1, the pin is pulled up; otherwise, the pin is pulled down.

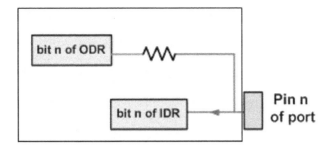

Figure 8-14: The Pull-up/Pull-down Resistor

Program 8-8 is similar to Program 8-7. But the pull-up resistor for PB10 is enabled. So, you can connect a key to the pin and there is no need to put an external pull-up resistor any more.

Program 8-8: Monitoring PB10

```c
#include <stm32f10x.h>

int main()
{
   RCC->APB2ENR |= 0xFC; //Enable GPIO ports clocks

   GPIOB->CRH = 0x44444844; //PB8-PB15 as inputs, pull-up for PB10
   GPIOB->ODR |= (1<<10); //set the bit 10 of ODR to pull-up

   GPIOC->CRH = 0x44344444; /* PC13 as output */

   while(1)
   {
      if((GPIOB->IDR & (1<<10)) != 0) /* is PB10 high */
           GPIOC->ODR |= (1 << 13); /* make PC13 high */
      else
           GPIOC->ODR &= ~(1 << 13); /* make PC13 low */
   }
}
```

The following program monitors PA2. If it is low, PC13 toggles. Otherwise, PC13 is low. You can connect PA2 to ground using a piece of wire and check the program.

Program 8-9: It toggles PC13 if PA2 is low

```c
#include <stm32f10x.h>

void delay_ms(uint16_t t);

int main()
{
   RCC->APB2ENR |= 0xFC; /* Enable GPIO ports clocks */
   GPIOC->CRH = 0x44344444; /* PC13 as output */

   GPIOA->CRL = 0x44444844; /* PA2 as input with pull-up */
   GPIOA->ODR |= (1<<2);    /* pull-up PA2 */
```

215

```
    while(1)
    {
       if((GPIOA->IDR&(1<<2)) == 0) /* is PA2 low? */
          GPIOC->ODR ^= (1<<13); /* toggle PC13 */
       else
          GPIOC->ODR &= ~(1<<13);
       delay_ms(500);
    }
}
//copy delay_ms from Program 8-1 to here
```

Review Questions

1. True or false. Upon power-up, the I/O pins are configured as output ports.
2. To set the direction of a port, we use _____ and _____ registers.
3. True or false. We use ODR to send data out to the pins.
4. True or false. We use IDR to bring data into the CPU from the pins.

Section 8.2: Seven-segment LED interfacing and programming

One of the popular output displays is seven-segment LED. The 7-seg LED can have common anode or common cathode. With common anode, the anode of the LED is driven by the positive supply voltage and the microcontroller drives the individual cathodes LOW for current to flow through LEDs to light up. In this configuration, the sink current capability of the microcontroller is critical. With common cathode, the cathode of the LED is grounded and microcontroller drives the individual anodes HIGH to light up the LED. In this configuration, the microcontroller pins must provide sufficient source current for each LED segment. In each of the configurations, if the microcontroller does not have sufficient drive or sink current capacity, we must add a buffer between the 7-seg LED and the microcontroller. The buffer for the 7-seg LED can be an IC chip or transistors.

The seven segments of LED are designated as a, b, c, d, e, f, and g. See Figure 8-15.

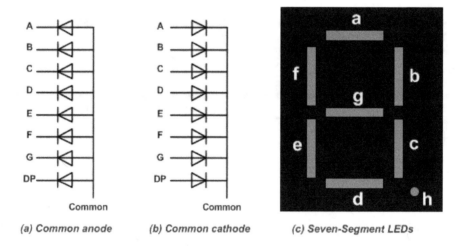

(a) Common anode   (b) Common cathode   (c) Seven-Segment LEDs

Figure 8-15: Seven-Segment

A byte of data should be sufficient to drive all of the segments. In the example below, segment a is assigned to bit D0, segment b is assigned to bit D1, and so on as shown in Table 8-2.

216

D7	D6	D5	D4	D3	D2	D1	D0
.	g	f	e	D	c	b	a

Table 8-2: Assignments of port pins to each segment of a 7-seg LED

The D7 bit is assigned to decimal point. One can create the following patterns for numbers 0 to 9 for the common cathode configuration:

Num	D7	D6	D5	D4	D3	D2	D1	D0	Hex value
	.	g	f	e	d	c	b	a	
0	0	0	1	1	1	1	1	1	0x3F
1	0	0	0	0	0	1	1	0	0x06
2	0	1	0	1	1	0	1	1	0x5B
3	0	1	0	0	1	1	1	1	0x4F
4	0	1	1	0	0	1	1	0	0x66
5	0	1	1	0	1	1	0	1	0x6D
6	0	1	1	1	1	1	0	1	0x7D
7	0	0	0	0	0	1	1	1	0x07
8	0	1	1	1	1	1	1	1	0x7F
9	0	1	1	0	1	1	1	1	0x6Γ

Table 8-3: Segment patterns for the 10 decimal digits for a common cathode 7-seg LED

In Figures 8-16 and 8-17 the connection for 2-digit 7-seg LED and STM32F103 is shown. The program 8-9 shows the code.

Notice since the same segment for both digit 1 and digit 2 are connected to the same I/O port pin, the common cathode of each digit must be driven separately so that only one digit is on at a time. The two digits are turned on alternatively. For example, if we want to display number 95 on the 7-seg LED, the following steps should be used:

1) Configure Port A as output port to drive the segments,
2) Configure Port B as output port to select the digit,
3) Write the pattern of numeral "9" from Table 8-3 to Port A,
4) Write one bit of Port B to activate the tens digit,
5) Delay for some time,
6) Write the pattern of numeral "5" from Table 8-3 to Port A,
7) Write one bit of Port B to activate the ones digit,
8) Delay for some time,
9) Repeat from step 3 to 8.

At low frequency of alternating digits, the display will appear to be flickering. To eliminate the flickering display, each digit should be turned on and off at least 60 times each second. From the example above, the delay for steps 5 and 8 should be 8 milliseconds or less. See Program 8-10.

1 second / 60 / 2 = 8 millisecond

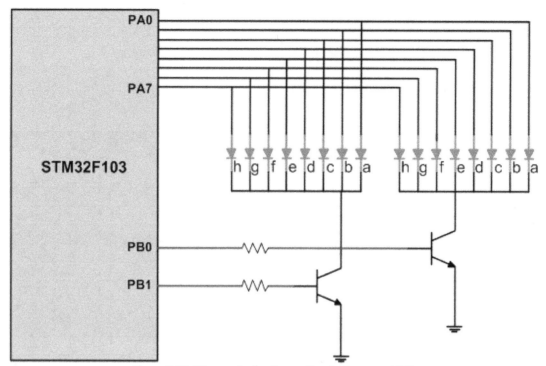

Figure 8-16: Microcontroller Connection to 7-segment LED

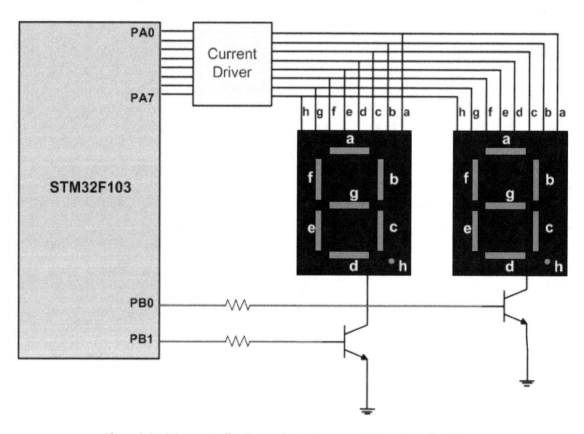

Figure 8-17: Microcontroller Connection to 7-segment LED with Buffer Driver

```
#include <stm32f10x.h>

void delay_ms(uint16_t t);

int main()
{
   const uint8_t sevenLookup[]={0x3F,0x06,0x5B,0x4F,0x66,0x6D,0x7D,0x07,0x7F,0x6F};
   RCC->APB2ENR |= 0xFC; /* Enable GPIO ports clocks */

   GPIOA->CRL = 0x33333333; /* PA0-PA7 as outputs */
   GPIOB->CRL = 0x44444433; /* PB0 and PB1 as outputs */

   while(1)
   {
      GPIOA->ODR = sevenLookup[9];
      GPIOB->ODR |= 1<<1; /* enable tens */
      delay_ms(10);
      GPIOB->ODR &= ~(1<<1); /* disable tens */

      GPIOA >ODR = scvcnLookup[5];
      GPIOB->ODR |= 1<<0; /* enable ones */
      delay_ms(10);
      GPIOB->ODR &= ~(1<<0); /* disable ones */
   }
}
```

Notice in Figure 8-17, a single pin is used to select each digit. That means if we want 4 digits, we must use a total of 12 pins. That is 8 pins for the segments a through g, decimal point, and 4 pins to select each digit. This might not be feasible in applications in which we have a limited number of microcontroller pins to spare. One solution is to use a decoder for the digit selection. For example, a 74LS138 decoder can be used for up to 8-digit 7-seg LED system with three select pins. Another approach is to use a 7-segment LED driver chip such as MAX7219, which only uses two interface pins. An additional advantage of MAX7219 is that the refreshing of the segments is handled by the driver chip itself. So, the microcontroller does not have to spend time refreshing the display and can concentrate on other important tasks. The MAX7219 is an SPI device and the vast majority of microcontrollers come with on-chip SPI serial communication feature, which we will discuss in a separate chapter.

Review Questions

1. In a common cathode 7-seg LED connection, to turn on a segment the microcontroller drives it (high, low).
2. True or false. In connecting the 7-seg LED directly to microcontroller, the refreshing of digits is done by microcontroller itself.
3. What is the disadvantage of letting microcontroller to do the refreshing of 7-seg LEDs?
4. List two advantages of using an IC chip such as MAX7219 chip?
5. In an application, we need 8 digits of 7-seg LED. How many pins of microcontroller will be used if we connect microcontroller to 7-seg directly (similar to Figure 2-16)? How about if we use 3-8 decoder for digit selection?

Section 8.3: Clock sources, Reset, and Power Supply Pins in STM32F10x

See Figure 8-1. In Section 8-1, we discussed to use GPIO pins. Now, we cover the other pins.

Power supply pins

VDD and VSS

The pins provide supply voltage to the chip. The operating voltage range is between 2 and 3.6V.

VDDA and VSSA

The pins provide supply voltage to the internal ADCs and DACs, as will be discussed in Chapter 13.

VBAT

The STM32F10x chips have an internal RTC (Real-Time Clock). If you connect a 3V battery to the VBAT pin, the RTC is powered with the battery when the main VDD power is off.

Oscillator pins

OSC_IN and OSC_OUT

The pins can be connected to an external crystal or clock. Figure 8-18 shows the clock sources fed to the CPU. The system clock is provided either from an external source named HSE (High Speed External clock) or from an internal RC named HSI (High Speed Internal clock). The clock circuit contains some dividers and a multiplier as well. The multiplier generates a high frequency from a low frequency wave. Thanks to the clock circuit the clock frequency is programmable. You may run high clock frequency for CPU intensive tasks and slow down the clock in other times to conserve energy. On startup, the internal 8MHz is selected by default. Then the program can change the clock source. The circuit monitors the external clock and in the case of failure, it automatically switches to internal clock and generates an interrupt. Figure 8-19 shows the RCC_CFGR register which is used to configure the clock source.

The clock source is configured in the system_stm32F10x.c file. There are some defines at the beginning of the file as shown below. By default, the external clock source with frequency of 72 MHz is selected:

```
/* #define SYSCLK_FREQ_HSE    HSE_VALUE */
/* #define SYSCLK_FREQ_24MHz  24000000 */
/* #define SYSCLK_FREQ_36MHz  36000000 */
/* #define SYSCLK_FREQ_48MHz  48000000 */
/* #define SYSCLK_FREQ_56MHz  56000000 */
#define SYSCLK_FREQ_72MHz  72000000
```

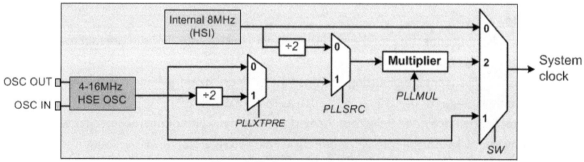

Figure 8-18: System Clock Sources

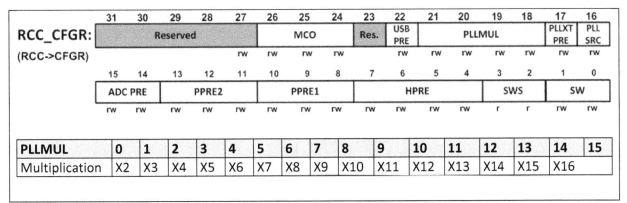

Figure 8-19: RCC_CFGR Register

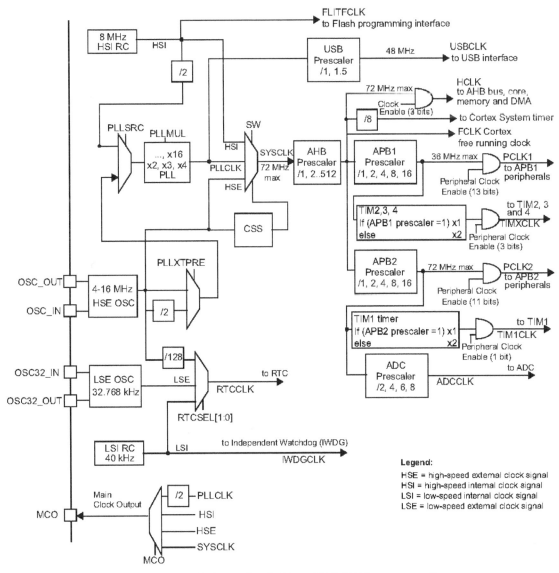

Figure 8-20: Clock Tree (Copied from STM32F103 user manual)

Figure 8-20 shows the complete clock tree of STM32F103. Figure 8-21 shows the internal architecture of the STM32F10x series. This figure shows the peripherals which are connected to APB1 and APB2. For more information, read *reset and clock control* (Chapter 7) of the STM32F10X reference manual (RM0008).

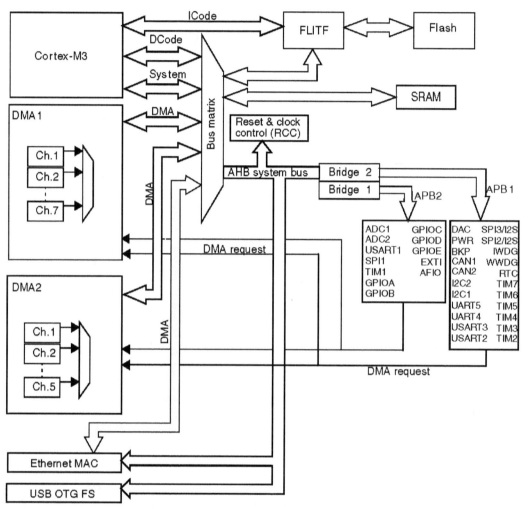

Figure 8-21: The Internal Architecture for STM32F10x

In the Blue Pill board, the OSC_IN and OSC_OUT pins are connected to an 8MHz crystal, as shown in Figure 8-22. You can use the same circuit if you design a board for STM32F10x chips.

OSC32_IN and OSC32_OUT

The internal RTC (Real-Time Clock) of STM32F10x needs a clock to work. The clock can be provided from an external source named LSE (Low Speed External clock) or from an internal RC named LSI (Low Speed Internal clock). To provide an external clock for the RTC, a 32.768 kHz crystal and 2 capacitors should be connected to the OSC32_IN and OSC32_OUT pins. See Figure 8-22.

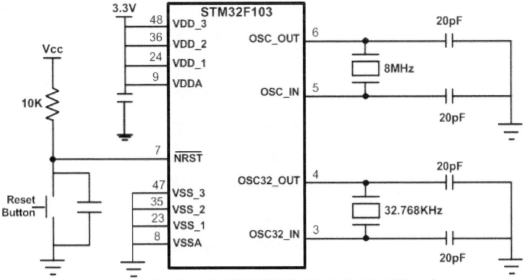

Figure 8-22: The Circuit for STM32F103 Vital Pins in the Blue Pill Board

NRST pin

In STM32F10x, the reset pin is named as NRST since it is active low. When a LOW pulse is applied to this pin, the microcontroller will reset and terminate all activities. After applying reset, contents of all registers and SRAM locations (except the RTC registers) will be initialized. The CPU will start executing the program from location 0x00004 when the Reset pin is released.

Figures 8-23a, 8-23b, and 8-23c show three ways of connecting the RESET pin. Figure 8-23b uses a momentary switch for reset circuitry. Putting the capacitor between the RESET pin and GND filters the noise during reset and working time. In the Blue Pill, the reset circuit is similar to Figure 8-23b. See Figure 8-23c. The diode protects the RESET pin from being powered by the capacitor when the power is off.

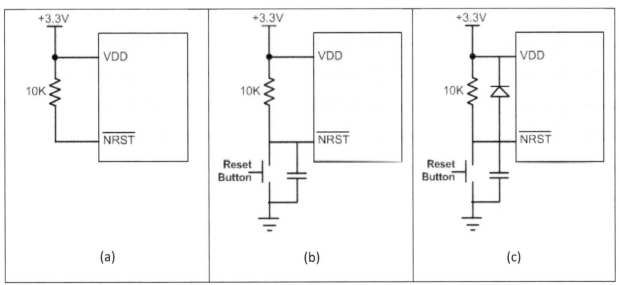

Figure 8-23: Reset Circuits

In addition to the RESET pin there are other sources of reset in STM32. For more information, see the reference manual.

Review Questions
1.   True or false. In STM32 family, programs can change the speed of the CPU.
2.   True or false. On startup, CPU works with the internal 8 MHz oscillator.
3.   Reset is an active-_____ (LOW, HIGH) pin.

Section 8.4: GPIO Programming in Assembly Language

To write Assembly programs for GPIOs you should be able to calculate the addresses for registers. See Examples 8-2 and 8-3.

Example 8-2

Find the physical address of the CRL, CRH, IDR, and ODR registers for PORTA.

Solution:

To find the physical address, we should add the offset to the base address. According to Figure 8-2, the Base address of the PORTA is 0x4001 0800. So,
The physical address location of the CRL for PORTA is 0x4001 0800 + 0000 = 0x4001 0800.
The physical address location of the CRH for PORTA is 0x4001 0800 + 0004 = 0x4001 0804.
The physical address location of IDR is 0x4001 0800 + 0008 = 0x4001 0808.
The physical address location of ODR is 0x4001 0800 + 000C = 0x4001 080C.

Example 8-3

Find the physical address for AHBENR, APB1ENR, and APB2ENR. According to the reference manual, the offsets of the registers are 0x14, 0x1C, and 0x18, respectively and the base address is 0x40021000.

Solution:

The physical address of AHBENR = 0x40021000 + 0x14 = 0x40021014
The address for APB1ENR = 0x40021000 + 0x1C = 0x4002101C
The physical address of APB2ENR = 0x40021000 + 0x18 = 0x40021018

Below, you see some of the programs of this chapter in Assembly language.

Program 8-1: Toggling Port A (Assembly)

```
RCC_APB2ENR EQU 0x40021018

GPIOA_CRL  EQU 0x40010800
GPIOA_CRH  EQU 0x40010804
GPIOA_IDR  EQU 0x40010808
```

```
GPIOA_ODR  EQU 0x4001080C

    EXPORT __main
    AREA MAIN, CODE, READONLY
__main
    LDR   R1,=RCC_APB2ENR
    LDR R0,[R1]              ;read the APB2ENR
    ORR R0,R0,#0xFC          ;enable the clocks for GPIOs
    STR R0,[R1]              ;write R0 to APB2ENR


    LDR R1,=GPIOA_CRL
    LDR R0,=0x33333333
    STR R0,[R1]              ;PA0 to PA7 as outputs


    LDR R1,=GPIOA_CRH
    LDR R0,=0x33333333
    STR R0,[R1]              ;PA8 to PA15 as outputs

LOOP
    LDR R1,=GPIOA_ODR
    LDR R0,=0x0000           ;ODR = 0x0000 (all pins low)
    STR R0,[R1]


    BL delay


    LDR R1,=GPIOA_ODR
    LDR R0,= 0xFFFF          ;ODR = 0xFFFF (all pins high)
    STR R0,[R1]


    BL delay
    B LOOP

delay
      LDR R0,= 48           ; R0 = 48, modify this value for different delays
d_L1 LDR R1,= 250000        ; R1 = 250, 000 (inner loop count)
d_L2 SUBS R1,R1,#1
      BNE   d_L2
      SUBS R0,R0,#1
      BNE d_L1
      BX LR


      END
```

Program 8-2: Toggling PA2 (Assembly)

```
RCC_APB2ENR EQU 0x40021018

GPIOA_CRL  EQU 0x40010800
GPIOA_CRH  EQU 0x40010804
GPIOA_IDR  EQU 0x40010808
GPIOA_ODR  EQU 0x4001080C

    EXPORT __main
    AREA MAIN, CODE, READONLY
```

```
__main
    LDR    R1,=RCC_APB2ENR
    LDR R0,[R1]
    ORR R0,R0,#0xFC          ;enable the clocks for GPIOs
    STR R0,[R1]

    LDR R1,=GPIOA_CRL
    LDR R0,=0x44444344
    STR R0,[R1]              ;PA2 as output

LOOP
    LDR R1,=GPIOA_ODR
    LDR R0,[R1]              ;R0 = ODR
    EOR R0,R0,#2_00000100    ;toggle bit 2
    STR R0,[R1]              ;ODR = R0

    BL delay
    B  LOOP

    ;copy delay from Program 8-1 (Assembly)
    END
```

Example 8-4

Find the physical address of the CRL, CRH, IDR, and ODR registers for PORTB.

Solution:

According to Figure 8-2, the Base address of the PORTB is 0x4001 0C00. So,
The physical address location of the CRL for PORTB is 0x4001 0C00 + 0000 = 0x4001 0C00.
The physical address location of the CRH for PORTB is 0x4001 0C00 + 0004 = 0x4001 0C04.
The physical address location of IDR is 0x4001 0C00 + 0008 = 0x4001 0C08.
The physical address location of ODR is 0x4001 0C00 + 000C = 0x4001 0C0C.

Example 8-5

Find the physical address of the CRL, CRH, BRR, and BSRR registers for PORTC.

Solution:

According to Figure 8-2, the Base address of the PORTC is 0x4001 1000. So,
The physical address location of the CRL for PORTC is 0x4001 1000 + 0000 = 0x4001 1000.
The physical address location of the CRH for PORTC is 0x4001 1000 + 0004 = 0x4001 1004.
The physical address location of BRR is 0x4001 1000 + 0014 = 0x4001 1014.
The physical address location of BSRR is 0x4001 1000 + 0010 = 0x4001 1010.

Program 8-3: Toggling PC13 (Assembly)

```
RCC_APB2ENR EQU 0x40021018

GPIOC_CRL  EQU 0x40011000
GPIOC_CRH  EQU 0x40011004
GPIOC_ODR  EQU 0x4001100C

  EXPORT __main
  AREA MAIN, CODE, READONLY
__main
  LDR  R1,=RCC_APB2ENR
  LDR R0,[R1]
  ORR R0,R0,#0xFC          ;enable the clocks for GPIOs
  STR R0,[R1]

  LDR R1,=GPIOC_CRH
  LDR R0,=0x44344444
  STR R0,[R1]              ;PC13 as output

LOOP
  LDR R1,=GPIOC_ODR
  LDR R0,[R1]             ;R0 = ODR
  EOR R0,R0,#0x2000       ;toggle bit 13
  STR R0,[R1]            ;ODR = R0

  BL delay
  B  LOOP
  ;copy delay from Program 8-1 (Assembly)
  END
```

Program 8-5: Toggling PC13 using BSRR and BRR (Assembly)

```
RCC_APB2ENR EQU 0x40021018

GPIOC_CRL  EQU 0x40011000
GPIOC_CRH  EQU 0x40011004
GPIOC_BRR  EQU 0x40011014
GPIOC_BSRR EQU 0x40011010

  EXPORT __main
  AREA MAIN, CODE, READONLY
__main
  LDR  R1,=RCC_APB2ENR
  LDR R0,[R1]
  ORR R0,R0,#0xFC          ;enable the clocks for GPIOs
  STR R0,[R1]

  LDR R1,=GPIOC_CRH
  LDR R0,=0x44344444
  STR R0,[R1]              ;PC13 as output
```

```
LOOP
   LDR R1,=GPIOC_BRR
   LDR R0,=0x2000          ;R0 = (1<<13); bit 13
   STR R0,[R1]             ;BRR = (1<<13) (make the pin low)

   BL delay    ;wait

   LDR R1,=GPIOC_BSRR
   LDR R0,=0x2000          ;R0 = (1<<13); bit 13
   STR R0,[R1]             ;BSRR = (1<<13) (make the pin high)

   BL delay    ;wait
   B  LOOP

   ;copy delay from Program 8-1

   END
```

Program 8-6: Reading from port B and writing to port A (Assembly)

```
RCC_APB2ENR EQU 0x40021018

GPIOA_CRL  EQU 0x40010800
GPIOA_CRH  EQU 0x40010804
GPIOA_IDR  EQU 0x40010808
GPIOA_ODR  EQU 0x4001080C

GPIOB_CRL  EQU 0x40010C00
GPIOB_CRH  EQU 0x40010C04
GPIOB_IDR  EQU 0x40010C08
GPIOB_ODR  EQU 0x40010C0C

   EXPORT __main
   AREA MAIN, CODE, READONLY
__main
   LDR   R1,=RCC_APB2ENR
   LDR R0,[R1]
   ORR R0,R0,#0xFC         ;enable the clocks for GPIOs
   STR R0,[R1]

   LDR R1,=GPIOA_CRL
   LDR R0,=0x33333333
   STR R0,[R1]             ;PA0 to PA7 as outputs

   LDR R1,=GPIOA_CRH
   LDR R0,=0x33333333
   STR R0,[R1]             ;PA8 to PA15 as outputs

   LDR R1,=GPIOB_CRL
   LDR R0,=0x44444444
   STR R0,[R1]             ;PB0 to PB7 as inputs

   LDR R1,=GPIOB_CRH
   LDR R0,=0x44444444
   STR R0,[R1]             ;PB8 to PB15 as inputs
```

```
LOOP
  LDR  R1,=GPIOB_IDR
  LDR  R0,[R1]              ;R0 = value of GPIOB_IDR
  LDR R1,=GPIOA_ODR
  STR R0,[R1]              ;GPIOA_ODR = R0
  B LOOP

  END
```

Program 8-8: Monitoring PB10 (Assembly)

```
RCC_APB2ENR EQU 0x40021018

GPIOB_CRL  EQU 0x40010C00
GPIOB_CRH  EQU 0x40010C04
GPIOB_IDR  EQU 0x40010C08
GPIOB_ODR  EQU 0x40010C0C

GPIOC_CRL  EQU 0x40011000
GPIOC_CRH  EQU 0x40011004
GPIOC_BRR  EQU 0x40011014
GPIOC_BSRR EQU 0x40011010

  EXPORT __main
  AREA MAIN, CODE, READONLY
__main
  LDR  R1,=RCC_APB2ENR
  LDR R0,[R1]
  ORR R0,R0,#0xFC          ;enable the clocks for GPIOs
  STR R0,[R1]

  LDR R1,=GPIOB_CRH
  LDR R0,=0x44444844
  STR R0,[R1]              ;PB8 to PB15 as inputs

  LDR R1,=GPIOB_ODR
  LDR  R0,[R1]
  ORR  R0,R0,#(1<<10)
  STR R0,[R1]              ;set bit 10 of ODR to pull-up PB10

  LDR R1,=GPIOC_CRH
  LDR R0,=0x44344444
  STR R0,[R1]              ;PC13 as output


LOOP
  LDR R1,=GPIOB_IDR
  LDR R0,[R1]              ;R0 = value of GPIOB_IDR
  TST R0,#(1<<10)         ;test bit 10
  BEQ LL
  LDR R1,=GPIOC_BSRR
  LDR R0,=0x2000          ;R0 = (1<<13); bit 13
  STR R0,[R1]              ;BSRR = (1<<13) (make the pin high)
  B   LOOP
LL
```

```
LDR R1,=GPIOC_BRR
LDR R0,=0x2000          ;R0 = (1<<13); bit 13
STR R0,[R1]             ;BRR = (1<<13) (make PC13 low)
B LOOP
END
```

Review Questions

1. Find the physical address of the CRL for port D.
2. Find the physical address of the IDR for port C.

Problems

Section 8-1

1. In STM32, how many pins each port has?
2. Upon reset, all the bits of ports are configured as _____ (input, output).
3. Write a program to toggle all the bits of PORTA continuously.
4. Write a program to toggle all the bits of PORTC continuously.
5. Write a program to get 8-bit data from PA0-PA7 and send it to PB0-PB7.
6. Write a program to get 8-bit data from PA0-PA7 and send it to PB8-PB15.
7. Write a program to monitor the PA0 bit. When it is HIGH, send 0x9999 to PORTB. If it is LOW, send 0x5555 to PORTB.
8. Write a program to monitor the PA0 bit. When it is HIGH, toggle PC13 every second. If it is LOW, toggle PC13 every 0.4 second.
9. Using BRR and BSRR, write a program to toggle PB1 and PB5, every second.
10. Using BRR and BSRR, write a program to toggle PC13 and PC14, every 0.5 second.
11. Write a program to monitor the PB5 and PB6 bits. When both of them are HIGH, make PC13 low; otherwise, make PC13 high. (Notice that in the Blue Pill the cathode pin of green LED is connected to PC13 and when the PC13 is low, the LED lights up.)
12. Write a program to monitor the PA5 and PA6 bits. When either of them is high, make PC13 low; otherwise, make PC13 high.

Section 8-2

13. See Figure 8-17. Write a program that displays 0 on the 7-segments. Then, increase the displayed value, every second and count up to 99.
14. See Figure 8-17. Write a program that displays 99 on the 7-segments. Then, decrease the displayed value, every second and count down to 0. When it reaches to zero make PC13 high.

Section 8-3

15. The STM32F103C8 package is a(n) _____-pin package.
16. The HSE crystal oscillator is connected to pins _____ and _____ .

Section 8-4

17. Find the address for BSRR register of port B.
18. What is the address for the IDR of port D?

Answer to Review Questions

Section 8-1
1. False
2. CRL and CRH
3. True
4. True

Section 8-2
1. High
2. True
3. The time and pins of microcontroller is wasted to scan the 7-segments.
4. (1) It refreshes the 7-segments, (2) it is connected to the microcontroller using SPI which uses less pins of the microcontroller.
5. 8 pins for data and 8 pins for selector; 8 pins for data and 3 pins for selector.

Section 8-3
1. True
2. True
3. Low

Section 8-4
1. 0x4001 1400
2. 0x4001 1008

Chapter 9: LCD and Keyboard Interfacing

In this chapter, we show interfacing to two real-world devices: LCD and Keyboard. They are widely used in different embedded systems.

Section 9.1: Interfacing to an LCD

This section describes the operation modes of the LCDs and then shows how to program and interface an LCD to the STM32F103.

LCD operation

In recent years the LCD is replacing LEDs (seven-segment LEDs or other multi-segment LEDs). This is due to the following reasons:

1. The declining prices of LCDs.
2. The ability to display numbers, characters, and graphics. This is in contrast to LEDs, which are limited to numbers and a few characters. (The new OLED panels are relatively much more expensive except the very small ones. But their prices are dropping. The interface and programming to OLED are similar to graphic LCD.)
3. Incorporation of the refreshing controller into the LCD itself, thereby relieving the CPU of the task of refreshing the LCD.
4. Ease of programming for both characters and graphics.
5. The extremely low power consumption of LCD (when backlight is not used).

In this chapter, we will limit the discussions to the character LCD modules. The graphic LCD modules will be discussed in a later chapter.

LCD module pin descriptions

For many years, the use of Hitachi HD44780 LCD controller dominated the character LCD modules. Even today, most of the character LCD modules still use HD44780 or a variation of it. The HD44780 controller has a 14 pin interface for the microprocessor. We will discuss this 14 pin interface in this section. The function of each pin is given in Table 9-1. (Pins 1 to 14 are for the controller and pins 15 and 16 are for the backlight.) Figure 9-1 shows the pin positions for various LCD modules.

Pin	Symbol	I/O	Description
1	VSS	--	Ground
2	VCC	--	+5V power supply
3	VEE	--	Power supply to control contrast
4	RS	I	RS = 0 to select command register, RS = 1 to select data register
5	R/W	I	R/W = 0 for write, R/W = 1 for read
6	E	I	Enable
7	DB0	I/O	The 8-bit data bus
8	DB1	I/O	The 8-bit data bus
9	DB2	I/O	The 8-bit data bus
10	DB3	I/O	The 8-bit data bus
11	DB4	I/O	The 4/8-bit data bus
12	DB5	I/O	The 4/8-bit data bus
13	DB6	I/O	The 4/8-bit data bus
14	DB7	I/O	The 4/8-bit data bus
15	LED+	--	Backlight LED power supply
16	LED-	--	Backlight LED power supply

Table 9-1: Pin Descriptions for LCD

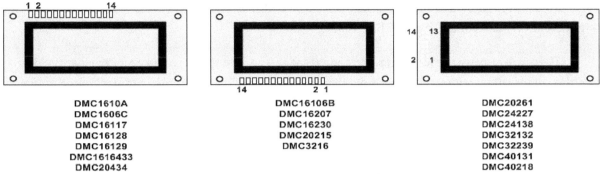

Figure 9-1: Pin Positions for Various LCDs from Optrex

VCC, VSS, and VEE: While VCC and VSS provide +5V power supply and ground, respectively, VEE is used for controlling the LCD contrast.

RS (Register Select): There are two registers inside the LCD and the RS pin is used for their selection as follows. If RS = 0, the instruction command code register is selected, allowing the user to send a command such as clear display, cursor at home, and so on (or query the busy status bit of the controller). If RS = 1, the data register is selected, allowing the user to send data to be displayed on the LCD (or to retrieve data from the LCD controller).

R/W (Read/Write): R/W input allows the user to write information into the LCD controller or read information from it. R/W = 1 when reading and R/W = 0 when writing.

E (Enable): The enable pin is used by the LCD to latch information presented to its data pins. When data is supplied to data pins, a pulse (Low-to-High-to-Low) must be applied to this pin in order for the LCD to latch in the data present at the data pins. This pulse must be a minimum of 230 ns wide, according to Hitachi datasheet.

D0–D7: The 8-bit data pins are used to send information to the LCD or read the contents of the LCD's internal registers. The LCD controller is capable of operating with 4-bit data and only D4-D7 are used. We will discuss this in more details later.

Code (Hex)	Command to LCD Instruction Register
1	Clear display screen
2	Return cursor home
6	Increment cursor (shift cursor to right)
F	Display on, cursor blinking
80	Force cursor to beginning of 1st line
C0	Force cursor to beginning of 2nd line
38	2 lines and 5x7 character (8-bit data, D0 to D7)
28	2 lines and 5x7 character (4-bit data, D4 to D7)

Table 9-2: Some commonly used LCD Command Codes

To display letters and numbers, we send ASCII codes for the letters A–Z, a–z, numbers 0–9, and the punctuation marks to these pins while making RS = 1.

There are also instruction command codes that can be sent to the LCD in order to clear the display, force the cursor to the home position, or blink the cursor. Table 9-2 lists some commonly used command codes. For detailed command codes, see Table 9-4.

Sending commands to LCDs

To send any of the commands to the LCD, make pins RS = 0, R/W = 0, and send a pulse (L-to-H-to-L) on the E pin to enable the internal latch of the LCD. The connection of an LCD to the microcontroller is shown in Figure 9-2.

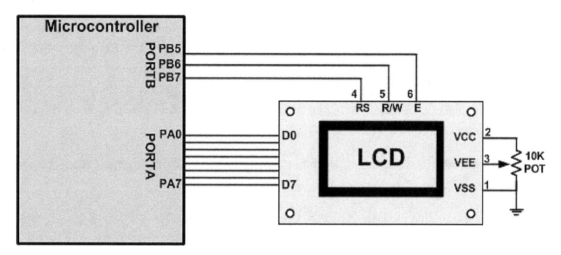

Figure 9-2: LCD Connection to Microcontroller

Notice the following for the connection in Figure 9-2:

1. The LCD's data pins are connected to PORTA of the microcontroller.
2. The LCD's RS pin is connected to Pin 5 of PORTB of the microcontroller.
3. The LCD's R/W pin is connected to Pin 6 of PORTB of the microcontroller.
4. The LCD's E pin is connected to Pin 7 of PORTB of the microcontroller.
5. Both Ports A and B are configured as output ports.

Sending data to the LCD

In order to send data to the LCD to be displayed, we must set pins RS = 1, R/W = 0, and also send a pulse (L-to-H-to-L) to the E pin to enable the internal latch of the LCD.

Because of the extremely low power feature of the LCD controller, it runs much slower than the microcontroller. The first two commands in Table 9-2 take up to 1.64 ms to execute and all the other commands and data take up to 40 us. After one command or data is written to the LCD controller, one must wait until the LCD controller is ready before issuing the next command/data, otherwise the second command/data will be ignored. An easy way (not as efficient though) is to delay the microcontroller for the maximal time it may take for the previous command. We will use this method in the following

234

examples. All the examples in this chapter use much more relaxed timing than the original HD44780 datasheet (See Table 9-4) to accommodate the variations of different LCD modules. You may want adjust the delay time for the LCD module you use.

Program 9-1: This program displays a message on the LCD using 8-bit mode and delay.

```c
/* The program shows "Hello" on the LCD. */
/* The pins are connected as shown in Figure 9-2. */
#include <stm32f10x.h>

void lcd_init(void);
void lcd_sendCommand(unsigned char cmd);
void lcd_sendData(unsigned char data);

void delay_us(uint16_t t);

#define LCD_RS  7    /* PB7 for RS */
#define LCD_RW  6    /* PB6 for RW */
#define LCD_EN  5    /* PB5 for Enable */

int main()
{
    RCC->APB2ENR |= 0xFC; /* Enable clocks for GPIO ports */
    GPIOA->CRL = 0x33333333; /* PA0-PA7 as outputs (for LCD data) */

    GPIOB->CRL = 0x33344444; /* PB7-PB5 as outputs */
    lcd_init();  /* initialize the LCD */

    lcd_sendData('H');
    lcd_sendData('e');
    lcd_sendData('l');
    lcd_sendData('l');
    lcd_sendData('o');

    while(1)   /* stay here forever */
    {
    }
}

/* The function puts value on the data pins and makes a H-to-L pulse on the enable */
void lcd_putValue(unsigned char value)
{
    GPIOA->BRR = 0xFF; /* clear PA0-PA7 */
    GPIOA->BSRR = value&0xFF; /* put value on PA0-PA7 */

    GPIOB->BSRR = (1<<LCD_EN); /* PB5 (EN) = 1 for H-to-L pulse */
    delay_us(2);               /* make EN pulse wider */
    GPIOB->BRR = (1<<LCD_EN); /* PB5 (EN) = 0 for H-to-L pulse */
    delay_us(100);            /* wait        */
}

/* The function sends a command to the LCD */
void lcd_sendCommand(unsigned char cmd)
{
    GPIOB->BRR = (1<<LCD_RS);       /* RS = 0 for command */
    lcd_putValue(cmd);
```

```
}

/* The function sends data to the lcd */
void lcd_sendData(unsigned char data)
{
   GPIOB->BSRR = (1<<LCD_RS);      /* RS = 1 for data */
   lcd_putValue(data);
}

void lcd_init()
{
   GPIOB->BRR = (1<<LCD_EN);        /* LCD_EN = 0 */
   delay_us(3000);                  /* wait 3ms */
   lcd_sendCommand(0x38);  /* init. LCD 2 line,5x7 matrix */
   lcd_sendCommand(0x0e);  /* display on, cursor on */
   lcd_sendCommand(0x01);  /* clear LCD */
   delay_us(2000);                  /* wait 2ms */
   lcd_sendCommand(0x06);  /* shift cursor right */
}

void delay_us(uint16_t t)
{
   volatile unsigned long l = 0;
   for(uint16_t i = 0; i < t; i++)
     for(l = 0; l < 6; l++)
     {
     }
}
```

Checking LCD busy flag

In the above program, the RW pin of the LCD is always 0 and we can connect it to ground. The above program used a time delay before issuing the next data or command. This allows the LCD a sufficient amount of time to get ready to accept the next data. However, the LCD has a busy flag. We can monitor the busy flag and send data when it is ready. This will speed up the process. To check the busy flag, we must read the command register (R/W = 1, RS = 0). The busy flag is the D7 bit of that register. Therefore, if R/W = 1, RS = 0. When D7 = 1 (busy flag = 1), the LCD is busy taking care of internal operations and will not accept any new information. When D7 = 0, the LCD is ready to receive new information.

Doing so requires switching the direction of the port connected to the data bus to input mode when polling the status register then switch the port direction back to output mode to send the next command. If the port direction is incorrect, it may damage the microcontroller or the LCD module. The next program example uses polling of the busy bit in the status register.

Program 9-2: This program displays a message on the LCD using 8-bit mode and checking the busy flag

```
/* The program shows "Hello" on the LCD */

#include <stm32f10x.h>

void lcd_init(void);
void lcd_sendCommand(unsigned char cmd);
```

```c
void lcd_sendData(unsigned char data);

void delay_us(uint16_t t);

#define LCD_RS  7
#define LCD_RW  6
#define LCD_EN  5

int main()
{
    RCC->APB2ENR |= 0xFC; /* Enable clocks for GPIO ports */
    GPIOA->CRL = 0x33333333; /* PA0-PA7 as outputs */
    GPIOB->CRL = 0x33344444; /* PB7-PB5 as outputs */
    lcd_init();

    lcd_sendData('H');
    lcd_sendData('e');
    lcd_sendData('l');
    lcd_sendData('l');
    lcd_sendData('o');

    while(1)    /* stay here forever */
    {
    }
}

void lcd_putValue(unsigned char value)
{
    GPIOA->BRR = 0xFF; /* clear PA0-PA7 */
    GPIOA->BSRR = value&0xFF; /* put value on PA0-PA7 */

    GPIOB->BSRR = (1<<LCD_EN); /* EN = 1 for H-to-L pulse */
    delay_us(2);               /* make EN pulse wider */
    GPIOB->BRR = (1<<LCD_EN);      /* EN = 0 for H-to-L pulse */
}

/* The function waits until the busy flag becomes low */
void lcd_waitForReady()
{
    GPIOB->ODR &= ~ (1<<LCD_RS);   /* RS = 0 for command */
    GPIOB->BSRR = (1<< LCD_RW); /* RW = 1 (read) */
    GPIOA->CRL = 0x44444444; /* PA0-PA7 as inputs */
    delay_us(1);

    unsigned char stat;
    do{
        GPIOB->BSRR = (1<<LCD_EN); /* EN = 1 for H-to-L pulse */
        delay_us(1);               /* make EN pulse wider */
        stat = GPIOA->IDR;    /* read status */
        GPIOB->BRR = (1<<LCD_EN);    /* EN = 0 for H-to-L pulse */
        delay_us(0);
    }while(stat&0x80);         /* repeat while busy */

    GPIOA->CRL = 0x33333333; /* PA0-PA7 as outputs */
    GPIOB->BRR = (1<< LCD_RW); /* RW = 0 (write) */
}
```

```
void lcd_sendCommand(unsigned char cmd)
{
    lcd_waitForReady();
    GPIOB->BRR = (1<<LCD_RS);        /* RS = 0 for command */
    lcd_putValue(cmd);
}

void lcd_sendData(unsigned char data)
{
    lcd_waitForReady();
    GPIOB->BSRR = (1<<LCD_RS);       /* RS = 1 for data */
    lcd_putValue(data);
}

void lcd_init()
{
    GPIOB->BRR = (1<<LCD_EN);        /* LCD_EN = 0 */
    delay_us(3000);            /* wait 3ms */
    lcd_sendCommand(0x38);     /* init. LCD 2 line,5x7 matrix */
    lcd_sendCommand(0x0e);     /* display on, cursor on */
    lcd_sendCommand(0x01);     /* clear LCD */
    delay_us(2000);            /* wait 2ms */
    lcd_sendCommand(0x06);     /* shift cursor right */
}

void delay_us(uint16_t t)
{
    volatile unsigned long l = 0;
    for(uint16_t i = 0; i < t; i++)
        for(l = 0; l < 6; l++)
        {
        }
}
```

LCD 4-bit Option

To save the number of microcontroller pins used by LCD interfacing, we can use the 4-bit data option instead of 8-bit. In the 4-bit data option, we only need to connect D7-D4 to microcontroller. Together with the three control lines, the interface between the microcontroller and the LCD module will fit in a single 8-bit port. See Figure 9-3. To save the number of pins, we can also connect the R/W pin of the LCD to ground and use time delays before sending the next command or data.

With 4-bit data option, the microcontroller needs to issue commands to put the LCD controller in 4-bit mode during initialization. This is done with command 0x20 in Program 9-3. After that, every command or data needs to be broken down to two 4-bit operations, upper nibble first. In Program 9-3, the lower nibble is extracted using **value & 0x0F** and the upper nibble is shifted into place by **(value>>4)&0x0F**.

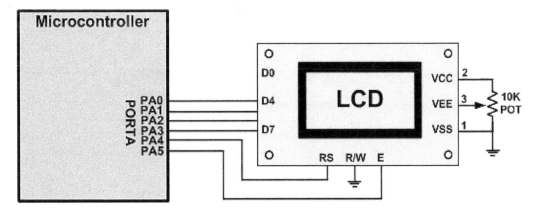

Figure 9-3: LCD Connection for 4-bit Data

Program 9-3: This program uses the 4-bit data option to show a message on the LCD.

```c
/* The program shows "The world is but one country" on the LCD */

#include <stm32f10x.h>

void lcd_init(void);
void lcd_sendCommand(unsigned char cmd);
void lcd_sendData(unsigned char data);
void lcd_print(char * str);

void delay_us(uint16_t t);

#define LCD_RS   4
#define LCD_EN   5

int main()
{
   RCC->APB2ENR |= 0xFC; /* Enable clocks for GPIO ports */
   GPIOA->CRL = 0x44333333; /* PA0-PA5 as outputs */

   lcd_init();

   lcd_print("Hello World!"); /* show "Hello World!" */

   while(1)   /* stay here forever */
   {
   }
}

void lcd_putValue(unsigned char value)
{
   GPIOA->BRR = 0x0F; /* clear PA0-PA3 */
   GPIOA->BSRR = (value>>4)&0x0F; /* put high nibble on PA0-PA3 */

   GPIOA->ODR |= (1<<LCD_EN); /* EN = 1 for H-to-L pulse */
   delay_us(1);              /* make EN pulse wider */
   GPIOA->ODR &= ~ (1<<LCD_EN);   /* EN = 0 for H-to-L pulse */
```

239

```c
    delay_us(100);                        /* wait        */

    GPIOA->BRR = 0x0F; /* clear PA0-PA3 */
    GPIOA->BSRR = value&0x0F; /* put low nibble on PA0-PA3 */

    GPIOA->ODR |= (1<<LCD_EN); /* EN = 1 for H-to-L pulse */
    delay_us(1);                  /* make EN pulse wider */
        GPIOA->ODR &= ~(1<<LCD_EN);           /* EN = 0 for H-to-L pulse */
        delay_us(100);                        /* wait        */
}

void lcd_sendCommand(unsigned char cmd)
{
    GPIOA->ODR &= ~ (1<<LCD_RS);    /* RS = 0 for command */
    lcd_putValue(cmd);
}

void lcd_sendData(unsigned char data)
{
    GPIOA->ODR |= (1<<LCD_RS);      /* RS = 1 for data */
    lcd_putValue(data);
}

void lcd_print(char * str)
{
  unsigned char i = 0;

    while(str[i] != 0) /* while it is not end of string */
    {
        lcd_sendData(str[i]); /* show str[i] on the LCD */
        i++;
    }
}

void lcd_init()
{
    GPIOA->ODR &= ~(1<<LCD_EN);      /* LCD_EN = 0 */
    delay_us(3000);            /* wait 3ms */
    lcd_sendCommand(0x33);  /* send 0x33 for init. */
    lcd_sendCommand(0x32);  /* send 0x32 for init. */
    lcd_sendCommand(0x28);  /* init. LCD 2 line,5x7 matrix */
    lcd_sendCommand(0x0e);  /* display on, cursor on */
    lcd_sendCommand(0x01);  /* clear LCD */
    delay_us(2000);            /* wait 2ms */
    lcd_sendCommand(0x06);  /* shift cursor right */
}

void delay_us(uint16_t t)
{
    volatile unsigned long l = 0;
    for(uint16_t i = 0; i < t; i++)
        for(l = 0; l < 6; l++)
        {
        }
}
```

LCD cursor position

In the LCD, one can move the cursor to any location in the display by issuing an address command. The next character sent will appear at the cursor position. For the two-line LCD, the address command for the first location of line 1 is 0x80, and for line 2 it is 0xC0. The following shows address locations and how they are accessed:

RS	R/W	DB7	DB6	DB5	DB4	DB3	DB2	DB1	DB0
0	0	1	A6	A5	A4	A3	A2	A1	A0

where $A_6A_5A_4A_3A_2A_1A_0$= 0000000 to 0100111 for line 1 and $A_6A_5A_4A_3A_2A_1A_0$ = 1000000 to 1100111 for line 2. See Table 9-3.

	DB7	DB6	DB5	DB4	DB3	DB2	DB1	DB0
Line 1 (min)	1	0	0	0	0	0	0	0
Line 1 (max)	1	0	1	0	0	1	1	1
Line 2 (min)	1	1	0	0	0	0	0	0
Line 2 (max)	1	1	1	0	0	1	1	1

Table 9-3: LCD Addressing Commands

The upper address range can go as high as 0100111 for the 40-character-wide LCD while for the 20-character-wide LCD the address of the visible positions goes up to 010011 (19 decimal = 10011 binary). Notice that the upper range 0100111 (binary) = 39 decimal, which corresponds to locations 0 to 39 for the LCDs of 40 × 2 size. Figure 9-4 shows the addresses of cursor positions for various sizes of LCDs. All the addresses are in hex. Notice the starting addresses for 4-line LCD are not in sequential order.

As an example of setting the cursor at the fourth location of line 1 we have the following:

```
lcd_sendCommand(0x83);
```

and for the sixth location of the second line we have:

```
lcd_sendCommand(0xC5);
```

Notice that the cursor location addresses are in hex and starting at 0 as the first location.

In Program 9-4, the lcd_gotoxy function moves the cursor to the desired location. The addresses for the first column of each row is stored in the firstCharAdr array and the address of each location is calculated by adding the column to the address of the first column of the row.

Program 9-4: This program uses the 4-bit data option to show a message on the LCD.

```
/* The program shows "The world is but one country" on the LCD */

#include <stm32f10x.h>

void lcd_init(void);
void lcd_sendCommand(unsigned char cmd);
void lcd_sendData(unsigned char data);
```

```c
void lcd_gotoxy(unsigned char x, unsigned char y);
void lcd_print(char * str);

void delay_us(uint16_t t);

#define LCD_RS    4
#define LCD_EN    5

int main()
{
    RCC->APB2ENR |= 0xFC; /* Enable clocks for GPIO ports */
    GPIOA->CRL = 0x44333333; /* PA0-PA5 as outputs */

    lcd_init();

    lcd_gotoxy(1,1);  /* move cursor to 1,1 */
    lcd_print("The world is but"); /* show "The world is but" */
    lcd_gotoxy(1,2);  /* move cursor to 1,2 */
    lcd_print("one country"); /* show "one country" */

    while(1)   /* stay here forever */
    {
    }
}

/* The function moves the LCD cursor to location x, y */
void lcd_gotoxy(unsigned char x, unsigned char y)
{
    const unsigned char firstCharAdr[]={0x80,0xC0,0x94,0xD4};
    lcd_sendCommand(firstCharAdr[y-1] + x - 1);
    delay_us(100);
}

/* copy lcd_putValue, lcd_sendCommand, lcd_sendData, lcd_init, lcd_print, and delay_us
from Program 9-3 */
```

LCD timing and data sheet

Figures 9-5 and 9-6 show timing diagrams for LCD write and read timing, respectively.

Notice that the write operation happens on the H-to-L transition of the E pin. The microcontroller must have data ready and stable on the data lines before the H-to-L transition of E to satisfy the setup time requirement.

The read operation is activated by the L-to-H pulse of the E pin. After the delay time, the LCD controller will have the data available on the data bus if the R/W line is high. The microcontroller should read the data from the data lines before lowering the E pulse.

Table 9-4 provides a more detailed list of LCD instructions.

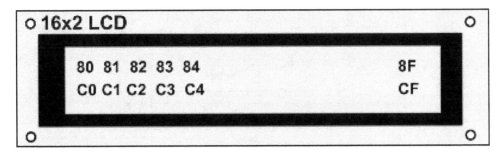

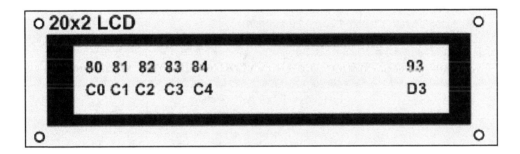

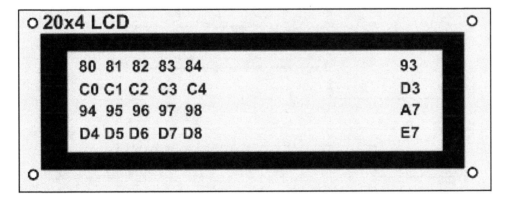

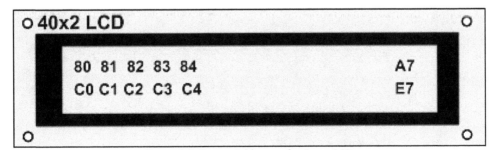

Figure 9-4: Cursor Addresses for Some LCDs

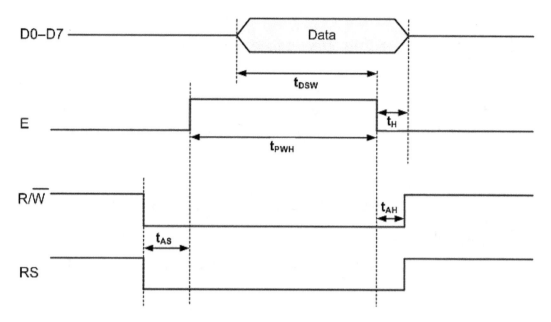

t_{PWH} = Enable pulse width = 230 ns (minimum)
t_{DSW} = Data setup time = 80 ns (minimum)
t_H = Data hold time = 10 ns (minimum)
t_{AS} = Setup time prior to E (going high) for both RS and R/W = 40 ns (minimum)
t_{AH} = Hold time after E has come down for both RS and R/W = 10 ns (minimum)

Figure 9-5: LCD Write Timing

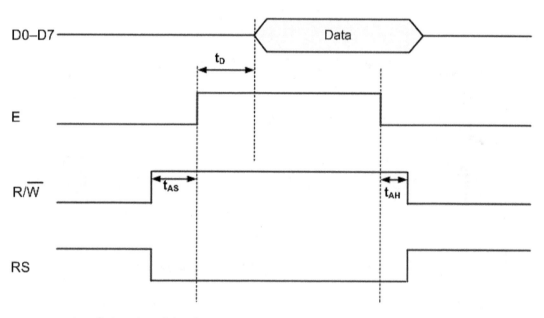

t_D = Data output delay time
t_{AS} = Setup time prior to E (going high) for both RS and R/W = 40 ns (minimum)
t_{AH} = Hold time after E has come down for both RS and R/W = 10 ns (minimum)

Note: Read requires an L-to-H pulse for the E pin.

Figure 9-6: LCD Read Timing

Instruction	RS	R/W	DB7	DB6	DB5	DB4	DB3	DB2	DB1	DB0	Description	Execution Time (Max)
Clear display	0	0	0	0	0	0	0	0	0	1	Clears entire display and sets DD RAM address 0 in address counter	1.64 ms
Return Home	0	0	0	0	0	0	0	0	1	-	Sets DD RAM address to 0 as address counter. Also returns display being shifted to original positions. DD RAM contents remain unchanged.	1.64 ms
Entry Mode Set	0	0	0	0	0	0	0	1	I/D	S	Sets cursor move direction and specifies shift of display. These operations are performed during data write and read.	40μs
Display On/Off Control	0	0	0	0	0	0	1	D	C	B	Sets On/Off of entire display (D), cursor On/Off (C), and blink of cursor position character (B).	40μs
Cursor or Display shift	0	0	0	0	0	1	S/C	R/L	-	-	Moves cursor and shifts display without changing DD RAM contents.	40μs
Function Set	0	0	0	0	1	DL	N	F	-	-	Sets interface data length (DL), number of display lines (L), and character font (F)	40μs
Set CG RAM Address	0	0	0	1	AGC						Sets CG RAM address. CG RAM data is sent and received after this setting.	40μs
Set DD RAM Address	0	0	1	ADD							Sets DD RAM address. DD RAM data is sent and received after this setting.	40μs
Read Busy Flag & Address	0	1	BF	AC							Reads Busy flag (BF) indicating internal operation is being performed and reads address counter contents.	40μs
Write Data CG or DD RAM	1	0	Write Data								Writes data into DD or CG RAM.	40μs
Read Data CG or DD RAM	1	1	Read Data								Reads data from DD or CG RAM.	40μs

Abbreviations:

DD RAM: Display data RAM

CG RAM: Character generator RAM

AGC: CG RAM address

ADD: DD RAM address, corresponds to cursor address

AC: address counter used for both DD and CG RAM addresses

I/D: 1 = Increment, 0: Decrement

S =1: Accompanies display shift

S/C: 1 = Display shift, 0: Cursor move

R/L: 1: Shift to the right, 0: Shift to the left

DL: 1 = 8 bits, 0 = 4 bits

N: 1 = 2-line, 0 = 1-line

F: 1 = 5 x 10 dots, 0 = 5 x 7 dots

BF: 1 = Internal operation, 0 = Can accept instruction

Table 9-4: List of LCD Instructions

Review Questions

1. The RS pin is an _____ (input, output) pin for the LCD.
2. The E pin is an _____ (input, output) pin for the LCD.
3. The E pin requires an _____ (H-to-L, L-to-H) transition to latch in information at the data pins of the LCD.
4. For the LCD to recognize information at the data pins as data, RS must be set to _____ (high, low).
5. Give the command codes for line 1, first character, and line 2, first character.

Section 9.2: Interfacing the Keyboard to the CPU

To reduce the microcontroller I/O pin usage, keyboards are organized in a matrix of rows and columns. The CPU accesses both rows and columns through ports; therefore, with two 8-bit ports, an 8 × 8 matrix of 64 keys can be connected to a microprocessor. When a key is pressed, a row and a column make a contact; otherwise, there is no connection between rows and columns. In PC keyboards, an embedded microcontroller in the keyboard takes care of the hardware and software interfacing of the keyboard. In such systems, it is the function of programs stored in the ROM of the microcontroller to scan the keys continuously, identify which one has been activated, and present it to the main CPU on the motherboard. In this section, we look at the mechanism by which the microprocessor scans and identifies the key. For clarity some examples are provided.

Scanning and identifying the key

Figure 9-7 shows a 4 × 4 matrix connected to two ports. The rows are connected to an output port and the columns are connected to an input port.

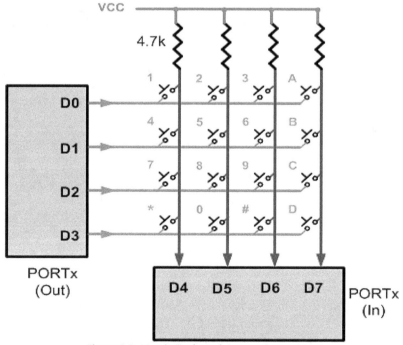

Figure 9-7: Matrix Keyboard Connection to Ports

All the input pins have pull-up resistor connected. If no key has been pressed, reading the input port will yield 1s for all columns. If all the rows are driven low and a key is pressed, the column of that key will read back a 0 since the key pressed shorted that column to the row that is driven low. It is the function of the microprocessor to scan the keyboard continuously to detect and identify the key pressed. How it is done is explained next.

Key press detection

To detect the key pressed, the microprocessor drives all rows low then it reads the columns. If the data read from the columns is D7–D4 = 1111, no key has been pressed and the process continues until a key press is detected. However, if one of the column bits has a zero, this means that a key was pressed. For example, if D7–D4= 1101, this means that a key in the D5 column has been pressed.

The following program detects whether any of the keys is pressed. If a key press is detected, the blue LED is turned on.

Program 9-5: This program turns on the blue LED when a key is pressed.

```
/* The program checks if any key is pressed or not */
/* It turns on the green led (PC13=0) if a key is pressed. */
/* PB15-PB12 as rows (outputs) */
/* PB6-PB9 as columns (inputs) */

#include <stm32f10x.h>

void delay_us(uint16_t t);

uint8_t keypad_kbhit(void);

int main()
{
   RCC->APB2ENR |= 0xFC; /* Enable clocks for GPIO ports */
   /* init keypad pins */
   GPIOB->CRL = 0x88444444; /* PB6 and PB7 as inputs */
   GPIOB->CRH = 0x33334488; /* PB15-PB12 as outputs, PB8 and PB9 as inputs */
   GPIOB->ODR |= (0xF<<6); /* pull-up PB9-PB6 */

   GPIOC->CRH = 0x44344444; /* PC13 as output */

   while(1)
   {
     if(keypad_kbhit())
       GPIOC->BRR = (1<<13); /* PC13 = low (LED on) */
     else
       GPIOC->BSRR = (1<<13); /* PC13 = high (LED off) */
   }
}

/* The function checks if a key is pressed or not */
/* Returns: 0: Not pressed, 1: Pressed */
uint8_t keypad_kbhit()
{
   GPIOB->BRR = (0x0F<<12); /* make all rows zeros */
```

```
    delay_us(1);
    if(((GPIOB->IDR>>6)&0xF) == 0xF) /* if all cols are high */
        return 0;
    else
        return 1;
}

void delay_us(uint16_t t)
{
    volatile unsigned long l = 0;
    for(uint16_t i = 0; i < t; i++)
        for(l = 0; l < 6; l++)
        {
        }
}
```

Key identification

After a key press is detected, the microprocessor will go through the process of identifying the key. Starting from the top row, the microprocessor drives one row low at a time; then it reads the columns. If the data read is all 1s, no key in that row is pressed and the process is moved to the next row. It drives the next row low, reads the columns, and checks for any zero. This process continues until a row is identified with a zero in one of the columns. The next task is to find out which column the pressed key belongs to. This should be easy since each column is connected to a separate input pin. Look at Example 9-1.

Example 9-1

From Figure 9-7, identify the row and column of the pressed key for each of the following.
(a) D3–D0 = 1110 for the row, D7–D4= 1011 for the column
(b) D3–D0 = 1101 for the row, D7–D4= 0111 for the column

Solution:

From Figure 9-7 the row and column can be used to identify the key.
(a)   The row belongs to D0 and the column belongs to D6; therefore, the key number 3 was pressed.
(b)   The row belongs to D1 and the column belongs to D7; therefore, the key B was pressed.

Figure 9-8 is the flowchart for the detection and identification of the key activation.

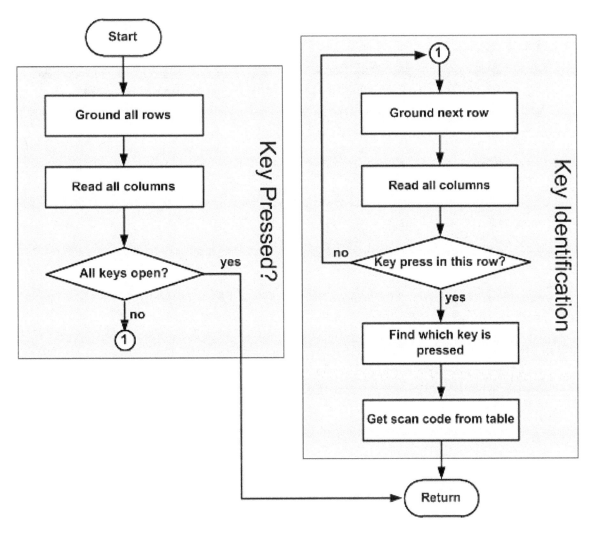

Figure 9-8: The Flowchart for Key Press Detection and Identification

Program 9-6 provides an implementation of the detection and identification algorithm in C language. We will exam it in details here. First for the initialization of the ports, Port 4 pins 3-0 are used for rows. The Port 4 pins 7-4 are used for columns. They are all configured as input digital pin to prevent accidental short circuit of two output pins. If output pins are driven high and low and two keys of the same column are pressed at the same time by accident, they will short the output low to output high of the adjacent pins and cause damages to these pins. To prevent this, all pins are configured as input pin and only one pin is configured as output pin at a time. Since only one pin is actively driving the row, shorting two rows will not damage the circuit. The input pins are configured with pull-up enabled so that when the connected keys are not pressed, they stay high and read as 1.

The key scanning function is a non-blocking function, meaning the function returns regardless of whether there is a key pressed or not. The function first drives all rows low and check to see if any key pressed. If no key is pressed, a zero is returned. Otherwise the code will proceed to check one row at a time by driving only one row low at a time and read the columns. If one of the columns is active, it will find out which column it is. With the combination of the active row and active column, the code will find out the key that is pressed and return a unique numeric code. The program below reads a 4x4 keypad and use the key code returned to set the tri-color LEDs. LED program is borrowed from P2-7 in Chapter 2.

Program 9-6: This program displays the pressed key on the tri-color LED.

```c
/* The program shows the pressed key on the LCD */
/* keypad rows: PB15-PB12 cols: PB9-PB6 */

#include <stm32f10x.h>
#include <stdio.h>

void lcd_init(void);
void lcd_sendCommand(unsigned char cmd);
void lcd_sendData(unsigned char data);
void lcd_gotoxy(unsigned char x, unsigned char y);
void lcd_print(char * str);

void delay_ms(uint16_t t);
void delay_us(uint16_t t);

#define LCD_RS   4
#define LCD_EN   5

uint8_t keypad_kbhit(void);
uint8_t keypad_getkey(void);


int main()
{
   RCC->APB2ENR |= 0xFC; /* Enable clocks for GPIO ports */
   GPIOA->CRL = 0x44333333; /* PA0-PA5 as outputs */

   lcd_init();

   lcd_gotoxy(1,1);  /* move cursor to 1,1 */
   lcd_print("The pressed key is: "); /* show the message */

   /* init keypad pins */
   GPIOB->CRL = 0x88444444; /* PB6 and PB7 as inputs */
   GPIOB->CRH = 0x33334488; /* PB15-PB12 as outputs, PB8 and PB9 as inputs */
   GPIOB->ODR |= (0xF<<6); /* pull-up PB9-PB6 */

   GPIOC->CRH = 0x44344444; /* PC13 as output */

   while(1)
   {
     lcd_gotoxy(1,2);       /* move cursor to 1,2 */
     if(keypad_kbhit() == 0)
       lcd_print(" ");
     else
       lcd_data(keypad_getkey()); /* show the pressed key */
   }
}

/* The function checks if a key is pressed or not */
/* Returns: 0: Not pressed, 1: Pressed */
```

```c
uint8_t keypad_kbhit()
{
   GPIOB->ODR &= ~(0x0F<<12); /* make all rows ones */
   delay_us(1);
   if(((GPIOB->IDR>>6)&0xF) == 0xF)
      return 0;
   else
      return 1;
}


/* The function returns the pressed key. */
/* It returns 0 if no key is pressed. */

uint8_t keypad_getkey()
{
   uint8_t keypadLookup[16]=
      {'1','2','3','A','4','5','6','B','7','8','9','C','*','0','#','D'};
   const uint32_t rowSelect[4]={
      0x1000E000, /* Row3-Row0 = 1110 */
      0x2000D000, /* Row3-Row0 = 1101 */
      0x4000B000, /* Row3-Row0 = 1011 */
      0x80007000};/* Row3-Row0 = 1011 */

   if(keypad_kbhit() == 0) /* if no key is pressed */
      return 0;

   for(int r = 0; r <= 3; r++) /* rows 0 to 3 */
   {
      GPIOB->BSRR = rowSelect[r]; /* ground row r and make the others high */
      delay_us(1);   /* wait for the columns to be updated */
      uint8_t cols = (GPIOB->IDR>>6)&0xF;

      switch(cols)
      {
         case 0x0E: return keypadLookup[r*4+0]; /* col0 is low */
         case 0x0D: return keypadLookup[r*4+1]; /* col1 is low */
         case 0x0B: return keypadLookup[r*4+2]; /* col2 is low */
         case 0x07: return keypadLookup[r*4+3]; /* col3 is low */
      }
   }

   return 0;
}
void delay_ms(uint16_t t)
{
   volatile unsigned long l = 0;
   for(uint16_t i = 0; i < t; i++)
      for(l = 0; l < 6000; l++)
      {
      }
}


/* Copy lcd_putValue, lcd_sendCommand, lcd_sendData, lcd_init, lcd_print, delay_us,
and lcd_gotoxy from Programs 3-3 and 3-4. */
```

Contact Bounce and Debounce

When a mechanical switch is closed or opened, the contacts do not make a clean transition instantaneously, rather the contacts open and close several times before they settle. This event is called contact bounce (see Figure 9-9). So, it is possible when the program first detects a switch in the keypad is pressed but when interrogating which key is pressed, it would find no key pressed. This is the reason we have a return 0 after checking all the rows. Another problem manifested by contact bounce is that one key press may be recognized as multiple key presses by the program. Contact bounce also occurs when the switch is released. Because the switch contacts open and close several times before they settle, the program may detect a key press when the key is released.

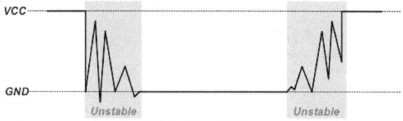

Figure 9-9: Switch contact bounces

For many applications, it is important that each key press is only recognized as one action. When you press a numeral key of a calculator, you expect to get only one digit. A contact bounce results in multiple digits entered with a single key press. A simple software solution is that when a transition of the contact state change is detected such as a key pressed or a key released, the software does a delay for about 10 – 20 ms to wait out the contact bounce. After the delay, the contacts should be settled and stable.

There are IC chips such as National Semiconductor's MM74C923 that incorporate keyboard scanning and decoding all in one chip. Such chips use combinations of counters and logic gates (no microprocessor) to implement the underlying concepts presented in Programs 9-5 and 9-6.

Review Questions

1. True or false. To see if any key is pressed, all rows are driven low.
2. If D7–D4 = 0111 is the data read from the columns, which column does the key pressed belong to?
3. True or false. Key press detection and key identification require two different processes.
4. In Figure 9-7, if the row has D3–D0 = 1110 and the columns are D7–D4 = 1110, which key is pressed?
5. True or false. To identify the key pressed, one row at a time is driven low.

Problems

Section 9-1

1. Describe the function of pins E, R/W, and RS in the LCD.
2. What is the difference between the VCC and VEE pins on the LCD?

3. "Clear LCD" is a _____ (command code, data item) and its value is ___ hex.
4. What is the hex value of the command code for "display on, cursor on"?
5. Give the state of RS, E, and R/W when sending a command code to the LCD.
6. Give the state of RS, E, and R/W when sending data character 'Z' to the LCD.
7. Which of the following is needed on the E pin in order for a command code (or data) to be latched in by the LCD?

   (a) H-to-L pulse   (b) L-to-H pulse
8. True or false. For the above to work, the value of the command code (data) must already be at the D0–D7 pins.
9. There are two methods of sending commands and data to the LCD: (1) 4-bit mode or (2) 8-bit mode. Explain the difference and the advantages and disadvantages of each method.
10. For a 16 × 2 LCD, the location of the last character of line 1 is 8FH (its command code). Show how this value was calculated.
11. For a 16 × 2 LCD, the location of the fourth character of line 2 is C3H (its command code). Show how this value was calculated.
12. Show the value (in hex) for the command code for the 10th location, line 1 on a 20 × 2 LCD. Show how you got your value.
13. Show the value (in hex) for the command code for the 20th location, line 2 on a 40 × 2 LCD. Show how you got your value.

Section 9-2

14. In reading the columns of a keyboard matrix, if no key is pressed, we should get all _____ (1s, 0s).
15. In the 4 × 4 keyboard interfacing, to detect the key press, which of the following is grounded?

   (a) all rows   (b) one row at time   (c) both (a) and (b)
16. In the 4 × 4 keyboard interfacing, to identify the key pressed, which of the following is grounded?

   (a) all rows   (b) one row at time   (c) both (a) and (b)
17. For the 4 × 4 keyboard interfacing (Figure 9-7), indicate the column and row for each of the following.

   (a) D3–D0 = 0111, D4-D7=1011   (b) D3–D0 = 1110, D4-D7= 1101
18. Indicate the steps to detect the key press.
19. Indicate the steps to identify the key pressed.
20. Indicate an advantage and a disadvantage of using an IC chip for keyboard scanning and decoding instead of using a microcontroller.

Answers to Review Questions

Section 9-1

1. Input
2. Input
3. H-to-L

4. High
5. 0x80 and 0xC0

Section 9-2

1. True
2. Column 3 (D7)
3. True
4. 1
5. True

Chapter 10: UART Serial Port Programming

Computers transfer data in two ways: parallel and serial. In parallel data transfers, often eight or more lines (wire conductors) are used to transfer data to a device that is only a few feet away. Although a lot of data can be transferred in a short amount of time by using many wires in parallel, the distance cannot be great. To transfer to a device located many meters away, the serial method is used. In serial communication, the data is sent one bit at a time, in contrast to parallel communication, in which the data is sent a byte or more at a time. As technology advances, the data rate of serial communication may exceed parallel communication while parallel communication still retains the disadvantages of the size and cost of cable and connector, and the crosstalk between the data lines at longer distance.

Serial communication of the STM32 is the topic of this chapter. The STM32 has serial communication capability built into it, thereby making possible fast data transfer using only a few wires.

Section 10.1: Basics of Serial Communication

When a microprocessor communicates with the outside world it usually provides the data in byte-sized chunks. For parallel transfer, 8-bit data is transferred at the same time. For serial transfer, 8-bit data is transferred one bit at a time. Figure 10-1 diagrams serial versus parallel data transfers.

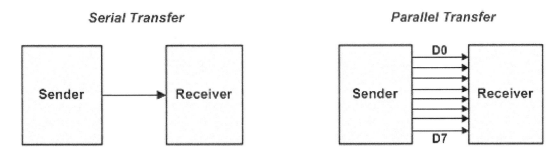

Figure 10-1: Serial vs. Parallel Data Transfer

The fact that in serial communication, a single data line is used instead of the 8-bit data line of parallel communication not only makes it much cheaper but also makes it possible for two computers located in two different cities to communicate.

For serial data communication to work, the byte of data must be grabbed from the 8-bit data bus of the microprocessor and converted to serial bits using a parallel-in-serial-out shift register; then it can be transmitted over a single data line. This also means that at the receiving end there must be a serial-in-parallel-out shift register to receive the serial data, pack it into a byte, and present it to the system at the receiving end. See Figures 10-2 and 10-3.

When the distance is short, the digital signal can be transferred as it is on a simple wire and requires no modulation. This is how PC keyboards transfer data between the keyboard and the motherboard. However, for long-distance data transfers using communication lines such as a telephone, serial data communication requires a modem to modulate the data (convert from 0s and 1s to audio tones) before putting it on the transmission media and demodulate (convert from audio tones to 0s and 1s) at the receiving end.

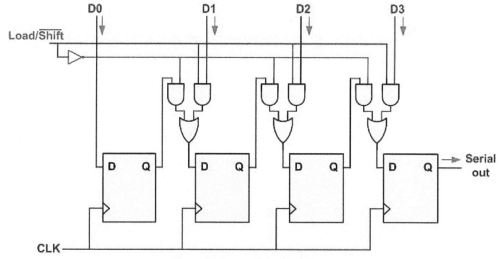

Figure 10-2: Parallel In Serial Out

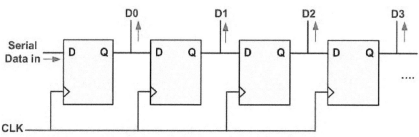

Figure 10-3: Serial In Parallel Out

Serial data communication uses two methods, asynchronous and synchronous. The synchronous method transfers the data with the clock and usually a block of data (characters) at a time while the asynchronous transfers without clock and usually a single byte at a time.

It is possible to write software to use either of these methods, but the programs can be tedious and long. For this reason, special IC chips are made by many manufacturers for serial data communications. These chips are commonly referred to as UART (universal asynchronous receiver-transmitter) and USART (universal synchronous-asynchronous receiver-transmitter). The COM port in the PC uses the UART. When this function incorporated into a microcontroller, it is often referred as SCI (Serial Communication Interface). The STM32 chips have some built-in USART, which is discussed in detail in this chapter.

Half- and full-duplex transmission

In data transmission, a duplex transmission is one in which the data can be transmitted and received. This is in contrast to a simplex transmission such as printers, in which the computer only sends data. Duplex transmissions can be half or full duplex. If data is transmitted one way at a time, it is referred to as *half duplex*. If the data can go both ways at the same time, it is *full duplex*. Of course, full duplex requires two wire conductors for the data lines (in addition to ground), one for transmission and one for reception, in order to transfer and receive data simultaneously. See Figure 10-4.

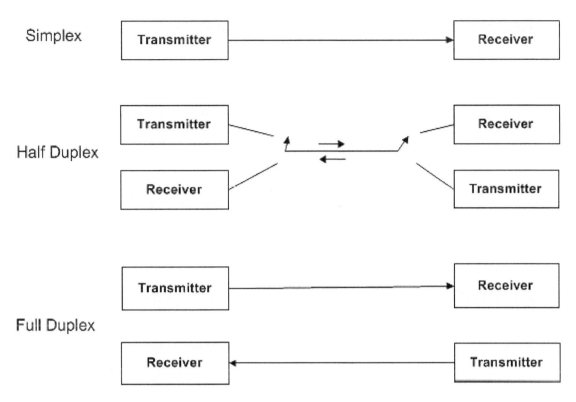

Figure 10-4: Simplex, Half-, and Full-Duplex Transfers

Asynchronous serial communication and data framing

The data coming in at the receiving end of the data line in a serial data transfer is all 0s and 1s; it is difficult to make sense of the data unless the sender and receiver agree on a set of rules, a *protocol*, on how the data is packed, how many bits constitute a character, and when the data begins and ends.

Start and stop bits

Asynchronous serial data communication is widely used for character-oriented transmissions. In the asynchronous method, each character, such as ASCII characters, is packed between start and stop bits. This is called *framing*. The start bit is always one bit but the stop bit can be one or two bits. The start bit is always a 0 (low) and the stop bit(s) is a 1 (high). For example, look at Figure 10-5 where the ASCII character "A", binary 0100 0001, is framed between the start bit and 2 stop bits. Notice that the LSB is sent out first.

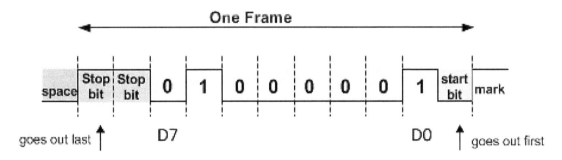

Figure 10-5: Framing ASCII "A" (0x41)

In Figure 10-5, when there is no data transfer, the signal stays at 1 (high), which is referred to as *mark*. The 0 (low) is referred to as *space*. Notice that the transmission begins with a start bit followed by D0, the LSB, then the rest of the bits until the MSB (D7), and finally, the 2 stop bits indicating the end of the character "A".

In asynchronous serial communications, peripheral chips can be programmed for data that is 5, 6, 7, or 8 bits wide. While in older systems ASCII characters were 7-bit, the modern systems usually send 8-bit data. The old Baud code uses 5- or 6-bit characters but they are rarely seen these days even though most of the hardware still supporting them. In some older systems, due to the slowness of the receiving mechanical device, 2 stop bits were used to give the device sufficient time to organize itself before transmission of the next byte. However, in modern PCs the use of 1 stop bit is common. Assuming that we are transferring a text file of ASCII characters using 1 stop bit, we have a total of 10 bits for each character since 8 bits are for the ASCII code, and 2 bits are for start and stop bits, respectively. Therefore, for each 8-bit character there are an extra 2 bits, or 12.5% overhead. (2/8 x100=12.5%)

Parity bit

In some systems in order to maintain data integrity, the parity bit of the character byte is included in the data frame. This means that for each character (7- or 8-bit, depending on the system) we have a single parity bit in addition to start and stop bits. The parity bit may be odd or even. In the case of an odd-parity the number of data bits, including the parity bit, has an odd number of 1s. Similarly, in an even-parity the total number of bits, including the parity bit, is even. For example, the ASCII character "A", binary 0100 0001, has 0 for the even-parity bit. UART chips allow programming of the parity bit for odd-, even-, and no-parity options, as we will see in the next section. If a system requires the parity, the parity bit is transmitted after the MSB, and is followed by the stop bit.

Data transfer rate

The rate of data transfer in serial data communication is stated in *bps* (bits per second). Another widely used terminology for bps is *Baudrate*. However, the baud and bps rates are not necessarily equal. This is due to the fact that baud rate is defined as number of signal changes per second. In modems, it is possible for each signal to transfer multiple bits of data. As far as the conductor wire is concerned, the baud rate and bps are the same, and for this reason in this book we use the terms bps and baud interchangeably.

Example 10-1

Calculate the total number of bits used in transferring 50 pages of text, each with 80 × 25 characters. Assume 8 bits per character and 1 stop bit.

Solution:

For each character a total of 10 bits is used, 8 bits for the character, 1 stop bit, and 1 start bit. Therefore, the total number of bits is 80 × 25 × 10 = 20,000 bits per page. For 50 pages, 1,000,000 bits will be transferred.

Example 10-2

Calculate the minimal time it takes to transfer the entire 50 pages of data in Example 10-1 using a baud rate of:

(a) 9600   (b) 57,600

Solution:

(a) 1,000,000 / 9600 = 104 seconds

(b) 1,000,000 / 57,600 = 17 seconds

Example 10-3

Calculate the time it takes to download a movie of 2 gigabytes using a telephone line. Assume 8 bits, 1 stop bit, no parity, and 57,600 baud rate.

Solution:

$2 \times 2^{30} \times 10 / 57,600 = 347,222$ seconds = ~4 days

---

RS232 and other serial I/O standards

To allow compatibility among data communication equipment made by various manufacturers, an interfacing standard called RS232 was proposed by the Electronics Industries Association (EIA) in 1960. It has several revisions through the years with an alphabet at the end to denote the revision number such as RS232C. RS stands for recommended standard. It was finally adopted as an EIA standard and renamed EIA232, later on TIA232 (Telecommunications Industry Association). In this book we refer to it simply as RS232. Today, RS232 is the most widely used serial I/O interfacing standard. However, since the standard was set long before the advent of the TTL logic family, the input and output voltage levels are not TTL compatible. In the RS232 at the receiver, a 1 is represented by −3 to −25 V, while the 0 bit is +3 to +25 V, making −3 to +3 undefined. For this reason, to connect any RS232 to a TTL-level chip (microprocessor or UART) we must use voltage converters such as MAX232 or MAX233 to convert the TTL logic levels to the RS232 voltage level, and vice versa. MAX232 and MAX233 IC chips are commonly referred to as line drivers. This is shown in Figures 10-6 and 10-7. The MAX232 has two sets of line drivers for transferring and receiving data, as shown in Figure 10-6. The line drivers used for TxD are called T1 and T2, while the line drivers for RxD are designated as R1 and R2. In many applications only one of each is used. Notice in MAX232 that the T1 line driver has a designation of T1in and T1out on pin numbers 11 and 14, respectively. The T1in pin is the TTL side and is connected to TxD of the USART, while T1out is the RS232 side that is connected to the RxD pin of the RS232 DB connector. The R1 line driver has a designation of

R1in and R1out on pin numbers 13 and 12, respectively. The R1in (pin 13) is the RS232 side that is connected to the TxD pin of the RS232 DB connector, and R1out (pin 12) is the TTL side that is connected to the RxD pin of the USART.

To convert CMOS voltage levels to RS232 and vice versa we can use MAX3232 which is pin compatible with MAX232 and works with supply voltages between 3V and 5.5V.

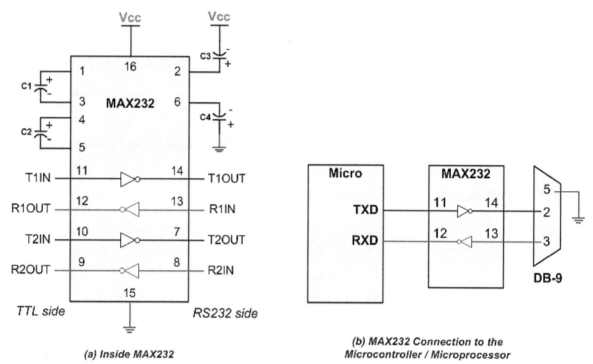

(a) Inside MAX232

(b) MAX232 Connection to the Microcontroller / Microprocessor

Figure 10-6: MAX232

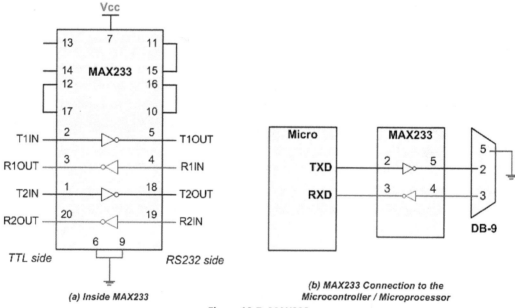

(a) Inside MAX233

(b) MAX233 Connection to the Microcontroller / Microprocessor

Figure 10-7: MAX233

MAX232 requires four capacitors of 1 μF. To save board space, some designers use the MAX233 chip from Maxim. The MAX233 performs the same job as the MAX232 but eliminates the need for capacitors. However, the MAX233 chip is much more expensive than the MAX232. See Figure 10-7 for MAX233 with no capacitor used.

RS232 pins

Table 10-1 provides the pins and their labels for the RS232 cable, commonly referred to as the DB-9 connector. The DB-9 male connector is shown in Figure 10-8.

Pin	Description
1	Data carrier detect (DCD)
2	Received data (RxD)
3	Transmitted data (TxD)
4	Data terminal ready (DTR)
5	Signal ground (GND)
6	Data set ready (DSR)
7	Request to send (RTS)
8	Clear to send (CTS)
9	Ring indicator (RI)

Table 10-1: RS232 Pins

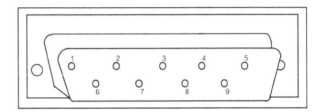

Figure 10-8: 9-Pin Male Connector

Data communication classification

Current terminology classifies data communication equipment as DTE (data terminal equipment) or DCE (data communication equipment). DTE refers to terminals and computers that send and receive data, while DCE refers to communication equipment, such as modems, that is responsible for transferring the data. Notice that all the RS232 pin function definitions of Table 10-1 are from the DTE point of view.

The simplest connection between two PCs (DTE and DTE) requires a minimum of three pins, TxD, RxD, and ground, as shown in Figure 10-9. Notice that the connection between two DTE devices, such as two PCs, requires pins 2 and 3 to be interchanged as shown in Figure 10-9. In looking at Figure 10-9, keep in mind that the RS232 signal definitions are from the point of view of DTE.

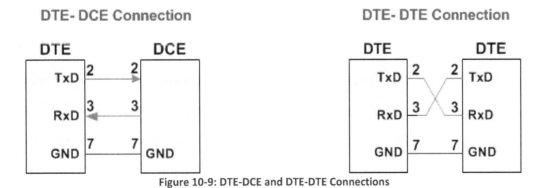

Figure 10-9: DTE-DCE and DTE-DTE Connections

Examining the RS232 handshaking signals

To ensure fast and reliable data transmission between two devices, the data transfer must be coordinated. Some of the pins of the RS-232 are used for handshaking signals. They are described below. Due to the fact that in serial data communication the receiving device may run out of room for more data at times, there must be a way to inform the sender to stop sending data. So, some of these handshaking lines may be used for flow control.

1. **DTR (data terminal ready)**: When the terminal (or a PC COM port) is turned on, after going through a self-test, it sends out signal DTR to indicate that it is ready for communication. If there is something wrong with the COM port, this signal will not be activated. This is an active-low signal and can be used to inform the modem that the computer is alive and kicking. This is an output pin from DTE (PC COM port) and an input to the modem.

2. **DSR (data set ready):** When a DCE (modem) is turned on and has gone through the self-test, it asserts DSR to indicate that it is ready to communicate. Therefore, it is an output from the modem (DCE) and an input to the PC (DTE). This is an active-low signal. If for any reason the modem cannot make a connection to the telephone, this signal remains inactive, indicating to the PC (or terminal) that it cannot accept or send data.

3. **RTS (request to send):** When the DTE device (such as a PC) has data to transmit, it asserts RTS to signal the modem. RTS is an active-low output from the DTE and an input to the modem.

4. **CTS (clear to send):** In response to RTS, when the modem is ready to accept data for transmission, it sends out signal CTS to the DTE (PC) to indicate that it can accept data from the PC now. This input signal to the DTE is used by the DTE to start transmission.

5. **CD (carrier detect, or DCD, data carrier detect):** The modem asserts signal DCD to inform the DTE (PC) that a valid carrier signal from the other modem has been detected and that contact between it and the other modem is established. Therefore, DCD is an output from the modem and an input to the PC (DTE).

6. **RI (ring indicator):** An output from the modem (DCE) and an input to a PC (DTE) indicates that the telephone is ringing. It goes on and off in synchronization with the ring tone. Of the six handshake signals, this is the least often used, due to the fact that modems take care of answering the phone. However, if in a given system the PC is in charge of answering the phone, this signal can be used.

From the above description, PC and modem communication can be summarized as follows: While signals DTR and DSR are used by the PC and modem, respectively, to indicate that they are alive and well, it is RTS and CTS that actually control the flow of data. When the PC wants to send data, it asserts RTS, and in response, if the modem is ready to accept the data, it sends back CTS. If not, the modem does not activate CTS, and the PC will have to wait until CTS goes active. RTS and CTS are also referred to as *hardware control flow signals*.

This concludes the description of the most important pins of the RS232 handshake signals plus TxD, RxD, and ground. Ground is also referred to as SG (signal ground). In the next section we will see serial communication programming for the microcontroller.

Review Questions

1. The transfer of data using parallel lines is _____ (more expensive, less expensive).
2. In communications between two PCs in New York and Dallas, we use _____ (serial, parallel) data communication.
3. In serial data communication, which method fits block-oriented data?
4. True or false. Sending data to a printer is duplex.
5. True or false. In duplex we must have two data lines.
6. The start and stop bits are used in the _____ (synchronous, asynchronous) method.
7. Assuming that we are transmitting letter "D", binary 100 0100, with odd-parity bit and 2 stop bits, show the sequence of bits transferred.
8. In Question 7, find the overhead due to framing.
9. Calculate the time it takes to transfer 400 characters as in Question 7 if we use 1200 bps. What percentage of time is used due to overhead?
10. True or false. RS232 is not TTL-compatible.

Section 10.2: Programming UART Ports

In this section we discuss the serial communication registers of the STM32F10x and show how to program them to transfer and receive data using asynchronous mode. STM32F103RB and STM32F103C8 have 3 on-chip USART ports. They are designated as USART1, USART2, and USART3. See Table 10-2.

	Bus	Address
USART1	APB2	0x40013800 – 0x40013BFF
USART2	APB1	0x40004400 – 0x400047FF
USART3	APB1	0x40004800 – 0x40004BFF
USART4	APB1	0x40004C00 – 0x40004FFF
USART5	APB1	0x40005000 – 0x400053FF

Table 10-2: USART Addresses

You can use a USB-to-Serial module (or cable) and connect the USART ports to the computer. One side of the USB-to-Serial module should be 3.3V logic level TxD and RxD pins and is connected to the USARTx pins on STM32 board. The other side is USB port connected to the PC USB port.

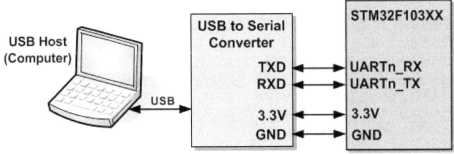

Figure 10-10: Connecting to the PC

WWW: Please download the step by step tutorial for connecting the STM32 boards to the PC using serial from https://NicerLand.com

Figure 10-11 shows the simplified block diagram of the USART units.

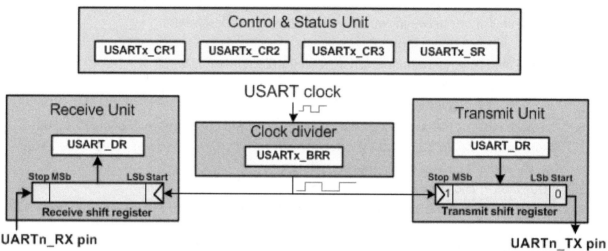

Figure 10-11: a Simplified Block Diagram of UARTn

In all microcontrollers, there are 3 groups of registers in USART peripherals:

1. **Configuration (Control) registers:** Before using the USART peripheral the configuration registers must be initialized. This sets some parameters of the communication including: Baud rate, word length, stop bit, interrupts (if needed). In STM32 microcontrollers, some of the configuration registers are: BRR, CR1, CR2, and CR3.

2. **Transmit and receive register:** To send data, we simply write to the transmit register. The USART peripheral sends out the contents of the transfer register through the serial transmit pin (TX). The received data is stored in the receive register. In STM32, the transfer and receive registers are named as USARTx_DR.

3. **Status register:** the status register contains some flags which show the state of sending and receiving data including: the transmitter sent out the entire byte, the transmitter is ready for another byte of data, the receiver received a whole byte of data, etc. The status register is named as USARTx_SR in STM32.

Register name	Offset	Description
USARTx_SR	0x0000	Status register
USARTx_DR	0x0004	Data register
USARTx_BRR	0x0008	Baud rate register
USARTx_CR1	0x000C	Control Register 1
USARTx_CR2	0x0010	Control Register 2
USARTx_CR3	0x0014	Control Register 3

Table 10-3: USART Registers

See Table 10-3. Next, we examine each of the registers and show how they are used in full-duplex serial data communication.

264

BRR Register and Baud rate

See Figure 10-12. The BRR register is used to set the baud rate.

(USARTx->BRR)

Figure 10-12: USART_BRR

To find the BRR value use the following formula:

$$BRR\ value = \frac{f_{pclkx}}{baud\ rate}$$

See Table 10-2 and Figures 8-20 and 8-21. USART1 uses the APB2 prescaler clock and the APB1 Prescaler provides the clock for USART2, 3, 4, and 5. APB1 and APB2 clocks are configured to 36MHz and 72MHz by Keil default startup code, respectively. See Example 10-4.

Example 10-4

In a system, the clock frequency for APB1 and APB2 are 36MHz and 72MHz, respectively. Find the BRR value for:

   (a) USART1 with baud rate of 57600
   (b) USART2 with baud rate of 9600

Solution:

   (a) USART1 is connected to APB2. So, BRR value = 72MHz/57600 = 1250
   (b) USART2 is connected to APB1. So, BRR value = 36MHz/9600 = 3750.

Using BRR you can make very different baud rates. But some of the baud rates are standard. Many devices (e.g. GSMs, Modems, and computers) use the standard baud rates. If you want to communicate with the devices you should use the standard baud rates. Some of the standard baud rates are: 1200, 2400, 4800, 9600, 19200, 38400, 57600, and 115200.

USART Control 1 (USARTx_CR1) register

The next important register in USART is the control register. We have 3 USART Control Registers. The most important among them are USART Control Register 1 (USARTx_CR1) and USART Control Register 2 (USARTx_CR2). They are 16-bit registers. The Control Register 1 is used to enable the serial port to send and receive data among other things. See Figure 10-13. Notice that the M bit (D12) of the CR1 register determines the framing of data by specifying the number of bits per character. In this textbook, we use the no parity option with a data size of 8 bits. The Control Register 2 can be used to set the number of stop bits. See Figure 10-14. In most applications, 1 stop bit is used (STOP=00). Since the default values of CR2 and CR3 are 0x0000, their default values are proper in most cases.

	15	14	13	12	11	10	9	8	7	6	5	4	3	2	1	0
USARTx_CR1:	Reserved		UE	M	WAKE	PCE	PS	PEIE	TXEIE	TCIE	RXNEIE	IDLEIE	TE	RE	RWU	SBK

(USARTx->CR1)

Figure 10-13: USARTx_CR1

Field	Bit	Description
UE	D13	0 = USART prescaler and outputs disabled. 1 = USART enabled
M	D12	Data format mode bit. We must use this to select 8-bit data frame size 0 = select 8-bit data frame and one start bit 1 = Select 9-bit data frame and one start bit
WAKE	D11	Wake-up condition bit. See the user manual 0 = Idle line wakeup 1 = Address mark wake-up
PCE	D10	Parity Control Enable bit. This will insert a parity bit right after the MSB bit. 0 = no parity bit 1 = parity bit
PS	D9	Parity select (used only if PE is one.) 0 = even parity bit 1 = odd parity bit
PEIE	D8	PE interrupt enable 0 = disabled 1 = A USART interrupt is generated whenever the PE flag of USART_SR is set.
TXEIE	D7	TXE interrupt enable 0 = disabled 1 = A USART interrupt is generated whenever the TXE flag of USART_SR is set.
TCIE	D6	Transmission complete interrupt enable 0 = disabled 1 = A USART interrupt is generated whenever the TC flag of USART_SR is set.
RXNEIE	D5	RXNE interrupt enable 0 = disabled 1 = A USART interrupt is generated whenever ORE or RXNE are set.
IDLEIE	D4	Idle interrupt enable 0 = disabled 1 = A USART interrupt is generated whenever the IDLE bit of USART_SR is set.
TE	D3	Transmitter enable (The bit enables the USART transmitter) 0 = transmitter is disabled 1 = transmitter is enabled
RE	D2	Receiver enable (The bit enables the USART receiver) 0 = disable the receiver 1 = enable the receiver
RWU	D1	Receiver wakeup 0 = Receiver in active mode 1 = Receiver in mute mode (See the user manual for more information)
SBK	D0	Send break (0 = do not send, 1 = send break character) See the user manual.
Note: In applications using the polling method we make the interrupt request bits all zeros.		

Table 10-4: USART_CR1

	15	14	13	12	11	10	9	8	7	6	5	4	3	2	1	0
USARTx_CR2:	Res.	LINEN	STOP		CLKEN	CPOL	CPHA	LBCL	Res.	LBDIE	LBDL	Res.	ADD			

(USARTx->CR2)

Field	Bit	Description
STOP	D12, D13	Number of stop bits 0 = 1 stop bit 1 = 0.5 stop bit 2 = 2 stop bits 3 = 1.5 stop bits

Note: The most important bits in this register are the stop bits. In most cases the number of stop bits is set to 1 stop bit. In asynchronous mode, we set all the other bits to zeros. So, the USARTx_CR2 is set to 0x00 in most cases. For more information about the other bits see the user manual.

Figure 10-14: USART_CR2

Example 10-5

(a)   What are the values of USART_CR1 and USART_CR2 needed to configure USART2 for 8 data bits (character size), no parity, and 1 stop bit? Enable both receive and transmit.

(b)   Write a program to set the values of USART_CR1 and USART_CR2 for this configuration

Solution:

(a)   UE, RE, and TE have to be 1 to enable receive and transmit. M should be 0 for 8-bit data, STOP should be 00 for one stop bit, and PCE should be 0 for no parity.

	UE	M	WAKE	PCE	PS	PEIE	TXEIE	TCIE	RXNEIE	IDLEIE	TE	RE	RWU	SBK
CR1:	1	0	0	0	0	0	0	0	0	0	1	1	0	0

(b)

Assembly:

```
LDR   R1, =0x40004400    ;address for USART2
LDR   R2, =0x200C        ; R2 = 0x200C
STS   R2,[R1,#0x0C]      ;CR1 = R2
MOV   R2,#0
STS   R2,[R1,#0x10]      ;CR2 = 0
```

C:

```
USART2->CR1 = 0x200C;
USART2->CR2 = 0;
```

USART Data Register

To transmit a byte of data we must place it in USART Data register. It must be noted that a write to this register initiates a transmission from the USART. In the same way, the received byte is placed in this register and must be retrieved by reading it before it is lost. Notice this is an 8-bit register.

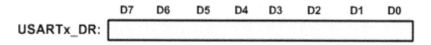

Figure 10-15: USART Data (USART_DR) register

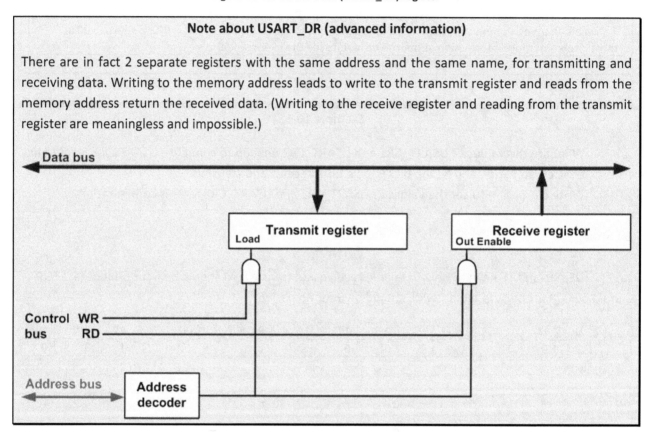

Status Register

See Figure 10-16. USARTn_SR is a 16-bit register. It is used to monitor the arrival of data among other things. Table 10-5 describes various bits of the USARTn_SR. Several of the SR register bits are widely used by the USART. We monitor (poll) the TC flag bit to make sure that all the bits of the last byte are transmitted. By the same logic, we monitor (poll) the RXNE flag to see if a byte of data is received. The transmitter is double buffered. That means there is a data register in addition to the shift register that shifts the bits out. While the shift register is shifting the last byte out, the program may write another byte of data to the Data Register to wait for the shift register to be ready. The transfer of data between the data register and the shift register is automatic and the program does not have to worry about it. The TXE flag indicates that the Data Register is empty and ready to accept another byte. The TC flag mentioned above actually indicates whether the shift register is empty or not. When the shift register is empty, the

Data Register must be empty too and the USART has no data to transmit. In Chapter 12, we will see how these flags are used with interrupts instead of polling.

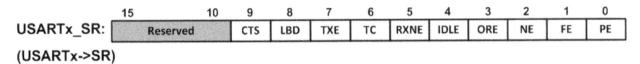

(USARTx->SR)

Figure 10-16: USART_SR

Field	Bit	Description
CTS	D9	When the CTS pin is changed the flag is set by hardware. For more information see the user manual.
LBD	D8	LIB break detection flag (See the user manual)
TXE	D7	Transmit data register Empty 0 = Data is not transferred to the shift register yet. 1 = The Data Register is empty and ready for the next byte.
TC	D6	Transmit Complete flag (The flag is set by hardware. To clear the flag, you can write a zero to it. If you read the USART_SR and then write to the USART_DR register, the flag will be automatically cleared.) 0 = Transmission is in progress (shift register is occupied) 1 = Transmission complete (both shift register and Data Register are empty)
RXNE	D5	Receive Data Register not empty flag. This indicates a byte has been received and is sitting in USART Data Register and ready to be picked up. 0 = No data is available in USART Data Register. 1 = Data is available in USART Data Register and ready to be picked up.
IDLE	D4	Idle line detected. See the user manual.
ORE	D3	Overrun error (The flag is set by hardware, if a new byte of data is received while RXNE is set.) 0 = No overrun 1 = Overrun error
NF	D2	Noise error Flag bit (The bit is set by hardware, if noise is detected on the received frame.) 0 = No noise 1 = Noise error
FE	D1	Framing Error bit (The bit is set by hardware, if de-synchronization, noise or a break character is detected.) 0 = No framing error 1 = Framing error
PF	D0	Parity flag error bit (The bit is set by hardware, if the received data has a parity error.) 0 = No parity error 1 = Parity error

Table 10-5: UART Status Register (UARTx_SR)

The TXE

To transmit a byte of data serially via the TX pin, we must write it into the USART Data Register (USARTx_DR). The transmit shift register is an internal register whose job is to get the data from the USART

Data Register (USARTx_DR), frame it with the start and stop bits, and send it out one bit at a time via the USARTx_TX pin. Notice that the transmit shift register is a parallel-in-serial-out shifter and is not accessible to the programmer. We can only write to the USART Data Register. Whenever the shifter is empty, it gets its new data from the USART Data Register and clears the USART Data Register immediately, so it does not send out the same data twice. When the shifter fetches the data from the USART Data Register, it sets the TXE flag to indicate it is empty and the USART Data Register is ready for the next character. We must check the TXE flag before we write another byte to the USART Data Register.

The importance of the TC

The TC flag indicates that both the Data Register and the shift register are empty and there is no data left for the transmitter to send. This bit is used when the program needs to know that all the data is sent before starting the next task. When TXE flag is set, the Data Register is empty but the last byte of data may still be in the shift register. If the program is to shut down the transmitter, the last byte of data will be lost. Checking TC ensures that all the data written to the USART is transmitted already.

The importance of the RXNE

The internal serial-in-parallel-out receive register receives data via the USARTx_RX pin. It gets rid of the start and stop bits and writes the received byte to the USART Data Register and makes RXNE high. We must check the RXNE flag to see if we need to pick up the received byte.

The GPIO pins used for UART TX and RX

In addition to the USART registers setup, we must also configure the I/O pins used for USART. The USART signals are digital. So, the RX pins need to be set as input pins and the TX pins need to be configured as alternate function output push/pull. See Figure 8-3. Table 10-4 lists the TX and RX pins for each USART in STM32F103.

	Bus	GPIO pin
USART1	TX	PA9
	RX	PA10
USART2	TX	PA2
	RX	PA3
USART3	TX	PB10
	RX	PB11

Table 10-6: TX and RX pins in STM32F103

Example 10-6

Configure the GPIO pins for using USART1.

Solution:

USART1_TX is available on PA9. It needs to be configured as alternate function output push/pull. According to Figure 8-3, the CNF9 bits need to be set to 10 and MODE9 bits should be 11 (or any values other than 00). So, CNF9:MODE9 should be set to 1011 (0x9 in hex). USART1_RX needs to be configured as digital input. It is a good practice to enable the pull-up resistor. The internal pull-up makes the line high if the wire between the microcontroller and the other device is cut.

```
GPIOA->ODR |= (1<<10);  //pull-up PA10
GPIOA->CRH = 0x444448B4; // RX1=input with pull-up, TX1=alt. func output
```

Steps for transmitting data

Here are the steps to configure a USART and transmit a byte of data:

1) Provide clock to USART and GPIO by writing 1 to the proper bits of APB1ENR and APB2ENR registers.
2) Configure the control register value for 1 stop bit, no parity, and 8-bit data size and enable the transmitter and receiver by writing 0x200C for the UARTn_CR1 register.
3) Set the baud rate for USART using the BRR register.
4) Select the alternate function output for UARTn_TX pin and configure UARTn_RX as input.
5) Write a byte to UARTn Data Register to be transmitted.
6) Monitor the TC bit of the Status Register (UARTn_SR) and wait for the transmission to complete. See Examples 10-7 and 10-8.

Example 10-7

Write a program that sends "Hi\n\r" through USART1 every second with baud rate of 9600 and 1 stop bit. Assume APB2 clock = 72MHz.

Solution:

```
#include "stm32f10x.h"

void delay_ms(uint16_t t);
void usart1_sendByte(unsigned char c);

int main()
{
   RCC->APB2ENR |= 0xFC | (1<<14); //enable GPIO clocks

   //USART1 init.
   GPIOA->ODR |= (1<<10);  //pull-up PA10
   GPIOA->CRH = 0x444448B4; // RX1=input with pull-up, TX1=alt. func output
   USART1->CR1 = 0x200C;
   USART1->BRR = 7500;        // 72MHz/9600bps = 7500

   while(1)
   {
      usart1_sendByte('H');
      usart1_sendByte('i');
      usart1_sendByte('\n'); //go to new line
      usart1_sendByte('\r'); //carrier return
```

```
      delay_ms(1000);        //wait 1 second
   }
}

void usart1_sendByte(unsigned char c)
{
   USART1->DR = c;

   while((USART1->SR&(1<<6)) == 0);  //wait until the TC flag is set
   USART1->SR &= ~(1<<6); //clear TC flag
}

void delay_ms(uint16_t t)
{
   volatile unsigned long l = 0;
   for(uint16_t i = 0; i < t; i++)
      for(l = 0; l < 6000; l++)
        {
        }
}
```

Example 10-8

Write a program that reports the state of PB5 through USART2 with baud rate of 9600 and 1 stop bit, continuously. If it is high, send "H"; otherwise, send "L". Assume APB1 clock = 36MHz.

Solution:

```
#include "stm32f10x.h"

void delay_ms(uint16_t t);
void usart2_sendByte(unsigned char c);

int main()
{
   RCC->APB1ENR |= (1<<17); //enable usart2 clock
   RCC->APB2ENR |= 0xFC; //enable GPIO clocks

   GPIOB->CRL = 0x44844444;  // enable pull-up resistor
   GPIOB->ODR |= (1<<5);     // pull-up PB5
   //USART2 init.
   GPIOA->CRL = 0x44448B44; //RX2=in, TX2=alt. func. Output
   GPIOA->ODR |= (1<<3); // pull-up PA3(RX2)
   USART2->CR1 = 0x200C;
   USART2->BRR = 3750; //36MHz/9600 = 3750

   while(1)
   {
      if((GPIOB->IDR&(1<<5)) == 0) //is PB5 low?
         usart2_sendByte('L');
      else
         usart2_sendByte('H');
```

```
   }
}

void usart2_sendByte(unsigned char c)
{
   USART2->DR = c;

   while((USART2->SR&(1<<6)) == 0);       //wait until the TC flag is set
   USART2->SR &= ~(1<<6); //clear TC flag
}
```

Steps for receiving data

Here are the steps to receive data through USART:

1) Provide clock to USART and GPIO by writing 1 to the proper bits of APB1ENR and APB2ENR registers.
2) Configure the control register value for 1 stop bit, no parity, and 8-bit data size and enable the transmitter and receiver by writing 0x200C for the UARTn_CR1 register.
3) Set the baud rate for USART using the BRR register.
4) Select the alternate function output for UARTn_TX pin and configure UARTn_RX as input.
5) Monitor the RXNE bit of the Status Register (UARTn_SR) until the bit is high.
6) Read the contents of USARTn_DR. to receive the next byte go to step 5. See Examples 10-9 through 10-11.

Example 10-9

Write a program to receive a byte of data through USART2 and send it back to the PC. Set the baud rate at 9600, 8-bit data, and 1 stop bit.

Solution:

```
#include "stm32f10x.h"

void usart2_sendByte(unsigned char c);
uint8_t usart2_recByte(void);

int main()
{
   RCC->APB1ENR |= (1<<17); //enable usart2 clock
   RCC->APB2ENR |= 0xFC; //enable GPIO clocks

   //USART2 init.
   GPIOA->CRL = 0x44448B44; //RX2=in, TX2=alt. func. Output
   GPIOA->ODR |= (1<<3); // pull-up PA3(RX2)
   USART2->CR1 = 0x200C;
   USART2->BRR = 3750; //36MHz/9600 = 3750

   while(1)
   {
```

273

```
      uint8_t receivedByte = usart2_recByte();
      usart2_sendByte(receivedByte);
   }
}


uint8_t usart2_recByte()
{
   while((USART2->SR&(1<<5)) == 0);       //wait until the RXNE flag is set

   return USART2->DR;
}

void usart2_sendByte(unsigned char c)
{
   USART2->DR = c;

   while((USART2->SR&(1<<6)) == 0);       //wait until the TC flag is set
   USART2->SR &= ~(1<<6); //clear TC flag
}
```

Example 10-10

Write a program to receive data serially through USART1. If the received data is "H", make PC13 high; If the received data is "L", make PC13 low. Set the baud rate at 9600, 8-bit data, and 1 stop bit.

Solution:

```
#include "stm32f10x.h"

uint8_t usart1_recByte(void);

int main()
{
   RCC->APB2ENR |= 0xFC | (1<<14); /* enable GPIO and usart1 clocks */

   /* USART1 init. */
   GPIOA->ODR |= (1<<10);  /* pull-up PA10 */
   GPIOA->CRH = 0x444448B4; /* RX1=input with pull-up, TX1=alt. func output */
   USART1->CR1 = 0x200C;
   USART1->BRR = 7500;       /* 72MHz/9600bps = 7500 */

   GPIOC->CRH = 0x44344444;

   while(1)
   {
      uint8_t c = usart1_recByte();
      switch(c)
      {
         case 'H':
         case 'h':
            GPIOC->ODR |= (1<<13);  /* PC13 = high */
            break;
```

```
            case 'L':
            case 'l':
                GPIOC->ODR &= ~(1<<13);   /* PC13 = low */
                break;
        }
    }
}

uint8_t usart1_recByte()
{
    while((USART1->SR&(1<<5)) == 0);        /* wait until the RXNE flag is set */

    return USART1->DR;
}
```

Example 10-11

Write a program to receive a character from USART3. If it is a small letter ('a' to 'z') change it to capital letters and send it back.

Solution:

```
#include "stm32f10x.h"

void usart3_sendByte(unsigned char c);
uint8_t usart3_recByte(void);

int main()
{
    RCC->APB1ENR |= (1<<18); //enable usart3 clock
    RCC->APB2ENR |= 0xFC; //enable GPIO clocks

    //USART3 init.
    GPIOB->CRH = 0x44448B44; //RX2=in, TX2=alt. func. Output
    GPIOB->ODR |= (1<<11); // pull-up PB11(RX3)
    USART3->CR1 = 0x200C;
    USART3->BRR = 3750; //36MHz/9600 = 3750

    while(1)
    {
        uint8_t c = usart3_recByte();

        if((c >= 'a')&&(c <= 'z')) //is a small letter?
            c = c - 32; // to capital letters

        usart3_sendByte(c);
    }
}

uint8_t usart3_recByte()
{
    while((USART3->SR&(1<<5)) == 0);        //wait until the RXNE flag is set
```

```
    return USART3->DR;
}

void usart3_sendByte(unsigned char c)
{
    USART3->DR = c;

    while((USART3->SR&(1<<6)) == 0);        //wait until the TC flag is set
    USART3->SR &= ~(1<<6); //clear TC flag
}
```

Baud rate error calculation

In calculating the baud rate, we have used the integer number for the BRR register values. By dropping the decimal fraction portion of the calculated values, we run the risk of introducing error into the baud rate. One way to calculate this error is to use the following formula:

$$Error = \frac{calculated\ value\ for\ BRR - integer\ part}{integer\ part} \times 100$$

See Example 10-12.

Example 10-12

Assume the clock source of 8 MHz is fed to USART Baud rate. Calculate the baud rate error for each of the following baud rates:

    a) 115200   b) 57600   c) 9600

Solution:

    a) BRR = 8MHz/115200 = 69.444 ≈69
        Error = (69.444 − 69) × 100 / 69 = 0.444× 100/69 = 0.64%
    b) BRR = 8M/57600 = 138.888 ≈ 139
        Error = (138.888 − 139) × 100/139 = 0.08%
    c) BRR = 8M/9600 = 833.333 ≈ 833
        Error = (833.333 − 833) × 100/833 = 0.0399%

Idle and break characters

In the user manual we see the mention of some terminology such as idle and break. The idle is when the USART output is high with no start bit. That is ten ones when the data frame size is 8-bit. The break character is when the USART sends out a low signal much longer than one frame, that is ten zeros when the data size is 8-bit. The break character is used to force a framing error at the receiver for the testing purpose and writing diagnostic software.

Review Questions
    1. Which register is used to set the USART baud rate?

2. Which register is used to enable the USART?
3. How do we know if the transmit buffer is not full before we load in another byte?
4. How do we know if a new byte has been received?

Problems

Section 10-1

1. For long distances, which is more expensive, parallel or serial data transfer?
2. True or false. 0- and 5-V digital pulses can be transferred on the telephone without being converted (modulated).
3. Show the framing of the letter ASCII 'Z' (0101 1010), no parity, 1 stop bit.
4. Calculate the overhead percentage if the data size is 7, 1 stop bit, and no parity bit.
5. True or false. The RS232 voltage specification is TTL compatible.
6. What is the function of the MAX232 chip?
7. State the absolute minimum number of signals needed to transfer data between two PCs connected serially. What are those signals?
8. If two PCs are connected through the RS232 without a modem, both are configured as a _____ (DTE, DCE) -to- _____ (DTE, DCE) connection.
9. Calculate the total number of bits transferred if 200 pages of ASCII data are sent using asynchronous serial data transfer. Assume a data size of 8 bits, 1 stop bit, and no parity. Assume each page has 80 x 25 of text characters.
10. In Problem 10, how long will the data transfer take if the baud rate is 9600?

Section 10-2

11. When is the RXNE flag bit raised?
12. When is the TC flag bit raised?
13. Write a program to transfer serially the letter 'Z' continuously at 9600 baud rate. Assume peripheral clock = 36 MHz.
14. Write a program to transmit serially the message "The earth is but one country and mankind its citizens" continuously at 57,600 baud rate.
15. Assume the clock source of 8 MHz is fed to USART Baud rate. Calculate the baud rate error for each of the following baud rates:
   a) 19200   b) 38400

Answer to Review Questions

Section 10-1

1. False, more expensive
2. Serial
3. Synchronous
4. False; it is simplex.
5. True

6. Asynchronous
7. With 100 0100 binary we have 1 as the odd-parity bit. The bits as transmitted in the sequence are:

1	2	3	4	5	6	7	8	9	10	11
0 (start bit)	0	0	1	0	0	0	1	1 (parity)	1 (first stop bit)	1 (second stop bit)

8. 4 bits
9. $400 \times 11 = 4400$ bits (total bits transmitted). $4400/1200 = 3.667$ seconds, $4/7 = 58\%$.
10. True

Section 10-2

1. USARTn_BRR
2. USARTn_CR1
3. The TXE flag from the UARTn_SR register goes high.
4. The RXNE flag from the UARTn_SR register goes high.

Chapter 11: STM32 ARM Timer Programming

In Section 11-0, the counter and timer concepts are reviewed. Section 11-1 covers the System Tick Timer which is available in all ARM Cortex microcontrollers. In Section 11-2, delays are made using 16-bit STM32 timers. Section 11-3 shows Output Compare mode. In Section 11-4, input edge-time mode is discussed and the pulse width and frequency measuring are covered. The event counter feature is studied in Section 11-5.

Section 11.0: Introduction to counters and timers

In the digital design course, you connected many flip flops (FFs) together to create up counter/down counter. For example, connecting 3 FFs together we can count up to 7 (000-111 in binary). This is called *3-bit counter*. The same way, to create a 4-bit counter (counting up to 15, or 0000-1111 in binary) we need 4 FFs. For 16-bit counter, we need 16 FFs and it counts up to $2^{16} - 1$. Figure 11-1 shows the T flip flop connection and pulse outputs for all three flip flops.

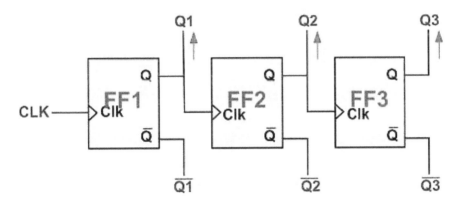

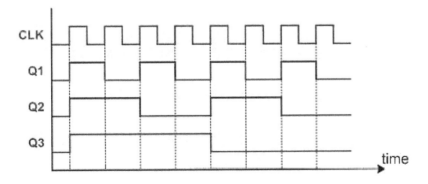

Figure 11-1: A 3-bit Counter

Regarding Figure 11-1, notice the following points:

1) The Q outputs give the down counter.
2) The $\bar{Q}$ (Q not) outputs give us up counter.
3) The frequency on Q3 is $\frac{1}{8}$ of the Clock fed to FF1.
4) We can use the circuit in Figure 11-1 to divide clock frequency.
5) We can use the circuit in Figure 11-1 to count the number of pulses fed to CLK pin of FF1.

An up counter begins counting from 0 and its value increases on each clock until it reaches its maximum value. Then, it overflows and rolls over to zero in the next clock. The following figure shows the stages which an 8-bit counter goes through.

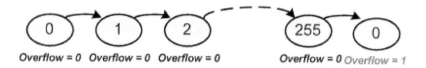

Figure 11-2: an 8-bit Up-Counter Stages

A down counter begins counting from its maximum value and decreases on each clock until it reaches to 0. Then, it underflows and rolls over to its maximum value in the next clock. The following figure shows the stages which an 8-bit down counter goes through.

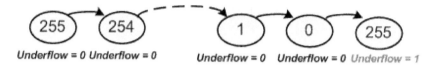

Figure 11-3: an 8-bit Down-Counter Stages

Counter Usages

Counters have different usages. Some of them are:

1. Counting events
2. Making delays (Using Counter as a Timer)
3. Measuring the time between 2 events

1. Counting events

You might need to count the number of cars going through a street or the number of spaghetti packages which produced in a factory. To do so, you can connect the output of a sensor to a counter, as shown in the following figure.

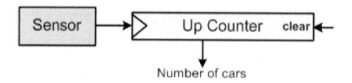

Figure 11-4: Counting Events Using a Counter

2. Making delays (Using Counter as a timer)

While controlling devices, it is a common practice to start or terminate a task when a desired amount of time elapsed. For example, a washing machine or an oven do each task for a determined amount of time. To do timing, we can connect a clock generator to a counter, and wait until a desired amount of time elapses. For example, in the following picture, the clock generator makes a 1 Hz signal and the counter increasing every second. The counter reaches to 60 after 60 seconds.

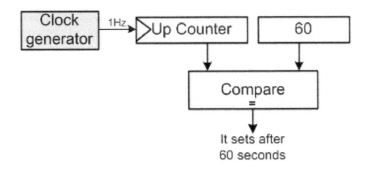

Figure 11-5: Using Counter as a Timer

3. Measuring the time between 2 events

You might need to measure the time between 2 events. For example, the amount of time it takes a marathon runner to go from the start to the finish point. In such cases we can use a circuit similar to the following:

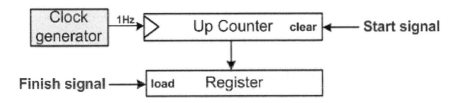

Figure 11-6: Capturing

The counter is cleared at the start. Then, it increases on each clock pulse. The value of the counter is loaded into another register when the runner passes the finish line.

Counters and Timers in microcontrollers

Nowadays, all the microcontrollers come with on-chip Timer/Counter. If the clock to the Timer comes from internal source such as PLL, XTAL, and RC, then it is called a *Timer*. If the clock source comes from external source, such as pulses fed to the CPU pin, then it is called a *Counter*. By Counter it is meant event-counter since it counts the event happening outside the CPU. In many microcontrollers, the Timers can be used as Timer or Counter.

Review Questions

1. With 5 FFs we can get maximum of _____ count.
2. With 5 FFs we can divide the frequency by maximum of _____.
3. When pulses are fed to a timer from the outside it is called _____.
4. When clocks pulses are fed to a timer from inside it is called _____.
5. If we need to divide a frequency by 500, we need _____ flip flops.

Section 11.1: System Tick Timer

Every ARM Cortex-M comes with a System tick timer. System tick timer allows the system to initiate an action on a periodic basis. This action is performed internally at a fixed rate without external signal. For example, in a given application we can use SysTick to read a sensor every 200 msec. SysTick is used widely by operating systems so that the system software may interrupt the application software periodically (often 10 ms interval) to monitor and control the system operations. The SysTick is a 24-bit down counter driven by the system clock. It counts down from an initial value to 0. When it reaches 0, in the next clock, it underflows and it raises a flag called COUNT and reloads the initial value and starts all over. We can set the initial value to a value between 0x000000 to 0xFFFFFF. See the following figure.

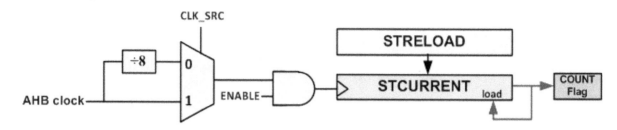

Figure 11-7: System Tick Timer Internal Structure

The down counter is named as STCURRENT (SysTick->VAL). The counter can receive clock from 2 different sources: the AHB clock (the clock which the CPU and all the peripherals work with it) or the AHB clock divided by 8. The clock source is chosen using the CLK_SRC bit of STCTRL (SysTick->CTRL) register. The clock is ANDed with the ENABLE bit of STCTRL register. So, it counts down when the ENABLE bit is set. The STCTRL register is shown in Figure 11-8.

SysTick Registers

Next, we will describe the SysTick registers. There are three registers in the SysTick module: SysTick Control and Status register, SysTick Reload Value register, and SysTick Current Value register.

The STCTRL (SysTick Control and Status) register is located at 0xE000E010. We use it to start the SysTick counter among other things.

	D31	.....	D17	D16	D15	...	D3	D2	D1	D0	
STCTRL:	Reserved			COUNT FLAG	Reserved			CLK SOURCE	TICKINT	ENABLE	0x010

SysTick->CTRL

Name	bit	Description
ENABLE	0	0: The counter is disabled, 1: enables SysTick to begin counting down
TICKINT	1	Interrupt Enable 0: Interrupt generation is disabled. 1: when SysTick counts to 0 an interrupt is generated
CLKSOURCE	2	Clock Source 0: AHB clock divided by 8 1: AHB clock
COUNTFLAG	16	Count flag 0: The SysTick has not counted down to zero since the last time this bit was read 1: The SysTick has counted down to zero Note: This flag is cleared by reading the STRCTRL or writing to STCURRENT.

Figure 11-8: STCTRL (System Tick Control)

ENABLE (D0): enables or disables the counter. When the *ENABLE* bit is set the counter initializes the STCURRENT with the value of the *STRELOAD* register and it counts down until it reaches to zero. Then, in the next clock, it underflows which sets the *COUNT* Flag to high and the counter reloads the STCURRENT with the value of the *STRELOAD* register and then the process is repeated. See the following figure.

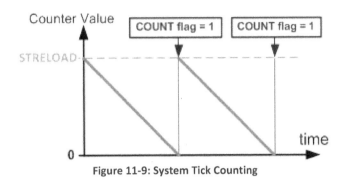

Figure 11-9: System Tick Counting

TICKINT (Tick Interrupt Enable, D1): If INTEN=1, an interrupt occurs when the COUNT flag is set. See Chapter 12.

CLKSOURCE (Clock Source D2): We have the choice of clock coming from AHB clock or AHB clock/4. If CLKSOURCE=0 then the clock comes from AHB clock/4. If CLKSOURCE=1, then the AHB clock provides the clock source to SysTick down counter.

COUNTFLAG (D16): Counter counts down from the initial value and when it reaches 0, in the next clock it underflows and the COUNT flag is set high. See Figure 11-9. The flag remains high until it is cleared by software. The flag can be cleared by reading the STCTRL register or writing to the STCURRENT register.

SysTick Reload Value Register (STRELOAD), offset 0x014

The STRELOAD (SysTick Reload Value) register is located at 0xE000E014. This is used to program the start value of SysTick down counter, the STCURRENT register. The STRELOAD should contain the value N − 1 for the COUNT to fire every N clock cycles because the counter counts down to 0. For example, if we need 1000 clocks of interval, then we make STRELOAD = 999. Although this is a 32-bit register, only the lower 24 bits are used. That means the highest value that can be loaded into this register is 0xFFFFFF or 16,777,216 decimal. See Figures 11-7 and 11-10.

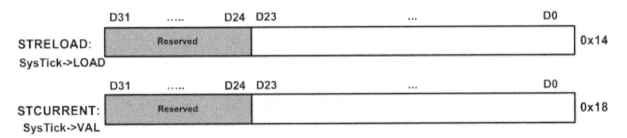

Figure 11-10: STRELOAD vs. STCURRENT

See Examples 11-1 through 11-6.

Example 11-1

In an ARM microcontroller system clock = 8 MHz. Calculate the delay which is made by the following function.

```
void delay() {
  SysTick->LOAD = 9;
  SysTick->CTRL = 5; /*Enable the timer and choose system clock as the clock source */

  while((SysTick->CTRL &0x10000) == 0) /*wait until the Count flag is set */
  { }
  SysTick->CTRL = 0; /*Stop the timer (Enable = 0) */
}
```

Solution:

The timer is initialized with 9. So, it goes through the following 10 stages:

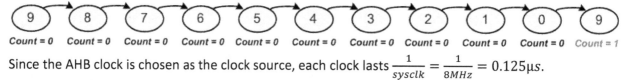

Count = 0   Count = 0   Count = 0   Count = 0   Count = 0   Count = 0   Count = 0   Count = 0   Count = 0   Count = 0   Count = 1

Since the AHB clock is chosen as the clock source, each clock lasts $\dfrac{1}{sysclk} = \dfrac{1}{8MHz} = 0.125\mu s$.

So, the program makes a delay of $10 \times 0.125\mu s = 1.25\mu s = 1250ns$.

Note: *the function call and the instructions execution take a few clock cycles as well. If you want to calculate the exact amount of delay, you should include this overhead, as well. But, in this book we do not consider it since most of the time it is negligible.*

Example 11-2

In an ARM microcontroller a clock with frequency of clk is fed to the sysTick timer. Calculate the delay which is made by the timer if the STRELOAD register is loaded with N.

Solution:

The timer is initialized with N. So, it goes through N+1 stages.
Since the system clock is chosen as the clock source, each clock lasts 1 / clk
So, the program makes a delay of (N + 1) × (1 / clk) = (N + 1) / clk.

Example 11-3

Using the System Tick timer, write a function that makes a delay of 1 ms. Assume AHB clock = 72 MHz.

Solution:

From the equation derived in Example 11-2
delay = (N + 1) / clk

$(N + 1)$ = delay × clk = 0.001 sec × 72 MHz = 72,000 ==> N = 72,000 − 1 = 71999

```
void delay1ms(void) {
   SysTick->LOAD = 71999;
   SysTick->CTRL = 0x5;    /* Enable the timer and choose sysclk as the clock source */

   while((SysTick->CTRL & 0x10000) == 0) /* wait until the COUNT flag is set */
   { }
   SysTick->CTRL = 0; /* Stop the timer (Enable = 0) */
}
```

Example 11-4

The AHB clock is 72MHz by default. What is the largest time delay possible if (a) AHB is selected as the clock source (b) AHB clock/8 is used as the clock source?

Solution:

a) The STRELOAD register has 24 bits and its biggest values is 2^{24}−1.
   delay = 2^{24} / 72M = 16,777,216 / 72,000,000 = 0.23301 Second.

b) Since AHB clock/8 is selected the clock source is 72MHz/8 = 9MHz.
   delay = 2^{24} / 9M = 16,777,216 / 9,000,000 = 1.864 Second.

Example 11-5

Using SysTick, toggle PC13 at 5Hz. The AHB clock is 72MHz by default.

Solution:

To make a wave with frequency of 5Hz, the pin is needed to toggle every 0.1 second.

delay = $(N + 1)$ / clk ➔ N + 1 = 0.1 x 72M ➔ N + 1 = 7,200,000 − 1 = 7,199,999.

```
#include <stm32f10x.h>

void delay(void);

int main()
{
   RCC->APB2ENR = 0xFC;     /* enable GPIO clocks */
   GPIOC->CRH = 0x44344444;       /* PC13 as output */

   while(1)
   {
      GPIOC->ODR ^= (1<<13); /* toggle PC13 */
```

```
      delay();
    }
}

void delay()
{
    SysTick->LOAD = 7199999;                    /* STRELOAD = 7,199,999 */
    SysTick->CTRL = 0x05;            /* Clock source = AHB clock, Enable = 1 */
    while((SysTick->CTRL&(1<<16)) == 0); /* wait until the COUNT flag is set */
    SysTick->CTRL = 0x00;               /* stop the SysTick timer */
}
```

Example 11-6

Using SysTick, make a 1 second delay. Toggle PC13 every second. The AHB clock is 72MHz by default.

Solution:

To make a delay of 1 second, it is needed to use AHB clock/8 as the clock source. In the case, the timer counts with frequency of 72MHz/8 = 9MHz.

delay = (N + 1) / clk ➜ N + 1 = delay x clk = 9M ➜ N = 9,000,000 − 1 = 8,999,999.

```
#include <stm32f10x.h>

void delay(void);

int main()
{
    RCC->APB2ENR = 0xFC;     /* enable GPIO clocks */
    GPIOC->CRH = 0x44344444;       /* PC13 as output */

    while(1)
    {
        GPIOC->ODR ^= (1<<13); /* toggle PC13 */
        delay();
    }
}

void delay()
{
    SysTick->LOAD = 9000000-1;               /* STRELOAD = 9,000,000-1 */
    SysTick->CTRL = 0x01;                /* Clock = AHB clock/8, Enable = 1 */
    while((SysTick->CTRL&(1<<16)) == 0); /* wait until the COUNT flag is set */
    SysTick->CTRL = 0x00;                /* stop the SysTick timer */
}
```

The System Tick Timer has a very simple structure and is the same across all the ARM Cortex chips regardless of who makes them. In contrast, STM32 has its own timers which are covered in the next section.

Review Questions
1. True or false. The highest number we can place in RELOAD register is _____.
2. Assume CPU frequency of 16MHz. Find the value for RELOAD register if we want 5 ms elapsed time.
3. The SysTick is _____-bit wide.
4. Which bit of STCTRL is used to enable the SysTick.
5. The SysTick is _____ (down or up) counter.

Section 11.2: Delay Generation with STM32 Timers

In this section we examine the timers for STM32F10x chip. We use the timers to create time delay. STM32F10x chips have three kinds of timers: basic timers, general-purpose timers, and advanced-control timers. The timers have the same basis. But there are some features in advanced-control and general-purpose which do not exist in basic timers. In STM32F103C8, Timers 2, 3, and 4 are general-purpose timers and Timer1 is advanced-control. In this book we concentrate on the general-purpose timers. For more information about the more advanced features of Timer1 read the reference manual. Figure 11-11 shows the simplified view of the timers.

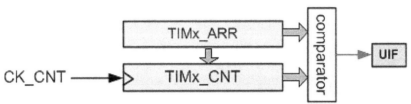

Figure 11-11: A Simplified View of TIM timers

CNT (Counter) Register

Each timer has a CNT (Counter) register. CNT is a 16-bit up/down counter. You can load a value into the CNT register or read its value. It is called TIMx_CNT in which x is the timer number. That means we have TIM1_CNT for TIM1 (timer 1) and TIM2_CNT for TIM2 (timer 2). When the clock is fed to TIMx_CNT, it keeps counting up (or counting down). Upon Reset TIMx_CNT=0000. See Figure 11-12.

Figure 11-12: TIMx_CNT

ARR (Auto Reload Register)

Each timer has an ARR register. It is a 16-bit register whose value is continuously compared with the TIMx_CNT register. See Figures 11-13.

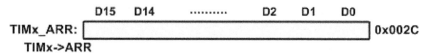

Figure 11-13: TIMx_ARR

Note: In STM32 microcontrollers, all the timer registers begin with TIMx. So, for simplicity, just consider the letters which come after TIM. For example, consider TIMx_CNT as CNT (Counter). That means, TIM1_CNT is Counter register for TIM1 and TIM2_ARR is the auto reload register for TIM2.

TIMx_CR1 (Control Register)

Each timer has its own CR1 register. The timer can count in 3 ways: counting up, counting down, and counting both up and down like a Yoyo! Using the CR1 register you choose the desired counting mode. See Figure 11-14. CMS and DIR bits are used to select the counting mode. The CEN (Counter Enable) bit of the TIMx_CR1 is used to enable/disable the counter. The counter counts when the CEN is 1 and it is stopped when the CEN is 0. See Example 11-7.

	15	14	...	10	9	8	7	6	5	4	3	2	1	0
TIMx_CR1:		Reserved			CKD		ARPE	CMS		DIR	OPM	URS	UDIS	CEN

(TIMx->CR1)

Name	bit	Description
CKD	9-8	Clock Division used by dead-time generators (for more info. see the manual)

ARPE	7	Auto Reload Preload Enable
		  0: ARR register is not buffered.
		  1: ARR is buffered.

CMS	6-5	Center-aligned Mode Selection When CMS=00, the timer counts up or down depending on the value of DIR bit. Otherwise, (CMS = 01, 10, or 11) the timer counts up and down, alternatively. | CMS	DIR	Counting mode	Counting event (interrupt flag set) | |---|---|---|---| | 00 | 0 | Counting up | When the counter reaches ARR | | 00 | 1 | Counting down | When the counter reaches 0 | | 01 | X | Count up and down | When the counter reaches 0 and ARR | | 10 | X | Count up and down | When the counter reaches 0 and ARR | | 11 | X | Count up and down | When the counter reaches 0 and ARR |

DIR	4	If the CMS bits are set to 00, the DIR bit chooses the direction of counting:
		  0: the CNT counter counts up.
		  1: the CNT counter counts down.

OPM	3	One Pulse Mode
		  0: the counter counts continuously
		  1: the counter stops at the next update event.

URS	2	Update request source
UDIS	1	Update Disable: We can mask (disable) generating any update events.
CEN	0	Counter Enable (0: The counter is disabled, 1: enable the counter to begin counting)

Figure 11-14: TIMx_CR1 Register

Example 11-7

Find the TIMx_CR1 value to: (a) count up continuously (b) count down continuously (c) stop counting.

Solution:

(a)

TIMx_CR1:	CKD	ARPE	CMS	DIR	OPM	URS	UDIS	CEN
	0	0	00	0	0	0	0	1

TIMx->CR1 = 0x01;

(b)

TIMx_CR1:	CKD	ARPE	CMS	DIR	OPM	URS	UDIS	CEN
	0	0	00	1	0	0	0	1

TIMx->CR1 = 0x11;

(c)

TIMx_CR1:	CKD	ARPE	CMS	DIR	OPM	URS	UDIS	CEN
	0	0	00	0	0	0	0	0

TIMx->CR1 = 0;

TIMx_SR (Status Register)

Figure 11-15 shows the TIMx_SR register. The bit 0 of TIMx_SR is used as UIF (Update Interrupt Flag). In counting up mode, when the value of CNT (counter) reaches ARR, UIF flag is set. The flag remains high until it is cleared by software. In the counting down mode, the flag is set when the counter reaches zero.

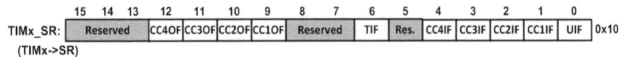

Figure 11-15: TIMx_SR (Status Register)

Counting Up

When TIMx_CNT counter register is counting up, it is compared with the contents of TIMx_ARR register. Whenever the contents of TIMx_CNT counter and TIMx_ARR register are equal, the UIF flag (Update Interrupt flag) goes up indicating there is a match and TIMx_CNT rolls over to zero. See Figure 10-16. A smaller value of the TIMx_ARR register leads the timer times out faster and the UIF flag sets sooner. In other words, delays can be made by setting the TIMx_ARR register and monitoring the UIF flag.

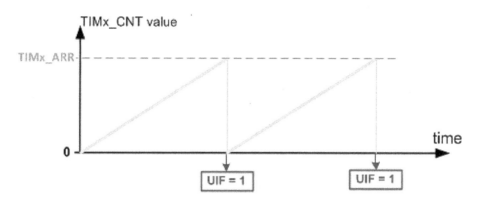

Figure 11-16: Counting up

Example 11-8

Assume TIM2_ARR = 5 and TIM2_CNT is counting up. (a) Explain when the UIF flag is raised. (b) How many clocks does it take until the UIF flag rises?

Solution:

As the timer counts up, it goes through the states of 0, 1, 2, 3, 4, and 5. Now since TIM2_CNT and TIM2_ARR match, it raises the UIF flag.

(b) When the counter starts counting, it goes through 6 states (ARR+1 states) until the flag rises.

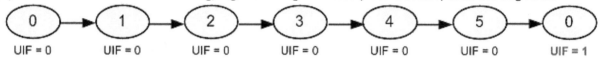

Enabling clock for TIMx

Before we can use any of the Timer Modules, we must enable the clock to them. For each timer module, there is a separate clock enable bit in the RCC_APB1ENR and RCC_APB2ENR registers. See Figures 8-7 and 8-8 in Chapter 8. See Example 11-9.

Example 11-9

Write an instruction that enables the clock for TIM2.

Solution:

According to Figure 8-7, bit 0 of RCC_APB1ENR is used for TIM2EN. So, to enable the clock for TIM2 we write the following:

```
RCC->APB1ENR |= (1<<0); //set bit 0 of RCC_APB1ENR
```

Notice that in STM32 there is a circuit between the APB prescaler and the timers. If the APB prescaler is set to 1, the same clock is fed to the timers. If the prescaler is set to any value other than 1, the APB clock is multiplied by 2 and then fed to the timers. See Figure 8-20 and Example 11-10.

Example 11-10

By default, the CPU clock frequency is 72MHz; the APB1 clock is set to 36MHz, and APB2 is 72MHz. Calculate the frequency of the clock that is fed to the timers.

Solution:

Since the CPU clock is 72MHz and APB1 clock is 36MHz, the prescaler for APB1 is set to 2 (other than 1). So, the APB2 clock is multiplied by 2 and fed to the timers which are connected to APB1 bus.

36MHz × 2 = 72MHz. According to Figure 8-7, timers TIM2,3,4,5,6,7,12,13, and 14 are connected to APB1. So, the clock for the timers is 72MHz.

By default, APB2 clock has the same frequency as the CPU. So, its prescaler is set to 1 and the timer clocks are the same as the APB2 clock. According to Figure 8-8, TIM1 and TIM8 are connected to APB2. So, a clock with frequency of 72MHz are fed to the timers.

Notice that the clocks for all timers are 72MHz by default, unless we change the APB prescalers.

Making delays using the TIM timer in up-counting mode

The steps to program the timer for TIMx_CNT are:

1) enable the clock to TIMx module,
2) load TIMx_ARR register with proper value,
3) clear UIF flag,
4) set the mode as up-counter timer and enable timer,
5) wait for UIF flag to go HIGH,
6) Stop the timer.

See Examples 11-11 through 11-13.

Example 11-11

Calculate the delay which is made by the following function. The clock frequencies are set by default.

```
void delay() {
   RCC->APB1ENR |= (1<<0);          /* enable TIM2 clock */
   TIM2->ARR = 71;
   TIM2->SR = 0; /* clear the UIF flag */
   TIM2->CR1 = 1; /* up counting */
   while((TIM2->SR & 1) == 0); /* wait until the UIF flag is set */
   TIM2->CR1 = 0; /*stop counting */
}
```

Solution:

Since the ARR is initialized with 71, it goes from 0 to 71 and in the next clock, it rolls over to 0 and the UIF flag rises. So, it goes through 72 stages.

Since the timer clock is 72MHz by default, each clock lasts $\frac{1}{72MHz}$.

So, the program makes a delay of $72 \times \frac{1}{72MHz} = 1\mu s$.

Example 11-12

Using TIM2 make a delay of 50μs. The clock frequencies are set by default.

Solution:

Delay = ARR+1 /72MHz ➔ ARR+1 = delay × 72MHz = 50 μs × 72MHz = 3600 ➔ ARR = 3600 − 1 = 3599.

```
void delay() {
    RCC->APB1ENR |= (1<<0);         /* enable TIM2 clock */
    TIM2->ARR = 3599;
    TIM2->CR1 = 1; /* up counting */
    while((TIM2->SR & 1) == 0); /* wait until the UIF flag is set */
    TIM2->CR1 = 0; /*stop counting */
    TIM2->SR = 0; /* clear the UIF flag */
}
```

Example 11-13

In Example 11-11, what is the largest delay, that we can make by changing the ARR value?

Solution:

TIM2_ARR is a 16-bit register. So, the biggest value we can load into the ARR is 65535.

Delay = ARR+1 /72MHz ➔ Delay = 65535+1 /72MHz = 65536 / 72M = 910 μs = 0.910ms

Prescaler

We saw in Example 11-13 that the largest time delay that we can make is 0.910ms. What if that is not enough? We can use the prescaler to increase the delay by reducing the period. See Figure 11-17. The prescaler is a 16-bit up-counter that sits between the clock source and the TIMx_CNT. It can be configured to divide the clock source by a number before feeding it to the TIMx_CNT. See Figure 11-17.

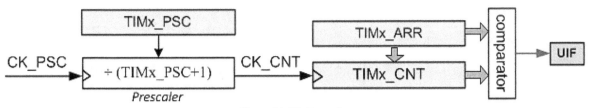

Figure 11-17: Prescaler

The prescaler contains a 16-bit up-counter and a comparator. The up-counter counts from 0 to the value of TIMx_PSC register. Then, in the next clock the compare match signal rises for 1 clock which makes the TIMx_CNT to count. The signal also triggers the load pin of the prescaler counter and the counter is initiated with 0. See Figure 11-18 and Example 11-14.

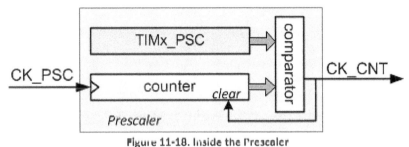

Figure 11-18. Inside the Prescaler

Example 11-14

Calculate the delay which is made by the following function. The clock frequencies are set by default.

```
void delay() {
    RCC->APB1ENR |= (1<<0);        /* enable TIM2 clock */
    TIM2->PSC = 7200-1;            /* PSC = 7199 */
    TIM2->ARR = 500-1;
    TIM2->SR = 0; /* clear the UIF flag */
    TIM2->CR1 - 1; /* up counting */
    while((TIM2->SR & 1) == 0); /* wait until the UIF flag is set */
    TIM2->CR1 = 0; /*stop counting */
}
```

Solution:

Since the prescaler is initialized with 7199, the clock is divided by 7200. 72MHz/7200 = 10KHz.

Each clock lasts $\frac{1}{10KHz} = 0.1ms$ and the program makes a delay of $500 \times 0.1\text{ms} = 50\text{ms}$.

Example 11-15

Using TIM2 write a program that toggles PC13, every second.

Solution:

```
#include <stm32f10x.h>
```

```
void delay(void);

int main()
{
    RCC->APB2ENR |= 0xFC;    /* enable GPIO clocks */
    RCC->APB1ENR |= (1<<0);          /* enable TIM2 clock */
    GPIOC->CRH = 0x44344444;         /* PC13 as output */

    while(1)
    {
        GPIOC->ODR ^= (1<<13); /* toggle PC13 */
        delay();
    }
}
void delay() {
    TIM2->PSC = 7200-1;              /* PSC = 7199 */
    TIM2->ARR = 10000-1;
    TIM2->SR = 0; /* clear the UIF flag */
    TIM2->CR1 = 1; /* up counting */
    while((TIM2->SR & 1) == 0); /* wait until the UIF flag is set */
    TIM2->CR1 = 0; /*stop counting */
}
```

Down counting mode

The timer counts down if the DIR bit of the TIMx_CR1 register is 1. In the counting down mode, the TIMx_CNT is loaded with the value of TIMx_ARR and begins counting down until it reaches zero. Then, the timer counter is reloaded with the value from TIMx_ARR and the UIF flag is set. See Figure 11-18.

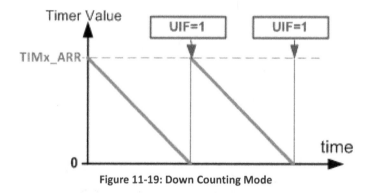

Figure 11-19: Down Counting Mode

Example 11-16

Calculate the delay which is made by the following function.

```
void delay()
{
    RCC->APB1ENR |= (1<<0);          /* enable TIM2 clock */
    TIM2->ARR = 999;
    TIM2->CR1 = 1; /* up counting */
    while((TIM2->SR & 1) == 0); /* wait until the UIF flag is set */
    TIM2->CR1 = 0; /*stop counting */
    TIM2->SR = 0; /* clear the UIF flag */
}
```

Solution:

The counter counts down from 999 to 0. Then in the next clock the TIM2_CNT is loaded with TIM2_ARR and the UIF flag is set. So, it takes 1000 clocks to rise the flag. Each clock lasts 1/72MHz and the function makes a delay of 1000×1/72MHz = 1/72K = 13.8μs.

One pulse mode vs. continuous

See Figure 11-14. The bit 3 of the TIMx_CR1 register is named OPM (One Pulse Mode). When the bit is 0, the timer is in continuous mode and the timer continues counting after each timeout as discussed above. But when the OPM bit is 1, the timer stops counting after timeout is reached. For example, when it is in up counting and one pulse modes, it counts from 0 to TIMx_ARR and then goes to zero just once and then the CEN bit of TIMx_CR1 is cleared causing the timer to stop.

In all the programs of this section, we stopped the timer by writing 0 to the TIMx_CR1 register. If we set the OPM bit to 1, then the timer stops automatically, when the UIF flag is set. See Example 11-17.

Example 11-17

Rewrite Example 11-14 using One Pulse Mode.

Solution:

```
void delay()
{
   RCC->APB1ENR |= (1<<0);        /* enable TIM2 clock */
   TIM2->PSC = 7200-1;            /* PSC = 7199 */
   TIM2->ARR = 500-1;
   TIM2->SR = 0; /* clear the UIF flag */
   TIM2->CR1 = 9; /* up counting and One Pulse Modes*/
   while((TIM2->SR & 1) == 0); /* wait until the UIF flag is set */
}
```
Notice that there is no need to stop the timer any more since it stops automatically.

Review Questions
1. True or false. TIMx_CNT is a 16-bit register.
2. True or false. Each timer has its own TIMx_CNT.
3. Which register is used to enable the clock to the TIM2?
4. Which register is used to select the counting direction for timers?

Section 11.3: Output Compare and TIM Channels

In the last section, we showed how to use timers to generate time delay. In this and following sections, we will examine the use of timers with the I/O pins. In this section, we will study the Output Compare feature of the timers. We examine the channels of TIMs, as well.

Programming Output Compare option

In some applications we need to control the digital pin output transition with precision timing. To do that, we use the Output Compare function of the timer. In the STM32, each of the TIMx module has 4 channels. See Figure 11-20. Each Channel has an input circuit for input capturing function and an output circuit for output compare function.

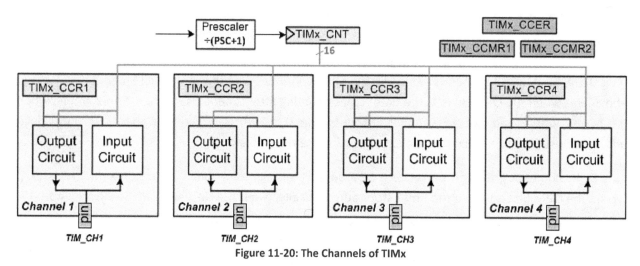

Figure 11-20: The Channels of TIMx

Register name	Offset	Description
TIMx_CCR1	0x0034	Capture/Compare Register 1
TIMx_CCR2	0x0038	Capture/Compare Register 2
TIMx_CCR3	0x003C	Capture/Compare Register 3
TIMx_CCR4	0x0040	Capture/Compare Register 4
TIMx_CCMR1	0x0018	Capture/Compare Mode Register 1
TIMx_CCMR2	0x001C	Capture/Compare Mode Register 2
TIMx_CCER	0x0020	Capture/Compare Enable Register

Table 11-1: Some TIMx Registers which are Used in Compare Mode

Capture/Compare Register (TIMx_CCRn)

Each channel has its own 16-bit register for the compare purpose. The registers are called Capture/Compare Register (TIMx_CCRn) and are designated as TIMx_CCR1 to TIMx_CCR4. See Figure 11-21. The 16-bit registers of TIMx_CCRn are readable and writable, which means we can initialize them to a desired value. After the initialization, the TIMx_CCRn content is compared with the value of TIMx_CNT after each clock cycle as TIM_CNT is counting. When the value of the CNT register and CCRn register match, the CCnOF flag is set high. See Figure 11-22. There is a wave generator in each channel. The wave generator can perform some actions such as toggling a pin, making a pin to go Low or High. We choose one of these options using the Capture/Compare Mode Registers (TIMx_CCMR1 and TIMx_CCMR2), which is discussed next.

Figure 11-21: TIMx_CCRn (Capture/Compare Register)

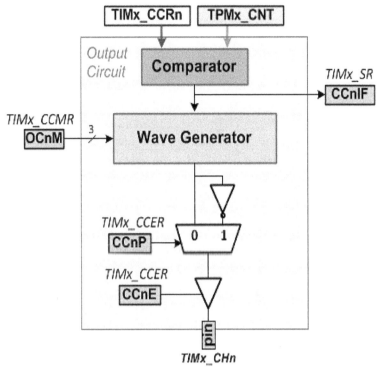

Figure 11-22: Output Circuit

Capture/Compare Mode Registers (TIMx_CCMR1 and TIM_CCMR2)

The mode selections for Output Compare of a given channel is done with TIMx Capture/Compare Mode Registers (TIMx_CCMR1 and TIMx_CCMR2). See Figure 11-23. There are 8 configuration bits for each channel. Bits D0-D7 of TIMx_CCMR1 is for channel 1, Bits D8-D15 of TIMx_CCMR1 is for channel 2, and so on for channel 3 and channel 4. Each channel has CCnS (Compare/Capture Selection) bits. The CCnS bits, are used to enable the input or output circuit of the channel. When CCnS = 00, the output circuit is enabled and the channel is in Compare mode. Otherwise, the input circuit is enabled and the channel is in Capture mode.

The OCnM bits select the behavior of the wave generator on compare matches (TIMx_CNT = TIMx_CCRn). See Figure 11-23. When the value of TIMx_CNT becomes equal to TIMx_CCRn, the wave generator might Set the pin to high, make the pin low, or toggle the pin. For example, if OCnM = 011, the wave generator toggles the GPIO pin on compare match. Figures 11-24 through 5-26 show the output pin in toggle, set, and clear modes.

	D15	D14	D13	D12	D11	D10	D9	D8	D7	D6	D5	D4	D3	D2	D1	D0	
TIMx_CCMR1:	OC2CE	OC2M			OC2PE	OC2FE	CC2S		OC1CE	OC1M			OC1PE	OC1FE	CC1S		0x18

(TIMx->CCMR1)

	D15	D14	D13	D12	D11	D10	D9	D8	D7	D6	D5	D4	D3	D2	D1	D0	
TIMx_CCMR2:	OC4CE	OC4M			OC4PE	OC4FE	CC4S		OC3CE	OC3M			OC3PE	OC3FE	CC3S		0x1C

(TIMx->CCMR2)

Name	Description
OCnCE	Output Compare n Clear Enable (For more information, see the reference manual.)
OCnM	Output Compare n Mode

OCnM	Mode	Description
000	Frozen	Compare match has no effect on the GPIO pin
001	Active on match	When CNT=CCRn, the wave generator makes its output high and the GPIO pin (TIMx_CHn) becomes active.
010	Inactive on match	When CNT=CCRn, the wave generator makes its output low and the GPIO pin (TIMx_CHn) changes to inactive level.
011	Toggle on match	When CNT=CCRn, it toggles the GPIO pin (TIMx_CHn).
100	Force inactive	It forces the GPIO pin to inactive level without considering the values of the TIMx_CNT and TIMx_CCRn registers.
101	Force active	It forces the GPIO pin to active level without considering the values of the TIMx_CNT and TIMx_CCRn registers.
110	PWM 1	It is discussed in the PWM chapter.
111	PWM 2	It is discussed in the PWM chapter.

Note: Using the CCnP of the CCER register, we can choose the GPIO pin to be active high or active low. If CCnP is 0, the pin is active high. Otherwise, it is active low. See Figure 11-22.

Name	Description
OCnPE	Output Compare n Preload Enable. (0: TIMx_CCRn is not buffered, 1: It is buffered.)
OCnFE	Output compare 1 Fast Enable (For more information see the reference manual)
CCnS	Compare/Capture n Selection The bit-field selects between the Compare and Capture modes. If CCnS = 00, the Compare (output) function of the channel is enabled. Otherwise, the Capture (input) function is selected.

Figure 11-23: CCMR1 and CCMR2

Compare/Capture Enable Register (TIMx_CCER)

See Figure 11-22. There is a multiplexer next to the wave generator. If CCnP is 0, the output of the wave generator goes directly to the GPIO pin. When CCnP is 1, the output signal of wave generator becomes inverted and then goes to the GPIO pin. So, if CCnP is 0, the pin becomes active high. This means that when the output of the output compare is active (high) the GPIO pin is high and when the output of the compare is inactive (low), the GPIO pin becomes low. The GPIO pin becomes active low when CCnP is 1. In the case, when the output signal of the wave generator becomes active, the pin becomes low.

See Figure 11-22. There is a buffer between the wave generator and the pin. We must set the CCnE bit to 1 to let the wave generator change the GPIO pin. See Figure 11-27.

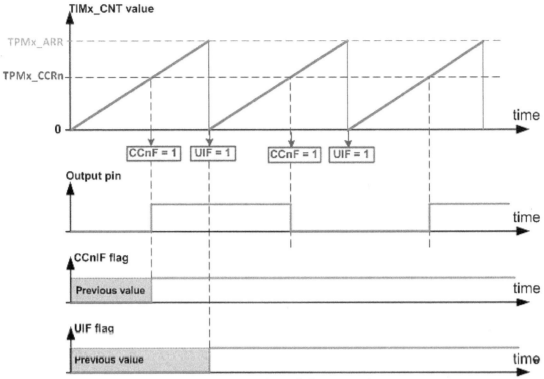

Figure 11-24: Toggle Mode (OCnM=011)

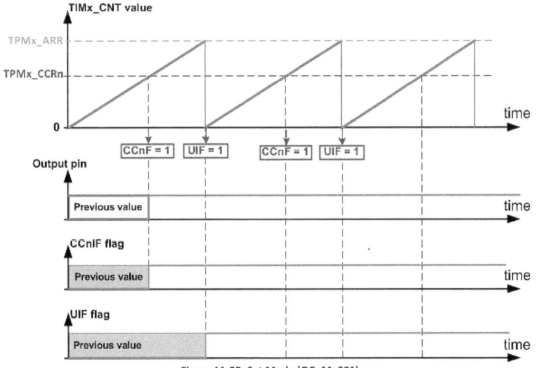

Figure 11-25: Set Mode (OCnM=001)

299

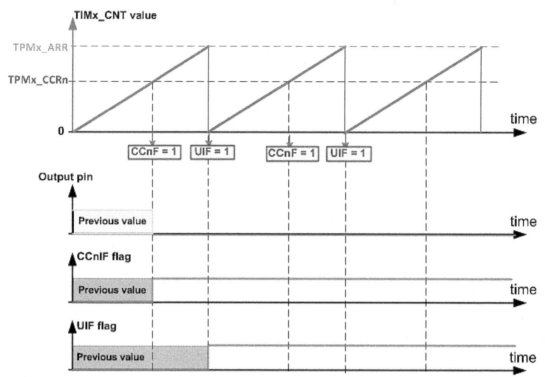

Figure 11-26: Clear Mode (OCnM=010)

	15	14	13	12	11	10	9	8	7	6	5	4	3	2	1	0	
TIMx_CCER:	Reserved		CC4P	CC4E	Reserved		CC3P	CC3E	Reserved		CC2P	CC2E	Reserved		CC1P	CC1E	0x20

(TIMx->CCER)

Name	Description
CCnP	Compare/Capture n output Polarity 0 (Active high): the output signal of the wave generator goes to the pin directly. 1 (Active low): the output signal of the wave generator is inverted.
CCnE	Compare/Capture n output Enable 0: the output buffer is disabled. 1: the output buffer is enabled.

Figure 11-27: CCER (Compare/Capture Enable Register)

Channel pins

Each channel has its own designated output pins. Table 11-2 shows the pin designations for Timer modules, in STM32F103.

Timer	Channel 1	Channel 2	Channel 3	Channel 4
TIM1	PA8	PA9	PA10	PA11
TIM2	PA0	PA1	PA2	PA3
TIM3	PA6	PA7	PB0	PB1
TIM4	PB6	PB7		

Table 11-2: Channel Pins

Example 11-18

Find the CCMR value to: (a) Set high channel 2 on compare match, (b) toggle channel 2 on compare match, (c) toggle channel 3 on compare match.

(a)

TIMx_CCMR1:	OC2CE	OC2M	OC2PE	OC2FE	CC2S	OC1CE	OC1M	OC1PE	OC1FE	CC1S
	0	001	0	0	00	0	000	0	0	00

TIMx->CCMR1 = 0x1000;

(b)

TIMx_CCMR1:	OC2CE	OC2M	OC2PE	OC2FE	CC2S	OC1CE	OC1M	OC1PE	OC1FE	CC1S
	0	011	0	0	00	0	000	0	0	00

TIMx->CCMR1 = 0x3000;

(c)

TIMx_CCMR2:	OC4CE	OC4M	OC4PE	OC4FE	CC4S	OC3CE	OC3M	OC3PE	OC3FE	CC3S
	0	000	0	0	00	0	011	0	0	00

TIMx->CCMR2 = 0x0030;

Example 11-19

Draw the wave generated by the following program. Calculate the frequency of the generated wave.

```c
#include <stm32f10x.h>

int main()
{
    RCC->APB2ENR |= 0xFC;      /* enable GPIO clocks */
    RCC->APB1ENR |= (1<<0);        /* enable TIM2 clock */
    GPIOA->CRL = 0x44444B44;       /* PA2: alternate func. output */

    TIM2->CCR3 = 200;
    TIM2->CCER = 0x1 << 8; /* CC3P = 0, CC3E = 1 */
    TIM2->CCMR2 = 0x0030;  /* toggle channel 3 */
    TIM2->ARR = 10000-1;
    TIM2->CR1 = 1;    /* start counting up */

    while(1)
    {
    }
}
```

Solution:

The timer counts up and when it toggles CH3 when CNT = CCR3. In the code, CCR3 is set to 200. So, CH3 toggles after 200 clocks. The timer counts up until the TIM3_CNT reaches TIM3_ARR. Then, the timer rolls over. The CH3 pin toggles again on the next compare match (when CNT is equal to CCR3). In the code, the frequency of the toggles is dependent to the value of ARR. If the ARR register had a smaller value, the timer rolled over faster and the frequency of the wave was higher.

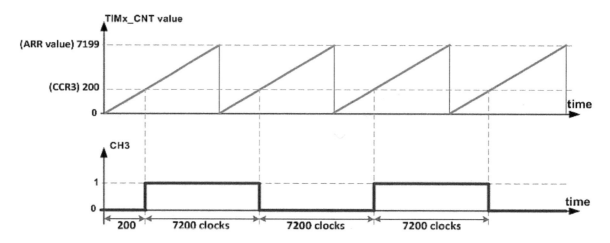

By default, the timer clock is 72MHz. So,

$$T_{Timer\ clock} = \frac{1}{72MHz} \Rightarrow T_{wave} = 2 \times 7200 \times \frac{1}{72M} = 200\mu s \Rightarrow F_{wave} = \frac{1}{T_{wave}} = \frac{1}{200\mu} = 5KHz$$

Example 11-20

Draw the waves generated by the following program.

```
#include <stm32f10x.h>

int main()
{
    RCC->APB2ENR |= 0xFC;      /* enable GPIO clocks */
    RCC->APB1ENR |= (1<<0);         /* enable TIM2 clock */
    GPIOA->CRL = 0x44444BB4;        /* PA2(CH3), PA1(CH2): alternate func. output */

    TIM2->CCR2 = 1000;
    TIM2->CCR3 = 3000;
    TIM2->CCER = (0x1<<8)|(0x1<<4); /*CC3P = 0, CC3E = 1, CC2E = 1 */
    TIM2->CCMR1 = 0x3000;    /* toggle channel 2 */
    TIM2->CCMR2 = 0x0030;    /* toggle channel 3 */
    TIM2->PSC = 7200-1;
    TIM2->ARR = 10000-1;
    TIM2->CR1 = 1;
```

302

```
    while(1)
    {
    }
}
```

Solution:

The prescaler is set to 7199. So, the timer counts with frequency of 72MHz/7200 = 10KHz.

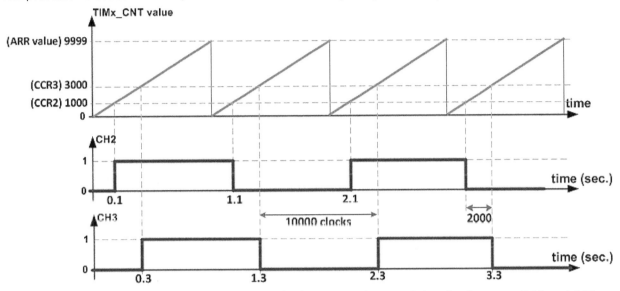

The program toggles PA2 and PA3, every second. There is a 0.2 second time lag between PA2 and PA3.

Timer Status Register (TIMx_SR)

The TIMx_SR contains the status flags for all the channels. This allows us to monitor the status of all the 4 channels with a single read of the register to see whether any given CCnIF (Compare/Capture n Interrupt Flag) flag has been raised. Notice from Figure 5-26, that we have D4 bit for Channel 4 flag and D1 bit for Channel 1 flag. Also notice that, the D0 bit of the TIMx_SR register is for UIF.

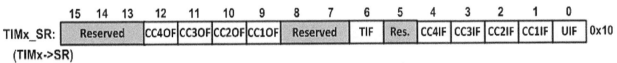

15	14	13	12	11	10	9	8	7	6	5	4	3	2	1	0	
	Reserved		CC4OF	CC3OF	CC2OF	CC1OF		Reserved		TIF	Res.	CC4IF	CC3IF	CC2IF	CC1IF	UIF	0x10

TIMx_SR:
(TIMx->SR)

Figure 11-28: TIMx_SR (Status Register)

Review Questions

1. True or false. Each Channel has its own designated pins.
2. True or false. To generate waves the pins must be configured as input pins.
3. True or false. Each Channel has two wave generators.

Section 11.4: Using Timer for Input Capturing

Input Capturing mode

If a timer channel is configured in input capturing mode, the I/O pin of the channel is used to capture the signal transition events. When an event occurs, the content of the TIMx_CNT timer counter is captured and saved in the CCRn register of the channel.

The Input Capture function is widely used for many applications. Among them are (a) recording the arrival time of an event, (b) pulse width measurement, and (c) period measurement.

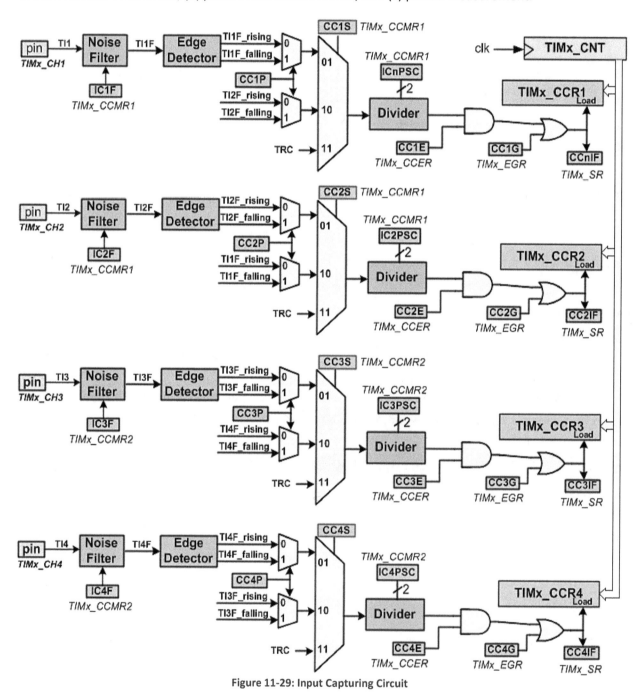

Figure 11-29: Input Capturing Circuit

Figure 11-29 shows the input capture circuit for the 4 channels of timers. The input signal is filtered by the Noise Filter. The Noise Filter makes sure the signal is stable for a given number of clocks and then informs the new signal. The edge detector detects the rising and falling edges of the signal. Using the CCnP bit of the CCER, we choose to capture on the rising or falling edge.

Using the CCnS bits, we choose the channel to be in input or output mode. We also use the bits to choose the input signal. For example, in channel 1, when CC1S = 01, it captures the TIMx_CH1 pin; when CC1S = 10, it captures the TIMx_CH2 pin.

The divider can be used to set the ratio of capturing. If ICnPSC = 00, it captures each time an edge is detected. If you like to capture after 4 edges you can set the ICnPSC to 10. See Figure 11-30.

	D15 D14 D13 D12	D11 D10	D9 D8	D7 D6 D5 D4	D3 D2	D1 D0	
TIMx_CCMR1: (TIMx->CCMR1)	IC2F	IC2PSC	CC2S	IC1F	IC1PSC	CC1S	0x18
	D15 D14 D13 D12	D11 D10	D9 D8	D7 D6 D5 D4	D3 D2	D1 D0	
TIMx_CCMR2: (TIMx->CCMR2)	IC4F	IC4PSC	CC4S	IC3F	IC3PSC	CC3S	0x1C

Name	Description
ICnF	Input Capture Filter: The filter samples the input signal with frequency of fsampling, N times. If the samples are the same informs the new level of signal. N and fsampling are defined using the ICnF bits.

ICnF	N	Fsampling	ICnF	N	Fsampling	ICnF	N	Fsampling	ICnF	N	Fsampling
0000	1	f_{DTS}	0100	6	$f_{DTS}/2$	1000	6	$f_{DTS}/8$	1100	8	$f_{DTS}/16$
0001	2	f_{CK_INT}	0101	8	$f_{DTS}/2$	1001	8	$f_{DTS}/8$	1101	5	$f_{DTS}/32$
0010	4	f_{CK_INT}	0110	6	$f_{DTS}/4$	1010	5	$f_{DTS}/16$	1110	6	$f_{DTS}/32$
0011	8	f_{CK_INT}	0111	8	$f_{DTS}/4$	1011	6	$f_{DTS}/16$	1111	8	$f_{DTS}/32$

CK_INT is the timer internal clock. The DTS clock is made from CLK_INT. There is a prescaler between CK_INT and DTS which is configured with TIMx_CR1 register. (See the manual.) |
| ICnPSC | Input Capture n Prescaler
00: no prescaler, capture is done each time an edge is detected on the capture input
01: capture is done once every 2 events
10: capture is done once every 4 events
11: capture is done once every 8 events |
| CCnS | Compare/Capture n Selection: The bit-field selects between the Compare and Capture modes. It also chooses the input signal.

CC1S	Mode	CC2S	Mode	CC3S	Mode	CC4S	Mode
00	Output Compare mode	00	Output Compare mode	00	Output Compare mode	00	Output Compare mode
01	Input from TIMx_CH1	01	Input from TIMx_CH2	01	Input from TIMx_CH3	01	Input from TIMx_CH4
10	Input from TIMx_CH2	10	Input from TIMx_CH1	10	Input from TIMx_CH4	10	Input from TIMx_CH3
11	Input from TRC	11	Input from TRC	11	Input from TRC	11	Input from TRC

Figure 11-30: TIMx_CCMR Registers in Input Mode

See Figures 11-29 and 11-31. The CCnE (Compare/Capture Enable) bit is used to enable capturing. To capture, the CCnE bit must be 1. The CCnG (Compare/Capture Generation) bit of the TIMx_EGR register can be used to make a capture. If we set the bit, one capture takes place and then the CCnG bit is cleared by hardware. See Figure 11-32.

Capture signal triggers the load pin of the TIMx_CCRn register, causing the TIMx_CCRn to be loaded with the value of the TIMx_CNTn register. On each capture, the CCnIF flag of the TIMx_SR register is also set to 1 and remains 1 until it is cleared by software. If we read the TIMx_CCRn register, the CCnIF flag is cleared automatically. See Figure 11-32. For each channel there is also a CCnOF (Compare/Capture n Overflow Flag) flag. The CCnOF flag is set if a new capture takes place while the previous captured value is unread.

Notice that the bits of CCMR registers differ in output Compare and input capture modes. Compare Figure 11-23 with Figure 11-30. In both modes the same bits are used as the CCnS bits. But when we set a channel to input capture mode, the rolls of the other bits of the channel changes as shown in Figure 11-30.

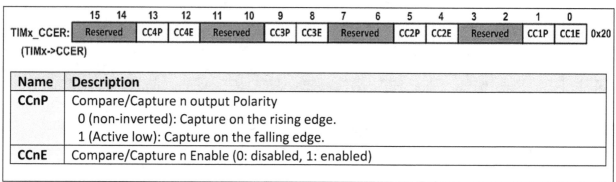

15 14	13	12	11 10	9	8	7 6	5	4	3 2	1	0	
TIMx_CCER: Reserved	CC4P	CC4E	Reserved	CC3P	CC3E	Reserved	CC2P	CC2E	Reserved	CC1P	CC1E	0x20

(TIMx->CCER)

Name	Description
CCnP	Compare/Capture n output Polarity
	0 (non-inverted): Capture on the rising edge.
	1 (Active low): Capture on the falling edge.
CCnE	Compare/Capture n Enable (0: disabled, 1: enabled)

Figure 11-31: CCER (Compare/Capture Enable Register)

15 14 ... 7	6	5	4	3	2	1	0	
TIMx_EGR: Reserved	TG	Res.	CC4G	CC3G	CC2G	CC1G	UG	0x14

(TIMx->EGR)

Figure 11-32: TIMx_EGR (Event Generation Register)

15 14 13	12	11	10	9	8 7	6	5	4	3	2	1	0	
TIMx_SR: Reserved	CC4OF	CC3OF	CC2OF	CC1OF	Reserved	TIF	Res.	CC4IF	CC3IF	CC2IF	CC1IF	UIF	0x10

(TIMx->SR)

Name	Description
CCnOF	Capture/Compare n Overcapture flag (0: No overcapture, 1: overcapture detected)
CCnIF	Capture/Compare n Input Flag (0: No capture occurred, 1: capture occurred)
UIF	Update Interrupt Flag (0: No interrupt occurred, 1: Update interrupt occurred)

Figure 11-33: TIMx_SR (Status Register)

Example 11-21

Find the CCMR1 and CCER values to configure channel 1 for capturing pin TIMx_CH1 on rising edge, and no division. The noise filter should accept signals after 4 timer clocks.

Solution:

TIMx_CCMR1:	IC2F	IC2PSC	CC2S	IC1F	IC1PSC	CC1S
	0000	00	00	0010	00	01

TIMx->CCMR1=0x0021;

TIMx_CCER	Res.	CC4P	CC4E	Res.	CC3P	CC3E	Res.	CC2P	CC2E	Res.	CC1P	CC1E
	00	0	0	00	0	0	00	0	0	00	0	1

TIMx->CCER=0x0001;

Steps to program the Input Capture function

Perform the following steps to measure the period of a periodic waveform based on the edge arrival time of the Input Capture function.

1) Enable the clock to the input pin GPIO port, and the TIMx module,
2) select the channel pin as input in the CRL/CRH registers,
3) set the timer prescaler value,
4) select the input signal using CCMR1 and CCMR2,
5) select the rising/falling edge and enable the capture using CCER,
6) set the value of TIMx_ARR register,
7) enable timer,
8) wait until the CCnIF bit is set in TIMx_SR register,
9) read the current counter value captured,
10) calculate the current counter value difference from the last value,
11) save the current value for next calculation,
12) repeat from step 8.

As shown in Figure 11-34, to measure the period of a signal we must measure the time between two falling edges or two rising edges. Program 11-1 measures the period of the square wave and sends through serial.

Program 11-1

```
/* this program measures the frequency of the wave and reports through usart1 */

#include <stm32f10x.h>
#include <stdio.h>

void usart1_init(void);
void usart1_sendByte(unsigned char c);
```

```c
void usart1_sendInt(unsigned int i);
void usart1_sendStr(char *str);

int main()
{
   RCC->APB2ENR |= (0xFC| (1<<14));      /* enable GPIO clocks and USART1 clock */
   RCC->APB1ENR |= (1<<0);        /* enable TIM2 clock */

   usart1_init();

   GPIOA->CRL = 0x44444844;        /* PA2(CH3): input pull-up */
   GPIOA->ODR |= (1<<2);

   TIM2->CCMR2 = 0x001;   /* Pin TIM2_CH3 as input for channel 3 */
   TIM2->CCER = 0x1 << 8; /*CC3P = 0 (rising), CC3E = 1 */
   TIM2->PSC = 7200-1;
   TIM2->ARR = 50000-1;
   TIM2->CR1 = 1;              /* start counting up */

   uint16_t t = 0, t0 = 0;

   while(1)
   {
      while((TIM2->SR &(1<<3)) == 0); /* wait until the CC3IF flag sets */

      t = TIM2->CCR3;        /* read the captured value */

      usart1_sendInt(t - t0); /* send the difference */
      usart1_sendStr("\n\r"); /* go to new line */

      t0 = t;
   }
}

void usart1_init()
{
   GPIOA->ODR |= (1<<10);   //pull-up PA10
   GPIOA->CRH = 0x444448B4; // RX1=input with pull-up, TX1=alt. func output
   USART1->CR1 = 0x200C;
   USART1->BRR = 7500;        // 72MHz/9600bps = 7500
}

void usart1_sendByte(unsigned char c)
{
   while((USART1->SR&(1<<6)) == 0);  //wait until the TC flag is set
   USART1->DR = c;
}

/* the function sends a zero-ending string through USART1 */
void usart1_sendStr(char *str)
{
   while(*str != 0)
   {
      usart1_sendByte(*str);
      str++;
   }
}
```

```
/* The function sends a number through USART1 */
void usart1_sendInt(unsigned int i)
{
    char str[10];
    sprintf(str,"%d",i);

    usart1_sendStr(str);
}
```

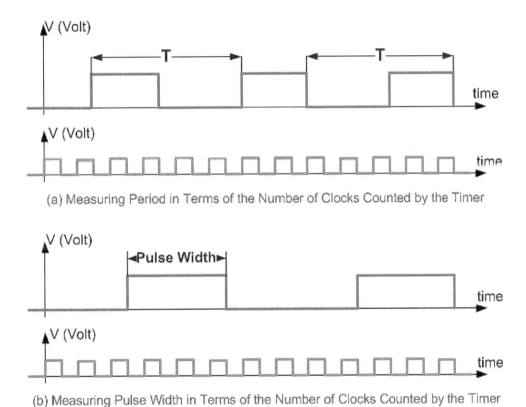

(a) Measuring Period in Terms of the Number of Clocks Counted by the Timer

(b) Measuring Pulse Width in Terms of the Number of Clocks Counted by the Timer

Figure 11-34: Measuring Period and Pulse Width

Review Questions

1. True or false. To measure the frequency of a signal, the time interval between a falling edge and a rising edge are needed.
2. True or false. If the time interval between two consecutive falling edges is measured, the frequency of the periodic signal can be calculated.
3. True or False. In STM32F1xx, the input capturer supports both rising and falling edge detection.
4. True or False. The Noise Filter can filter transient signals and noises.

Section 11.5: Using Timer as a Counter

See Figure 11-35. The TIMx_CH1, TIMx_CH2, and TIMx_ETR pins can be used as inputs for timers. In the case, the TIM_CNT counter is clocked by the pulses fed to the pins. You can also connect an encoder to TIMx_CH1 and TIMx_CH2 pins and the value of the TIMx_CNT changes as the encoder rotates.

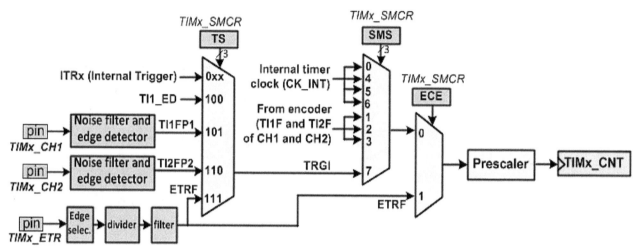

Figure 11-35: Timers Clock Sources

The ECE, SMS, and TS bits of the TIMx_SMCR register choose the clock source for the timers. When ECE is 1, the timer counts the pulses of TIMx_ETR pin. Otherwise, the SMS bits choose the clock among the following chooses: internal clock (SMS=0, 4, 5, or 6), the encoder input (SMS=1, 2, or 3), and TRGI (mode 7). The TS bits choose the source TRGI signal. See Figures 11-35 and 11-37.

Using TIMx_CH1 and TIMx_CH2 as Clock Source

See Figure 11-36. To use TIMx_CH1 or TIMx_CH2 as the external clock you should choose the edge using the CCER register and initialize the SMCR register with proper value. See Examples 11-22 and 11-23.

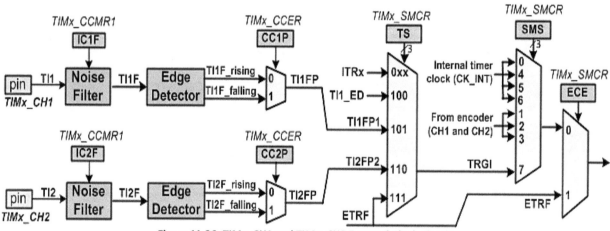

Figure 11-36: TIMx_CH1 and TIMx_CH2 External Clock Circuit

15	14	13	12	11	10	9	8	7	6	5	4	3	2	1	0
ETP	ECE	ETPS			ETF			MSM	TS			Res.	SMS		

TIMx_SMCR:
(TIMx->SMCR)

Name	Description
ETP	External Triger (ETR) Polarity The bit selects the ETR polarity. (0: rising edge, 1: falling edge.)
ECE	External Clock Enable (0: The SMS bits select the clock source, 1: the TIMx_CNT is clocked by the TIMx_ETR signal)
ETPS	External trigger (ETR) Prescaler The bits can be used to initialize the ETR prescaler. The ETR signal frequency must be less than ¼ of INT_CK. If the ETR input signal is fast, you can use the prescaler to slow down. 00: Prescaler off 01: ETRP frequency is divided by 2 10: ETRP frequency is divided by 4 11: ETRP frequency is divided by 8
ETF	External trigger filter The filter works the same way as noise filter in input circuits of channels.
MSM	Master/Slave Mode (See the reference manual.)
TS	Trigger Selection: It selects the input for the TRGI signal. See Figure 11-35. 000 to 011: Internal Triggers (See the reference manual.) 100: TI1 Edge detector (TI1F_ED) 101: Filtered Timer Input 1 (TI1FP1) from TIMx_CH1 110: Filtered Timer Input 2 (TI2FP2) from TIMx_CH2 111: External Trigger Input (ETRF) from TIMx_ETR pin
SMS	Slave Mode Selection

SMS	Mode	Description
000	Timer mode	clocked by the internal clock
001	Encoder mode 1	counter counts up/down on TI2FP1 edge depending on TI1FP2 level. (See the manual.)
010	Encoder mode 2	counter counts up/down on TI1FP2 edge depending on TI2FP1 level. (See the manual.)
011	Encoder mode 3	counter counts up/down on both TI1FP2 and TI2FP2 edges depending on the level of the other.
100	Reset mode	The counter is reset and the registers are updated on each rising edge of TRGI signal. (See the manual.)
101	Gated mode	As long as the TRGI signal is high, the timer counts. The timer stops counting when the TRGI signal is low.
110	Trigger mode	at the rising edge of the TRGI signal, it starts counting.
111	External clock mode 1	Counts the rising edges of TRGI signal.

Figure 11-37: SMCR (Slave Mode Control Register)

Example 11-22

Find the SMCR value to choose (a) TIMx_CH1 (b) TIMx_CH2 as the clock source for the counter.

Solution:

(a) ECE = 0, SMS = 7 (111 in binary), TS = 101

TIMx_SMCR	ETP	ECE	ETPS	ETF	MSM	TS	Res.	SMS
	0	0	00	0000	0	101	0	111

TIMx_SMCR = 0x57;

(b) ECE = 0, SMS = 7 (111 in binary), TS = 110

TIMx_SMCR	ETP	ECE	ETPS	ETF	MSM	TS	Res.	SMS
	0	0	00	0000	0	110	0	111

TIMx_SMCR = 0x67;

Example 11-23

Write a program that counts the input pulses of TIMx_CH2 on rising edge and sends the value of counter through USART1.

Solution:

To count on the rising edge, the CC2P bit of CCER should be 0.

```c
#include <stm32f10x.h>
#include <stdio.h>

void usart1_init(void);
void usart1_sendByte(unsigned char c);
void usart1_sendStr(char *str);
void usart1_sendInt(unsigned int i);

void delay_ms(uint16_t t);

int main()
{
  RCC->APB2ENR |= (0xFC| (1<<14));      /* enable GPIO clocks and USART1 clock */
  RCC->APB1ENR |= (1<<0);          /* enable TIM2 clock */

  usart1_init();           /* initialize the usart1 */

  GPIOA->CRL = 0x44444484;        /* PA1(CH2): input pull-up */
  GPIOA->ODR |= (1<<1);

  TIM2->CCMR1 = 0x0000;  /* no filter */
  TIM2->CCER = 0; /* CC2P = 0 (rising) */
  TIM2->SMCR = 0x67;       /* TIM2_CH2 as clock source */
  TIM2->ARR = 50000-1; /* count from 0 to 49999 then roll over to 0 */
  TIM2->CR1 = 1;    /* start counting up */
```

```
   while(1)
   {
      usart1_sendInt(TIM2->CNT);   /* send the counter value through serial */
      usart1_sendStr("\n\r"); /* go to new line */
      delay_ms(100);
   }
}

/* copy usart1_init, usart1_sendByte, usart1_sendStr, and usart1_sendInt from Program
11-1 to here */

void delay_ms(uint16_t t)
{
   for(int i = 0; i < t; i++)
   {
      for(volatile uint16_t a = 0; a < 6000; a++)
      {}
   }
}
```

Example 11-24

A clock pulse is fed into pin TIM2_CH1(PA0). Write a program that toggles PC13 every 100 pulses.

Solution:

```
#include <stm32f10x.h>
#include <stdio.h>

int main()
{
   RCC->APB2ENR |= 0xFC;    /* enable GPIO clocks and USART1 clock */
   RCC->APB1ENR |= (1<<0);            /* enable TIM2 clock */

   GPIOA->CRL = 0x44444448;        /* PA0(CH1): input pull-up */
   GPIOA->ODR |= (1<<0);

   GPIOC->CRH = 0x44344444;        /* PC13 as output */

   TIM2->CCMR1 = 0x0000;  /* no filter */
   TIM2->CCER = 0x1 << 1; /* CC0P = 1 (falling) */
   TIM2->SMCR = 0x57;              /* TIM2_CH1 as clock source */
   TIM2->ARR = 100-1;
   TIM2->CR1 = 1;

   while(1)
   {
      if((TIM2->SR&1) != 0)
      {
         TIM2->SR = 0;
         GPIOC->ODR ^= (1<<13);
      }
   }
}
```

Note: In the above program, ARR is set to 99. So, after 99 clocks the TIM2_CNT becomes equal to ARR and in the next clock the UIF is set and TIM2_CNT rolls over to zero. The code monitors the UIF flag and toggles PC13 when the flag is set.

___

Using TIMx_ETR as input clock

To use the TIMx_ETR pin, you can set the ECE bit of the SMCR register to 1. As an alternative choice, the counter is clocked by TIMx_ETR if you set both TS and SMS fields of the SMCR register to 7.

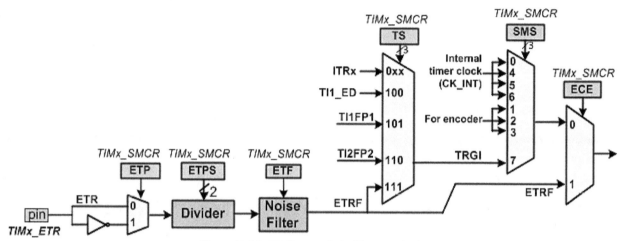

Figure 11-38: ETR (External Clock) Input Block

Review Questions
1. True or false. The Timer can also be used as counter.
2. True or false. STM32 timer can only count the falling edges.
3. True or false. To use the timer as counter, we must configure the SMS bits in gated mode.

Problems

Section 11-1
1. True or false. SysTick is a 24-bit timer.
2. True or false. In STM32, SysTick can be used as a counter.
3. Using SysTick, make a delay of 3ms. The AHB clock is 72MHz.
4. Using SysTick, make a delay of 70ms. The AHB clock is 72MHz.

Section 11-2
5. True or false. Each timer has its own status register.
6. True or false. Each timer has its own counter.
7. True or false. Timer 2 can only count down.
8. True or false. In One pulse mode, the timer stops counting one clock after starting counting.
9. Using Timer 1, write a program to generate a time delay of 20 μs.
10. Using Timer 3, write a program to generate a time delay of 50 μs.
11. Using Timer 2, write a program to generate a time delay of 20 ms.

12. Using Timer 3, write a program to generate a time delay of 50 ms.

Section 11-3

13. True or false. In STM32F103, each timer has only one output channel.
14. True or false. In STM32F103, each channel has a wave generator.
15. Using Timer 2, write a program to generate a square wave with frequency of 2MHz.
16. Using Timer 2, write a program to generate a square wave with frequency of 15KHz.

Section 11-4

17. True or false. STM32F103 captures the value of TIMx_CNT into the TIMx_SR register.
18. True or false. Each channel has an input capture circuit.
19. Find the value for TIM2_CCER register, if you want to capture on falling edges using channel 2.
20. Find the CCMR1 value to capture the TIMx_CH2 pin using channel 2 with no prescale. Configure the noise filter so that the new signal is accepted after 4 internal clocks.

Section 11-5

21. Write a program that counts the falling edges on the TIM2_CH2 pin and sends the counter value to the PC.
22. We want to pack each 9 eggs in a packet. Write a program that counts the rising edges on the TIM2_CH1 pin. When the value of counter reaches 8, set the PB3 pin high for 10ms and then make the pin low.

Answers to Review Questions

Section 11-0

1. 31
2. 32
3. event counter
4. Timer
5. 9

Section 11.1

1. 0xFFFFFF
2. 1/16MHz=62.5 nsec. Now, 5 msec/62.5nsec=80,000. Therefore, RELOAD-80,000 − 1 =79,999
3. 24
4. The D0 of STCTRL (the Enable)
5. Down counter

Section 11.2

1. True
2. True
3. RCC_APB1ENR
4. TIMx_CR1

Section 11.3

1. True
2. False
3. False

Section 11.4

1. False
2. True
3. True
4. True

Section 11.5

1. True
2. False
3. False

Chapter 12: Interrupt and Exception Programming

This chapter examines the interrupts in ARM. We also discuss sources of hardware interrupts in STM32F10x chips. In Section 12.1 we discuss the concept of interrupts in the ARM CPU, and then we look at the interrupt assignment of the ARM Cortex-M. The SysTick interrupt is covered in Section 12.2. The interrupt for I/O ports are discussed in Section 12.3. Section 12.4 examines the interrupt for USART. Timers' interrupts are explored in Section 12.5. The interrupt priority is discussed in Section 12.6. Section 12.7 examines the NVIC interrupt controller and discusses the Thread and Handler mode in Cortex-M.

Section 12.1: Interrupts and Exceptions in ARM Cortex-M

In this section, first we examine the difference between polling and interrupt and then describe the various interrupts of the ARM Cortex.

Interrupts vs. polling

A single microprocessor can serve several devices. There are two ways to do that: interrupts or polling. In the *interrupt* method, whenever any device needs service, the device notifies the CPU by sending it an interrupt signal. Upon receiving an interrupt signal, the CPU interrupts whatever it is doing and serves the device. The program associated with the interrupt is called the *interrupt service routine* (ISR) or *interrupt handler*. In *polling*, the CPU continuously monitors the status of a given device; when the status condition is met, it performs the service. After that, it moves on to monitor the next device until each one is serviced. See Figure 12-1.

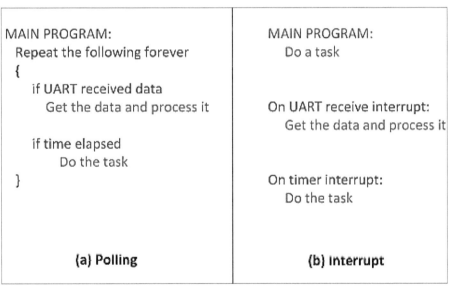

Figure 12-1: Polling vs. Interrupts

Although polling can monitor the status of several devices and serve each of them as certain conditions are met, it is not an efficient use of the CPU time. The polling method wastes much of the CPU's time by polling devices when they do not need service. So in order to avoid tying down the CPU, interrupts are used. For example, in Timer we might wait until a determined amount of time elapses, and while we were waiting we cannot do anything else. That is a waste of the CPU's time that could have been used to perform some useful tasks. In the case of the Timer, if we use the interrupt method, the CPU can go about doing other tasks, and when the COUNT flag is raised the Timer will interrupt the CPU to let it know that the time is elapsed. See Figure 12-1.

Interrupt service routine (ISR)

For every interrupt there must be a program associated with it. When an interrupt occurs, this program is executed to perform certain service for the interrupt. This program is commonly referred to as an *interrupt service routine* (ISR). The interrupt service routine is also called the *interrupt handler*. When an interrupt occurs, the CPU runs the interrupt service routine. Now the question is how the ISR gets executed?

As shown in Figure 12-2, in the ARM CPU there are pins that are associated with hardware interrupts. They are input signals into the CPU. When the signals are triggered, CPU pushes the PC register onto the stack and loads the PC register with the address of the interrupt service routine. This causes the ISR to get executed.

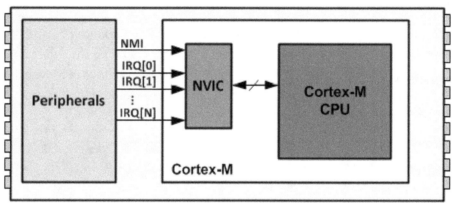

Figure 12-2: NVIC in ARM Cortex-M

As can be seen from Table 12-1, for every interrupt there are four bytes of memory allocated in the interrupt vector table. These four memory locations provide the addresses of the interrupt service routine for which the interrupt was invoked.

Interrupt Vector Table

Since there is a program (ISR) associated with every interrupt and this program resides in memory (RAM or ROM), there must be a look-up table to hold the addresses of these ISRs. This look-up table is called *interrupt vector table*. In the ARM, the lowest 1024 bytes (256 × 4 = 1024) of memory space are set aside for the interrupt vector table and must not be used for any other function. Table 12-1 provides a list of interrupts and their designated functions as defined by ARM Cortex-M products. Of the 256 interrupts, some are used for software interrupts and some are for hardware IRQ interrupts.

NVIC (nested vector interrupt controller) In ARM Cortex-M

In the ARM Cortex series, we have Cortex-A, Cortex-R and Cortex-M. Currently only the Cortex-M has an on-chip interrupt controller called NVIC (Nested Vector Interrupt Controller). See Figure 12-2. This allows some degree of standardization among the ARM Cortex-Mx (M0, M1, M3, and M4) family members. The classical ARM chips and Cortex-A and Cortex-R series do not have this NVIC interrupt controller, therefore ARM manufacturers' implementation of the interrupt handling varies. This chapter

318

focuses on the interrupts for ARM Cortex-M series. It must be noted that there are substantial differences between the ARM Cortex-M series and classical ARM versions as far as interrupt handling are concerned. The study of classical ARM and ARM Cortex A and R series interrupts are left to the reader since they are used for high performance systems using complex OS and real-time system.

Interrupt and Exception assignments in ARM Cortex-M

The NVIC of the ARM Cortex-M has room for the total of 255 interrupts and exceptions. The interrupt numbers are also referred to INT type (or INT #) in which the type can be from 1 to 255 or 0x01 to 0xFF. That is INT 01 to INT 255 (or INT 0x01 to INT 0xFF.) The NVIC in ARM Cortex-M assigns the first 15 interrupts for internal use. The memory locations 0-3 are used to store the value to be loaded into the stack pointer when the device is coming out of reset. See Table 12-1.

Interrupt #	Interrupt	Memory Location (Hex)
	Stack Pointer initial value	0x00000000
1	Reset	0x00000004
2	NMI	0x00000008
3	Hard Fault	0x0000000C
4	Memory Management Fault	0x00000010
5	Bus Fault	0x00000014
6	Usage Fault (undefined instructions, divide by zero, unaligned memory access, ...)	0x00000018
7	Reserved	0x0000001C
8	Reserved	0x00000020
9	Reserved	0x00000024
10	Reserved	0x00000028
11	SVCall	0x0000002C
12	Debug Monitor	0x00000030
13	Reserved	0x00000034
14	PendSV	0x00000038
15	SysTick	0x0000003C
16	IRQ for peripherals	0x00000040
17	IRQ for peripherals	0x00000044
...	...	...
255	IRQ for peripherals	0x000003FC

Table 12-1: Interrupt Vector Table for ARM Cortex-M

The predefined Interrupts (INT 0 to INT 15)

The followings are the first 15 interrupts in ARM Cortex-M:

Reset

When the device is coming out of reset, the ARM Cortex-M loads the program counter from memory location 0x00000004 and the SP register is initialized with the contents of addresses 0x00000000 to 0x00000003.

Non-maskable interrupt

As shown in Figure 12-2, there are pins in the ARM chip that are associated with hardware interrupts. They are IRQs (interrupt request) and NMI (nonmaskable interrupt). IRQ is an input signal into the CPU, which can be masked (ignored) and unmasked through the use of software using the I bit of the CPSR register. However, NMI, which is also an input signal into the CPU, cannot be masked by software, and for this reason it is called a *nonmaskable interrupt*. ARM Cortex-M NVIC has embedded "INT 02" into the ARM CPU to be used only for NMI. Whenever the NMI pin is activated, the CPU will go to memory location 0x0000008 to get the address of the interrupt service routine (ISR) associated with NMI. Memory locations 0x00000008, 0x00000009, 0x0000000A, and 0x0000000B contain the 4 bytes of address associated with the ISR belonging to NMI.

Exceptions (Faults)

There is a group of interrupts belongs to the category referred to as *fault* or *exception*. Internally, they are invoked by the microprocessor whenever there are conditions (exceptions) that the CPU is unable to handle. One such situation is divide-by-zero. Since the result is undefined, and the CPU has no way of handling it, it automatically invokes the usage fault interrupt. Cortex-M supports the following faults:

Hard Fault

The hard fault is an exception that occurs when the CPU having difficulties executing the ISR for any of the exceptions. One common cause of hard fault is trying to write to the registers of a peripheral before the clock is enabled for that peripheral.

Memory Management Fault

The memory manager unit fault is used for protection of memory from unwanted access. An example of memory management exception fault is when the access permission in MPU is violated by attempting to write into a region of memory designated as read-only. In an ARM chip with an on-chip MMU, the page fault can also be mapped into the memory management fault.

Bus Fault

The bus fault is an exception that occurs when there is an error in accessing the buses. This can be due to memory access problem during the fetch stage of an instruction or reading and writing to data section of memory. For example, if you try to access memory address location that has not been mapped to a memory chip or peripheral device the Bus Fault exception will occur.

Usage Fault

The ARM Cortex-M chip has implemented the divide-by-zero, unaligned memory access, undefined instruction, and so on as part of the Usage Fault exception.

Whenever an invalid instruction is executed, the CPU will go to memory location 0x00000018 to get the address of the ISR to handle the situation. The undefined instruction fault is part of the *Usage Fault* exceptions. An attempt to divide a number by zero is another Usage Fault. Since the result of dividing a number by zero is undefined, and the CPU has no way of handling such a result, it automatically invokes the exception.

SVCall

An ISR can be called upon as a result of the execution of SVC (supervisor call) instruction. This is referred to as a *software interrupt* since it was invoked from software, not from a fault exception, external hardware, or any peripheral IRQ interrupt. Whenever the SVC instruction is executed, the CPU will go to memory location 0x0000002C to get the address of the ISR associated with SVC. The SVC is widely used by the operating system to call the OS kernel functions and services that can be provided only by the privileged access mode of the OS. In many systems, the API and function calls needed by various User applications are handled by the SVCall to make sure the OS is protected. In the classical ARM literature, SVC was called SWI (software interrupt), but the ARM Cortex-M has renamed it as SVC. Again it must be noted that the SVC is an ARM Cortex-M instruction and can be used like any other ARM instruction.

PendSV (pendable service call)

The PendSV (pendable service call) can be used to do the same thing as the SVC to get the OS services. However, the SVC is an instruction and is executed right away just like all ARM instructions. The PendSV is an interrupt and can wait until NVIC has time to service it when other urgent higher priority interrupts are being taken care. Examine the concept of nested interrupt and pending interrupts at end of this section to see how NVIC handles multiple pending interrupts.

Debug Monitor

In executing a sequence of instructions, there is a need to examine the contents of the CPU's registers and system memory. This is often done by executing the program one instruction at a time and then inspecting registers and memory. This is commonly referred to as *single-stepping*, or performing a trace. ARM has designated INT 12, debug monitor, specifically for implementation of single-stepping.

SysTick

In the multitasking OS we need a real time interrupt clock to notify the CPU that it needs to service the task. The clock tick happens at a regular interval and is used mainly by the OS system. The SysTick in ARM Cortex is designed for this purpose.

IRQ Peripheral interrupts

An ISR can be launched as a result of an event at the peripheral devices such as timer timeout or analog-to-digital converter (ADC) conversion complete. The largest number of the interrupts in the ARM Cortex-M belongs to this category. Notice from Table 12-1 that ARM Cortex-M NVIC has set aside the first 15 interrupts (INT 1 to INT 15) for internal use and exceptions and is not available to chip designer. The Reset, NMI, undefined instructions, and so on are part of this group of exceptions. The rest of the interrupts can be used for peripherals. Many of the INT 16 to INT 255 are used by the chip manufacturer to be assigned to various peripherals such as timers, ADC, Serial COM, external hardware interrupts, and so on. There is no standard in assigning the INT 16 to INT 255 to the peripherals. Different manufacturers assign different interrupts to different peripherals. Each peripheral device has a group of special function registers that must be used to access the device for configuration. For a given peripheral interrupt to take effect, the interrupt for that peripheral must be enabled. The special function registers for that device provide the way to enable the interrupts.

context saving in task switching

Most of the interrupts are asynchronous, that means they may happen any time in the middle of program execution. When the interrupt is acknowledged and the interrupt service routine is launched, the interrupt service routine will need some CPU resource, mainly the CPU registers, to execute the code. In order not to corrupt the register content of the program that was running before interrupt occurs, these CPU registers need to be preserved. This saving of the CPU contents before switching to interrupt handler is called context switching (or context saving). The use of the stack as a place to save the CPU's contents is tedious and time consuming. It takes time to save all the registers. In executing an interrupt service routine, each task generally needs some key registers such as PC (R15), LR (R14), and CPSR (flag register), in addition to some working registers. For that reason, the ARM Cortex-M automatically saves the registers of CPSR, PC, LR, R12, R3, R2, R1, and R0 on stack when an interrupt is acknowledged. See Figure 12-3. If the interrupt service routine needs to use more registers than those preserved, the program has to save the content before using the other registers. The choice of the registers automatically saved adheres to the ARM Architecture Procedure Call Standard (AAPCS) so that an interrupt handler may be written as a plain C function without the need of any special provision.

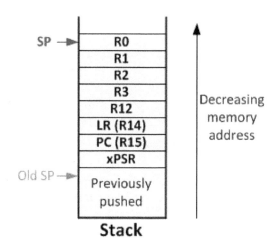

Figure 12-3: ARM Cortex-M Stack Frame upon Interrupt

Processing interrupts in ARM Cortex-M

When the ARM Cortex-M processes any interrupt (from either Fault Exceptions or peripheral IRQs), it goes through the following steps:

1. The Current processor status register (CPSR) is pushed onto the stack and SP is decremented by 4, since CPSR is a 4-byte register.
2. The current PC (R15) is pushed onto the stack and SP is decremented by 4.
3. The current LR (R14) is pushed onto the stack and SP is decremented by 4.
4. The current R12 is pushed onto the stack and SP is decremented by 4.
5. The current R3 is pushed onto the stack and SP is decremented by 4.
6. The current R2 is pushed onto the stack and SP is decremented by 4.
7. The current R1 is pushed onto the stack and SP is decremented by 4.
8. The current R0 is pushed onto the stack and SP is decremented by 4.

9. The CPU goes into the Handler Mode (details will be described later). LR is loaded with a number with bit 31-5 all 1s.

10. The INT number (type) is multiplied by 4 to get the address of the location within the vector table to fetch the program counter of the interrupt service routine (interrupt handler).

11. From the memory locations pointed to by this new PC, the CPU starts to fetch and execute instructions belonging to the ISR program.

12. When one of the return instructions is executed in the interrupt service routine, the CPU recognizes that it is in the Handler Mode from the value of the LR. It then restores the registers saved when entering ISR including the program counter from the stack and makes the CPU run the code where it left off when interrupt occurred. See Figure 12-4.

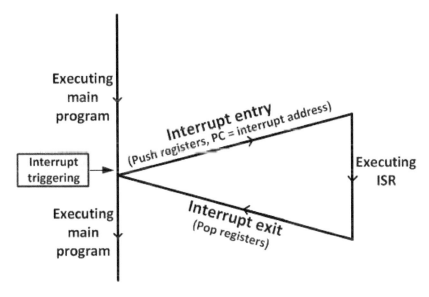

Figure 12-4: Main Program gets Interrupted

Difference between interrupt and a subroutine call

If the execution of an interrupt saves the program counter of the following instruction and jumps indirectly to the subroutine associated with the interrupt, what is the difference between that and a BL instruction, which also saves the program counter and jumps to the desired subroutine (procedure)? The differences can be summarized as follows:

1. A "BL" instruction can take an argument and jump to any location within the 4-gigabyte address range of the ARM CPU, but "INT" goes to a fixed memory location in the interrupt vector table to get the address of the interrupt service routine.

2. A "BL" instruction is used by the programmer in the sequence of instructions in the program but an externally activated hardware interrupt can come in at any time, requesting the attention of the CPU.

3. A "BL" instruction cannot be masked (disabled), but "INT#" belonging to externally activated hardware interrupts can be masked.

4. A "BL" instruction automatically saves only PC of the next instruction on the stack, while "INT#" saves SP, R12, R3–R0, CPSR (flag register) in addition to PC of the next instruction.

323

5. An interrupt puts the CPU in the Handler Mode while the "BL" instruction does not change the CPU execution mode.

6. When returning from the end of the subroutine that has been called by the "BL" instruction, the PC is restored to the address of the next instruction after the "BL" instruction. When returning from the interrupt handler, the CPU will restore the registers saved when the CPU entered into ISR (the CPSR, R15, R14, R12, R3–R0 registers) from the top of stack.

Vector Table Implementation

The vector table can be found in the start-up header file of your compiler. For example, when you make a new project in Keil for STM32F10x chips, the startup_stm32f10x_md.s is added to your project. See Figure 12-5 and compare it with Table 12-1. Using DCD, 4 bytes of memory is allocated for each vector. The names Reset_Handler, SVC_Handler, SysTick_Handler, and so on are names of the interrupt service routines (ISRs). For example, if you want to write an interrupt handler for SysTick, you should write a function (in Assembly or C) and name it SysTick_Handler. When the SysTick interrupt is generated, the program counter is loaded with the address of the SysTick_Handler function. So, the SysTick_Handler is executed. Next, we write a simple program for the SysTick Interrupt.

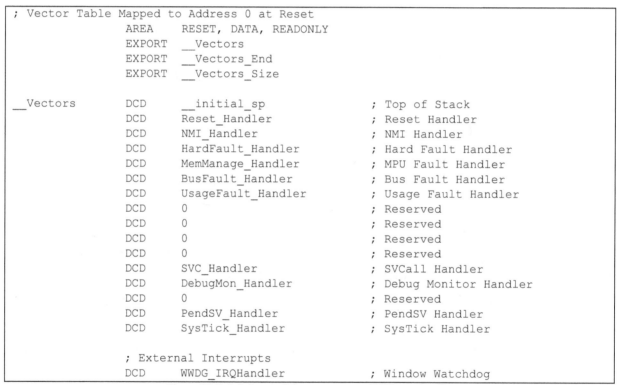

```
; Vector Table Mapped to Address 0 at Reset
                AREA    RESET, DATA, READONLY
                EXPORT  __Vectors
                EXPORT  __Vectors_End
                EXPORT  __Vectors_Size

__Vectors       DCD     __initial_sp            ; Top of Stack
                DCD     Reset_Handler           ; Reset Handler
                DCD     NMI_Handler             ; NMI Handler
                DCD     HardFault_Handler       ; Hard Fault Handler
                DCD     MemManage_Handler       ; MPU Fault Handler
                DCD     BusFault_Handler        ; Bus Fault Handler
                DCD     UsageFault_Handler      ; Usage Fault Handler
                DCD     0                       ; Reserved
                DCD     0                       ; Reserved
                DCD     0                       ; Reserved
                DCD     0                       ; Reserved
                DCD     SVC_Handler             ; SVCall Handler
                DCD     DebugMon_Handler        ; Debug Monitor Handler
                DCD     0                       ; Reserved
                DCD     PendSV_Handler          ; PendSV Handler
                DCD     SysTick_Handler         ; SysTick Handler

                ; External Interrupts
                DCD     WWDG_IRQHandler         ; Window Watchdog
```

Figure 12-5: A Snippet of the startup_stm32f10x_md.s File

Review Questions
1. True or false. When any interrupt is activated, the CPU jumps to a fixed and unique address.
2. There are _____ bytes of memory in the interrupt vector table for each interrupt.
3. How many bytes of memory are used by the interrupt vector table, and what are the beginning and ending addresses of the table for the first 256 interrupts?
4. The program associated with an interrupt is also referred to as _____.

5. What is the function of the interrupt vector table?
6. What memory locations in the interrupt vector table hold the address for INT 16 ISR?
7. The ARM Cortex-M has assigned INT 2 to NMI. Can that be changed?
8. Which interrupt is assigned to divide error exception handling?

Section 12.2: SysTick Programming and Interrupt

The 15th Cortex predefined interrupt is SysTick. The predefined interrupts are implemented the same way in all Cortex microcontrollers. So, SysTick can be used in the same way in ARM Cortexes.

As discussed in Chapter 11, the SysTick is a 24-bit down counter driven by the system clock. It counts down from an initial value to 0. When it reaches 0, in the next clock, it underflows and it raises the COUNT flag and reloads the initial value and starts all over. See Figure 12-5.

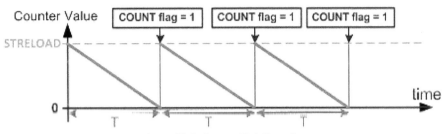

Figure 12-6: System Tick Counting

See Figure 12-6. If TICINT=1, when COUNTFLAG is set, it generates an interrupt. TICKINT is D1 of the STCTRL register, as shown in Figure 12-7.

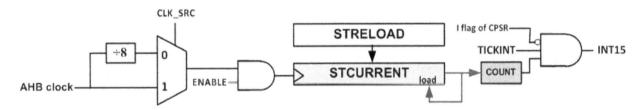

Figure 12-7: SysTick Internal Structure

	D31	.....	D17	D16	D15	...	D3	D2	D1	D0	
STCTRL:	Reserved			COUNT FLAG	Reserved			CLK SOURCE	TICKINT	ENABLE	0x010

SysTick->CTRL

Name	bit	Description
ENABLE	0	0: The counter is disabled, 1: enables SysTick to begin counting down
TICKINT	1	Interrupt Enable 0: Interrupt generation is disabled. 1: when SysTick counts to 0 an interrupt is generated
CLKSOURCE	2	Clock Source (0: AHB clock divided by 8, 1: AHB clock)
COUNTFLAG	16	Count flag

Figure 12-8: SysTick Control and status register (STCTRL)

325

The SysTick interrupt can be used to initiate an action on a periodic basis. For example, in a given application we can use SysTick to read a sensor every 200 msec. SysTick is used widely for an operating system so that the system software may interrupt the application software periodically (often at 10 ms interval) to monitor and control the system operations.

Writing the first Interrupt Program

The Program 12-1 uses the SysTick to toggle the LED of PC13 every second. This program is similar to the program of Example 11-6. But it uses the interrupt.

In the program, the COUNTFLAG is raised every 72,000,000 system clocks (which is 1 second) and the timer is loaded with 9,000,000 − 1 automatically.

In the program, the TICKINT bit of the SysTick->Ctrl is set. So, whenever the COUNTFLAG is set, an interrupt is generated. This causes the CPU to push the values of registers CPSR, PC (R15), R14 (LR), R12, R3, R2, R1, and R0 onto the stack and load LR with 0xFFFFFFF9 (to show that it is in interrupt handler mode) and then the mode changes to handler mode and the PC is loaded with the address of the SysTick_Handler routine and the routine begins executing.

When the execution of the SysTick_Handler ends, the BX LR instruction is executed. But since the LR register contains 0xFFFFFFF9, the CPU pops the values of the registers including the program counter from the stack, and runs the code where it left off when interrupt occurred.

Notice that the COUNTFLAG is cleared automatically by hardware and there is no need to clear COUNTFLAG in the interrupt handler.

Program 12-1: SysTick interrupt

```c
#include <stm32f10x.h>

void SysTick_Handler()
{
    GPIOC->ODR ^= (1<<13); /* toggle PC13 */
}

int main()
{
  RCC->APB2ENR = 0xFC; /* enable GPIO clocks */
  GPIOC->CRH = 0x44344444; /* PC13 as output */

  SysTick->LOAD = 9000000-1;    /* STRELOAD = 72,000,000/8 -1 */
  SysTick->CTRL = 0x03;         /* Clock = AHB clock/8, TickInt enable, Enable = 1 */

  while(1)
  {
  }
}
```

Review Questions
1. Which interrupt is assigned to SysTick?
2. We use register _____ to enable the interrupt associated with SysTick.
3. True or false. We use NVIC registers to enable SysTick interrupt.

Section 12.3: STM32 I/O Port Interrupt Programming

In Chapter 8, we showed how to use GPIO ports for simple I/O. We also showed a simple program getting (polling) an input switch and placing it on LED. In this section, we show how to program the interrupt capability of the I/O ports. However, before we do that, we need to examine the NVIC and vector table for the STM32F10x chips. As discussed in Section 12-1, INT 16 to INT 255 are for peripheral interrupts and each vendor implements them as they please. Table 12-2 shows interrupt assignment in STM32F10x.

INT#	IRQ#	Vector location	Acronym	Device
1-15	None	0000 0000 to 0000 003C		CPU Exception (set by ARM) See Table 12-1.
16	0	0000 0040	WWDG	Window Watchdog interrupt
17	1	0000 0044	PVD	PVD through EXTI line detection interrupt
18	2	0000 0048	TAMPER	Tamper interrupt
19	3	0000 004C	RTC	RTC global interrupt
20	4	0000 0050	FLASH	Flash global interrupt
21	5	0000 0054	RCC	RCC global interrupt
22	6	0000 0058	EXTI0	EXTI line 0 interrupt
23	7	0000 005C	EXTI1	EXTI line 1 interrupt
24	8	0000 0060	EXTI2	EXTI line 2 interrupt
25	9	0000 0064	EXTI3	EXTI line 3 interrupt
26	10	0000 0068	EXTI4	EXTI line 4 interrupt
27	11	0000 006C	DMA1_Channel1	DMA1 Channel1 global interrupt
28	12	0000 0070	DMA1_Channel2	DMA1 Channel2 global interrupt
29	13	0000 0074	DMA1_Channel3	DMA1 Channel3 global interrupt
30	14	0000 0078	DMA1_Channel4	DMA1 Channel4 global interrupt
31	15	0000 007C	DMA1_Channel5	DMA1 Channel5 global interrupt
32	16	0000 0080	DMA1_Channel6	DMA1 Channel6 global interrupt
33	17	0000 0084	DMA1_Channel7	DMA1 Channel7 global interrupt
34	18	0000 0088	ADC1_2	ADC1 and ADC2 global interrupt
35	19	0000 008C	CAN1_TX	CAN1 TX interrupts
36	20	0000 0090	CAN1_RX0	CAN1 RX0 interrupts
37	21	0000 0094	CAN1_RX1	CAN1 RX1 interrupt
38	22	0000 0098	CAN1_SCE	CAN1 SCE interrupt
39	23	0000 009C	EXTI9_5	EXTI Line[5:9] interrupts
40	24	0000 00A0	TIM1_BRK	TIM1 Break interrupt
41	25	0000 00A4	TIM1_UP	TIM1 Update interrupt
42	26	0000 00A8	TIM1_TRG_COM	TIM1 Trigger and Commutation interrupts
43	27	0000 00AC	TIM1_CC	TIM1 Capture Compare interrupt
44	28	0000 00B0	TIM2	TIM2 global interrupt
45	29	0000 00B4	TIM3	TIM3 global interrupt

46	30	0000 00B8	TIM4	TIM4 global interrupt
47	31	0000-00BC	I2C1_EV	I2C1 event interrupt
48	32	0000-00C0	I2C1_ER	I2C1 error interrupt
49	33	0000-00C4	I2C2_EV	I2C2 event interrupt
50	34	0000-00C8	I2C2_ER	I2C2 error interrupt
51	35	0000-00CC	SPI1	SPI1 global interrupt
52	36	0000-00D0	SPI2	SPI2 global interrupt
53	37	0000-00D4	USART1	USART1 global interrupt
54	38	0000-00D8	USART2	USART2 global interrupt
55	39	0000-00DC	USART3	USART3 global interrupt
56	40	0000-00E0	EXTI15-10	EXTI Line[15:10] interrupts
57	41	0000-00E4	RTCAlarm	RTC alarm through EXTI line interrupt
58	42	0000-00E8	USBWakeup	USB wakeup from suspend through EXTI line interrupt
59	43	0000-00EC	TIM8_BRK	TIM8 Break interrupt
60	44	0000-00F0	TIM8_UP	TIM8 Update interrupt
61	45	0000-00F4	TIM8_TRG_COM	TIM8 Trigger and Commutation interrupts
62	46	0000-00F8	TIM8_CC	TIM8 Capture Compare interrupt
63	47	0000-00FC	ADC3	ADC3 global interrupt
64	48	0000-0100	FSMC	FSMC global interrupt
65	49	0000-0104	SDIO	SDIO global interrupt
66	50	0000-0108	TIM5	TIM5 global interrupt
67	51	0000-010C	SPI3	SPI3 global interrupt
68	52	0000-0110	UART4	UART4 global interrupt
69	53	0000-0114	UART5	UART5 global interrupt
70	54	0000-0118	TIM6	TIM6 global interrupt
71	55	0000-011C	TIM7	TIM7 global interrupt
72	56	0000-0120	DMA2_Channel1	DMA2 Channel1 global interrupt
73	57	0000-0124	DMA2_Channel2	DMA2 Channel2 global interrupt
74	58	0000-0128	DMA2_Channel3	DMA2 Channel3 global interrupt
75	59	0000-012C	DMA2_Channel4_5	DMA2 Channel4 and DMA2 Channel5 global interrupts

Table 12-2: IRQ assignment in STM32F10x Chips

In Cortex-M series NVIC is used to serve the interrupts. So, the NVIC registers are used to enable/disable the interrupts and set the priority of the interrupts.

NVIC has some interrupt set enable registers (ISER). Each bit of the registers is for an entry of the interrupt vector table. So, each register covers 32 peripheral interrupts. For example, register ISER[0] controls the enable the interrupts for IRQ0 to IRQ31, ISER[1] for IRQ32 to IRQ63, and so on. See Figure 12-10. Notice, we have an array for all of the ISER registers. The array is referred to as ISER[0], ISER[1], and so on. The STM32F10x chips have a total of 60 peripheral interrupts and only ISER[0] and ISER[1] are used.

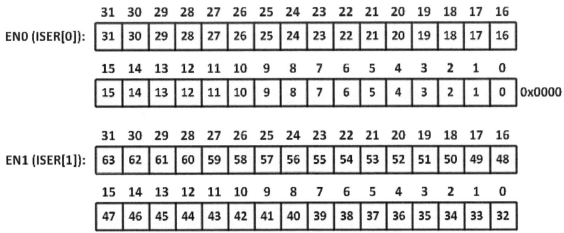

Figure 12-9: Interrupts 0–31 Set Enable (EN0)

As we can see in Table 12-2, the external line 1 interrupt (EXTI1) is assigned to IRQ7. Therefore, to enable the interrupt associated with EXTI1 in Vector table, we need the following:

```
NVIC->ISER[0] = 1<<7;    /* enable IRQ7 (bit 7 of ISER[0]) */
```

The interrupts can be enabled using the following function, as well:

```
void NVIC_EnableIRQ(IRQn_Type IRQn);
```

The function is defined in the *core_cm3.h* file which is included in the *stm32f10x.h* header file. To enable an interrupt using this function, the IRQ number of the interrupt should be passed as the argument to the function. For example, the following statement enables EXTI interrupt:

```
NVIC_EnableIRQ(7);
```

Since the IRQ numbers of all the interrupts are defined in the *stm32f10x.h*, we can use their names instead of their numbers. For example, to enable EXTI1 interrupt the following can be used as well:

```
    NVIC_EnableIRQ(EXTI1_IRQn);
```

In the above instruction EXTI1_IRQn is made of EXTI1 (the acronym for external line 1) and _IRQn. For any interrupts, you can simply check their acronyms in Table 12-2 and add _IRQn to them. For more information, open the *stm32f10x.h* file and find "typedef enum IRQn" in the file.

To disable interrupts there are other registers: ICER0 to ICER3. Again because STM32F10x devices have 60 IRQs, only ICER[0] and ICER[1] are used. See the Figure 12-10.

Each interrupt can be disabled by writing a 1 to the corresponding bit in the ICER registers. Writing 0 to the ICER registers has no effect on their values. For example, the following instruction disables USART1 interrupt, keeping the other interrupts unchanged:

```
NVIC->ICER[1] = 1<<5;    /*disable bit 37 (bit 5 of ICER[1]) to disable USART1 Int. */
```

The interrupts can be disabled using the following function, as well:

```
void NVIC_DisableIRQ(IRQn_Type IRQn);
```

For example, the following instruction disables the UART0 interrupt:

```
NVIC_DisableIRQ(UART0_IRQn);
```

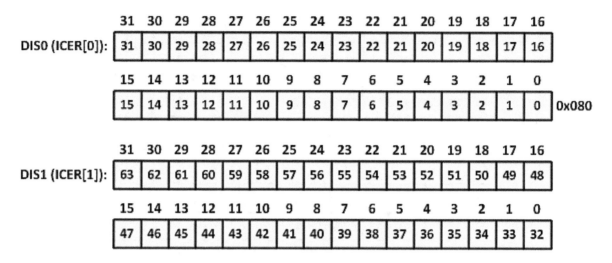

Figure 12-10: Interrupts 0–31 Clear Enable (DIS0)

In fact, each bit of the ISER register together with its peer in the ICER register is connected to a J-K Flip-Flop, as shown below:

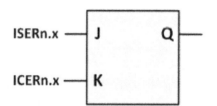

Figure 12-11: Enabling and Disabling an Interrupt

About the I (Interrupt) flag

The CPSR register contains the I flag. On reset the flag is 0. When the I flag is 1, all the interrupts (except the NMI interrupt) are disabled. Global interrupt enable/disable allows us with a single instruction to mask all interrupts during the execution of critical tasks such as manipulating a common pointer shared by multiple threads. In ARM Cortex-M, we do the global enable/disable of interrupts with assembly language instructions of CPSID I (Change processor state-disable interrupts) and CPSIE I (Change processor state-enable interrupts). In Keil C language, we use pseudo-functions:

```
__enable_irq();     /* Enable interrupt Globally */
        and

__disable_irq();    /* Disable interrupt Globally */
```

It is a good idea to disable all interrupts during the initialization of the program and enable interrupts after all the initializations are complete.

GPIO Interrupts (EXTIn)

In STM32 chips, all the GPIO pins are connected as shown in Figure 12-12. Pins PAn, PBn, PCn, …, and PGn are multiplexed using EXTIn field of the AFIO_EXTICR (External Interrupt Configuration Register).

The bits n of the EXTI_RTSR (Rising Trigger Selection Register) and EXTI_FTSR (Falling Trigger Selection Register) registers are used to choose if the interrupt occurs in falling/rising edges. If the bit n of the RTSR register is 1, the interrupt occurs on rising edges. If the bit n of the FTSR register is 1, the interrupts are generated on the falling edges. We can also set the bit n of both RTSR and FTSR registers to generate interrupt on both edges.

Whenever the desired edge is detected the bit n of the EXTI_PR is set. Now, if the bit n of the EXTI_IMR register is 1, an EXTIn interrupt is generated.

See Table 12-2. The interrupt vectors IRQ6 to IRQ10 are for EXTI0 to EXTI4. Notice that EXTI5 to EXTI9 use the same interrupt vector and they use IRQ23 (EXTI9_5). In the same way, EXTI10 to EXTI15 use IRQ40 (EXTI15_10).

Figures 12-13 to 12-17 show the registers used in the EXTI circuit in detail.

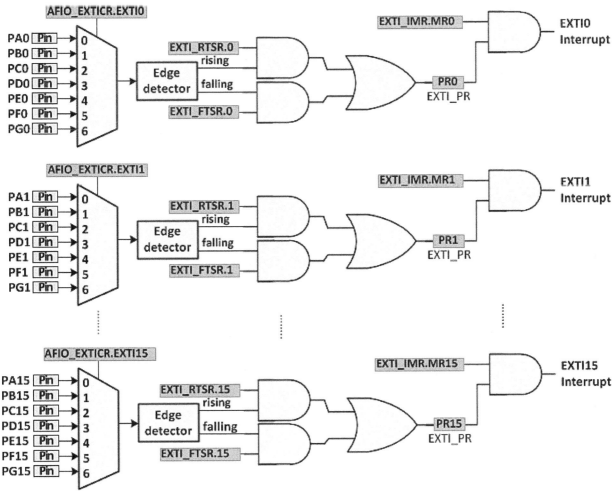

Figure 12-12: External Line Interrupt (EXTI) Circuits

331

AFIO_EXTICR Registers

	D31 ..... D16	D15 D14 D13 D12	D11 D10 D9 D8	D7 D6 D5 D4	D3 D2 D1 D0	
AFIO_EXTICR1: AFIO->EXTICR[0]	Reserved	EXTI3	EXTI2	EXTI1	EXTI0	0x08
AFIO_EXTICR2: AFIO->EXTICR[1]	Reserved	EXTI7	EXTI6	EXTI5	EXTI4	0x0C
AFIO_EXTICR3: AFIO->EXTICR[2]	Reserved	EXTI11	EXTI10	EXTI9	EXTI8	0x10
AFIO_EXTICR4: AFIO->EXTICR[3]	Reserved	EXTI15	EXTI14	EXTI13	EXTI12	0x14

EXTIn: It selects the input pin for line EXTIn
0000: PAn, 0001: PBn, 0010: PCn, 0011: PDn, 0100: PEn, 0101: PFn, 0110: PGn

Figure 12-13: AFIO_EXTICRx (External Interrupt Configuration Register x)

EXTI_RTSR: EXTI->RTSR

D31 ..... D20	D19	D18	D17	D16	D15	D14	D13	D12	D11	D10	D9	D8	D7	D6	D5	D4	D3	D2	D1	D0	
Reserved	TR19	TR18	TR17	TR16	TR15	TR14	TR13	TR12	TR11	TR10	TR9	TR8	TR7	TR6	TR5	TR4	TR3	TR2	TR1	TR0	0x08

TRn: Rising trigger event configuration bit of line n (0: rising trigger disabled, 1: rising trigger enabled)

Figure 12-14: RTSR (Rising Trigger Selection Register)

EXTI_FTSR: EXTI->FTSR

D31 ..... D20	D19	D18	D17	D16	D15	D14	D13	D12	D11	D10	D9	D8	D7	D6	D5	D4	D3	D2	D1	D0	
Reserved	TR19	TR18	TR17	TR16	TR15	TR14	TR13	TR12	TR11	TR10	TR9	TR8	TR7	TR6	TR5	TR4	TR3	TR2	TR1	TR0	0x0C

TRn: Falling trigger event configuration bit of line n (0: falling trigger disabled, 1: falling trigger enabled)

Figure 12-15: FTSR (Falling Trigger Selection Register)

EXTI_PR: EXTI->PR

D31 ..... D20	D19	D18	D17	D16	D15	D14	D13	D12	D11	D10	D9	D8	D7	D6	D5	D4	D3	D2	D1	D0	
Reserved	PR19	PR18	PR17	PR16	PR15	PR14	PR13	PR12	PR11	PR10	PR9	PR8	PR7	PR6	PR5	PR4	PR3	PR2	PR1	PR0	0x14

PRn: Pending for line n (0: Not trigger is pending, 1: a trigger is pending)
Note: to bits are cleared by writing 1 to the bits

Figure 12-16: PR (Pending Register)

EXTI_IMR: EXTI->IMR

D31 ..... D20	D19	D18	D17	D16	D15	D14	D13	D12	D11	D10	D9	D8	D7	D6	D5	D4	D3	D2	D1	D0	
Reserved	MR19	MR18	MR17	MR16	MR15	MR14	MR13	MR12	MR11	MR10	MR9	MR8	MR7	MR6	MR5	MR4	MR3	MR2	MR1	MR0	0x00

MRn: mask for external interrupt (EXTI) line n (0: interrupt is masked (disabled), 1: unmasked(enabled))

Figure 12-17: IMR (Interrupt Mask Register)

Steps to use EXTI interrupt

To use the EXTI interrupt, the following steps must be taken:

1. Enable the clocks for the GPIOs and the AFIO,
2. Configure the interrupt input pin as input using the CRL and CRH registers
3. Select the interrupt input pin using AFIO_EXTICR,
4. Configure the RTSR and FTSR registers to choose the interrupt edge,
5. Set the EXTI_IMR register to enable the interrupt for the line,
6. Enable the EXTI interrupt using the ISER register or the NVIC_EnableIRQ function,
7. Clear the PR flag in the interrupt routine.

See Example 12-1 and Program 12-2. In program 12-2, PB4 is selected as the input for external interrupt. On each falling edge, an interrupt is generated. The interrupt service routine toggles PC13. So, if you connect a momentary switch to PB4, PC13 toggles when the switch is pressed. At the same time, the main program toggles PA2, every second. To check the program, you can also connect PA2 to PB4 using a wire.

Example 12-1

Find the value for AFIO_EXTICR registers to select PB4 for the external interrupt.

Solution:

The pin 4 of the ports are connected to line 4. So, the EXTI4 field of the EXTICR2 register should be configured. See Figure 12-13. To select PORTB, the EXTI4 must be initialized with 0001.

AFIO_EXTICR2 (AFIO->EXTICR[1])	Res.	EXTI7	EXTI6	EXTI5	EXTI4
		0000	0000	0000	**0001**

Program 12-2: Using PB4 external interrupt

```
#include <stm32f10x.h>

void delay_ms(uint16_t t);

void EXTI4_IRQHandler()     /* interrupt handler for EXTI4 */
{
   EXTI->PR = (1<<4);       /* clear the Pending flag */
   GPIOC->ODR ^= 1<<13;     /* toggle PC13 */
}

int main()
{
   RCC->APB2ENR |= (0xFC | 1); /* Enable clocks for GPIO ports and AFIO */

   GPIOB->CRL = 0x44484444; /* PB4 as input */
   GPIOB->ODR |= (1<<4);    /* pull-up PB4 */
   GPIOC->CRH = 0x44344444; /* PC13 as output */

   AFIO->EXTICR[1] = 1<<0; /* EXTI4 = 1 (selects PB4 for line 4) */

   EXTI->FTSR = (1<<4);     /* int. on falling edge */

   EXTI->IMR = (1<<4); /* enable interrupt EXTI4 */
   NVIC_EnableIRQ(EXTI4_IRQn);    /* enable the EXTI4 interrupt */

   GPIOA->CRL = 0x44444344; /* PA2 as output */

   while(1)
   {
```

```
        GPIOA->ODR ^= (1<<2); /* toggle PA2 */
        delay_ms(1000);          /* wait 1 second */
    }
}
/* copy the implementation of delay_ms from Program 8-1 */
```

Clearing the interrupt flag

It is critical to clear the interrupt flag in the interrupt handler. Otherwise the interrupt appears as if it is still pending and the interrupt handler will be executed again and again forever and the program hangs.

Notice that, to clear the pending interrupt flag, the program writes a 1 to the location of the flag in the EXTI_PR register. Writing a zero to the flag has no effect. So, to clear the interrupt flag of external interrupt line 4 (EXTI4), the following statement is used:

```
EXTI->PR = (1<<4); /* clear the flag for EXTI4 and do not change the other flags */
```

Review Questions

1. IRQ0 is assigned to INT number____.
2. We use _____in C to enable the interrupts globally.
3. True or false. The I/O ports in STM32 support both falling and rising edge trigger interrupts.
4. True or false. A separate interrupt vector is assigned to each pin of the STM32 chips.

Section 12.4: USART Serial Port Interrupt Programming

In Chapter 10, we showed the programming of USARTs in STM32F10x using polling. This chapter shows how to do the same thing using interrupt. Using interrupt frees up the CPU from having to poll the status of USART.

USART Interrupt Programming to receive data

Example 10-9 showed how USART2 receives data by polling the RXNE status flag. The disadvantage with that program is that it ties down the CPU while polling the status flag. We can modify it to make it an interrupt driven program. Examining the USARTx_CR1 (USARTx Control Register 1), we see bit 5 allows us to enable the receiver interrupt. If the receiver interrupt for USART is enabled when a byte is received, the receiver RXNE flag is directed to NVIC and that causes the interrupt handler associated with the USART to be executed. In the USART handler we must read the received character. Reading the received character from the data register clears the RXNE flag. See Figure 12-14 and Table 12-9.

From Table 12-2 we see IRQ37 is assigned to USART1. We enable the receiver interrupt in USART1 as follow:

```
USART1->CR1 = 0x2024;              /* enable receive and receive interrupt*/
NVIC_Enable_IRQ(USART1_IRQ);       /* enable USART1 interrupt */
```

	15	14	13	12	11	10	9	8	7	6	5	4	3	2	1	0
USARTx_CR1:	Reserved		UE	M	WAKE	PCE	PS	PEIE	TXEIE	TCIE	RXNEIE	IDLEIE	TE	RE	RWU	SBK

(USARTx->CR1)

Figure 12-18: USARTx_CR1 (USART Control Register)

334

Field	Bit	Description
UE	D13	0 = USART prescaler and outputs disabled. 1 = USART enabled
M	D12	Data format mode bit. We must use this to select 8-bit data frame size 0 = select 8-bit data frame and one start bit 1 = Select 9-bit data frame and one start bit
WAKE	D11	Wake-up condition bit. See the user manual 0 = Idle line wakeup 1 = Address mark wake-up
PCE	D10	Parity Control Enable bit. This will insert a parity bit right after the MSB bit. 0 = no parity bit 1 = parity bit
PS	D9	Parity select (used only if PE is one.) 0 = even parity bit 1 = odd parity bit
PEIE	D8	PE interrupt enable 0 = disabled 1 = A USART interrupt is generated whenever the PE flag of USART_SR is set.
TXEIE	D7	TXE interrupt enable 0 = disabled 1 = A USART interrupt is generated whenever the TXE flag of USART_SR is set.
TCIE	D6	Transmission complete interrupt enable 0 = disabled 1 = A USART interrupt is generated whenever the TC flag of USART_SR is set.
RXNEIE	D5	RXNE interrupt enable 0 = disabled 1 = A USART interrupt is generated whenever ORE or RXNE are set.
IDLEIE	D4	Idle interrupt enable 0 = disabled 1 = A USART interrupt is generated whenever the IDLE bit of USART_SR is set.
TE	D3	Transmitter enable (The bit enables the USART transmitter) 0 = transmitter is disabled 1 = transmitter is enabled
RE	D2	Receiver enable (The bit enables the USART receiver) 0 = disable the receiver 1 = enable the receiver
RWU	D1	Receiver wakeup 0 = Receiver in active mode 1 = Receiver in mute mode (See the user manual for more information)
SBK	D0	Send break (0 = do not send, 1 = send break character) See the user manual.

Table 12-3: USART_CR1

Program 12-3, is similar to the program of Example 10-10. It receives ASCII codes through USART1. When a character is received, the interrupt handler reads the character. If the received character is L, it makes PC13 Low. If it is H, it makes PC13 high.

```
#include "stm32f10x.h"

/* The program receives a character through USART1. */
/* If the character is L, it makes the PC13 Low. If it is H, PC13 becomes High. */

void USART1_IRQHandler()   /* USART1 interrupt routine */
{
   uint8_t c = USART1->DR; /* get received data */

   if((c == 'H')||(c == 'h'))
      GPIOC->ODR |= (1<<13);        /* make PC13 high */
   else
      if((c == 'L')||(c == 'l'))
         GPIOC->ODR &= ~(1<<13);   /* make PC13 low */
}

int main()
{
   RCC->APB2ENR |= 0xFC|(1<<14); /* enable GPIO and usart1 clocks */

   GPIOC->CRH = 0x44344444; /* PC13 as output */

   /* USART1 init. */
   GPIOA->ODR |= (1<<10);   /* pull-up PA10 */
   GPIOA->CRH = 0x444448B4; /* RX1=input with pull-up, TX1=alt. func output */
   USART1->CR1 = 0x2024;    /* receive int. enable, receive enable */
   USART1->BRR = 7500;         /* 72MHz/9600bps = 7500 */

   NVIC_EnableIRQ(USART1_IRQn);   /* USART1 IRQ enable */

   while(1)
   {
   }
}
```

Sending data Using USART Interrupt

See Figure 12-18 and Table 12-3. The TCIE bit of USARTx_CR1 is used to enable the transmit interrupt. If the transmitter is ready to send a new byte and the TCIE bit is set, an interrupt is generated. In the interrupt service routine, you load the USARTx_DR register with the next data to be sent. See Program 12-4.

```
/* The program reports the state of PB5 through USART1. */
/* If PB5 is L, it sends L. Otherwise, it sends H. */

#include "stm32f10x.h"

void USART1_IRQHandler()   /* USART1 interrupt routine */
{
   USART1->SR &= ~(1<<6); /* clear TC flag */
```

```
  if((GPIOB->IDR & (1<<5)) != 0) /* is PB5 high? */
     USART1->DR = 'H';      /* send 'H' */
  else
     USART1->DR = 'L';      /* send 'L' */
}

int main()
{
  RCC->APB2ENR |= 0xFC|(1<<14); /* enable GPIO and usart1 clocks */

  GPIOB->CRL = 0x44844444; /* PB5 as input */
  GPIOB->ODR |= (1<<5);    /* pull-up PB5 */

  /* USART1 init. */
  GPIOA->ODR |= (1<<10);   /* pull-up PA10 */
  GPIOA->CRH = 0x444448B4; /* RX1=input with pull-up, TX1=alt. func. output */
  USART1->CR1 = 0x2048;    /* TC int. enable, transmit enable */
  USART1->BRR = 7500;      /* 72MHz/9600bps = 7500 */

  NVIC_EnableIRQ(USART1_IRQn);   /* USART1 IRQ enable */

  while(1)
  {
  }
}
```

Notice that there is only a single interrupt for both receiver and transmitter of USARTs. If we want to implement both transmitter and receiver interrupts, then we have to test the TC and RXNE bits in register USARTx_CR1 to see which one caused the interrupt.

Review Questions

1. In STM32F103, Which IRQ is assigned to USART1?
2. True or false. There is only one interrupt for both Receiver and Transmitter.
3. We use register _____ to enable the interrupt associated with USART.
4. True or false. Upon Reset, the USART1 interrupt is enabled and ready to go.

Section 12.5: Timer Interrupt Programming

In Chapter 11, we showed how to program the timers. In those programming examples, we used polling to see if a timeout event occurred. In this section, we give interrupt-based version of those programs.

Examine the programs in Section 11.2 of Chapter 11. Notice, we could run those programs only one at a time since we have to monitor the timer flag continuously. By using interrupt, we can run several of timer programs all at the same time. To do that, we need to enable the timer interrupt using the UIE (Update Interrupt Enable) in TIMx_DIER (DMA/Interrupt Enable Register).

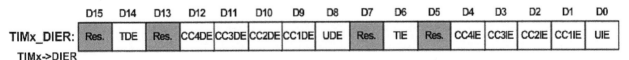

	D15	D14	D13	D12	D11	D10	D9	D8	D7	D6	D5	D4	D3	D2	D1	D0
TIMx_DIER:	Res.	TDE	Res.	CC4DE	CC3DE	CC2DE	CC1DE	UDE	Res.	TIE	Res.	CC4IE	CC3IE	CC2IE	CC1IE	UIE

TIMx->DIER

Figure 12-19: TIM_DIER (DMA/Interrupt Enable Register)

Field	Bit	Description
UIE	D0	Update interrupt enable (0: disabled, 1: enabled)
		If UIE=1, whenever UIF is set an interrupt is generated.
CCnIE	D1-D4	Capture/Compare n Interrupt Enable (0: disabled, 1: enabled)
		If the bit is set, an interrupt is generated when the CCnIF flag is set.
TIE	D6	Trigger Interrupt Enable (0: disabled, 1: enabled)
UDE	D8	Update DMA request Enable
CCnDE	D9-D12	Capture/Compare n DMA request Enable (0: disabled, 1: enabled)
TDE	D14	Trigger DMA request Enable

See Table 12-2. Notice that IRQ44, IRQ45, and IRQ46 are assigned to TIM2, TIM3, and TIM4, respectively.

In Program 12-5, the main program toggles PB5 continuously. The TIM2 interrupt, toggles PC13, every second.

Program 12-5: Toggling the green LED using the Timer interrupt

```
/* This program toggles PC13 using interrupt, every second. */
#include <stm32f10x.h>

void TIM2_IRQHandler()
{
   TIM2->SR = 0; /* clear UIF flag */
   GPIOC->ODR ^= (1<<13); /* toggle PC13 */
}

int main()
{
   RCC->APB2ENR |= 0xFC;    /* enable GPIO clocks */
   RCC->APB1ENR |= (1<<0);        /* enable TIM2 clock */
   GPIOC->CRH = 0x44344444;       /* PC13 as output */

   TIM2->PSC = 7200-1;              /* PSC = 7199 */
   TIM2->ARR = 10000-1;
   TIM2->SR = 0;      /* clear the UIF flag */
   TIM2->CR1 = 1;     /* up counting */
   TIM2->DIER = (1<<0);     /* enable UIE interrupt */
   NVIC_EnableIRQ(TIM2_IRQn);     /* enable TIM2 interrupt */

   GPIOB->CRL = 0x44344444;        /* PB5 as output */

   while(1)
   {
     GPIOB->ODR ^= (1<<5); /* toggle PB5 */
   }
}
```

Review Questions

1. In STM32F10x chips, which IRQ is assigned to TIM2?
2. True or false. There is only one interrupt for all the TIM2 through TIM4.
3. We use register _____to enable the interrupts associated with TIM2.
4. True or false. Upon Reset, TIM3 interrupt is enabled and ready to go.

Section 12.6: Interrupt Priority, nested interrupts, and latency

The next topic in this chapter is the concept of priority for exceptions and IRQs. What happens if two interrupts want the attention of the CPU at the same time? Which has priority? In the ARM Cortex-M the Reset, NMI and Hard Fault exceptions have fixed priority levels and are set by the ARM itself. Among the Reset, NMI and Hard Fault, the Reset has the highest priority. See Table 12-6. The lower priority value means the higher priority. As you can see from Table 12-6, the NMI and Hard Fault have lower priority than Reset, meaning if all three of them are activated at the same time, the Reset will be executed first.

The priority levels for the rest of the exceptions and IRQs are set to zero by default and the programmer can change their priority levels. When the priority levels of some interrupts are the same, the interrupts with lower interrupt number have higher priority. For example, if both interrupts IRQ5 and IRQ6 are generated at the same time while they have the same priority level, then IRQ5 is served first.

Int. #	Interrupt	Priority Level
	Stack Pointer initial value	
1	Reset	-3 Highest
2	NMI	-2
3	Hard Fault	-1
4	Memory Management Fault	Programmable
5	Bus Fault	Programmable
6	Usage Fault (undefined instructions, divide by zero, unaligned memory access, ...)	Programmable
7	Reserved	Programmable
8	Reserved	Programmable
9	Reserved	Programmable
10	Reserved	Programmable
11	SVCall	Programmable
12	Debug Monitor	Programmable
13	Reserved	Programmable
14	PendSV	Programmable
15	SysTick	Programmable
16	IRQ for peripherals	Programmable
17	IRQ for peripherals	Programmable
...	...	Programmable
255	IRQ for peripherals	Programmable

Table 12-4: Interrupt Priority for ARM Cortex-M

The IPR registers and Priority Grouping in ARM Cortex

The priority of an IRQ is assigned in one of the interrupt priority registers called *IPRx (Interrupt PRiorityx)* in NVIC. If we do not assign a priority to an IRQ, by default, it has priority 0. Each IRQ uses one byte in an interrupt priority register. Therefore, each interrupt priority register holds priorities for four IRQs. For example, IPR0 holds the priorities of IRQ0, IRQ1, IRQ2 and IRQ3. In the same way, the priorities of IRQ4, IRQ5, IRQ6 and IRQ7 are assigned in IPR1. For 60 IRQs, 15 interrupt priority registers are used. The STM32F10x device uses only the 4 most significant bits of the byte in the interrupt priority register. With 4 bits, there can be 16 different priorities, 0 to 15. The lower the number the higher the priority is. See Figure 12-20.

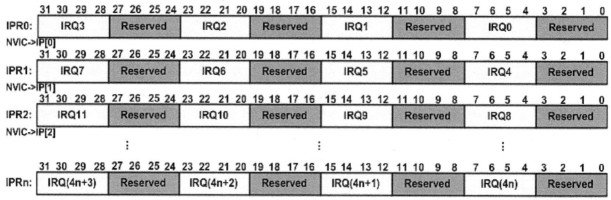

Figure 12-20: IPRn Registers

For example, if we want to set the Timer 3 interrupt priority to 2, first we need to find out the IRQ number of Timer 2 interrupt, which is 45. To locate the register for IRQ45, we will divide 45 by 4, which results in a quotient of 11 and remainder of 1. The byte that holds the priority of Timer 3 is byte 1 of IPR11. To get to byte 1, we need to shift the priority 8 bits (8 x 1) to the left and to get the 4 most significant bits, we need to shift it 4 more bits to the left. The statement will look like:

```
NVIC->IP[IRQn/4] |= PRIO << (8 * (IRQn % 4) + 6);
```
    or

```
NVIC->IP[11] |= 2 << (8 + 6);
```

This is tedious and error prone. To ease the calculation of finding the correct bits of the correct register to set the priority, the CMSIS has a macro NVIC_SetPriority defined in core_cm3.h for programmers to set the priority of an IRQ. It would be easier to use the macro:

```
NVIC_SetPriority (TIM3_IRQn, 2);
```

Configuring the priorities for INT4 to INT15

The SHPR1 (System Handler Priority Register 1), SHPR2, and SHPR3 registers are used to configure the system exceptions of 4 to 15. You can explore them by reading the ARM Cortex-M data sheet. For example, the priority of SysTick is controlled in the most significant byte of System Handler Priority 3 register (SHPR3) of System Control Block (SCB->SHP[2]).

Interrupt inside an interrupt handler (nested interrupt)

What happens if the ARM is executing an ISR belonging to an interrupt and another interrupt is activated? In such cases, a higher priority interrupt can preempt a lower priority interrupt. The CPU stops the execution of the lower priority interrupt handler and launches the higher priority interrupt handler. Then, resumes the execution of the lower priority interrupt handler when the execution of the higher priority interrupt handler is finished. The ARM Cortex-M allows only the higher priority interrupts to preempt the lower priority interrupt service routine. The programmer is responsible to assign the proper priority to each IRQ to determine whether an interrupt may preempt the other's interrupt handler. The NVIC in ARM Cortex-M has the ability to capture the pending interrupts and keeps track of each one until all are serviced.

Program 12-6 illustrates two interrupts with different priority. In this example, delay function is called in the interrupt handler to demonstrate the preemption by higher priority interrupt. (In real work, it is a bad practice to call delay function in interrupt handler.) TIM2 is programmed to interrupt at 1 second interval. In the interrupt handler, the green LED (PC13) is turned on for 500 ms. TIM3 is programmed to interrupt at 200ms interval and in its interrupt handler, the PB5 is low for 50 ms. Please, connect an LED to the PB5 so that the LED turns on when the pin is low and check the program. Since TIM2 has higher priority, you will observe that PB5 is not blinking when TIM2 interrupt is running (when the green LED is on). Now change the priority of the TIM3 to be higher than TIM2 by changing the following line from

```
NVIC_SetPriority(TIM3_IRQn,3); /* Priority level = 3 */
```
to

```
NVIC_SetPriority(TIM3_IRQn,1); /* Priority level = 1 */
```
You will see that the PB5 is blinking all the time so is the green LED (PC13) because the TIM3 (PB5) preempts TIM2 interrupt handler (PC13).

Program 12-6: Interrupt priority demonstration

```
/* In the program, Timer2 and Timer3 are used to generate interrupts. */
/* Timer2 interrupt is generated every second. Its interrupt vector takes around 0.5
second to be executed and during its execution PC13 is low. */
/* Timer3 interrupt is generated every 0.2 second. Its interrupt vector takes around
50ms to be executed and during its execution PB5 is low. */
/* Timer2 and Timer3 have the priority level of 2 and 3, respectively. */

#include <stm32f10x.h>

void delay_ms(uint16_t t);

void TIM2_IRQHandler()
{
    TIM2->SR = 0; /* clear UIF flag */
    GPIOC->ODR &= ~(1<<13); /* PC13 = low */
    delay_ms(500);   /* wait 0.5 sec */
    GPIOC->ODR |= (1<<13); /* PC13 = high */
}

void TIM3_IRQHandler()
```

```
{
   TIM3->SR = 0;      /* clear UIF flag */
   GPIOB->ODR &= ~(1<<5); /* PB5 = low */
   delay_ms(50); /* wait 50 ms */
   GPIOB->ODR |= (1<<5); /* PB5 = high */
}

int main()
{
   RCC->APB2ENR |= 0xFC;    /* enable GPIO clocks */
   RCC->APB1ENR |= (1<<0)|(1<<1);          /* enable TIM2 and TIM3 clocks */
   GPIOB->CRL = 0x44344444;      /* PB5 as output */
   GPIOC->CRH = 0x44344444;      /* PC13 as output */

   TIM2->PSC = 7200-1;      /* PSC = 7199 */
   TIM2->ARR = 10000-1;
   TIM2->SR = 0;      /* clear the UIF flag */
   TIM2->CR1 = 1;     /* up counting */
   TIM2->DIER = (1<<0);      /* enable UIE interrupt */
   NVIC_EnableIRQ(TIM2_IRQn);      /* enable TIM2 interrupt */
   NVIC_SetPriority(TIM2_IRQn,2); /* Priority level = 2 */

   TIM3->PSC = 7200-1;      /* PSC = 7199 */
   TIM3->ARR = 2000-1;
   TIM3->CR1 = 1;     /* up counting */
   TIM3->DIER = (1<<0);      /* enable UIE interrupt */
   NVIC_EnableIRQ(TIM3_IRQn);      /* enable TIM3 interrupt */
   NVIC_SetPriority(TIM3_IRQn,3); /* Priority level = 3 */

   while(1)
   {
   }
}

void delay_ms(uint16_t t)
{
   for(int i = 0; i < t; i++)
   {
      for(volatile uint16_t a = 0; a < 6000; a++)
      {}
   }
}
```

Interrupt latency

The time from the moment the event that triggers an interrupt signal to the moment the CPU starts to execute the ISR code is called the interrupt latency. This latency depends on whether the source of the interrupt is an internal (e.g., exceptions) or external hardware (e.g., peripheral hardware IRQ) interrupt. The duration of interrupt latency can also be affected by the type of the instruction which the CPU was executing when the interrupt occurs. It takes longer in cases where the instruction being executed lasts for many instruction cycles compared to the instructions that last for only one instruction cycle time. In the ARM Cortex-M, we also have extra clocks added to the latency due to the fact that it saves the content of registers CPSR, PC, LR, R12, and R0-R3 on stack. See your ARM Cortex-M manual for the timing data sheet.

Another source of the interrupt latency is the interrupt priority. As mentioned earlier, when several interrupts occur at the same time, the interrupt with the highest priority is acknowledged first. All other interrupts have to wait.

Review Questions
1. In ARM, which interrupt has the highest priority?
2. True or false. Upon Reset, all the IRQs have the same priority.
3. We use register _____ to modify the interrupt priority of IRQ8.
4. To assign priority to IRQ21, we need to program the IPR__ register.

Section 12.7: ARM Cortex-M Processor Modes

In this section we examine various operation modes in ARM Cortex-M.

ARM Cortex Thread (application) and Handler (exception) modes

In comparing the traditional ARM7 with ARM Cortex-M series we see some major changes in the ARM Cortex-M series. Among the changes are the CPU modes, stack, interrupt processing and many new instructions. These changes are meant to make the ARM Cortex-M systems to run programs faster and more efficiently. We have examined some of these changes in this chapter since the vast majority of them are related to the interrupt execution. The ARM Cortex-M can run in one of the two modes at any given time. They are: (1) Thread (Application) mode and (2) Handler (Exception) mode. The differences can be stated as follows:

1. When the ARM Cortex-M is powered on and coming out of reset, it automatically goes to the Thread mode. The Thread mode is the mode that vast majority of the applications programs are executed in. The CPU spends most of its time in Thread mode and gets interrupted only to execute ISR for exception faults or peripheral IRQs.

2. The ARM Cortex-M switches to Handler mode only when an exception fault (of course other than the Reset) or an IRQ interrupt from a peripheral is activated to get the attention of the CPU to execute an ISR (interrupt handler). Upon returning from ISR, the CPU automatically changes from Handler mode back to Thread mode. It must be noted that of all the exceptions and IRQs in the Table 12-1, only the Reset forces the CPU into Thread mode and the rest are executed in Handler mode.

A big advantage of having Handler mode is that when returning from Handler mode, the CPU will pop the stack and restore the registers saved during entry to Handler mode. With this interrupt handlers are written just like any other functions.

There are two Stacks in ARM Cortex (Case study)

The classical ARM has a single stack pointer (R13) to be used to point to RAM area for the purpose of stack. With a multi-threaded operating system, every thread should have their own stack so does the operating system itself. It is much more efficient to have separate stack pointers for the system and the thread. The ARM Cortex-M has two stack pointer registers. They are called PSP (processor stack pointer) and MSP (main stack pointer). Threads running in Thread mode should use the process stack and the kernel and exception handlers should use the main stack.

343

The bit 1, ASP (active stack pointer), of the special function register called CONTROL register gives the option of choosing MSP or PSP for stack pointer, in Thread mode. Upon Reset the ASP=0, meaning that R13 is the Main Stack pointer (MSP) and its value come from the first 4 bytes of the interrupt vector table starting at 0x00000000 address location. By making the ASP=1, the R13 is the same as PSP (processor stack pointer). Next, we examine the privilege levels in ARM Cortex-M.

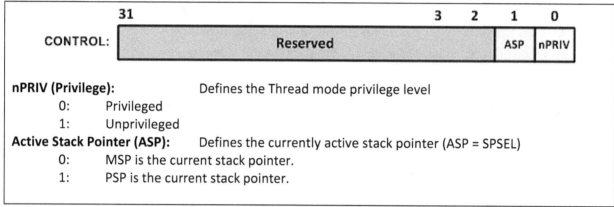

Figure 12-21: CONTROL Register in ARM Cortex-M4

Processor Mode	Software	Stack Usage
Thread	Applications	MSP or PSP
Handler	ISR for Exceptions and IRQs	MSP
Note: In Thread mode, use bit 1 of the Control register to select MSP or PSP for stack pointer.		

Table 12-5: Processor Modes and Stack Usage in ARM Cortex-M

Privileged and Unprivileged levels in ARM Cortex-M (Case study)

The ARM Cortex-M series has a new feature that did not exist in the previous ARM products. This new feature is called privileged level. There are two privilege levels in ARM Cortex-M. They are called Privileged and Unprivileged. The Privileged level in ARM Cortex-M can be used to limit the CPU access to special registers and protected memory area to prevent the system from getting corrupted due to error in coding or malicious user. Here is summary of the Privileged level software:

1. Privileged level software has access to all registers including the special function registers for interrupts.
2. Privileged level software has access to every region of memory.
3. Privileged level software has access to system timer, NVIC, and system resources.
4. The Privileged level software can execute all the ARM Cortex-M instructions including the MRS, MSR, and CPS.
5. The Handlers for fault exceptions and IRQs can be executed only in Privileged level.
6. Only the Privileged software can access the CONTROL register to see whether execution is in Privileged or Unprivileged mode. In Unprivileged mode one can switch from Unprivileged level to Privileged level by using SVC instruction.

Here is summary of the Unprivileged level software:

1. Unprivileged level software has no access to some registers such the special function registers for interrupts.
2. Unprivileged level software has limited access to some regions of memory.
3. Unprivileged level software is blocked from accessing system timer, NVIC, and system control block and resources.
4. The Unprivileged level software cannot execute some of the ARM instructions such as CPS. It has limited access to the MRS and MSR instructions.
5. While Handler mode is always executed in the Privileged level, the Thread mode software can be executed in Privileged or Unprivileged level. The bit 0 of the special a function register called CONTROL register gives the option of running the software in Privileged or Unprivileged mode.
6. In Unprivileged mode, one can use SVC instruction to make a supervisor call to switch from Unprivileged level to Privileged level.

Processor Mode	Software	Privilege level
Thread	Applications	Privileged and Unprivileged
Handler	ISR for Exceptions and IRQs	Always Privileged
Note: In Thread mode, use bit 0 of the CONTROL register to select Privileged or Unprivileged		

Table 12-6: Privileged level Execution and Processor Modes in ARM Cortex-M

Mode	Privilege	Stack Pointer	Typical Example usage
Handler	Privileged	Main	Exception Handling
Handler	Unprivileged	Any	Reserved since Handler is always Privileged
Thread	Privileged	Main	Operating system kernel
Thread	Privileged	Process	
Thread	Unprivileged	Main	
Thread	Unprivileged	Process	Application threads

Table 12-7: Processor Mode, Privilege, and Stack in ARM Cortex

Special Function register in ARM Cortex (Case study)

Beside the traditional general-purpose registers of R0–R15, the ARM Cortex has many new special function registers. These registers are widely used in programs written for the Cortex-M based embedded systems. See Figure 12-22.

While the general-purpose registers of R0–R15 can be accessed using the MOV, LDR, and STR instructions, these new special function registers can be accessed only with the two new instructions MSR and MRS. To manipulate (clear or set) the bits of special function registers, first we must use the MSR to move them to a general-purpose register and after changing their values they are moved back by using MRS instruction. Table 12-8 shows special function registers.

Review Questions
1. True or false. When a Reset pin is activated, the ARM CPU wakes up in Thread mode.
2. How many processor modes are there in the ARM Cortex-M? Give their names
3. True or false. When an interrupt comes in from exception fault or IRQ, the ARM CPU switches to Handler mode automatically.

Register name	Privilege Usage
MSP (main stack pointer)	Privileged
PSP (processor stack pointer)	Privileged or Unprivileged
PSR (Processor status register)	Privileged
APSR (application processor status register)	Privileged or Unprivileged
ISPR (interrupt processor status register)	Privileged
EPSR (execution processor status register)	Privileged
PRIMASK (Priority Mask register)	Privileged
FAULTMASK (fault mask register)	Privileged
BASEPRI (base priority register)	Privileged
CONTROL (control register)	Privileged
Note: We must use MSR and MRS instructions to access the above registers	

Table 12-8: Special function registers of ARM Cortex-M

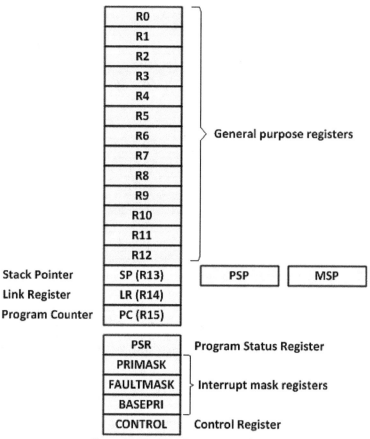

Figure 12-22: ARM Cortex-M Registers

Problems

Section 12.1

1. True or False. The polling method wastes the CPU's time.
2. What memory area is assigned to the interrupt vector table?
3. How many bytes are set aside for each interrupt vector?
4. Describe how the division-by-zero is handled in ARM Cortex.

5. Describe the fault interrupts in ARM.
6. What is SVCall?
7. How does the CPU find the address of interrupt handler?
8. How does the CPU distinguish between a subroutine return and an interrupt return?

Section 12.2
9. How can we enable the interrupt for the SysTick?
10. True or False. When the interrupt routine is executed the COUNTFLAG is cleared by hardware.

Section 12.3
11. What is the state of the I bit upon power-on reset, and what does it mean?
12. With a single instruction, show how to disable all the interrupts.
13. Using the external interrupt, write a program that counts the fall edges of PB4. Put the counted value on PA0-7 pins.
14. Initialize the external interrupt for PB5 so that it generates an interrupt on both edges. Then monitor PB5 in the interrupt routine. If PB5 is low, make PC13 low. Otherwise, make it high.

Section 12.4
15. Write a program to transmit serially the letter 'Z' continuously at 9600 baud rate.
16. Write a program to transmit letters 'A' and 'B' continuously at 9600 baud rate. (Hint: toggle a variable in the interrupt handler and use its value to decide if you should send 'A' or 'B'.)

Section 12.5
17. True or false. For each timer, there is a unique address in the interrupt vector table.
18. Show how to enable the TIM4 update interrupt (UIE).
19. Using the TIM3 interrupt, write a program to create a square wave of 60 kHz on pin PB4.
20. Using the TIM4 interrupt, write a program to create a square wave of 3 kHz on pin PC13.

Section 12.6
21. Explain what happens if both TIM3 and TIM4 are activated at the same time.
22. Explain what happens if an interrupt is generated while the CPU is serving another interrupt.
23. True or false. In the ARM, an interrupt inside an interrupt is not allowed.
24. What happens if two interrupts with the same priority level are triggered at the same time?

Section 12.7
25. True or False. The interrupt routines are executed in Handler mode.
26. True or False. The main function is executed in Handler mode.

Answer to Review Questions

Section 12.1
1. True
2. 4
3. 1K byte beginning at 00000000 and ending at 000003FFH

4. Interrupt service routine (ISR) or interrupt handler
5. To hold the starting address of each ISR
6. 0x00000040, 41, 42, and 43
7. No; it is internally embedded into the NVIC.
8. INT 6

Section 12.2

1. INT15
2. TICKINT
3. False. NVIC_EN0 register is used for IRQs (external interrupts) and SysTick is not part of them

Section 12.3

1. INT16
2. __enable_irq();
3. True
4. False

Section 12.4

1. IRQ37
2. True
3. USARTx_CR1
4. False

Section 12-5

1. IRQ44
2. False
3. TIM2_DIER
4. False

Section 12-6

1. Reset
2. False
3. IPR1 (IP[1])
4. IPR5 (IP[5])

Section 12.7

1. True
2. 2. Thread and Handler
3. True

Chapter 13: ADC, DAC, and Sensor Interfacing

This chapter explores more real-world devices such as ADCs (analog-to-digital converters), DACs (digital-to-analog converters), and sensors. In Section 13.1, we describe analog-to-digital converter (ADC) chips. We will program the ADC module of the STM32F1xx chip in Section 13.2. In Section 13.3, we show the interfacing of sensors and discuss the issue of signal conditioning. The characteristics and programming of DAC chips are discussed in Section 13.4.

Section 13.1: ADC Characteristics

This section will explore ADC generally. First, we describe some general aspects of the ADC itself, then focus on the functionality of some important pins in ADC.

ADC devices

Analog-to-digital converters are among the most widely used devices for data acquisition. Digital computers use binary (discrete) values, but in the physical world everything is analog (continuous). Temperature, pressure (wind or liquid), humidity, and velocity are a few examples of physical quantities that we deal with every day. A physical quantity is converted to electrical (voltage, current) signals using a device called a *transducer*. Transducers used to generate electrical outputs are also referred to as *sensors*. Sensors for temperature, velocity, pressure, light, and many other natural physical quantities produce an output that is voltage (or current). Therefore, we need an analog-to-digital converter to translate the analog signals to digital numbers so that the microcontroller can read and process the numbers. See Figures 13-1 and 13-2.

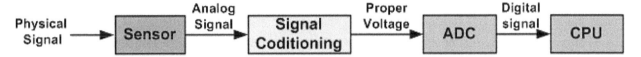

Figure 13-1: Microcontroller Connection to Sensor via ADC

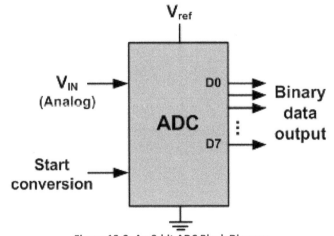

Figure 13-2: An 8-bit ADC Block Diagram

349

Some of the major characteristics of the ADC

Resolution

The ADC has *n*-bit resolution, where *n* can be 8, 10, 12, 16, or even 24 bits. Higher-resolution ADCs provide a smaller step size, where *step size* is the smallest change that can be discerned by an ADC. Some widely used resolutions for ADCs are shown in Table 13-1. Although the resolution of an ADC chip is decided at the time of its design and cannot be changed, we can control the step size with the help of what is called V_{ref}. This is discussed below.

n-bit	Number of steps	Step size
8	256	5V /256 = 19.53 mV
10	1024	5V /1024 = 4.88 mV
12	4096	5V /4096 = 1.2 mV
16	65,536	5V /65,536 = 0.076 mV
Note: V_{ref} = 5V		

Table 13-1: Resolution versus Step Size for ADC (Vref = 5V)

Vref

Vref is an input voltage used for the reference voltage. The voltage connected to this pin, along with the resolution of the ADC chip, determine the step size. For an 8-bit ADC, the step size is Vref / 256 because it is an 8-bit ADC, and 2 to the power of 8 gives us 256 steps. See Table 13-1. For example, if the analog input range needs to be 0 to 4 volts, Vref is connected to 4 volts. That gives 4 V / 256 = 15.62 mV for the step size of an 8-bit ADC. In another case, if we need a step size of 10 mV for an 8-bit ADC, then V_{ref} = 2.56 V, because 2.56 V / 256 = 10 mV. For the 10-bit ADC, if the V_{ref} = 5V, then the step size is 4.88 mV as shown in Table 13-1. Tables 13-2 and 13-3 show the relationship between the V_{ref} and step size for the 8- and 10-bit ADCs, respectively. In some applications, we need the differential reference voltage where $V_{ref} = V_{ref\,(+)} - V_{ref\,(-)}$. Often the $V_{ref\,(-)}$ pin is connected to ground and the $V_{ref\,(+)}$ pin is used as the V_{ref}.

V_{ref} (V)	V_{in} in Range (V)	Step Size (mV)
5.00	0 to 5	5 / 256 = 19.53
4.00	0 to 4	4 / 256 = 15.62
3.00	0 to 3	3 / 256 = 11.71
2.56	0 to 2.56	2.56 / 256 = 10
2.00	0 to 2	2 / 256 = 7.81
1.28	0 to 1.28	1.28 / 256 = 5
1.00	0 to 1	1 / 256 = 3.90
Note: In an 8-bit ADC, step size is V_{ref}/256		

Table 13-2: Vref Relation to V_{in} Range for an 8-bit ADC

V_{ref} (V)	V_{in} Range (V)	Step Size (mV)
5.00	0 to 5	5 / 1024 = 4.88
4.96	0 to 4.096	4.096 / 1024 = 4
3.00	0 to 3	3 / 1024 = 2.93
2.56	0 to 2.56	2.56 / 1024 = 2.5
2.00	0 to 2	2 / 1024 = 2
1.28	0 to 1.28	1.28 / 1024 = 1.25
1.024	0 to 1.024	1.024 / 1024 = 1
Note: In a 10-bit ADC, step size is V_{ref}/1024		

Table 13-3: Vref Relation to Vin Range for an 10-bit ADC

Conversion time

In addition to resolution, conversion time is another major factor in selecting an ADC. *Conversion time* is defined as the time it takes the ADC to convert the analog input to a digital number. The conversion time is dictated by the clock source connected to the ADC in addition to the method used for data conversion and technology used in the fabrication of the ADC.

Digital data output

In an 8-bit ADC we have an 8-bit digital data output of D0–D7, while in the 10-bit ADC the data output is D0–D9. To calculate the output voltage, we use the following formula:

$$D_{OUT} = V_{IN} / StepSize$$

where D_{out} = digital data output (in decimal), V_{in} = analog input voltage, and step size (resolution) is the smallest change, which is $V_{ref}/256$ for an 8-bit ADC.

Figure 13-3 shows a simple 2-bit ADC. In the circuit, the voltage between Vref(+) and Vref(-) is divided into 4 since resistors have the same values. As a result, the step size is $(V_{ref(+)} - V_{ref(-)}) / 4$.

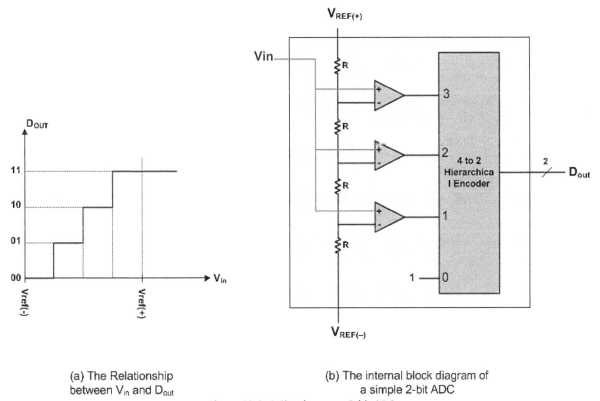

(a) The Relationship
between V_{in} and D_{out}

(b) The internal block diagram of
a simple 2-bit ADC

Figure 13-3: A Simultaneous 2-bit ADC

If V_{in} is below step size all the comparators send out zeros. When V_{in} is between step size and step size × 2, the lowest comparator sends out 1 and the encoder gives 01.

If V_{in} is between step size × 2 and step size × 3, the second comparator and the first comparator sends out 1. Since the encoder is hierarchical priority, it sends out the highest value in cases that more than 1 input is high. As a result, 2 (10 in binary) will be sent out.

When V_{in} is bigger than step size × 3, the third comparator becomes high and 3 will be sent out.

See Example 13-1. This data is brought out of the ADC chip either one bit at a time (serially), or in one chunk, using a parallel line of outputs. This is discussed next.

Example 13-1

For a given 8-bit ADC (e.g. ADC0848), we have V_{ref} = 2.56 V. Calculate the D0–D7 output if the analog input is: (a)1.7 V, and (b) 2.1 V.

Solution:

Since the step size is 2.56/256 = 10 mV, we have the following.
(a)D_{OUT} = 1.7V/10 mV = 170 in decimal, which gives us 10101011 in binary for D7–D0.
(b)D_{OUT} = 2.1V/10 mV = 210 in decimal, which gives us 11010010 in binary for D7–D0.

Parallel versus serial ADC

The ADC chips are either parallel or serial. In parallel ADC, we have 8 or more pins dedicated to bringing out the binary data, but in serial ADC we have only one pin for data out. The D0–D7 data pins of the 8-bit ADC provide an 8-bit parallel data path between the ADC chip and the CPU. In the case of the 16-bit parallel ADC chip, we need 16 pins for the data path. In order to save pins, many 12- and 16-bit ADCs use pins D0–D7 to send out the upper and lower bytes of the binary data. In recent years, for many applications where space is a critical issue, using such a large number of pins for data is not feasible. For this reason, serial devices such as the serial ADC are becoming widely used. While the serial ADCs use fewer pins and their smaller packages take much less space on the printed circuit board, more CPU time is needed to get the converted data from the ADC because the CPU must get data one bit at a time, instead of in one single read operation as with the parallel ADC. ADC0848 is an example of a parallel ADC with 8 pins for the data output, while the MAX1112 is an example of a serial ADC with a single pin for D_{out}. Figures 13-4 and 13-5 show the block diagram for ADC0848 and MAX1112, respectively.

Analog input channels

Many data acquisition applications need more than one analog input for ADC. For this reason, we see ADC chips with 2, 4, 8, or even 16 channels on a single chip. Multiplexing of analog inputs is widely used as shown in the ADC848 and MAX1112. In these chips, we have 8 channels of analog inputs, allowing us to monitor multiple quantities such as temperature, pressure, flow, and so on. Nowadays, some ARM microcontroller chips come with 16-channel on-chip ADC.

Start conversion and end-of-conversion signals

For the conversion to be controlled by the CPU, there are needs for start conversion (SC) and end-of-conversion (EOC) signals. When SC is activated, the ADC starts converting the analog input value of V_{in} to a digital number. The amount of time it takes to convert varies depending on the conversion method. When the data conversion is complete, the end-of-conversion signal notifies the CPU that the converted data is ready to be picked up.

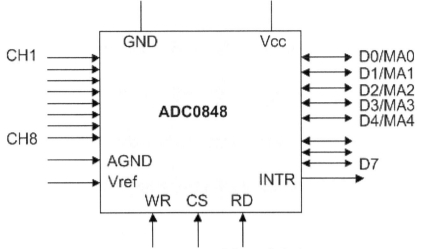

Figure 13-4: ADC0848 Parallel ADC Block Diagram

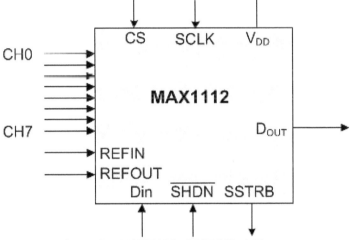

Figure 13-5: MAX1112 Serial ADC Block Diagram

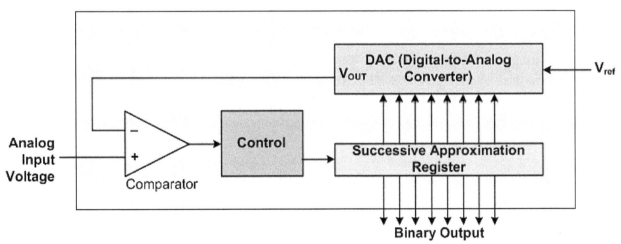

Figure 13-6: Successive Approximation ADC

Successive Approximation ADC

Successive Approximation is a widely used method of converting an analog input to digital output. It has three main components: (a) successive approximation register (SAR), (b) comparator, and (c) control unit. See Figure 13-6.

The successive approximation register is loaded with only the most significant bit set at the start. An internal digital-to-analog converter converts the value of SAR to an analog voltage which is used to compare to the input voltage. If the input voltage is higher, the bit is kept. If the voltage is lower, the bit is cleared. The next bit is tried and the DAC and compare are exercised. This process is repeated for all bits of the SAR. Assuming a step size of 10 mV, the 8-bit successive approximation ADC will go through the following steps to convert an input of 1 Volt:

(1) It starts with binary number 10000000. Since 128 × 10 mV = 1.28 V is greater than the 1 V input, bit 7 is cleared (dropped).

(2) 01000000 gives us 64 × 10 mV = 640 mV and bit 6 is kept since it is smaller than the 1 V input.

(3) 01100000 gives us 96 × 10 mV = 960 mV and bit 5 is kept since it is smaller than the 1 V input,

(4) 01110000 gives us 112 × 10 mV = 1120 mV and bit 4 is dropped since it is greater than the 1 V input.

(5) 01101000 gives us 108 × 10 mV = 1080 mV and bit 3 is dropped since it is greater than the 1 V input.

(6) 01100100 gives us 100 × 10 mV = 1000 mV = 1 V and bit 2 is kept since it is equal to input. Even though the answer is found it does not stop.

(7) 011000110 gives us 102 × 10 mV = 1020 mV and bit 1 is dropped since it is greater than the 1 V input.

(8) 01100101 gives us 101 × 10 mV = 1010 mV and bit 0 is dropped since it is greater than the 1 V input.

Notice that the Successive Approximation method goes through all the steps even if the answer is found in one of the earlier steps. The advantage of the Successive Approximation method is that the conversion time is fixed since it has to go through all the steps.

Review Questions
1. Give two factors that affect the step size calculation.
2. The ADC0848 is a(n) _____-bit converter.
3. True or false. While the ADC0848 has 8 pins for Dout, the MAX1112 has only one Dout pin.
4. Find the step size for an 8-bit ADC, if Vref = 1.28 V.
5. For question 4, calculate the output if the analog input is: (a) 0.7 V, and (b) 1 V.

Section 13.2: ADC Programming with STM32F1xx

Because the ADC is widely used in data acquisition, in recent years an increasing number of microcontrollers have on-chip ADC modules. In this section, we discuss the ADC feature of STM32F1xx and show how it is programmed.

The STM32F1xx chips have two or three ADC modules which support up to 18 ADC channels. These ADC modules have 12-bit resolution. Figure 13-7 shows a simplified block diagram of STM32F1xx chips.

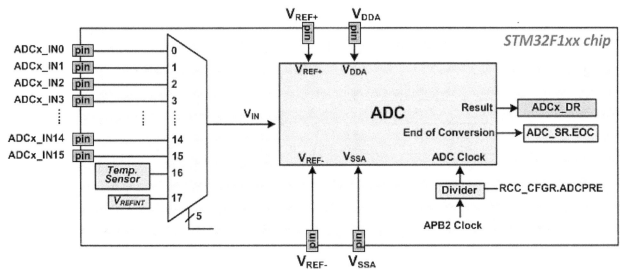

Figure 13-7: Simplified Block Diagram of ADC in STM32F1xx

Table 13-4 lists some registers of ADC modules. In this section, we examine some of these registers and show how to program the ADC.

Register name	Offset	Description
ADCx_SR	0x0000	Status Register
ADCx_DR	0x004C	Regular Data Register
ADCx_CR1	0x0004	Control Register 1
ADCx_CR2	0x0008	Control Register 2
ADCx_SMPR1	0x000C	Sample time register 1
ADCx_SMPR2	0x0010	Sample time register 2
ADCx_SQR1	0x002C	Regular Sequence register 1
ADCx_SQR2	0x0030	Regular Sequence register 2
ADCx_SQR3	0x0034	Regular Sequence register 3

Table 13-4: Some of the ADC Registers

Enabling Clock to ADC

First thing we need to do is to enable the clock to the ADC module. Bits 9, 10, and 15 of RCC_APB2ENR register is used to enable the clock to ADC1, ADC2, and ADC3. See Figure 13-8.

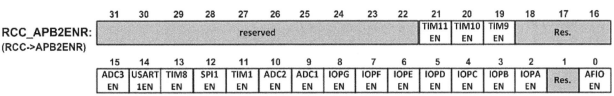

Figure 13-8: RCC_APB2ENR

The ADC clock can be 12 MHz at most. So, there is a divider between the APB2 clock (PCLK2) and the ADC which makes a lower clock frequency from the APB2 clock. The divider is configured by the ADCPRE bits of the RCC_CFGR register. See Figures 13-7 and 13-9. By default, the startup file configures the RCC_CFGR so that the ADC clock is 12MHz when the CPU frequency is 72MHz.

RCC_CFGR:
(RCC->CFGR)

31	30	29	28	27	26	25	24	23	22	21	20	19	18	17	16
		Reserved				MCO		Res.	USB PRE		PLLMUL			PLLXT PRE	PLL SRC
				rw	rw	rw	rw		rw	rw	rw	rw	rw	rw	rw

15	14	13	12	11	10	9	8	7	6	5	4	3	2	1	0
ADC PRE		PPRE2			PPRE1			HPRE				SWS		SW	
rw	rw	rw	rw	rw	rw	rw	rw	rw	rw	rw	rw	r	r	rw	rw

Field	Description
MCO	Microcontroller Clock Output
USBPRE	USB Prescaler
PLLMUL	PLL Multiplication factor
PLLXTPRE	LSB of division factor PREDIV1
PLLSRC	PLL entry clock source
ADCPRE	ADC Prescaler: the bits are used to set the frequency of ADC clock. <table><tr><th>ADCPRE value</th><th>ADC Clock Frequency</th></tr><tr><td>00</td><td>PCLK2 (APB2 clock) divided by 2</td></tr><tr><td>01</td><td>PCLK2 (APB2 clock) divided by 4</td></tr><tr><td>10</td><td>PCLK2 (APB2 clock) divided by 6</td></tr><tr><td>11</td><td>PCLK2 (APB2 clock) divided by 8</td></tr></table>
PPRE2	APB2 clock prescaler
PPRE1	APB1 clock prescaler
HPRE	AHB clock prescaler
SWS	System clock Switch Status
SW	System clock Switch

Figure 13-9: RCC_CFGR (Clock Configuration Register)

ADC Data Register (ADCx_DR)

Upon the completion of conversion, the binary result is placed in the ADCx_DR register. It is a 16-bit register. Since the resolution of ADCs are 12-bit, only 12 bits of the ADCx_DR register are used. The Align bit of the ADCx_CR2 is used to make the result right-justified or left-justified.

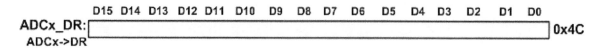

Figure 13-10: ADCx_DR (Data Register)

ADC Control Register 2 (ADCx_CR2)

Figure 13-11 shows the bits of ADCx_CR2 register and their usages. In this section we will focus more on the function of the following bits:

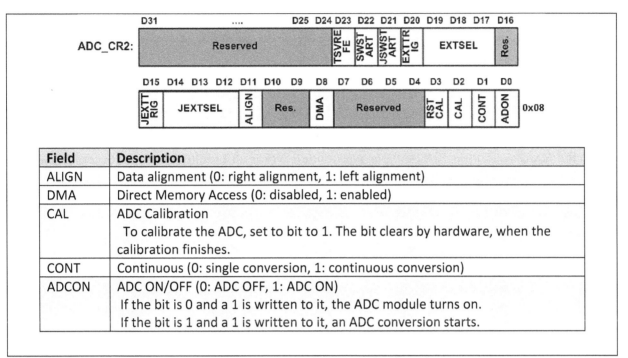

Figure 13-11: ADC CR2 (ADC Control Register 2)

Field	Description
ALIGN	Data alignment (0: right alignment, 1: left alignment)
DMA	Direct Memory Access (0: disabled, 1: enabled)
CAL	ADC Calibration To calibrate the ADC, set to bit to 1. The bit clears by hardware, when the calibration finishes.
CONT	Continuous (0: single conversion, 1: continuous conversion)
ADCON	ADC ON/OFF (0: ADC OFF, 1: ADC ON) If the bit is 0 and a 1 is written to it, the ADC module turns on. If the bit is 1 and a 1 is written to it, an ADC conversion starts.

Align bit operation

The STM32F1xx chips have 12-bit ADCs, which means that the result is 12 bits long. In STM32F1xx, a 16-bit register is dedicated to the ADC result, but only 12 of the 16 bits are used and 4 bits are unused. You can select the position of used bits in the bytes. If you set the Align bit in ADCx_CR2 register to high, the result bits will be left-justified; otherwise, the result bits will be right-justified. See Figure 13-12.

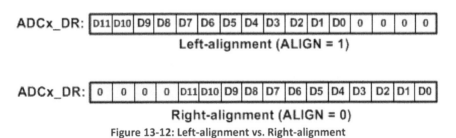

Figure 13-12: Left-alignment vs. Right-alignment

CAL bit (Calibration)

The STM32F1xx ADC can do self-calibration. If you set the CAL bit to 1, the internal calibration circuit, measures the internal capacitors and sets some internal variables. This increases the accuracy of the ADC. Once the calibration is finished the CAL bit is cleared by hardware. It is recommended to do calibration after each power up.

CONT bit

Using the CONT (continuous) bit you can choose between the single and continuous modes. In the single mode, the ADC converts the selected input from analog to digital and then stops.

In the continuous mode, the ADC module converts the input(s) from analog to digital and when the conversion finishes it starts another conversion. The conversion takes place continuously until the CONT bit is cleared by software.

ADCON bit

The bit is used for both waking up the ADC module and starting the conversion. The value of the bit shows if the ADC module is ON or OFF. On reset, the bit is 0 and ADC module is OFF. When we write one to the ADCON bit for the first time, the ADC module wakes up and the ADCON bit becomes one.

If we write one to the ADCON bit when the ADCON bit is one, the ADC starts conversion. Notice that if we want to start conversion, we should not change the other bits of the register at the same time; otherwise, the conversion does not start.

Notice that when you turn on the ADC module, it takes a while until the ADC module becomes stable and gives accurate result. The time is called stabilization time (t_{STAB}) and it is less than 1 microsecond. So, there should be a delay of 1 microsecond between the power-up and the first conversion.

ADC Sequence Registers (ADCx_SQR1, ADCx_SQR2, and ADCx_SQR3)

See Figure 13-7. There is a multiplexer between the input pins and the ADC. Using the ADCx_SQRn registers, we list the inputs which should be converted to digital. See Figure 13-13. In single mode, we store the input channel in the SQ1 bits of ADCx_SQR3 and leave all the other bits of ADCx_SQR registers zeros.

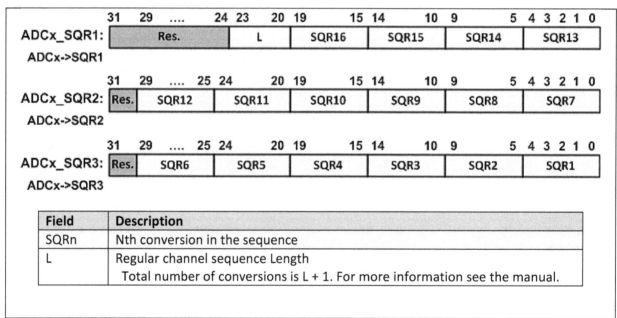

Figure 13-13: Sequence Registers (ADCx_SQR1, ADCx_SQR2, and ADCx_SQR3)

We can also give a list of inputs to ADC module. In the case, the ADC module converts the inputs one after another. To do so, you can write the list of inputs in the ADCx_SQR registers. For more information, see the reference manual.

ADC Sample time registers (ADCx_SMPR1 and ADCx_SMPR2)

Before converting, ADC samples the input voltage in a capacitor. In STM32F1xx chips, you can configure the sample time for each channel separately using the ADCx_SMPR1 and ADCx_SMPR2. See Figure 13-14.

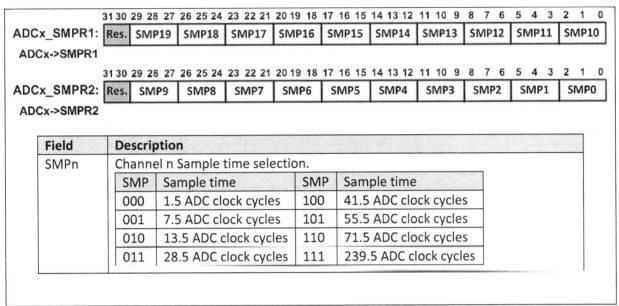

Figure 13-14: Sequence Registers (ADCx_SQR1, ADCx_SQR2, and ADCx_SQR3)

ADC Status Register (ADCx_SR)

Figure 13-15 shows the ADC status register (ADCx_SR). The EOC (end of conversion) bit of the register is set when the conversion is finished. The EOC bit is cleared by software. The EOC flag is also automatically cleared if we read the ADCx_DR register.

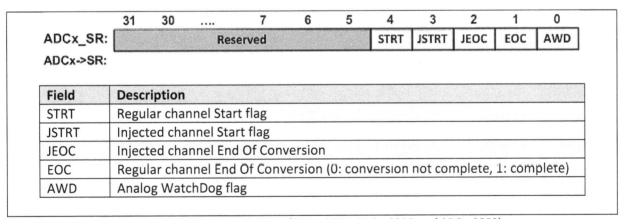

Figure 13-15: Sequence Registers (ADCx_SQR1, ADCx_SQR2, and ADCx_SQR3)

Programming the ADC module using Polling

To program the A/D converter, the following steps should be taken:

1. Enable the clock for the ADC module and the GPIO pins using the APB2ENR register,
2. Make the pin for the selected ADC channel an analog input pin.
3. Turn on the ADC module since it is disabled upon power-on reset to save power.

4. Initialize the ADCx_SMPR registers to select a proper sample time for the channels you use.
5. Wait 1 microsecond to make sure the ADC module is stable,
6. start conversion by writing a one to the ADCON bit,
7. Wait for the conversion to be completed by polling the EOC bit in the ADCx_SR register,
8. When the EOC bit is HIGH, read the ADCx_DR register to get the digital data output,
9. If you want to read the selected channel again, go back to step 6.

Program 13-1 illustrates the steps for ADC conversion shown above. Figure 13-16 shows the hardware connection of Program 13-1.

Program 13-1: This program configures PA1(ADC_IN1) as the ADC input and sends the result through usart1

```
#include <stm32f10x.h>
#include <stdio.h>

void usart1_init(void);
void usart1_sendByte(unsigned char c);
void usart1_sendStr(char *str);
void usart1_sendInt(unsigned int i);

void delay_ms(uint16_t t);
void delay_us(uint16_t t);

int main()
{
   RCC->APB2ENR |= 0xFC|(1<<9)|(1<<14); /* enable clocks for GPIO, ADC1 and usart1 */

   GPIOA->CRL = 0x44444404;        /* PA1(ADC_IN1) as analog input */
   ADC1->CR2 = 1;    /* ADON = 1 (power-up) */

   ADC1->SMPR2 = 1<<3; /* SMP1 = 001 (set sample time for IN1 to 7.5 clock cycles) */

   usart1_init(); /* initialize the usart1 */

   delay_us(1);  /* wait 1us to make sure the adc module is stable */

   while(1)
   {
      ADC1->SQR3 = 1;        /* choose channel 1 as the input */
      ADC1->CR2 = 1; /* ADON = 1 (start conversion) */
      while((ADC1->SR&(1<<1)) == 0); /* wait until the EOC flag is set */

      usart1_sendInt(ADC1->DR);   /* read ADC1_DR and send its value through serial */
      usart1_sendStr("\n\r");            /* go to new line */

      delay_ms(10);          /* wait 10ms */
   }
}

/* copy usart1_init, usart1_sendByte, usart1_sendStr, and usart1_sendInt
from Program 11-1 */
```

```
void delay_ms(uint16_t t)
{
    for(int i = 0; i < t; i++)
    {
        for(volatile uint16_t a = 0; a < 6000; a++)
        {}
    }
}

void delay_us(uint16_t t)
{
    for(int i = 0; i < t; i++)
    {
        for(volatile uint16_t a = 0; a < 6; a++)
        {}
    }
}
```

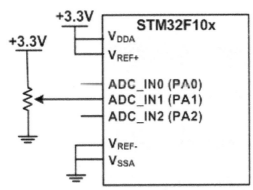

Figure 13-16: ADC Connection for Program 13-1

Programming A/D converter using interrupts

See Figure 13-17. The bits EOCIE (EOC interrupt Enable), AWDIE (AWD Interrupt Enable), and JEOCIE (JEOC Interrupt Enable) are used to enable the interrupts for EOC, AWD, and JEOC flags, respectively.

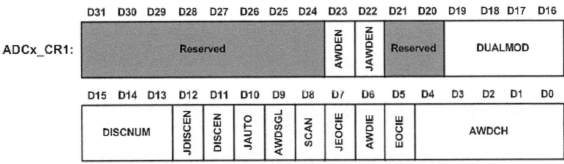

Figure 13-17: ADCx_CR1 (ADC Control Register 1)

Upon completion of conversion, the EOC flag is set to HIGH. If EOCIE = 1, on each end of conversion, an interrupt is generated and forces the CPU to jump to the ADC interrupt handler. Then, you can read the converted value in the interrupt routine. Program 13-2 shows how to read ADC using interrupts. In the program the CONT bit of ADC_CR2 is set. So, the ADCON bit is triggered only once and the ADC converts continuously. Whenever the conversion is completed the interrupt routine is executed which reads the result of conversion and copies to a global variable. The main program sends the value of the global variable using USART1 to the PC.

Program 13-2: This program reads PA1(ADC_IN1) using interrupt and sends the result through usart1

```c
/* This program configures PA1(ADC_IN1) as the ADC input and sends the result through
usart1 */

#include <stm32f10x.h>
#include <stdio.h>

void usart1_init(void);
void usart1_sendByte(unsigned char c);
void usart1_sendStr(char *str);
void usart1_sendInt(unsigned int i);

void delay_ms(uint16_t t);
void delay_us(uint16_t t);

volatile uint16_t adcResult;

void ADC1_2_IRQHandler()
{
   adcResult= ADC1->DR; /* read conversion result */
}

int main()
{
   RCC->APB2ENR |= 0xFC|(1<<9)|(1<<14); /* enable clocks for GPIO, ADC1 clock, and
usart1 */

   GPIOA->CRL = 0x44444404;  /* PA1(ADC_IN1) as analog input */
   ADC1->CR2 = 1;    /* ADON = 1 (power-up) */

   ADC1->SMPR2 = 7<<3; /* SMP1 = 111 (239.5 ADC clock cycles) */

   delay_us(1); /* wait 1us to make sure the adc module is stable */

   usart1_init(); /* initialize the usart1 */

   ADC1->CR1 = (1<<5); /* EOCIE = 1 */

   ADC1->SQR3 = 1;   /* choose channel 1 as the input */
   ADC1->CR2 |= 1<<1;      /* CONT = 1 (Convert continuously) */
   ADC1->CR2 |= 1<<0;      /* ADCON = 1 (start conversion) */

   NVIC_EnableIRQ(ADC1_2_IRQn); /* enable ADC1_2 interrupt */

   while(1)
   {
     usart1_sendInt(adcResult);  /* send the adc result through serial */
     usart1_sendStr("\n\r");     /* go to new line */

     delay_ms(10);            /* wait 10ms */
   }
```

```
}
/* copy usart1_init, usart1_sendByte, usart1_sendStr, and usart1_sendInt
from Program 11-1 */
/* copy delay_ms and delay us from Program 13-1 */
```

Vref in trainer boards

In the STM32F10x chips with 48-pin (or smaller) packages, Vref+ and Vref- are internally connected to V_{DDA} and V_{SSA}, respectively and there is no separate Vref pin. So, in the Blue pill, Vref is 3.3V since V_{DDA} is connected to +3.3V. See Example 13-2.

Example 13-2

Assuming that Vref = 3.3 V, (a) for a STM32 ADC calculate the step size, (b) find the relation between Vin and the output.

Solution:

(a) Because the ADC is 12-bit, the step size is 3.3/4096 = 805.6μV
(b) output − Vin/Step size = Vin/805.6μ = 1000000Vin/805.6

ADC hardware considerations

For digital logic signals a small variation in voltage level has no effect on the output. For example, 0.2 V is considered LOW, since in TTL logic, anything less than 0.5 V will be detected as LOW logic. That is not the case when we are dealing with analog voltage. See Example 13-3.

Example 13-3

For an 12-bit ADC, we have Vref = 3.3 V. Calculate the D0–D11 output if the analog input is: (a) 0.2 V, and (b) 0 V. How much is the variation between (a) and (b)?

Solution:

Because the step size is 3.3/4096 = 805.6μV, we have the following:
(a) Dout = 0.2 V/805.6μV = 248 in decimal, which gives us 11111000 in binary.
(b) Dout = 0 V/805.6μV = 0 in decimal, which gives us 0 in binary.

The difference is 11111000, which is 8 bits!

When you design a circuit using microcontrollers you can use some techniques to reduce the impact of ADC supply voltage and Vref variation on the accuracy of ADC output. Next, we examine two of the most widely used techniques.

Connecting a capacitor between V_{DDA} and GND

The V_{DDA} pin provides the supply for analog ADC circuitry. To get a better accuracy of ADC we must provide a stable voltage source to the V_{DDA} pin. Figure 13-18 shows how to use capacitors to achieve this.

Connecting a capacitor between V_{REF+} and V_{ref-}

If the microcontroller has the Vref pin, connect two capacitors between the V_{REF+} pin and V_{REF-} to make the Vref voltage more stable and increase the precision of ADC. See Figure 13-18.

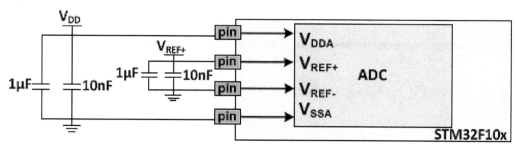

Figure 13-18: ADC Hardware Considerations

Review Questions

1. The ADC in STM32F10x is _____ bit.
2. In STM32F10x, the highest number we can get for the ADC output is _____ in decimal.
3. Assume VREF+ = 3.3V. Find the ADC output in decimal if V_{in} of analog input is 1.9V.
4. In STM32F10x, which register provides the ADC output converted data?

Section 13.3: Sensor Interfacing and Signal Conditioning

This section will show how to interface sensors to the microcontroller. We examine some popular temperature sensors and then discuss the issue of signal conditioning. Although we concentrate on temperature sensors, the principles discussed in this section are the same for other types of sensors such as light and pressure sensors.

Temperature sensors

Transducers convert physical data such as temperature, light intensity, flow, and speed to electrical signals. Depending on the transducer, the output produced is in the form of voltage, current, resistance, or capacitance. For example, temperature is converted to electrical signals using a transducer called a *thermistor*. A thermistor responds to temperature change by changing resistance, but its response is not linear, as seen in Table 13-5 and Figure 13-19.

Temperature ('C)	Tf (K ohms)
0	29.490
25	10.000
50	3.893
75	1.700
100	0.817

Table 13-5: Thermistor Resistance vs. Temperature

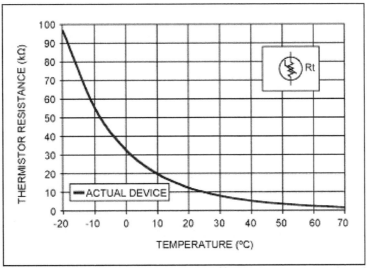

Figure 13-19: Thermistor (Copied from http://www.maximintegrated.com)

The resistance of a thermistor is typically modeled by Steinhart-Hart equation and requires a logarithmic amplifier to produce a linear output. The complexity associated with the circuit for such nonlinear devices has led many manufacturers to market a linear temperature sensor. Simple and widely used linear temperature sensors include the LM34 and LM35 series from National Semiconductor (now part of TI Corp.) They are discussed next.

LM34 and LM35 temperature sensors

The sensors of the LM34 series are precision integrated-circuit temperature sensors whose output voltage is linearly proportional to the Fahrenheit temperature. See Figure 13-20. The LM34 requires no external calibration because it is internally calibrated. It outputs 10 mV for each degree of Fahrenheit temperature.

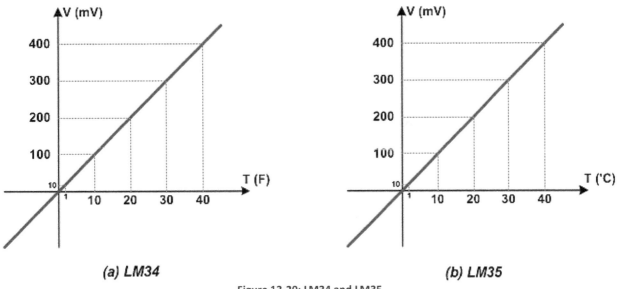

Figure 13-20: LM34 and LM35

365

The LM35 series sensors are similar to LM34 series sensors except that the output voltage is linearly proportional to the Celsius (centigrade) temperature. It outputs 10 mV for each degree of centigrade temperature. See Figure 13-20.

Signal conditioning

The common transducers produce an output in the form of voltage, current, charge, capacitance, or resistance. In order to perform A-to-D conversion on these signals, they need to be converted to voltage unless the transducer output is already voltage. In addition to the conversion to voltage, the signal may also need gain and offset adjustment to achieve optimal dynamic range. A low-pass analog filter is often incorporated in the signal conditioning circuit to eliminate the high frequency to avoid aliasing. Figure 13-21 shows a block diagram of the input of a data acquisition system.

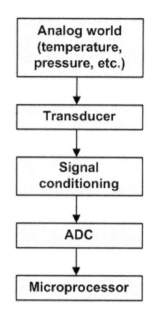

Figure 13-21: Getting Data to the CPU

Interfacing the LM34/LM35 to the STM32 Microcontroller

The A/D of STM32F10x microcontroller has 12-bit resolution with a maximum of 4096 steps, and the LM34 produces 10 mV for every degree of temperature change. The maximum operating temperature of the LM34 is 300 degrees F, so the highest output will be 3000 mV (3.00 V), which is below 3.3V of V_{ref}. The LM34/35 can be connected to the microcontroller as shown in Figure 13-22. See Example 13-4.

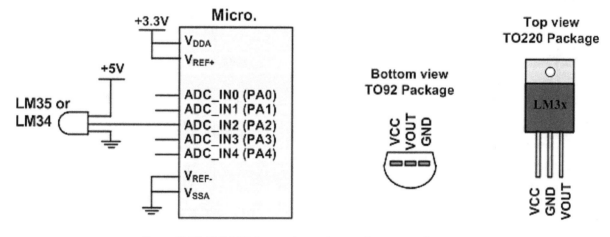

Figure 13-22: LM34/35 Connection to ARM and Its Pin Configuration

Example 13-4

Assuming that Vref = 3.3 V, find the relation between temperature and the ADC result.

Solution:

The step size is 3.3/4096 and for LM35/LM34, we have V = 0.01 × temperature. So,

$$adcResult = \frac{V_{in}}{stepSize} = \frac{0.01 \times temp}{\frac{3.3}{4096}} \Rightarrow adcResult = \frac{4096 \times temp}{330} \Rightarrow temp = \frac{330 \times adcResult}{4096}$$

Reading and displaying temperature

Programs 13-3 shows code for reading LM35 (or LM34) and displaying temperature. Notice that the LM34 (or LM35) is connected to ADC_IN2 pin.

Program 13-3: Measuring temperature using LM35/LM34

```c
#include <stm32f10x.h>
#include <stdio.h>

void usart1_init(void);
void usart1_sendByte(unsigned char c);
void usart1_sendStr(char *str);
void usart1_sendInt(unsigned int i);

void delay_ms(uint16_t t);
void delay_us(uint16_t t);

int main()
{
    RCC->APB2ENR |= 0xFC|(1<<9)|(1<<14);  /* enable clocks for GPIO, ADC1 clock, and
usart1 */

    GPIOA->CRL = 0x44444044;        /* PA2(ADC_IN2) as analog input */
    ADC1->CR2 = 1;     /* ADON = 1 (power-up) */

    ADC1->SMPR2 = 1<<6; /* SMPR2.SMP2 = 001 */

    delay_us(1); /* wait 1us to make sure the adc module is stable */

    usart1_init(); /* initialize the usart1 */

    while(1)
    {
        ADC1->SQR3 = 2;        /* choose channel 2 as the input */
        ADC1->CR2 = 1; /* ADON = 1 (start conversion) */
        while((ADC1->SR&(1<<1)) == 0); /* wait until the EOC flag is set */

        unsigned int temp = ADC1->DR*330/4096; /* read ADC1_DR and calculate temp. */
        usart1_sendInt(temp);         /* send temperature through serial */
        usart1_sendStr("\n\r");       /* go to new line */

        delay_ms(10);          /* wait 10ms */
    }
}
```

```
/* copy usart1_init, usart1_sendByte, usart1_sendStr, and usart1_sendInt
from Program 11-1 */
/* copy delay_ms and delay_us from Program 13-1 */
```

Review Questions

1. True or false. The transducer is connected to signal conditioning circuitry before its signal is sent to the ADC.
2. The LM35 provides ____ mV for each degree of _____ (Fahrenheit, Celsius) temperature.
3. The LM34 provides ____ mV for each degree of _____ (Fahrenheit, Celsius) temperature.
4. What is the temperature if the ADC output is 63?

Section 13.4: DAC Programming

This section will discuss the fundamentals of a DAC (digital-to-analog converter). Then we demonstrate how to generate a sawtooth wave and a sine wave. The analog conversion output can be observed on the oscilloscope at DAC output pins.

Digital-to-analog (DAC) converter

The digital-to-analog converter (DAC) is a device widely used to convert digital signals to analog signals. In this section we discuss the basics of a DAC.

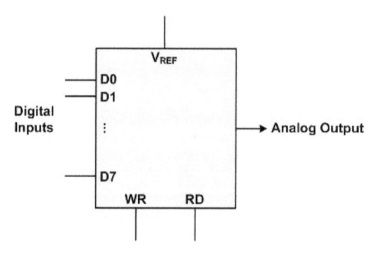

Figure 13-23: DAC Block Diagram

Recall from your digital electronics book the two methods of making a DAC: binary weighted and R/2R ladder. The vast majority of integrated circuit DACs, including the DAC0808 discussed in this section use the R/2R method since it can achieve a much higher degree of precision. The first criterion for selecting a DAC is its resolution, which is a function of the number of bits of the digital input. The common ones are 8, 10, and 12 bits. The number of digital input bits decides the resolution of the DAC since the number of analog output levels is equal to 2^n, where n is the number of digital input bits. Therefore, the 8-bit DAC such as the DAC0808 provides 256 discrete voltage (or current) levels of output. See Figure 13-23.

Similarly, the 12-bit DAC provides 4096 discrete voltage levels. Although there are 16-bit DACs, they are much more expensive.

368

DAC0808

In the DAC0808, the digital inputs are converted to current (I_{OUT}). By connecting a resistor to the I_{OUT} pin, we convert the conversion result current to voltage. The total current provided by the I_{OUT} is a function of the binary numbers at the D0–D7 inputs of the DAC0808 and the reference current (I_{ref}), and is as follows:

$$I_{OUT} = I_{ref} \times \left(\frac{D7}{2} + \frac{D6}{4} + \frac{D5}{8} + \frac{D4}{16} + \frac{D3}{32} + \frac{D2}{64} + \frac{D1}{128} + \frac{D0}{256}\right) = \frac{I_{ref} \times Data}{256}$$

where D0 is the LSB, D7 is the MSB for the inputs, and I_{ref} is the reference input current that must be applied to pin 14. The I_{ref} current is generally set to 2.0 mA. Figure 13-24 shows the generation of current reference (setting I_{ref} = 2 mA) by using the standard 5-V power supply and 5K ohm resistors.

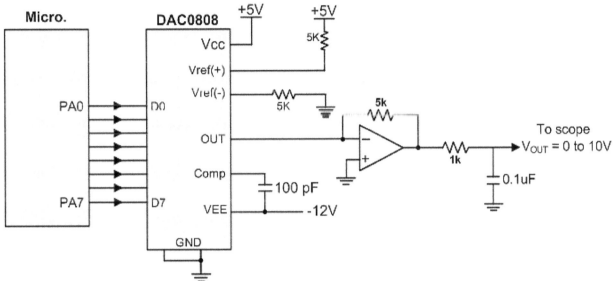

Figure 13-24: Microcontroller Connection to DAC0808

Some also use the Zener diode reference voltage device (LM336), which overcomes fluctuations associated with the power supply voltage. Now assuming that I_{ref} = 2 mA, if all the input bits to the DAC are high, the maximum output current is 1.99 mA (verify this for yourself).

Converting I_{out} to voltage in DAC0808

We connect the output pin I_{OUT} to a resistor, convert this current to voltage, and monitor the output on the scope. However, in real life this can cause inaccuracy since the input resistance of the load where it is connected will also affect the output voltage. For this reason, the I_{ref} current output is buffered by connecting it to an op amp such as the 741 with R_f = 5K ohms for the feedback resistor. Assuming that R = 5K ohms, by changing the binary input, the output voltage changes as shown in Example 13-5.

Example 13-5

Assuming that R = 5K and I_{ref} = 2 mA, calculate V_{out} for the following binary inputs:
(a) 10011001 binary (0x99)   (b) 11001000 (0xC8)

Solution:

(a) I_{out} = 2 mA (153/255) = 1.195 mA and V_{out} = 1.195 mA × 5K = 5.975 V

(b) I_{out} = 2 mA (200/256) = 1.562 mA and V_{out} = 1.562 mA × 5K = 7.8125 V

Generating a stair-step ramp

In order to generate a stair-step ramp, you can set up the circuit in Figure 13-24 and load Program 13-4 on the microcontroller chip. To see the result wave, connect the output to an oscilloscope. Figure 13-25 shows the output.

Program 13-4

```c
#include <stm32f10x.h>

void delay_us(uint16_t t);

int main()
{
    RCC->APB2ENR |= 0xFC;           /* enable clocks for GPIOs */
    GPIOA->CRL = 0x33333333;        /* PA0-PA7 as outputs */

    int8_t n = 0;
    while(1)
    {
        GPIOA->ODR = n;     /* copy n into PORTA to be converted */
        n ++;            /* increment n */
        delay_us(10);
    }
}

void delay_us(uint16_t t)
{
    for(int i = 0; i < t; i++)
    {
        for(volatile uint16_t a = 0; a < 6; a++)
        {}
    }
}
```

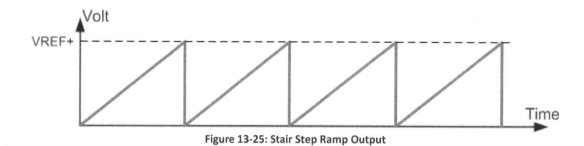

Figure 13-25: Stair Step Ramp Output

DAC of STM32F10x chips (Only for high density series)

Some series of STM32F10x chips including the high-density STM32F103 chips, have on-chip DACs. The on-chip DAC is 12-bit. For more information, see the reference manual.

Review Questions

1. In a DAC, input is _____ (digital, analog) and output is _____ (digital, analog).
2. DAC0808 is a(n) ____-bit D-to-A converter.
3. The output of DAC808 is in _____ (current, voltage).

Problems

Section 13.1

1. True or false. The output of most sensors is analog.
2. True or false. A 10-bit ADC has 10-bit digital output.
3. True or false. ADC0848 is an 8-bit ADC.
4. True or false. MAX1112 is a 10-bit ADC.
5. True or false. An ADC with 8 channels of analog input must have 8 pins, one for each analog input.
6. True or false. For a serial ADC, it takes a longer time to get the converted digital data out of the chip.
7. True or false. ADC0848 has 4 channels of analog input.
8. True or false. MAX1112 has 8 channels of analog input.
9. True or false. ADC0848 is a serial ADC.
10. True or false. MAX1112 is a parallel ADC.
11. Which of the following ADC sizes provides the best resolution?
    (a) 8-bit   (b) 10-bit   (c) 12-bit   (d) 16-bit   (e) They are all the same.
12. In Question 11, which provides the smallest step size?
13. Calculate the step size for the following ADCs, if Vref is 5 V:
    (a) 8-bit   (b) 10-bit   (c) 12-bit   (d) 16-bit
14. In an 8-bit ADC with Vref = 1.28 V, find the Vin for the following outputs:
    (a) D7–D0 = 11111111   (b) D7–D0 = 10011001   (c) D7–D0 = 1101100
15. In the ADC0848, what should the Vref value be if we want a step size of 5 mV?
16. In an 8-bit ADC with Vref = 2.56 V, find the Vin for the following outputs:
    (a) D7–D0 = 11111111   (b) D7–D0 = 10011001   (c) D7–D0 = 01101100

Section 13.2

17. True or false. The STM32F10x chips have on-chip A/D converters.
18. True or false. A/D of the STM32F103 is 8-bit ADC.
19. True or false. STM32F103 has 5 channels of analog input.
20. True or false. The unused ADC pins of the STM32F10x can be used for I/O pins.
21. True or false. Upon power-on reset, the ADC module of the STM32F103 is turned on and ready to go.
22. True or false. All the STM32F103 chips have Vref+ and Vref- pins.

23. In the ADC of STM32F10x, what happens to the converted analog data? How do we know that the ADC is ready to provide us the data?

24. For the ADC of STM32F103, find the step size for each of the following Vref. Consider GND as Vref-.

    (a) Vref = 1.024 V   (b) Vref = 2.048 V   (c) Vref = 2.56 V

25. In the STM32F10x, what should the Vref value be if we want a step size of 0.5 mV?

26. In the STM32F10x, what should the Vref value be if we want a step size of 0.3 mV?

27. With a step size of 0.5 mV, what is the analog input voltage if all outputs are 1?

28. With Vref = 1.024 V, find the Vin for the following outputs:

    (a) D11–D0 = 010011111111 (b) D11–D0 = 000010011000 (c) D11–D0 = 000011010000

29. With Vref = 2.56 V, find the Vin for the following outputs:

    (a) D11–D0 = 111111111111 (b) D11–D0 = 001000000001 (c) D11–D0 = 001100110000

30. How do we start conversion in STM32F10x chips?

31. How do we recognize the end of conversion in STM32F10x chips?

32. Which registers are used to select the analog channel to be converted?

33. Give the name of the register used to enable the ADC interrupts.

Section 13.3

34. What does it mean when a given sensor is said to have a linear output?

35. The LM35 sensor produces _____ mV for each degree of temperature.

36. What is signal conditioning?

Section 13.4

37. True or false. DAC0808 is 8-bit.

38. Find the number of discrete voltages provided by the n-bit DAC for the following:

    (a) n = 8 (b) n = 10 (c) n = 12

39. For DAC0808, if Iref = 2 mA, show how to get the Iout of 1.99 when all inputs are HIGH.

40. Find the Iout for the following inputs. Assume Iref = 2 mA for DAC0808.

    (a) 10011001   (b) 11001100   (c) 11101110

    (d) 00100010   (e) 00001001   (f) 10001000

41. To get a smaller step, we need a DAC with _____ (more, fewer) digital inputs.

Answers to Review Questions

Section 13.1

1. Number of steps and V_{ref} voltage

2. 8

3. True

4. 1.28 V / 256 = 5 mV

5.

    (a) 0.7 V / 5 mV = 140 in decimal and D7–D0 = 10001100 in bin

    (b) 1 V / 5 mV = 200 in decimal and D7–D0 = 11001000 in binary.

Section 13.2

1. 12
2. 4095
3. Step size is 3.3V / 4096 = 0.8056 mV and 1.9V / 0.8056mV = 2,358 in decimal or 0x936.
4. ADCn_DR

Section 13.3

1. True
2. 10, Celsius
3. 10, Fahrenheit
4. Temperature = 63 × 330 / 4096 = 5

Section 13.4

1. Digital, Analog
2. 8
3. current

Chapter 14: Relay, Optoisolator, and Stepper Motor Interfacing

Microcontrollers are widely used in motor control. We also use relays and optoisolators in motor control. This chapter discusses motor control and shows ARM interfacing with relays, optoisolators, and stepper motors.

Section 14.1: Relays and Optoisolators

This section begins with an overview of the basic operations of electromechanical relays, solid-state relays, reed switches, and optoisolators. Then we describe how to interface them to the ARM. We use the C language programs to demonstrate their control.

Electromechanical relays

A *relay* is an electrically controllable switch widely used in industrial controls, automobiles, and appliances. It allows the isolation of two separate sections of a system with two different voltage sources. For example, a +5 V system can be isolated from a 120 V system by placing a relay between them. One such relay is called an *electromechanical* (or *electromagnetic*) *relay* (EMR) as shown in Figure 14-1. The EMRs have three components: the coil, spring, and contacts.

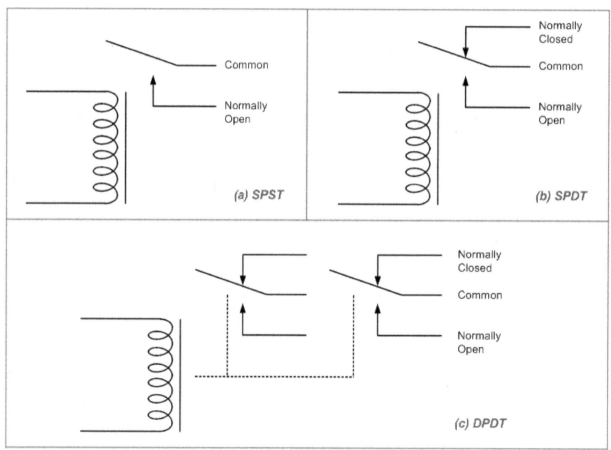

Figure 14-1: Relay Diagrams

In Figure 14-1, a digital +5 V on the left side can control a 12 V motor on the right side without any physical contact between them. When current flows through the coil, a magnetic field is created around the coil (the coil is energized), which causes the armature to be attracted to the coil. The armature's contact acts like a switch and closes or opens the circuit. When the coil is not energized, a spring pulls the armature to its normal state of open or closed. In the block diagram for electromechanical relays (EMR) we do not show the spring, but it does exist internally. There are all types of relays for all kinds of applications. In choosing a relay the following characteristics need to be considered:

1. The contacts can be normally open (NO) or normally closed (NC). In the NC type, the contacts are closed when the coil is not energized. In the NO type, the contacts are open when the coil is unenergized.

2. There can be one or more contacts. For example, we can have SPST (single pole, single throw), SPDT (single pole, double throw), and DPDT (double pole, double throw) relays.

3. The voltage and current needed to energize the coil. The voltage can vary from a few volts to 50 volts, while the current can be from a few mA to 20 mA. The relay has a minimum voltage, below which the coil will not be energized. This minimum voltage is called the "pull-in" voltage. In the datasheets for relays we might not see current, but rather coil resistance. The V/R will give you the pull-in current. For example, if the coil voltage is 5 V, and the coil resistance is 500 ohms, we need a minimum of 10 mA (5 V/500 ohms = 10 mA) pull-in current.

4. The maximum DC/AC voltage and current that can be handled by the contacts. This is in the range of a few volts to hundreds of volts, while the current can be from a few amps to 40 A or more, depending on the relay. Notice the difference between this voltage/current specification and the voltage/current needed for energizing the coil. The fact that one can use such a small amount of voltage/current on one side to handle a large amount of voltage/current on the other side is what makes relays so widely used in industrial controls. Examine Table 14-1 for some relay characteristics.

Part No.	Contact Form	Coil Volts	Coil Ohms	Contact Volts	Current
106462CP	SPST-NO	5 VDC	500	100 VDC	0.5 A
138430CP	SPST-NO	5 VDC	500	100 VDC	0.5 A
106471CP	SPST-NO	12 VDC	1000	100 VDC	0.5 A
138448CP	SPST-NO	12 VDC	1000	100 VDC	0.5 A
129875CP	DPDT	5 VDC	62.5	30 VDC	1 A

Table 14-1: Selected DIP Relay Characteristics (www.Jameco.com)

Driving a relay

Digital systems and microcontroller pins lack sufficient current to drive the relay. While the relay's coil needs around 10 mA to 50mA to be energized, the STM32F10x microcontroller's pin can provide a maximum of 25 mA current. For this reason, we place a driver, such as the ULN2803, or a transistor between the microcontroller and the relay as shown in Figure 14-2. In the circuit we can turn the lamp on and off by setting and clearing the PA8.

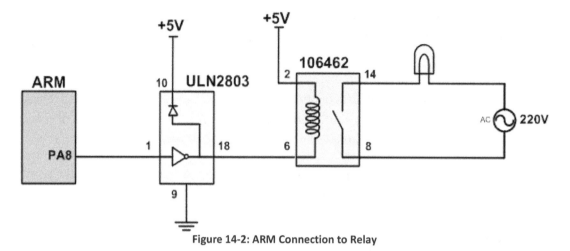

Figure 14-2: ARM Connection to Relay

Program 14-1 turns the lamp shown in Figure 14-2 on and off by energizing and de-energizing the relay every second.

Program 14-1: The program toggles PA8 every second

```c
/* The Program toggles PA8 every second and the lamp turns on and off. */

#include <stm32f10x.h>

void delay_ms(uint16_t t);

int main()
{
   RCC->APB2ENR |= 0xFC; /* Enable GPIO ports clocks */

   GPIOA->CRH = 0x44444443; /* PA8 as output */

   while(1)
   {
     GPIOA->ODR ^= (1<<8);/* toggle PA8 */
     delay_ms(1000);        /* wait 1 second */
   }
}

void delay_ms(uint16_t t)
{
   volatile unsigned long l = 0;
   for(uint16_t i = 0; i < t; i++)
     for(l = 0; l < 6000; l++)
     {
     }
}
```

Solid-state relay

Another widely used relay is the solid-state relay. See Table 14-2.

Part No.	Contact Style	Control Volts	Contact Volts	Contact Current
143058CP	SPST	4-32 VDC	240 VAC	3 A
139053CP	SPST	3-32 VDC	240 VAC	25 A
162341CP	SPST	3-32 VDC	240 VAC	10 A
172591CP	SPST	3-32 VDC	60 VAC	2 A
175222CP	SPST	3-32 VDC	60 VAC	4 A
176647CP	SPST	3-32 VDC	120 VAC	5 A

Table 14-2: Selected Solid-State Relay Characteristics (www.Jameco.com)

In this relay, there is no coil, spring, or mechanical contact switch. The entire relay is made out of semiconductor materials. Because no mechanical parts are involved in solid-state relays, their switching response time is much faster than that of electromechanical relays. Another advantage of the solid-state relay is its greater life expectancy. The life cycle for the electromechanical relay can vary from a few hundred thousand to a few million operations. Wear and tear on the contact points can cause the relay to malfunction after a while. Solid-state relays, however, have no such limitations. Extremely low input current and small packaging make solid-state relays ideal for microcontroller and logic control switching. They are widely used in controlling pumps, solenoids, alarms, and other power applications. Some solid-state relays have a phase control option, which is ideal for motor-speed control and light-dimming applications. Figure 14-3 shows control of a fan using a solid-state relay (SSR).

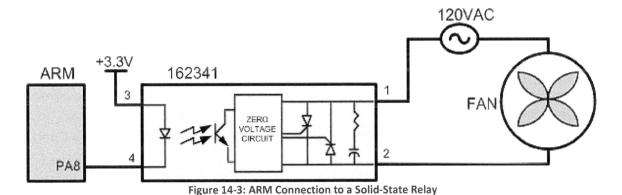

Figure 14-3: ARM Connection to a Solid-State Relay

Reed switch

Another popular switch is the reed switch. When the reed switch is placed in a magnetic field, the contact is closed. When the magnetic field is removed, the contact is forced open by its spring. See Figure 14-4. The reed switch is ideal for moist and marine environments where it can be submerged in fuel or water. Reed switches are also widely used in dirty and dusty atmospheres because they are tightly sealed.

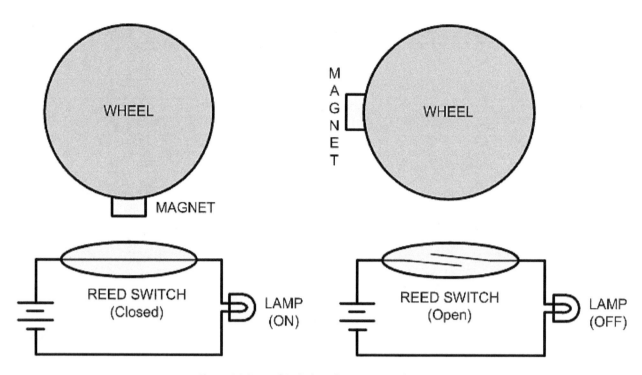

Figure 14-4: Reed Switch and Magnet Combination

Optoisolator

In some applications we use an optoisolator (also called optocoupler) to isolate two parts of a system. An example is driving a motor. Motors can produce what is called *back EMF*, a high-voltage spike produced by a sudden change of current as indicated in the formula V = Ldi/dt. In situations such as printed circuit board design, we can reduce the effect of this unwanted voltage spike (called *ground bounce*) by using decoupling capacitors (see Appendix A). In systems that have inductors (coil winding), such as motors, a decoupling capacitor or a diode will not do the job. In such cases we use optoisolators. An optoisolator has an LED (light-emitting diode) transmitter and a photosensor receiver, separated from each other by a gap. When current flows through the diode, it transmits a signal light across the gap and the receiver produces the same signal with the same phase but a different current and amplitude. See Figure 14-5. Optoisolators are also widely used in communication equipment such as modems. This device allows a computer to be connected to a telephone line without risk of damage from high voltage of telephone line. The gap between the transmitter and receiver of optoisolators prevents the electrical voltage surge from reaching the system.

Interfacing an optoisolator

The optoisolator comes in a small IC package with four or more pins. There are also packages that contain more than one optoisolator. When placing an optoisolator between two circuits, we must use two separate voltage sources, one for each side, as shown in Figure 14-6. Unlike relays, no drivers need to be placed between the microcontroller/digital output and the optoisolators.

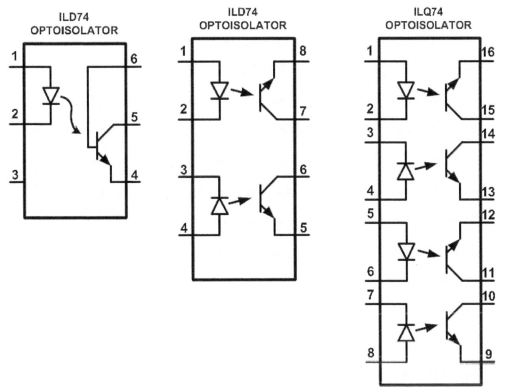

Figure 14-5: Optoisolator Package Examples

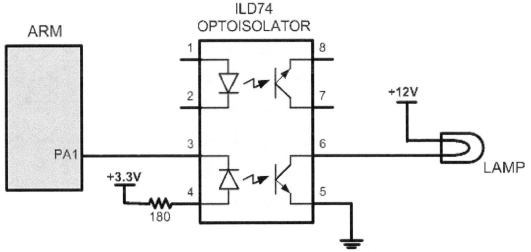

Figure 14-6: Controlling a Lamp via an Optoisolator

Review Questions

1. Give one application where would you use a relay.
2. Why do we place a driver between the microcontroller and the relay?
3. What is an NC relay?
4. Why are relays that use coils called electromechanical relays?
5. What is the advantage of a solid-state relay over EMR?
6. What is the advantage of an optoisolator over an EMR?

Section 14.2: Stepper Motor Interfacing

This section begins with an overview of the basic operation of stepper motors. Then we describe how to interface a stepper motor to the ARM. Finally, we use C language programs to demonstrate control of the rotation of stepper motor.

Stepper motors

A *stepper motor* is a widely used device that translates electrical pulses into mechanical movement. In applications such as dot matrix printers and robotics, the stepper motor is used for position control. Stepper motors commonly have a permanent magnet *rotor* (also called the *shaft*) surrounded by a stator (see Figure 14-7).

There are also steppers called *variable reluctance stepper motors* that do not have a permanent magnet rotor. The most common stepper motors have four stator windings that are paired with a center-tapped common as shown in Figure 14-8.

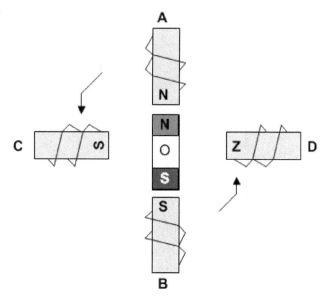

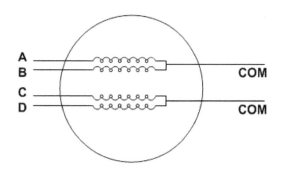

Figure 14-7: Stator Winding Configuration

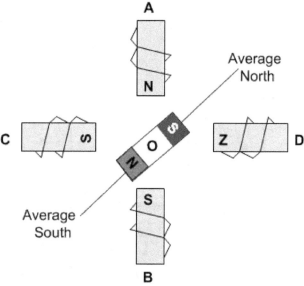

Figure 14-8: Rotor Alignment

This type of stepper motor is commonly referred to as a four-phase or unipolar stepper motor. The center tap allows a change of current direction in each of two coils when a winding is grounded, thereby resulting in a polarity change of the stator. Notice that while a conventional motor shaft runs freely, the stepper motor shaft moves in a fixed repeatable increment, which allows it to move to a precise position. This repeatable fixed movement is possible as a result of basic magnetic theory where poles of the same polarity repel and opposite poles attract. The direction of the rotation is dictated by the stator poles. The stator poles are determined by the current sent through the wire coils. As the direction of the current is changed, the polarity is also changed causing the reverse motion of the rotor. The stepper motor discussed here has a total of six leads: four leads representing the four stator windings and two commons for the center-tapped leads. As the sequence of power is applied to each stator winding, the rotor will rotate. There are several widely used sequences, each of which has a different degree of precision. Table 14-3 shows a two-phase, four-step stepping sequence.

	Step #	Winding A	Winding B	Winding C	Winding D	Counter
Clockwise	1	1	0	0	1	Clockwise
	2	1	1	0	0	
	3	0	1	1	0	
	4	0	0	1	1	

Table 14-3: Normal Four-Step Sequence

Note that although we can start with any of the sequences in Table 14-3, once we start, we must continue in the proper order. For example, if we start with step 3 (0110), we must continue in the sequence of steps 4, 1, 2, and so on.

Step angle

How much movement is associated with a single step? This depends on the internal construction of the motor, in particular the number of teeth on the stator and the rotor. The step angle is the minimum degree of rotation associated with a single step. Various motors have different step angles. Table 14-4 shows some step angles for various motors. In Table 14-4, notice the term steps per revolution. This is the total number of steps needed to rotate one complete rotation or 360 degrees (e.g., 180 steps × 2 degrees = 360).

Step Angle	Step per Revolution
0.72	500
1.8	200
2.0	180
2.5	144
5.0	72
7.5	48
15	24

Table 14-4: Stepper Motor Step Angles

It must be noted that perhaps contrary to one's initial impression, a stepper motor does not need more terminal leads for the stator to achieve smaller steps. All the stepper motors discussed in this section have four leads for the stator winding and two COM wires for the center tap. Although some manufacturers set aside only one lead for the common signal instead of two, they always have four leads for the stators. See Example 14-1. Next we discuss some associated terminology in order to understand the stepper motor further.

Example 14-1

Describe the ARM connection to the stepper motor of Figure 14-9.

Solution:

381

The following steps show the ARM connection to the stepper motor and its programming:

1. Use an ohmmeter to measure the resistance of the leads. This should identify which COM leads are connected to which winding leads.
2. The common wire(s) are connected to the positive side of the motor's power supply. In many motors, +5 V is sufficient.
3. The four leads of the stator winding are controlled by four bits of the ARM port (PA0–PA3). Because the microcontroller lacks sufficient current to drive the stepper motor windings, we must use a driver such as the ULN2003 (or ULN2803) to energize the stator. Instead of the ULN2003, we could have used transistors as drivers, as shown in Figure 14-11. However, notice that if transistors are used as drivers, we must also use diodes to take care of inductive current generated when the coil is turned off. One reason that using the ULN2003 is preferable to the use of transistors as drivers is that the ULN2003 has an internal diode to take care of back EMF.

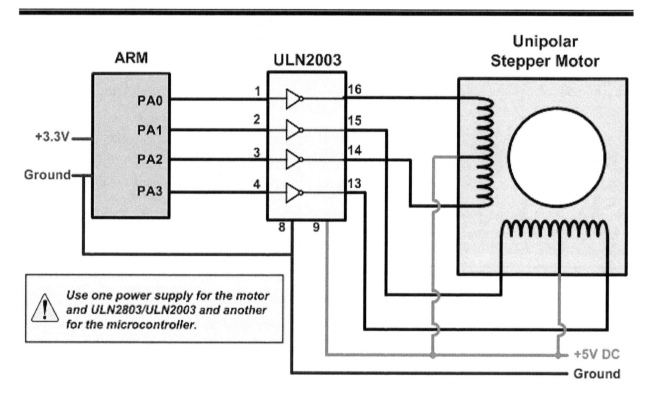

Figure 14-9: ARM Connection to Stepper Motor

Steps per second and RPM relation

The relation between RPM (revolutions per minute), steps per revolution, and steps per second is as follows.

$$Step\ per\ second = \frac{RPM \times Steps\ per\ revolution}{60}$$

The 4-step sequence and number of teeth on rotor

The switching sequence shown earlier in Table 14-3 is called the 4-step switching sequence because after four steps the same two windings will be "ON". How much movement is associated with these four steps? Therefore, in a stepper motor with 200 steps per revolution, the rotor has 50 teeth because 4 × 50 = 200 steps are needed to complete one revolution. This leads to the conclusion that the minimum step angle is always a function of the number of teeth on the rotor. In other words, the smaller the step angle, the more teeth the rotor has. See Example 14-2.

Example 14-2

Give the number of times the four-step sequence in Table 14-3 must be applied to a stepper motor to make an 80-degree move if the motor has a 2-degree step angle.

Solution:

A motor with a 2-degree step angle has the following characteristics:

Step angle: 2 degrees
Steps per revolution: 180
Number of rotor teeth: 45
Movement per 4-step sequence: 8 degrees

To move the rotor 80 degrees, we need to send 10 consecutive 4-step sequences, because 10 × 4 steps × 2 degrees = 80 degrees.

Looking at Example 14-2, one might wonder what happens if we want to move 45 degrees, because the steps are 2 degrees each. To provide finer resolutions, all stepper motors allow what is called an 8-step switching sequence. The 8-step sequence is also called half-stepping, because in the 8-step sequence each step is half of the normal step angle. For example, a motor with a 2-degree step angle can be used as a 1-degree step angle if the sequence of Table 14-5 is applied.

	Step #	Winding A	Winding B	Winding C	Winding D	Counter
Clockwise	1	1	0	0	1	**Clockwise**
	2	1	0	0	0	
	3	1	1	0	0	
	4	0	1	0	0	
	5	0	1	1	0	
	6	0	0	1	0	
	7	0	0	1	1	
	8	0	0	0	1	

Table 14-5: Half-Step 8-Step Sequence

Motor speed

The motor speed, measured in steps per second (steps/s), is a function of the switching rate. Notice in Example 14-1 that by changing the length of the time delay loop, we can achieve various rotation speeds.

Holding torque

The following is a definition of holding torque: "With the motor shaft at standstill or zero rpm condition, the amount of torque, from an external source, required to break away the shaft from its holding position. This is measured with rated voltage and current applied to the motor." The unit of torque is ounce-inch (or kg-cm).

Wave drive 4-step sequence

In addition to the 8-step and the 4-step sequences discussed earlier, there is another sequence called the *wave drive 4-step sequence*. It is shown in Table 14-6.

Clockwise	Step #	Winding A	Winding B	Winding C	Winding D	Counter Clockwise
⬇	1	1	0	0	0	⬆
	2	0	1	0	0	
	3	0	0	1	0	
	4	0	0	0	1	

Table 14-6: Wave Drive 4-Step Sequence

Notice that the 8-step sequence of Table 14-5 is simply the combination of the wave drive 4-step and normal 4-step normal sequences shown in Tables 14-6 and 14-3, respectively. Experimenting with the wave drive 4-step sequence is left to the reader.

Unipolar versus bipolar stepper motor interface

There are three common types of stepper motor interfacing: universal, unipolar, and bipolar. They can be identified by the number of connections to the motor. A universal stepper motor has eight, while the unipolar has six and the bipolar has four. The universal stepper motor can be configured for all three modes, while the unipolar can be either unipolar or bipolar. Obviously the bipolar cannot be configured for universal nor unipolar mode. Table 14-7 shows selected stepper motor characteristics.

Part No.	Step Angle	Drive System	Volts	Phase Resistance	Current
151861CP	7.5	unipolar	5 V	9 ohms	550 mA
171601CP	3.6	unipolar	7 V	20 ohms	350 mA
164056CP	7.5	bipolar	5 V	6 ohms	800 mA

Table 14-7: Selected Stepper Motor Characteristics (www.Jameco.com)

Figure 14-10 shows the basic internal connections of all three types of configurations.

Unipolar stepper motors can be controlled using the basic interfacing shown in Figure 14-11, whereas the bipolar stepper requires H-Bridge circuitry. Bipolar stepper motors require a higher operational current than the unipolar; the advantage of this is a higher holding torque.

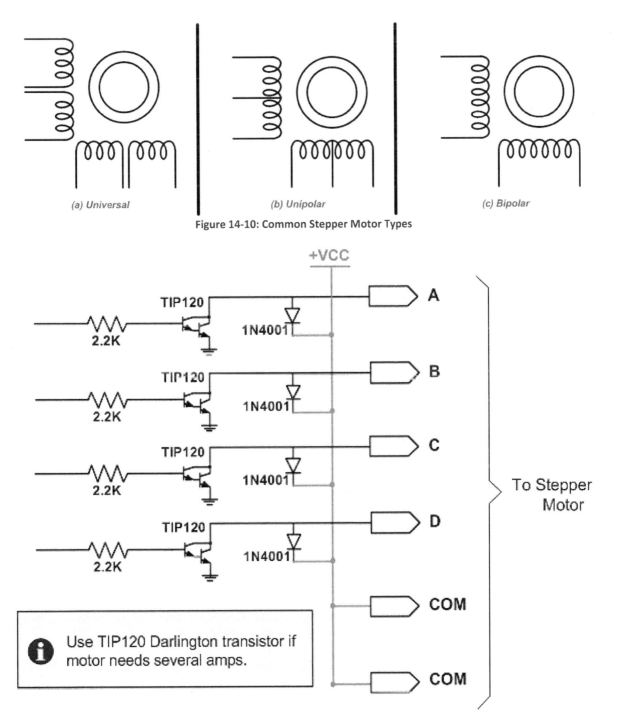

Figure 14-10: Common Stepper Motor Types

(a) Universal

(b) Unipolar

(c) Bipolar

Figure 14-11: Using Transistors for Stepper Motor Driver

Using transistors as drivers

Figure 14-11 shows an interface to a unipolar stepper motor using transistors. Diodes are used to reduce the back EMF spike created when the coils are energized and de-energized, similar to the electromechanical relays discussed earlier. TIP transistors can be used to supply higher current to the motor. Table 14-8 lists the common industrial Darlington transistors. These transistors can accommodate higher voltages and currents.

385

NPN	PNP	V$_{CEO}$ (volts)	I$_C$ (amps)	hfe (common)
TIP110	TIP115	60	2	1000
TIP111	TIP116	80	2	1000
TIP112	TIP117	100	2	1000
TIP120	TIP125	60	5	1000
TIP121	TIP126	80	5	1000
TIP122	TIP127	100	5	1000
TIP140	TIP145	60	10	1000
TIP141	TIP146	80	10	1000
TIP142	TIP147	100	10	1000

Table 14-8: Darlington Transistor Listing

Controlling stepper motor via optoisolator

In the first section of this chapter we examined the optoisolator and its use. Optoisolators are widely used to isolate the stepper motor's EMF voltage and keep it from damaging the digital/microcontroller system. This is shown in Figure 14-12. See Program 14-2.

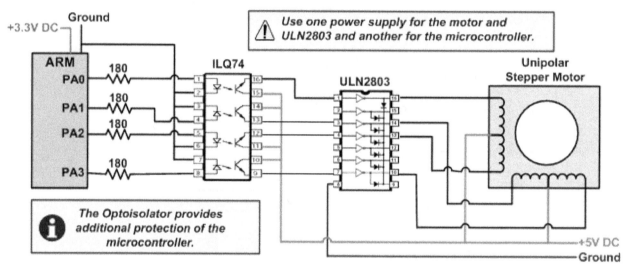

Figure 14-12: Controlling Stepper Motor via Optoisolator

Program 14-2: Controlling a stepper motor

```
/* The program monitors PB0. It rotates clockwise if it is high. Otherwise, it rotates
counter clockwise. */
#include <stm32f10x.h>

void delay_ms(uint16_t t);

int main()
{
    const uint8_t steps[4] ={0x09, 0x0C, 0x06, 0x03};  /* Table 14-3 */
    RCC->APB2ENR |= 0xFC; /* Enable GPIO ports clocks */

    GPIOA->CRL = 0x44443333; /* PA0-PA3 as outputs */
    GPIOB->CRL = 0x44444448; /* PB0 as input with pull-up */
    GPIOB->ODR |= (1<<0);    /* pull-up PB0 */
```

```
    uint8_t n = 0;
    while(1)
    {
        GPIOA->ODR = steps[n];        /* go to next step */

        if((GPIOB->IDR&(1<<0)) != 0) /* is PB0 high */
        {
            if(n >= 3)
                n = 0;
            else
                n++;        /* clockwise */
        }
        else
        {
            if(n == 0)
                n = 3;
            else
                n--; /* counter clockwise */
        }

        delay_ms(100);        /* wait 100 ms */
    }
}

/* copy delay_ms from Program 8-1 to here */
```

Review Questions

1. Give the 4-step sequence of a stepper motor if we start with 0110.
2. A stepper motor with a step angle of 5 degrees has _____ steps per revolution.
3. Why do we put a driver between the microcontroller and the stepper motor?

Problems

Section 14.1

1. True or false. The minimum voltage needed to energize a relay is the same for all relays.
2. True or false. The minimum current needed to energize a relay depends on the coil resistance.
3. Give the advantages of a solid-state relay over an EMR.
4. True or false. In relays, the energizing voltage is the same as the contact voltage.
5. Find the current needed to energize a relay if the coil resistance is 1200 ohms and the coil voltage is 5 V.
6. Give two applications for an optoisolator.
7. Give the advantages of an optoisolator over an EMR.
8. Of the EMR and solid-state relay, which has the problem of back EMF?
9. True or false. The greater the coil inductance, the worse the back EMF voltage.
10. True or false. We should use the same voltage sources for both the coil voltage and the contact voltage.

Section 14.2

11. If a motor takes 90 steps to make one complete revolution, what is the step angle for this motor?

12. Calculate the number of steps per revolution for a step angle of 7.5 degrees.

13. Finish the normal 4-step sequence clockwise if the first step is 0011 (binary).

14. Finish the normal 4-step sequence clockwise if the first step is 1100 (binary).

15. Finish the normal 4-step sequence counterclockwise if the first step is 1001 (binary).

16. Finish the normal 4-step sequence counterclockwise if the first step is 0110 (binary).

17. What is the purpose of the ULN2003 placed between the microcontroller and the stepper motor? Can we use that for 3A motors?

18. Which of the following cannot be a sequence in the normal 4-step sequence for a stepper motor?

(a) 0xCC   (b) 0xDD   (c) 0x99   (d) 0x33

19. What is the effect of a time delay between issuing each step?

20. In Question 19, how can we make a stepper motor go faster?

Answers to Review Questions

Section 14.1

1. With a relay we can use a 5 V digital system to control 12 V–220 V devices such as horns and appliances.

2. Because microcontroller/digital outputs lack sufficient current to energize the relay, we need a driver.

3. When the coil is not energized, the contact is closed.

4. When current flows through the coil, a magnetic field is created around the coil, which causes the armature to be attracted to the coil.

5. It is faster and needs less current to get energized.

6. It is smaller and can be connected to the microcontroller directly without a driver.

Section 14.2

1. 1100, 0110, 0011, 1001 for clockwise; and 1001, 0011, 0110, 1100 for counterclockwise

2. 72

3. The microcontroller pins do not provide sufficient current to drive the stepper motor.

Chapter 15: PWM and DC Motor Control

This chapter discusses the topic of PWM (pulse width modulation) and shows ARM interfacing with DC motors. The characteristics of DC motors are discussed along with their interfacing to the ARM. We use C programming examples to create PWM pulses.

Section 15.1: DC Motor Interfacing and PWM

This section begins with an overview of the basic operation of the DC motors. Then we describe how to interface a DC motor to the ARM. Finally, we use C language programs to demonstrate the concept of pulse width modulation (PWM) and show how to control the speed and direction of a DC motor.

DC motors

A direct current (DC) motor is a widely used device that translates electrical current into mechanical movement. In the DC motor we have only + and − leads. Connecting them to a DC voltage source moves the motor in one direction. By reversing the polarity, the DC motor will rotate in the opposite direction. One can easily experiment with the DC motor. For example, some small fans used in many motherboards to cool the CPU are run by DC motors. While a stepper motor moves in discrete steps of 1 to 15 degrees, the DC motor moves continuously. In a stepper motor, if we know the starting position, we can easily count the number of steps the motor has moved and calculate the final position of the motor. This is not possible in a DC motor. The maximum speed of a DC motor is indicated in RPM and is given in the data sheet. The DC motor has two types of RPM: no-load and loaded. The manufacturer's data sheet gives the no-load RPM. The no-load RPM can be from a few thousand to tens of thousands. The RPM is reduced when moving a load and it decreases as the load is increased. For example, a drill turning a screw has a much lower RPM speed than when it is in the no-load situation. DC motors also have voltage and current ratings. The nominal voltage is the voltage for that motor under normal conditions, and can be from 1 to 150 V, depending on the motor. As we increase the voltage, the RPM goes up. The current rating is the current consumption when the nominal voltage is applied with no load, and can be from 25 mA to a few amps. As the load increases, the RPM is decreased, unless the current or voltage provided to the motor is increased, which in turn increases the torque. With a fixed voltage, as the load increases, the current (power) consumption of a DC motor is increased. If we overload the motor it will stall, and that can damage the motor due to the heat generated by high current consumption.

See Table 15-1 for selected DC motors.

Part No.	Nominal Volts	Volt Range	Current	RPM	Torque
154915CP	3 V	1.5–3 V	0.070 A	5,200	4.0 g-cm
154923CP	3 V	1.5–3 V	0.240 A	16,000	8.3 g-cm
177498CP	4.5 V	3–14 V	0.150 A	10,300	33.3 g-cm
181411CP	5 V	3–14 V	0.470 A	10,000	18.8 g-cm

Table 15-1: Selected DC Motor Characteristics (http://www.Jameco.com)

Unidirectional control

Figure 15-1 shows the DC motor clockwise (CW) and counterclockwise (CCW) rotations.

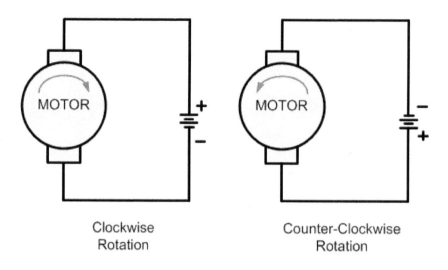

Clockwise Rotation   Counter-Clockwise Rotation

Figure 15-1: DC Motor Rotation (Permanent Magnet Field)

Bidirectional control

With the help of relays, transistor circuit or some specially designed chips we can change the direction of the DC motor rotation. Figures 15-2 through 15-4 show the basic concepts of the H-Bridge control of DC motors.

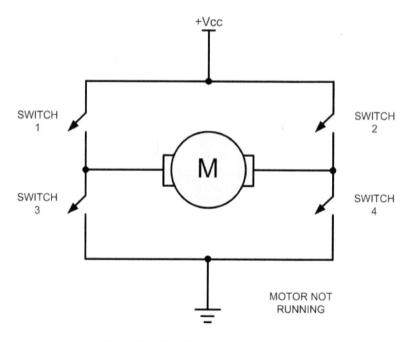

Figure 15-2: H-Bridge Motor Configuration

Figure 15-2 shows the connection of an H-Bridge using simple switches. All the switches are open, which does not allow the motor to turn.

Figure 15-3 shows the switch configuration for turning the motor in one direction. When switches 1 and 4 are closed, current is allowed to pass through the motor.

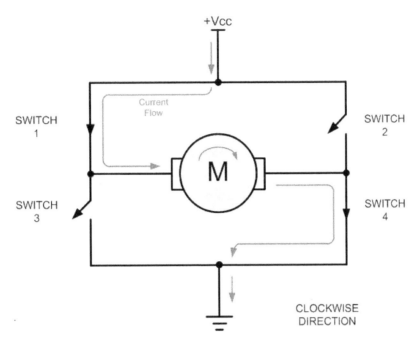

Figure 15-3: H-Bridge Motor Clockwise Configuration

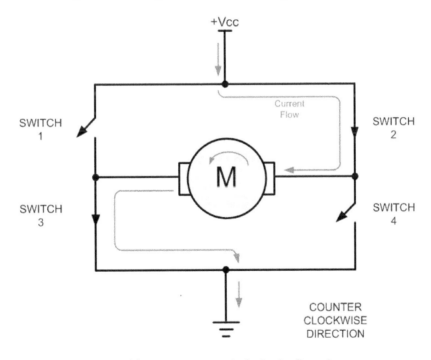

Figure 15-4: H-Bridge Motor Counterclockwise Configuration

Figure 15-4 shows the switch configuration for turning the motor in the opposite direction from the configuration of Figure 15-3. When switches 2 and 3 are closed, current is allowed to pass through the motor.

Figure 15-5 shows an invalid configuration. Current flows directly to ground, creating a short circuit. The same effect occurs when switches 1 and 3 are closed or switches 2 and 4 are closed.

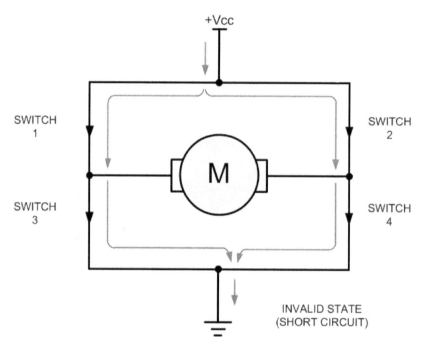

+Vcc

SWITCH 1

SWITCH 2

SWITCH 3

SWITCH 4

M

INVALID STATE
(SHORT CIRCUIT)

Figure 15-5: H-Bridge in an Invalid Configuration

Table 15-2 shows some of the logic configurations for the H-Bridge design.

Motor Operation	SW1	SW2	SW3	SW4
Off	Open	Open	Open	Open
Clockwise	Closed	Open	Open	Closed
Counterclockwise	Open	Closed	Closed	Open
Invalid	Closed	Closed	Closed	Closed

Table 15-2: Some H-Bridge Logic Configurations for Figure 15-2

H-Bridge control can be created using relays, transistors, or a single IC solution such as the L298. When using relays and transistors, you must ensure that invalid configurations do not occur.

Although we do not show the relay control of an H-Bridge, Example 15-1 shows a simple program to operate a basic H-Bridge.

Example 15-1

A switch is connected to pin PTD7. Using relays make the H-Bridge in Table 15-2 and write the proper program. We must perform the following:
(a) If PTD7 = 0, the DC motor moves clockwise.
(b) If PTD7 = 1, the DC motor moves counterclockwise.

Solution 1 (Using SPST Relays):

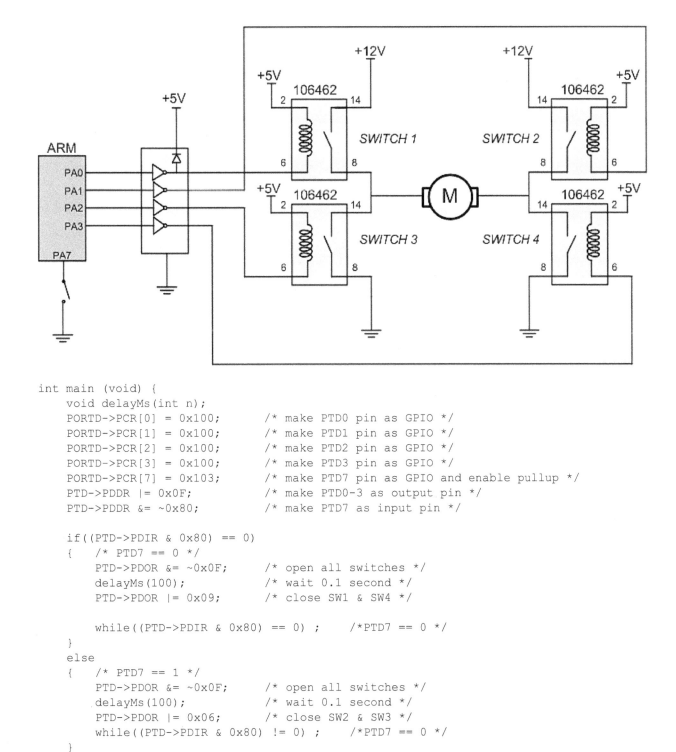

```
int main (void) {
    void delayMs(int n);
    PORTD->PCR[0] = 0x100;       /* make PTD0 pin as GPIO */
    PORTD->PCR[1] = 0x100;       /* make PTD1 pin as GPIO */
    PORTD->PCR[2] = 0x100;       /* make PTD2 pin as GPIO */
    PORTD->PCR[3] = 0x100;       /* make PTD3 pin as GPIO */
    PORTD->PCR[7] = 0x103;       /* make PTD7 pin as GPIO and enable pullup */
    PTD->PDDR |= 0x0F;           /* make PTD0-3 as output pin */
    PTD->PDDR &= ~0x80;          /* make PTD7 as input pin */

    if((PTD->PDIR & 0x80) == 0)
    {   /* PTD7 == 0 */
        PTD->PDOR &= ~0x0F;      /* open all switches */
        delayMs(100);            /* wait 0.1 second */
        PTD->PDOR |= 0x09;       /* close SW1 & SW4 */

        while((PTD->PDIR & 0x80) == 0) ;    /*PTD7 == 0 */
    }
    else
    {   /* PTD7 == 1 */
        PTD->PDOR &= ~0x0F;      /* open all switches */
        delayMs(100);            /* wait 0.1 second */
        PTD->PDOR |= 0x06;       /* close SW2 & SW3 */
        while((PTD->PDIR & 0x80) != 0) ;    /*PTD7 == 0 */
    }
}
```

Solution 2 (Using SPDT Relays):

The H-bridge can also be made using two SPDT relays as shown in the following figure.

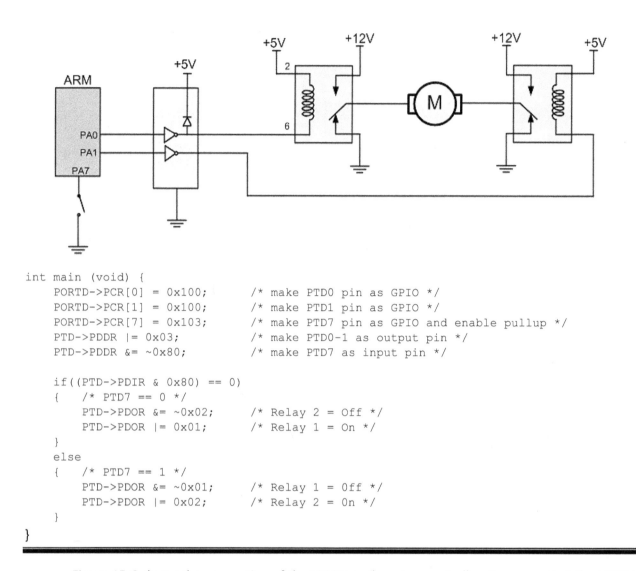

```c
int main (void) {
    PORTD->PCR[0] = 0x100;      /* make PTD0 pin as GPIO */
    PORTD->PCR[1] = 0x100;      /* make PTD1 pin as GPIO */
    PORTD->PCR[7] = 0x103;      /* make PTD7 pin as GPIO and enable pullup */
    PTD->PDDR |= 0x03;          /* make PTD0-1 as output pin */
    PTD->PDDR &= ~0x80;         /* make PTD7 as input pin */

    if((PTD->PDIR & 0x80) == 0)
    {   /* PTD7 == 0 */
        PTD->PDOR &= ~0x02;     /* Relay 2 = Off */
        PTD->PDOR |= 0x01;      /* Relay 1 = On */
    }
    else
    {   /* PTD7 == 1 */
        PTD->PDOR &= ~0x01;     /* Relay 1 = Off */
        PTD->PDOR |= 0x02;      /* Relay 2 = On */
    }
}
```

Figure 15-6 shows the connection of the L298N to the microcontroller. Be aware that the L298N will generate heat during operation. For sustained operation of the motor, use a heat sink.

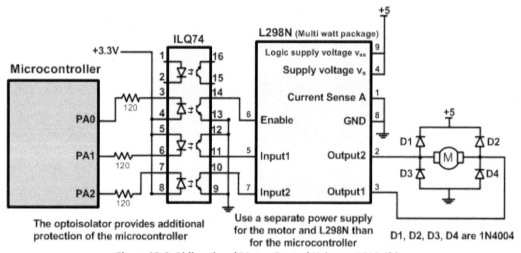

Figure 15-6: Bidirectional Motor Control Using an L298 Chip

Pulse width modulation (PWM)

The speed of the motor depends on three factors: (a) load, (b) voltage, and (c) current. For a given fixed load, we can maintain a steady speed by using a method called pulse width modulation (PWM). By changing (modulating) the width of the pulse applied to the DC motor we can increase or decrease the amount of power provided to the motor, thereby increasing or decreasing the motor speed. Notice that, although the voltage has a fixed amplitude, it has a variable duty cycle. That means the wider the pulse, the higher the speed. PWM is so widely used in DC motor control that many microcontrollers come with an on-chip PWM circuitry. In such microcontrollers all we have to do is load the proper registers with the values of the high and low portions of the desired pulse, and the rest is taken care of by the microcontroller. This allows the microcontroller to do other things. For microcontrollers without on-chip PWM circuitry, we must create the various duty cycle pulses using software, which prevents the microcontroller from doing other things. The ability to control the speed of the DC motor using PWM is one reason that DC motors are preferable over AC motors. AC motor speed is dictated by the AC frequency of the voltage applied to the motor and the frequency is generally fixed. As a result, we cannot control the speed of the AC motor when the load is increased. As will be shown later, we can also change the DC motor's direction and torque. See Figure 15-7 for PWM comparisons.

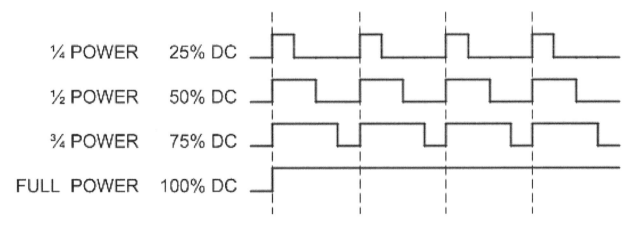

Figure 15-7: Pulse Width Modulation Comparison

DC motor control with optoisolator

The optoisolator is indispensable in many motor control applications. Figures 15-8 and 15-9 show the connections to a simple DC motor using a bipolar and a MOSFET transistor. Notice that the microcontroller is protected from EMI created by motor brushes by using an optoisolator and a separate power supply.

Figures 15-8 and 15-9 show optoisolators for single directional motor control, and the same principle should be used for most motor applications. Separating the power supplies of the motor and logic will reduce the possibility of damage to the control circuit. Figure 15-8 shows the connection of a bipolar transistor to a motor. Protection of the control circuit is provided by the optoisolator. The motor and the microcontroller use separate power supplies. The separation of power supplies also allows the use of high-voltage motors. Notice that we use a decoupling capacitor across the motor; this helps reduce the EMI created by the motor. The motor is switched on by clearing bit PTD0.

395

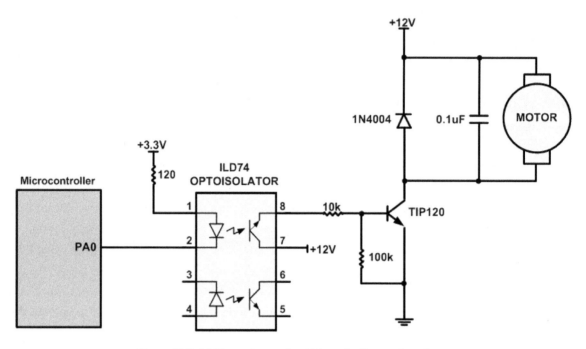

Figure 15-8: DC Motor Connection Using a Darlington Transistor

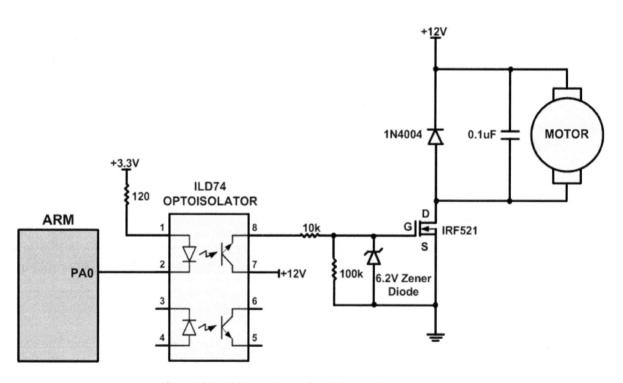

Figure 15-9: DC Motor Connection Using a MOSFET Transistor

Figure 15-9 shows the connection of a MOSFET transistor. The optoisolator protects the microcontroller from EMI. The Zener diode is required for the transistor to reduce gate voltage below the rated maximum value.

Review Questions

1. True or false. The permanent magnet field DC motor has only two leads for + and − voltages.
2. True or false. As with a stepper motor, one can control the exact angle of a DC motor's move.
3. Why do we put a driver between the microcontroller and the DC motor?
4. How do we change a DC motor's rotation direction?
5. What is stall in a DC motor?
6. The RPM rating given for the DC motor is for _____ (no-load, loaded).

Section 15.2: Programming PWM in STM32

In the first section of this chapter we showed how to use the CPU itself to create the PWM outputs. In this section, we use the timers built-in PWM features to generate PWM waves and relieve the CPU to do other important things.

In STM32, the PWM (Pulse Width Modulation) is incorporated into the Timer. To program the PWM features of the ARM STM32 chip, we must understand the Timer topics covered in Chapter 11 since PWM is subset of the Timer. In this section, we examine the PWM features and show how to program them.

The CR1 register and the TIM counting

As discussed in Chapter 11, the TIM_CR1 register has control on the counting of the timer. See Figures 15-11 and 15-12.

See Figure 15-13. The timer can count in one of the following modes:

1) **Count Up (CMS=00 and DIR = 0):** The TIMx_CNT counts up from the 0 value until it reaches the value of ARR register. Then, in the next clock, the CNT is cleared to zero and count-up starts again.

2) **Count Down (CMS=00 and DIR = 1):** The TIMx_CNT is initialized with the value of ARR and it counts down until it reaches 0. Then, in the next clock, the CNT is reloaded with the ARR value and count-down starts again.

3) **Count Up-Down (CMS = 01, 10, or 11):** It counts up from 0 until it reaches the ARR value. After reaching the ARR value, the DIR bit of CR1 is set, and in the next clock it counts down to 0. And upon reaching 0, the DIR bit is cleared and it repeats the process. Figure 15-10 shows the up-down counting when TIMx_ARR is 3. The timer is in Up-Down (Center-Aligned) mode, if the CMS bits have values 01, 10, or 11. The timer counts the same way for the 3 values. But the CCnIF flag rises differently as will be discussed.

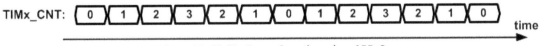

TIMx_CNT: 0 1 2 3 2 1 0 1 2 3 2 1 0   time

Figure 15- 10: Up-Down Counting when ARR=3

	15	14	...	10	9	8	7	6	5	4	3	2	1	0
TIMx_CR1: (TIMx->CR1)		Reserved			CKD		ARPE	CMS		DIR	OPM	URS	UDIS	CEN

Name	bit	Description
CKD	9-8	Clock Division used by dead-time generators (for more info. see the manual)
ARPE	7	Auto Reload Preload Enable 0: ARR register is not buffered. 1: ARR is buffered.
CMS	6-5	Center-aligned Mode Selection When CMS=00, the timer counts up or down depending on the value of DIR bit. Otherwise, (CMS = 01, 10, or 11) the timer counts up and down, alternatively.

CMS	DIR	Counting mode	Counting event (interrupt flag set)
00	0	Counting up	When the counter reaches ARR
00	1	Counting down	When the counter reaches 0
01	X	Count up and down	When the counter reaches 0 and ARR
10	X	Count up and down	When the counter reaches 0 and ARR
11	X	Count up and down	When the counter reaches 0 and ARR

Name	bit	Description
DIR	4	If the CMS bits are set to 00, the DIR bit chooses the direction of counting: 0: the CNT counter counts up. 1: the CNT counter counts down.
OPM	3	One Pulse Mode 0: the counter counts continuously 1: the counter stops at the next update event.
URS	2	Update request source
UDIS	1	Update Disable: We can mask (disable) generating any update events.
CEN	0	Counter Enable (0: The counter is disabled, 1: enable the counter to begin counting)

Figure 15-11: TIMx_CR1

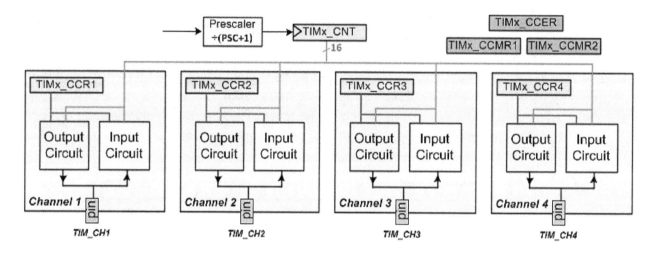

Figure 15-12: The Channels of TIMx

398

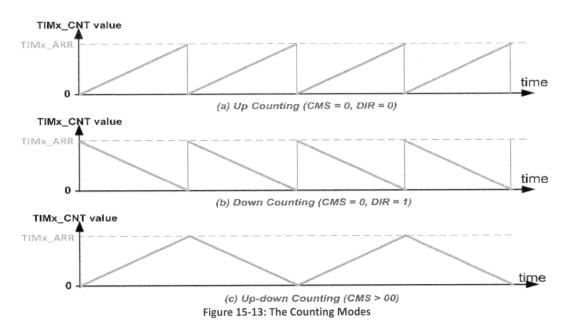

Figure 15-13: The Counting Modes

In Chapter 11, we used TIMx_CCMR (TIMx Capture/Compare Mode Register) to program the Output Control or Input Capture features of the Timer. As it was shown, the OCnM bits are used to configure the Output Compare wave generator. We use the same bits to choose the PWM features. See Figure 15-14.

	D15	D14	D13	D12	D11	D10	D9	D8	D7	D6	D5	D4	D3	D2	D1	D0	
TIMx_CCMR1:	OC2CE		OC2M		OC2PE	OC2FE		CC2S	OC1CE			OC1M	OC1PE	OC1FE		CC1S	0x18
(TIMx->CCMR1)																	

	D15	D14	D13	D12	D11	D10	D9	D8	D7	D6	D5	D4	D3	D2	D1	D0	
TIMx_CCMR2:	OC4CE		OC4M		OC4PE	OC4FE		CC4S	OC3CE			OC3M	OC3PE	OC3FE		CC3S	0x1C
(TIMx->CCMR2)																	

Name	Description		
OCnCE	Output Compare n Clear Enable (For more information, see the reference manual.)		
OCnM	Output Compare n Mode		
	OCnM	**Mode**	
	000	Frozen	Compare match has no effect on the GPIO pin
	001	Active on match	The output activates when CNT is equal to CCRn.
	010	Inactive on match	The output becomes inactive when CNT=CCRn.
	011	Toggle on match	The output toggles when CNT is equal to CCRn.
	100	Force inactive	It forces the GPIO pin to inactive level.
	101	Force active	It forces the GPIO pin to active level.
	110	PWM 1	The output is active when CNT is less than CCRn.
	111	PWM 2 (inverted)	The output is inactive when CNT is less than CCRn.
OCnPE	Output Compare n Preload Enable. (0: TIMx_CCRn is not buffered, 1: It is buffered.)		
OCnFE	Output compare n Fast Enable (For more information see the reference manual)		
CCnS	Compare/Capture n Selection (00: output compare, otherwise: input capture)		

Figure 15-14: CCMR1 and CCMR2

For PWM, we can use the options of Center-aligned (up-down counting) or Edge-aligned (up counting or down counting). Each is discussed next.

PWM in Up-Counting mode

In PWM1 mode (OCnM=110), when TIMx_CNT is less than TIMx_CCRn, the PWM pin is active. Otherwise, the PWM pin is inactive. See Figure 15-15. In PWM 1, when TIMx_CCRn is zero the duty cycle of the generated wave is 0%. The duty cycle of the generated wave increases when the value of CCRn increases. When TIMx_CCRn is bigger than TIMx_ARR, the duty cycle is 100%. See Figure 15-16.

See Figure 15-15 and compare PWM2 with PWM1. In PWM2 the outputs are inverted. We can also invert the output using the CCnP bit of the CCER register. For simplicity, we concentrate on PWM1 with CCnP = 0.

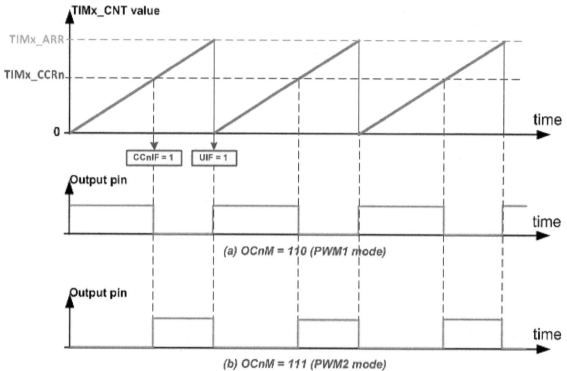

Figure 15-15: PWM in Up-Counting Mode

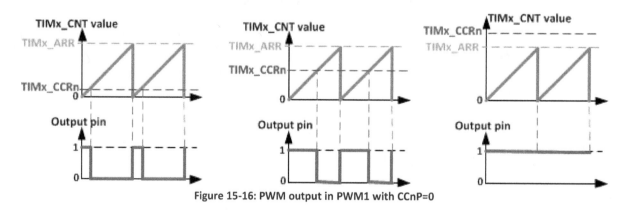

Figure 15-16: PWM output in PWM1 with CCnP=0

400

The PWM output duty cycle and frequency

Figure 15-17 shows the output waveform when OCnM = 110 (PWM1), TIMx_ARR = 8, and TIMx_CCRn = 5. The output is set on counter overflow (reload) and it is cleared on compare match. The CNT is reloaded with 0 after ARR + 1 clocks and the output is set to HIGH for CCRn clocks. So, the duty cycle can be calculated using the following formula:

$$duty\ cycle = \frac{TIMx\_CCRn}{TIMx\_ARR + 1} \times 100$$

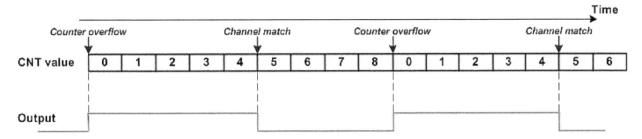

Figure 15-17: The PWM output for ARR = 8, CCRn= 5, OCnM = 110 (PWM1)

When OCnM = 111 (PWM2), the output is inverted and the duty cycle is:

$$Duty\ Cycle = 100 - (\frac{TIMx\_CCRn}{TIMx\_ARR + 1} \times 100)$$

In the mode, the timer counts from 0 to ARR and then rolls over. So, the frequency of the output is 1 / (ARR+1) of the frequency of timer clock. The frequency of the timer clock can be selected using the prescaler. So, the frequency of the output can be calculated as follows:

$$F_{generated\ wave} = \frac{\frac{F_{timer\ clock}}{prescaler}}{ARR + 1} = \frac{F_{timer\ clock}}{(ARR + 1) \times (PSC + 1)}$$

See Examples 15-2 through 15-5.

Example 15-2

Find the frequency (F) and pulse width (DC, duty cycle) of a PWM if TIMx_ARR=999 and TIMx_CCRn=250. Assume OCnM = 110 (PWM1), no prescaler, and TIMx clock frequency of 72MHz.

Solution:

Frequency=72M/(999+1)=72KHz=72000Hz.
The Duty Cycle is [TIMx_CCR/(TIMx_ARR +1)] × 100 = (250/1000) × 100 = 25%.

Example 15-3

Assume the TIMx Module clock frequency is 72MHz. Using no prescaler, find the value of the TIMx_ARR register if we want the PWM output Frequency of (a) 5KHz, (b) 10KHz, and (c) 25KHz.

Solution:

(a) ARR+1=72MHz/5KHz = 14400 ➔ ARR = 14399.
(b) ARR+1=72MHz/10KHz = 7200 ➔ ARR = 7199.
(c) ARR+1=72MHz/25KHz = 2880 ➔ ARR = 2879.

Example 15-4

In a PWM application, we need the PWM output frequency of 100Hz. (a) Using the TIMx Module frequency of 72MHz, find out the value of the TIMx_ARR register. (b) find the TIMx_CCRn value, to generate a wave with duty cycle of 30%.

Solution:

(a) TIMx_ARR = (72MHz / 100Hz) − 1 = 720,000 − 1 = 719,999. This is not acceptable since it is larger than 65,535, the maximum value the TIMx_ARR register can hold. So, we use prescaler:

(ARR + 1) × (PSC + 1) = (72MHz / 100Hz) = 720,000. With PSC = 719, ARR becomes 1000 − 1 = 999.

(b) duty cycle = [CCRn /(ARR+1)]×100 ➔ 30 = [CCRn /1000]×100 ➔ CCRn = 300

The update signal, the shadow registers and preloading (buffering)

when the wave generator is working, if we want to change the values of TIMx_CCRn or TIMx_ARR, the changes should be synchronized with the wave generator. Otherwise, unacceptable waves might be generated. To prevent this, there are shadow registers for TIMx_CCRn and TIMx_ARR registers.

The preloading for TIMx_ARR can be enabled using the ARPE (Auto Reload Preloading Enable) bit of CR1. To enable preloading for CCRn registers, there is an OCnPE bit for each CCRn in the CCMRx registers.

If preloading is enabled, the values of the shadow registers are used to generate waves. When we write to ARR/CCRn registers, the values sit in the registers. The registers will be loaded to the shadow registers at the end of generating a wave cycle, when the update signal is raised. See Figure 15-18.

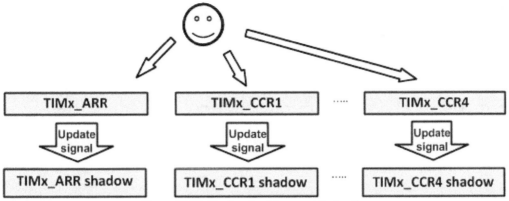

Figure 15-18: Shadow Registers and Preloading

When preloading is enabled, it is needed to generate an update signal before turning on the timer, in order to initialize the shadow registers with the values of the registers. If the UG bit of TIMx_EGR is set, an update signal is generated and then hardware clears the UG bit automatically. See Figure 15-19.

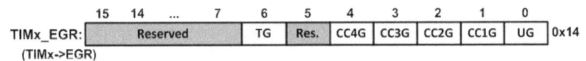

Figure 15-19: TIMx_EGR (Event Generating Register)

Program 15.1 generates a PWM wave with duty cycle of 30% and frequency of 100Hz on TIM2_CH1 (PA0). You can use an oscilloscope to observe the waveform. The register values of Program 15-1 are from Example 15-4.

Program 15-1: Using TIM2 to generate a 100Hz wave with duty cycle of 30%

```c
#include <stm32f10x.h>

int main()
{
    RCC->APB2ENR |= 0xFC;   /* enable GPIO clocks */
    RCC->APB1ENR |= (1<<0);        /* enable TIM2 clock */
    GPIOA->CRL = 0x4444444B;       /* PA0: alternate func. output */

    TIM2->CCER = 0x1 << 0; /* CC1P = 0, CC1E = 1 */
    TIM2->CCMR1 = 0x0068;  /* OC1M=PWM1, OC1PE=1 */
    TIM2->CR1 = 0x80; /* Auto reload preload enable */

    TIM2->PSC = 720-1;      /* prescaler = 720 */
    TIM2->ARR = 1000-1;     /* ARR = 999 */
    TIM2->CCR1 = 1000;      /* duty cycle = (300/1000)*100 */

    TIM2->EGR = 1; /* UG = 1 (generate update) */
    TIM2->CR1 |= 0x01; /* timer enable (CEN = 1) */

    while(1)
    {
    }
}
```

Program 15-2 is based on Program 11-1. But the duty cycle of the generated wave changes gradually from 0% to 100%. So, if you connect an LED between PA0 and ground, the light of the LED increases, gradually.

In Program 15-2, if you change PSC from 720 to 7200 and ARR from 1000 to 10000, the output wave becomes slow enough that you can observe the generated wave with naked eyes.

Program 15-2: Changing the duty cycle from 0% to 100%

```
#include <stm32f10x.h>

void delay_ms(uint16_t t);

int main()
{
   RCC->APB2ENR |= 0xFC;    /* enable GPIO clocks */
   RCC->APB1ENR |= (1<<0);          /* enable TIM2 clock */
   GPIOA->CRL = 0x4444444B;         /* PA0: alternate func. output */

   TIM2->CCER = 0x1 << 0; /* CC1P = 0, CC1E = 1 */
   TIM2->CCMR1 = 0x0068;   /* OC1M=PWM1, OC1PE=1 */
   TIM2->CR1 = 0x80; /* Auto reload preload enable */

   TIM2->PSC = 720-1;        /* prescaler = 720 */
   TIM2->ARR = 1000-1;       /* ARR = 999 */

   TIM2->EGR = 1; /* UG = 1 (generate update) */
   TIM2->CR1 |= 0x01; /* timer enable (CEN = 1) */

   while(1)
   {
     /* change the dc from 0 to 100.0% */
     for(uint16_t d = 0; d <= 1000; d+= 20)
     {
        TIM2->CCR1 = d;
        delay_ms(50);
     }
   }
}

void delay_ms(uint16_t t)
{
   for(int i = 0; i < t; i++)
   {
     for(volatile uint16_t a = 0; a < 6000; a++)
     {}
   }
}
```

PWM in Down-Counting mode

In PWM1 mode (OCnM=110), when TIMx_CNT is more than TIMx_CCRn, the PWM pin is inactive. Otherwise, the PWM pin is active. See Figure 15-20. In PWM 1, the duty cycle of the generated wave decreases when the value of CCRn increases. When TIMx_CCRn is bigger than TIMx_ARR, the duty cycle is 0%. The down-counting and up-counting modes are referred to as edge-aligned and they make similar waves. In contrast, the up-down counting mode is called center-aligned which will be discussed next.

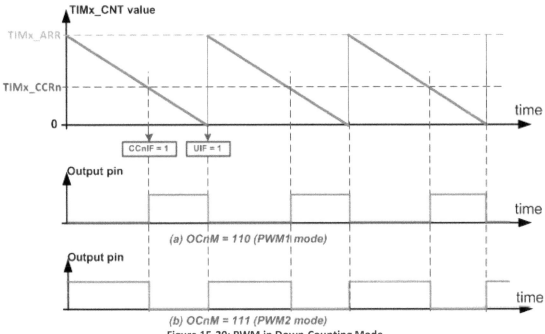

Figure 15-20: PWM in Down-Counting Mode

Up-Down Counting (Center aligned mode)

If we set the CMS bits in CR1 register to any values other than 00, then the timer counts in up-down counting mode and the output is Center-Aligned PWM. The counter will count up from 0 to the value in TIMx_ARR register then turn around and count down to 0. See Figure 15-21.

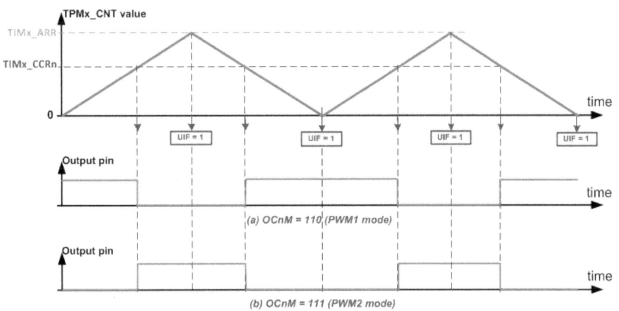

Figure 15-21: Center aligned mode

In PWM1, the output is high, as long as the TIMx_CNT value is less than TIMx_CCRn, and it is high when the TIMx_CNT is more than TIMx_CCRn. In other words, the timer counts up and when TIMx_CNT becomes equal to TIMx_CCRn, the output pin becomes low. The timer continues counting until it reaches ARR. Then it turns around and counts down. When TIMx_CNT becomes equal to TIMx_CCRn, the output pin becomes high. The timer continues counting until it reaches 0.

The PWM output duty cycle and frequency

See Figure 15-21. the period of the pulse is 2 × ARR. The same way, the pulse width = 2 × CCRn. Figure 15-22 shows the output when ARR = 7 and CCRn = 4. The output is set on compare match when counting down, and is cleared on compare match when counting up. The output is HIGH for CCRn×2 clocks and each cycle takes CCRn × 2 clocks. As a result, the duty cycle is:

$$duty\ cycle = \frac{TIMx\_CCRn \times 2}{TIMx\_ARR \times 2} \times 100 = \frac{TIMx\_CCRn}{TIMx\_ARR} \times 100$$

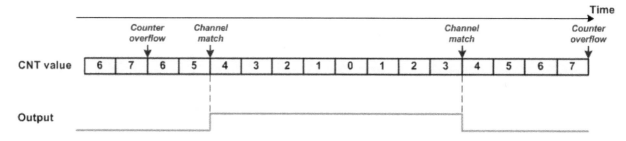

Figure 15-22: The PWM output for ARR = 7, CCRn= 4, OCnM = 110 (non-inverted)

When OCnM = 111, the output is inverted and the duty cycle is:

$$duty\ cycle = 100 - (\frac{TIMx\_CCRn}{TIMx\_ARR} \times 100)$$

The frequency of the generated wave is:

$$F_{generated\ wave} = \frac{F_{timer\ clock}/prescaler}{2 \times ARR} = \frac{F_{timer\ clock}}{2 \times ARR \times (1 + PSC)}$$

See the following examples.

Example 15-5

Find the CR1 value, to enable ARR preloading and the up-down counting mode.

Solution:

TIMx_CR1:	CKD	ARPE	CMS	DIR	OPM	URS	UDIS	CEN
	0	1	01	0	0	0	0	1

TIMx_CR1 = 0xA1;

Example 15-6

Using PWM1 mode, write a program that makes a wave with frequency of 1kHz and duty cycle of 70%.

Solution:

$2*ARR*(1+PSC) = F_{timer\ clock}/F_{generated\ wave} = 72M/1K = 72000 \rightarrow ARR*(1+PSC) = 36000$

We can choose many different values for ARR and PSC. We choose ARR = 36000 and no prescaler.

Duty cycle/100 = CCRn/ARR $\rightarrow$ 70/100 = CCRn/36000 $\rightarrow$ CCRn = 25200

In the following program, Channel 2 of TIM3 is used.

```
#include <stm32f10x.h>

int main()
{
    RCC->APB2ENR |= 0xFC;    /* enable GPIO clocks */
    RCC->APB1ENR |= (1<<1);  /* enable TIM3 clock */
    GPIOA->CRL = 0xB4444444; /* PA7: alternate func. output */

    TIM3->CCER = 0x1 << 4;  /* CC2P = 0, CC2E = 1 */
    TIM3->CCMR1 = 0x6800;   /* OC2M=PWM1, OC2PE=1 */
    TIM3->CR1 = 0xA0;   /* Auto reload preload enable, up down counting mode */

    TIM3->PSC = 0;    /* prescaler = 1 */
    TIM3->ARR = 36000;     /* ARR = 36000 */
    TIM3->CCR2 = 25200;    /* duty cycle = (25200/36000)*100 */

    TIM3->EGR = 1; /* UG = 1 (generate update) */
    TIM3->CR1 |= 0x01; /* timer enable (CEN = 1) */

    while(1)
    {
    }
}
```

In the above program, if you change the PSC to 1000, you can see the generated wave with naked eyes.

Difference between CMS=01, CMS=10, and CMS=11

In the 3 modes the timer counts in the same way. But the CCnIF (Compare/Capture Interrupt flag) is set differently:

- CMS = 01: In this mode, the CCnIF is set only when the timer is counting down.
- CMS = 10: In the mode, the CCnIF flag is set when the timer is counting up.
- CMS = 11: In the mode, the CCnIF flag is set both when the timer is counting up or down.

Edge-aligned vs. center-aligned mode

See Figure 11-19. In both figures the bold vertical blue lines are repeated periodically. In the edge-aligned mode, the left edge of the pulse is always on the bold blue line while in center-aligned mode, the center of the pulse is always fixed on the bold line. In other words, in edge-aligned mode, the phase of the wave is different for different duty cycles, while it remains unchanged in the center-aligned mode. For driving motors, it is preferable to use center-aligned rather than edge-aligned.

In edge-aligned mode, the frequency of the generated wave is twice that of the center-aligned mode. Thus, edge-aligned mode is preferable when we need to generate waves with higher frequencies.

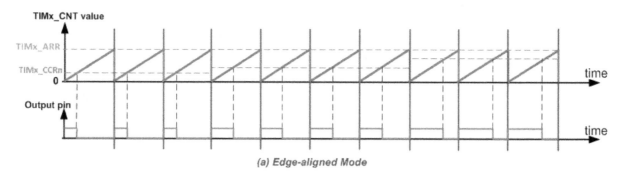

(a) Edge-aligned Mode

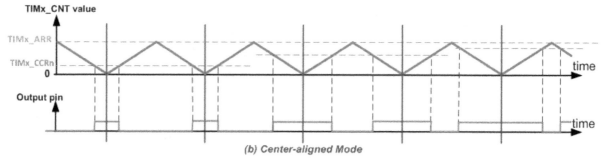

(b) Center-aligned Mode

Figure 15-23: Edge-aligned vs. Center-aligned Mode

Review Questions

1. We use _____ register to set the PWM output Period/Frequency.
2. We use _____ register to set the PWM output pulse width.
3. True or false. In STM32, the PWM module uses the timer registers to set the frequency and duty cycle.

Section 15.3: DC Motor Control Using PWM

As discussed in Section 15.1, you can control the speed of DC motors using PWM. See Example 15-7.

Example 15-7

Using the circuit of Figure 15-8 or Figure 15-9, write a program that gradually increases the speed of a DC motor by changing the duty cycle from 40% to 100%.

Solution:

408

For driving motors, it is preferable to use center-aligned mode. In the program we use TIM2_CH1 (PA0). DC motors have smoother move when the frequency of the wave is higher. But transistors have switching speed limits and if you switch transistors faster, they heat more. In the following program, the frequency of the wave is 2kHz. But you can change the PSC value to increase or decrease the frequency.

```
#include <stm32f10x.h>

void delay_ms(uint16_t t);

int main()
{
   RCC->APB2ENR |= 0xFC;    /* enable GPIO clocks */
   RCC->APB1ENR |= (1<<0);        /* enable TIM2 clock */
   GPIOA->CRL = 0x4444444B;       /* PA0: alternate func. output */

   TIM2->CCER = 0x1 << 0; /* CC1P = 0, CC1E = 1 */
   TIM2->CCMR1 = 0x0068;   /* OC1M=PWM1, OC1PE=1 */
   TIM2->CR1 = 0xA0; /* Auto reload preload enable, center-aligned mode */

   TIM2->PSC = 18-1; /* prescaler = 18 */
   TIM2->ARR = 1000; /* ARR = 1000 */

   TIM2->EGR = 1; /* UG = 1 (generate update) */
   TIM2->CR1 |= 0x01; /* timer enable (CEN = 1) */

   /* change the dc from 40 to 100.0% */
   for(uint16_t d = 400; d <= 1000; d+= 4)
   {
      TIM2->CCR1 = d;
      delay_ms(15);
   }

   while(1)
   {
   }
}

void delay_ms(uint16_t t)
{
   for(int i = 0; i < t; i++)
   {
      for(volatile uint16_t a = 0; a < 6000; a++)
      {}
   }
}
```

Speed control and L298N driver

When you drive a motor using L298 or L293, you can connect the enable pin of the chips to the microcontroller and control the speed using PWM. The speed of motor increases when the duty cycle increases. See Figure 15-6.

Controlling a 3-wheel or 4-wheel robot (Case Study)

Now you can make 3-wheel robots and control its speed and direction using L298 and PWM! L298 can drive two DC motors. So, you can connect two motors to L298 and control their speeds and directions, separately. The robot, moves forward when both wheels rotate forward, and the robot moves backward when both wheels rotate backward. The robot can also turn left and right if the wheels rotate in different directions or with different speeds.

Review Questions

1. True or false. We can control the speed of DC motors using PWM.
2. True or false. We cannot control the direction and speed of DC motors at the same time.

Problems

Section 15.1

1. True or false. DC motors move exactly 90 degrees per second.
2. True or false. Current dissipation of a DC motor is proportional to the load.
3. True or false. The RPM of a DC motor is the same for no-load and loaded.
4. The RPM given in data sheets is for _____ (no-load, loaded).
5. What is the advantage of DC motors over AC motors?
6. What is the advantage of stepper motors over DC motors?
7. True or false. Higher load on a DC motor slows it down if the current and voltage supplied to the motor are fixed.
8. What is PWM, and how is it used in DC motor control?
9. A DC motor is moving a load. How do we keep the RPM constant?
10. What is the advantage of placing an optoisolator between the motor and the microcontroller?

Section 15.2

11. Using PWM1 and up-counting mode, write a program that generates a wave with frequency of 62.5 kHz and duty cycle of 60%. Timer internal clock is 72 MHz.
12. Using PWM1 and up-counting mode, write a program that generates a wave with frequency of 46.875 kHz and duty cycle of 70%. The timer clock is 72 MHz.
13. Using PWM1 and up-counting mode, write a program that generates a wave with frequency of 1953 Hz and duty cycle of 20%. The timer clock is 72 MHz.
14. Using PWM1 and up-counting mode, write a program that generates a wave with frequency of 15.25 Hz and duty cycle of 10%. The timer clock is 72 MHz.
15. Using PWM1 and up-counting mode, write a program that generates a wave with frequency of 1960 Hz and duty cycle of 20%. The timer clock is 72 MHz.
16. Using PWM1 and center-aligned mode, write a program that generates a wave with frequency of 1.96 kHz and duty cycle of 95%. The timer clock is 72 MHz.
17. Using PWM1 and center-aligned mode, write a program that generates a wave with frequency of 61.3 Hz and duty cycle of 19%. The timer clock is 72 MHz.
18. Using PWM1 and center-aligned mode, write a program that generates a wave with frequency of 245 Hz and duty cycle of 82%. The timer clock is 72 MHz.

Answers to Review Questions

Section 15.1
1. True
2. False
3. Because microcontroller/digital outputs lack sufficient current to drive the DC motor.
4. By reversing the polarity of voltages connected to the motor leads
5. The DC motor is stalled if the load is beyond what it can handle.
6. No-load

Section 15.2
1. TIMx_ARR
2. TIMx_CCRn
3. True

Section 15.3
1. True
2. False

Chapter 16: I2C Protocol and RTC Interfacing

This chapter covers I2C bus interfacing and programming. Section 16.1 examines the I2C bus protocol. Section 16.2 shows the inner working of I2C module in STM32F10x chips. The DS3231 RTC and its I2C interfacing and programming are covered in Section 16.3.

Section 16.1: I2C Bus Protocol

The IIC (Inter-Integrated Circuit) is a bus interface connection incorporated into many devices such as sensors, RTC, and EEPROM. The IIC is also referred to as I2C or I square C in many technical literatures. In this section, we examine the signals of the I2C bus and focus on I2C terminology and protocols.

I2C Bus

The I2C bus was originally started by Philips, but in recent years has become a widely used standard adopted by many semiconductor companies. I2C is ideal to attach low-speed peripherals to a motherboard or embedded system or anywhere that a reliable communication over a short distance is required. As we will see in this chapter, I2C provides a connection-oriented communication with acknowledgement. I2C devices use only 2 pins for data transfer, instead of the 8 or more pins used in traditional parallel buses. These two signals are called SCL (Serial Clock) which synchronize the data transfer between two chips, and SDA (Serial Data). This reduction of communication pins reduces the package size and power consumption drastically, making them ideal for many applications in which space is a major concern. These two pins, SDA, and SCL, make the I2C a 2-wire interface. In some application notes, I2C is referred to as Two-Wire Serial Interface (TWI).

I2C line electrical characteristics

I2C devices use only 2 bidirectional open-drain pins for data communication. To implement I2C, a 4.7k ohm pull-up resistor for each of bus lines is needed (see Figure 16-1). This implements a wired-AND which is needed to implement I2C protocols. It means that if one or more devices pull the line to low (zero) level, the line state is zero. The level of line will be 1 only if none of devices pull the line to low level.

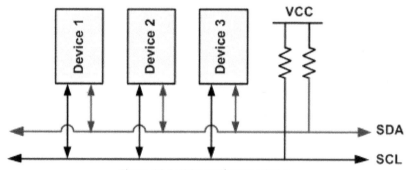

Figure 16-1: I2C Bus Characteristics

I2C Nodes

In I2C protocol, more than 100 devices can share an I2C bus. Each of these devices is called a *node*. In I2C terminology, each node can operate as either master or slave. Master is a device that generate the Clock for the system, it also initiates and terminates a transmission. Slave is a node that receives the clock and is addressed by the master. In I2C, both master and slave can receive or transmit data. So, there are 4 modes of operation for each node. They are: master transmitter, master receiver, slave transmitter and slave receiver. Notice that each node can have more than one mode of operation at different times but it has only one mode of operation at any given time. See Example 16-1

Example 16-1

Give an example to show how a device (node) can use more than one mode of operation.
Solution:

If you connect a microcontroller to an EEPROM with I2C, the microcontroller does master transmit operation to write to EEPROM and master receive operation to read from EEPROM

In next sections, you will see that a node can do the operations of master and slave at different time.

Bit Format

I2C is a synchronous serial protocol; each data bit transferred on the SDA line is synchronized by a high to low pulse of clock on SCL line. According to I2C protocols the data line cannot change when the clock line is high, it can change only when the clock line is low. See Figure 16-2. STOP and START condition are the only exceptions to this rule.

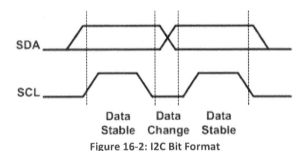

Figure 16-2: I2C Bit Format

START and STOP conditions

As we mentioned before, I2C is a connection-oriented communication protocol, it means that each transmission is initiated by a START condition and is terminated by STOP condition. Remember that the START and STOP conditions are generated by the master.

STOP and START conditions must be distinguished from bits of address or data and that is why they do not obey the bit format rule that we mentioned before.

START and STOP conditions are generated by keeping the level of the SCL line to high and then changing the level of the SDA line. START condition is generated by a high-to-low change in SDA line when SCL is high. STOP condition is generated by a low-to-high change in SDA line when SCL is high. See Figure 16-3.

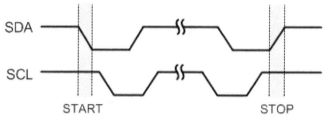

Figure 16-3: START and STOP Conditions

The bus is considered busy between each pair of START and STOP conditions and no other master tries to take control of the bus when it is busy. If a master, which has the control of the bus, wishes to initiate a new transfer and does not want to release the bus before starting the new transfer, it issues a new START condition between a pair of START and STOP condition. It is called REPEATED START condition or simply RESTART condition. See Figure 16-4.

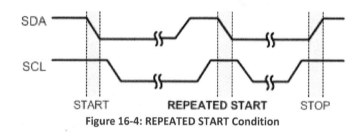

Figure 16-4: REPEATED START Condition

Example 16-2 shows why REPEATED START condition is necessary.

Example 16-2

Give an example to show when a master must use REPEATED START condition. What will happen if the master does not use it?

Solution:

If you connect two microcontrollers (uA and uB) and an EEPROM with I2C, and the uA wants to display the sum of the contents at address 0x34 and 0x35 of EEPROM, it has to use REPEATED START condition. Let's see what may happen if the uA does not use REPEATED START condition. uA transmit a START condition, reads the content of address 0x34 of EEPROM and transmit a STOP condition to release the bus. Before uA reads the contents of address 0x35, the uB seize the bus and change the contents of address 0x34 and 0x35 of EEPROM. Then uA reads the content of address 0x35, adds it to last content of address 0x34 and display the result to LCD. The result on the LCD is neither the sum of old values of address 0x34 and 0x35 nor the sum of the new values of address 0x34 and 0x35 of EEPROM!

Message format in I2C

In I2C, each address or data to be transmitted must be framed in 9-bit long. The first 8 bits are put on SDA line by the transmitter and the 9th bit is the acknowledgement by the receiver or it may be NACK (negative acknowledge). Notice that the clock is always generated by the master, regardless of it being transmitter or receiver. To allow acknowledge, the transmitter release the SDA line during the 9th clock so the receiver can pull the SDA line low to indicate an ACK. If the receiver doesn't pull the SDA line low, it is considered as NACK. See Figure 16-5.

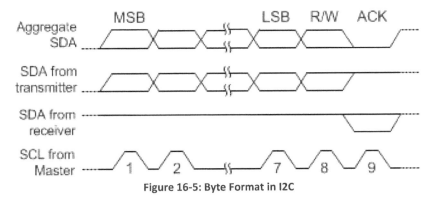

Figure 16-5: Byte Format in I2C

In I2C, each byte may contain either address or data. Also notice that: **START condition + slave address byte + one or more data byte + STOP condition** together form a complete data transfer. Next we will study slave address and data byte formats and how to combine them to make a complete transmission.

Address Byte Format

Like any other bytes, all address bytes transmitted on the I2C bus are nine bits long. It consists of seven address bits, one READ/WRITE control bit and an acknowledge bit. (See Figure 16-6)

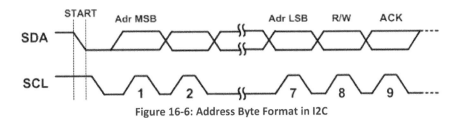

Figure 16-6: Address Byte Format in I2C

Slave address bits are used to address a specific slave device on the bus. 7-bit address let the master to address maximum of 128 slaves on the bus. Although address 0000 000 is reserved for general call and all address of the format 1111 xxx are reserved in many devices. There are 8 more reserved addresses. That means 111 = (128-1-8-8) device can share an I2C bus. In I2C bus the MSB of the address is transmitted first. The I2C bus also supports 10-bit address where the address is split into two frames at the beginning of the transmission. For the rest of the discussion, we will focus on 7-bit address only.

The 8th bit in the address byte is READ/WRITE control bit. If this bit is set, the master will read the next byte from the slave, otherwise, the master will write the next byte on the bus to the slave. When a slave detects its address on the bus, it knows that it is being addressed and it should acknowledge in the ninth clock cycle by pulling SDA to low. If the addressed slave is not ready or for any reason does not want to respond to the master, it should leave the SDA line high in the 9th clock cycle. It is considered as NACK. In case of NACK, the master can transmit a STOP condition to terminate the transmission, or a REPEATED START condition to initiate a new transmission.

Example 16-3 shows how a master says that it wants to write to a slave.

Example 16-3

Show how a master initiates a write to a slave with address 1001101?
Solution:

The following actions are performed by the master:
1) The master put a high to low pulse on SDA while SCL is high to generate a start condition to start the transmission
2) The master transmits 1001101 0 into the bus. The first seven bits (1001101) indicates the slave address and the 8th bit (0) indicates Write operation and the master will write the next byte (data) into the slave.

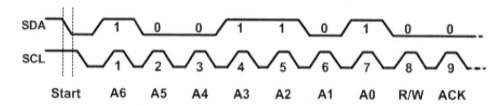

An address byte consisting of a slave address and a READ is called SLA+R while an address byte consisting of a slave address and a WRITE is called SLA+W.

As we mentioned before, address 0000 000 is reserved for general call. It means that when a master transmit address 0000 000 all slaves respond by changing the SDA line to zero for one clock cycle for an ACK and wait to receive the data byte. It is useful when a master wants to transmit the same data byte to all slaves in the system. Notice that the general call address cannot be used to read data from slaves because no more than one slave is able to write to the bus at a given time. Also, not all the devices respond to a general call.

Data Byte Format

Like other bytes, data bytes are 9 bits long too. The first 8 bits are a byte of data to be transmitted and the 9th bit, is for ACK. If the receiver has received the last byte of data and does not wish to receive more data, it may signal a NACK by leaving the SDA line high. The master should terminate the transmission with a STOP after a NACK appears. In data bytes, like address bytes, MSB is transmitted first.

Combining Address and Data Bytes into a Transmission

In I2C, normally, a transmission is started by a START condition, followed by an address byte (SLA+R/W), one or more data bytes and finished by a STOP condition. Figure 16-7 shows a typical data transmission. Try to understand each element in the figure. (See Example 16-4)

416

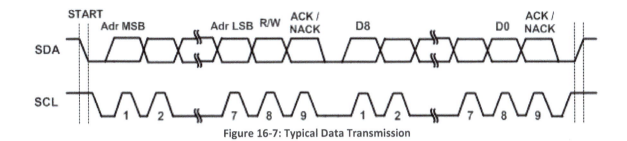

Figure 16-7: Typical Data Transmission

Example 16-4

Show how a master writes data value 1111 0000 to a slave with an address 1001 101?

Solution:

The following actions are performed by the master:

1) The master put a high to low transition on SDA while SCL is high to generate a START condition to start the transmission

2) The master transmits 1001 101 0 on the bus. The first seven bits (1001 101) indicates the slave address and the 8th bit (0) indicates a Write operation and say that the master will write the next byte (data) into the slave.

3) The slave pulls the SDA line low at the 9th clock pulse to signal an ACK to say that it is ready to receive data

4) After receiving the ACK, the master will transmit the data byte (1111 0000) on the SDA line. (MSB first)

5) When the slave device receives the data, it leaves the SDA line high to signal NACK and inform the master that the slave received the last data byte and does not need any more data

6) After receiving the NACK, the master will know that no more data should be transmitted. The master changes the SDA line when the SCL line is high to transmit a STOP condition and then releases the bus.

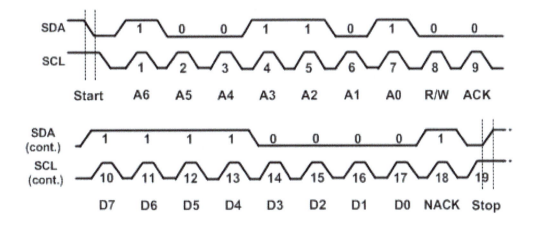

Clock stretching

      One of the features of the I2C protocol is clock stretching. It is used by a slow slave device to synchronize with the master. If an addressed slave device is not ready to process more data, it will stretch

417

the clock by holding the clock line (SCL) low after receiving (or sending) a bit of data so the master will not be able to raise the clock line (because devices are wire-ANDed) and will wait until the slave releases the SCL line to show it is ready for the next bit. See Figure 16-8. Clock stretching can be used to slow down the clock for each bit or it can be used to temporarily halt the clock at the end of a byte while the receiver is processing the data.

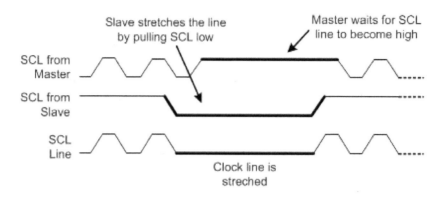

Figure 16-8: Clock Stretching

Arbitration

I2C protocol supports multi-master bus system. It doesn't mean that more than one master can use the bus at the same time. Each master waits for the current transmission to finish and then start to use the bus. But it is possible that two or more masters initiate a transmission at about the same time. In this case the arbitration happens.

Each master has to check the level of the bus and compare it with the levels it is driving; if it doesn't match, that master has lost the arbitration, and will switches to slave mode. In the case of arbitration, the winning master will continue the transmission. Notice that neither the bus is corrupted nor the data is lost. See Example 16-5

Example 16-5

If two master A and B start at about the same time, what happens if master A wants to write to slave 0010 000 and master B wants to write to slave 0001 111?

Solution:

Master A will lose the arbitration in the third clock because the SDA line is different from output of master A at the third clock. Master A switches to slave mode and stops driving the bus after losing the arbitration.

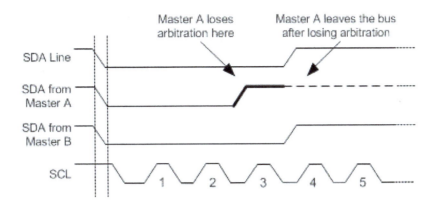

Multi-byte burst write

Burst mode writing is an effective means of loading data into consecutive memory locations. It is supported in I2C, SPI, and many other serial protocols. In burst mode, we provide the address of the first memory location, followed by the data for that location. From then on, consecutive bytes are written to consecutive memory locations. In this mode, the I2C device internally increments the address location as long as STOP condition is not detected. The following steps are used to send (write) multiple bytes of data in burst mode for I2C devices.

1. The master generates a START condition.
2. The master transmits the slave address followed by a zero bit (for write).
3. The master transmits the memory address of the first location.
4. The master transmits the data for the first memory location and from then on, the master simply provides consecutive bytes of data to be placed in consecutive memory locations in the slave.
5. The master generates a STOP condition.

Figure 16-9 shows how to write 0x05, 0x16, and 0x0B to 3 consecutive locations starting from location 00001111 of slave 1111000.

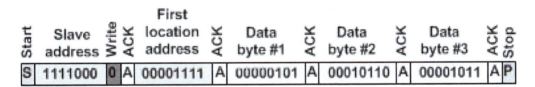

Figure 16-9: Multi-byte Burst Write

Multi-byte burst read

Burst mode reading is an effective means of bringing out the contents of consecutive memory locations. In burst mode, we provide the address of the first memory location only. From then on, contents are brought out from consecutive memory locations. In this mode, the I2C device internally increments the address location as long as STOP condition is not detected. The following steps are used to get (read) multiple bytes of data using burst mode for I2C devices.

419

1. The master generates a START condition.
2. The master transmits the slave address followed by a zero bit (for writing the memory address).
3. The master transmits the memory address of the first memory location.
4. The master generates a RESTART condition to switch the bus direction from write to read.
5. The master transmits the slave address followed by a one bit (for read).
6. The master clocks the bus 8 times and the slave device provides the data for the first location.
7. The master provides an ACK.
8. The master reads the consecutive locations and provides an ACK for each byte.
9. The master gives a NACK for the last byte received to signal the slave that the read is complete.
10. The master generates a STOP condition.

Figure 16-10 shows how to read three consecutive locations starting from location 00001111 of slave number 1111000.

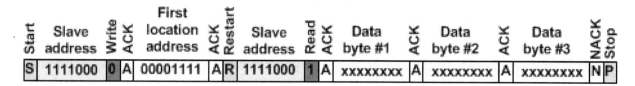

Figure 16-10: Multi-byte Burst Read

Review Questions
1. True or false. I2C protocol is ideal for short distance.
2. How many bits are there in a frame? Which bit is for acknowledgement?
3. True or false. START and STOP conditions are generated when the SDA is high.
4. What is the name of the procedure a slow slave device uses to synchronize with a fast master?
5. True or false. After arbitration of two masters, both of them must start transmission from beginning.

Section 16.2: I2C Programming in STM32F10x

The STM32 chips come with on-chip I2C modules. In this section, we examine the registers and features of I2C module. See Table 16-1.

Register name	Offset	Description
I2C_CR1	0x00	I2C Control Register 1
I2C_CR2	0x04	I2C Control Register 2
I2C_OAR1	0x08	I2C Own Address Register 1
I2C_OAR2	0x0C	I2C Own Address Register 2
I2C_DR	0x10	I2C Data Register
I2C_SR1	0x14	I2C Status Register 1
I2C_SR2	0x18	I2C Status Register 2
I2C_CCR	0x1C	Clock Control Register
I2C_TRISE	0x20	I2C Trise

Table 16-1: Some of the I2C Registers

I2C_CR1 (Control Register 1)

See Figure 16-11. To enable the I2C module you should set the PE bit of I2C_CR1. But before enabling the module, you should configure the module.

The I2C_CR1 register can be used to generate START and STOP conditions, as well. To generate START, set the START bit. The hardware, clears the START bit and tries to generate START condition. If the START condition is generated, the SB flag of I2C_SR1 sets. Similarly, to generate STOP condition, set the STOP bit.

When the device is in slave mode, by default, it stretches the SCL clock until the address or data is provided by software. If you set the NOSTRETCH bit, the hardware does not stretch the clock any more. But sends the previous contents of data register (I2C_DR) if no new data is provided and the OVR flag sets.

Using the ACK bit of I2C_CR1, we choose to send ACK or NACK. If the ACK bit is set, the device sends ACK when needed. Otherwise, NACK will be sent. The hardware automatically clears the ACK bit when an ACK is sent.

	D15	D14	D13	D12	D11	D10	D9	D8	D7	D6	D5	D4	D3	D2	D1	D0
I2Cx_CR1:	SW RST	Res.	ALERT	PEC	POS	ACK	STOP	START	NOST RETCH	ENGC	ENPEC	EN ARP	SMB TYPE	Res.	SMBU S	PE

Field	Bit	Descriptions
SWRST	15	Software Reset
ALERT	13	SMBus alert (It is used in SMBus.)
PEC	12	Packet Error Checking
POS	11	PEC Position
ACK	10	Acknowledge enable (0: No Ack, 1: ACK) The bit is set/cleared by software to send ACK/NACK. Hardware clears the bit after sending ACK.
STOP	9	Stop generation If software sets the bit, the I2C hardware generates a Stop condition and switches to slave mode (The MSL bit of I2C_SR2 is cleared.)
START	8	Start generation When software sets the bit, the I2C interface generates a Start condition and switches to master mode (MSL is set). If it is already in master mode, setting the START bit generates a repeated start.
NOSTRETCH	7	No clock stretching in slave mode (0: clock stretching enabled, 1: disabled) Software can set the bit to disable clock stretching.
ENGC	6	General Call Enable (0: General call disabled, 1: enabled (Addr. 0 is ACKed))
ENPEC	5	PEC enable (0: PEC calculation disabled, 1: PEC calculation enabled)
ENARP	4	ARP enable (It is used in SMBus.)
SMBTYPE	3	SMB Type (It is used in SMBus.)
SMBUS	1	SMB Bus (0: I2C mode, 1: SMBus mode)
PE	0	Peripheral Enable (0: peripheral disable, 1: enable)

Figure 16-11: I2C_CR1 (Control Register 1)

I2C_CR2 (Control Register 2)

See Figure 16-12. The FREQ field of the I2C_CR2 gives the peripheral clock frequency fed to the I2C module. The FREQ value is in megahertz and it must be between 2 (for 2MHz) and 50 (for 50MHz). For example, if the PCLK is 32MHz, we set FREQ to 32.

The I2C_CR2 is also used to enable/disable the I2C interrupts and DMA.

	D15	D14	D13	D12	D11	D10	D9	D8	D7	D6	D5	D4	D3	D2	D1	D0
I2Cx_CR2:	Reserved			LAST	DMA EN	ITBUF EN	ITEVT EN	ITERR EN	Reserved		FREQ					

Field	Bit	Descriptions
LAST	12	DMA Last Transfer
DMAEN	11	DMA request Enable (0: disabled, 1: enabled)
ITBUFEN	10	Buffer Interrupt Enable (0: Interrupt disabled, 1: enabled) If ITBUFEN is set, an interrupt is generated when TxE or RxNE flags of I2C_SR1 are set.
ITEVTEN	9	Event Interrupt Enable (0: Interrupt disabled, 1: enabled) If ITEVTEN is set, an interrupt is generated when any of the event flags (SB, ADDR, ADD10, STOPF, or BTF) are set.
ITERREN	8	Error Interrupt Enable (0: Interrupt disabled, 1: enabled) If ITERREN is set, an interrupt is generated when any of the error flags (BERR, ARLO, AF, OVR, PECERR, TIMEOUT, or SMBALERT) are set.
FREQ	5-0	PCLK (Peripheral Clock) Frequency

Figure 16-12: I2C_CR2 (Control Register 2)

I2C_CCR (Clock Control Register)

In the early I2C, the baud rate was 100kHz. Then Fast mode (Fm), fast mode plus, and High Speed are introduced. Table 16-2 lists the different I2C grades with their speeds. STM32F10x chips support standard and fast mode speeds. The CCR register is used to set the baud rate for I2C when the module is in master mode. See Figure 16-13. Table 16-3 summarizes the relation between the I2C_CCR fields and the duty cycle of SCL and the baud rate for I2C.

I2C grade	Baud rate
Standard	100Kbps
Fast	Up to 400Kbps
Fast plus	Up to 1Mbps
High Speed	Up to 3.2Mbps

Table 16-2: I2C Grades

	D15	D14	D13	D12	D11	D10	D9	D8	D7	D6	D5	D4	D3	D2	D1	D0
I2Cx_CCR:	F/S	DUTY	Reserved		CCR											

Field	Bit	Descriptions
F/S	15	Master mode selection (0: Standard mode, 1: Fast mode)
DUTY	14	SCL clock duty cycle in Fast mode
CCR	11-0	Clock Control in master mode

Figure 16-13: I2C_CCR (Clock Control Register)

F/S	DUTY	Duty cycle for SCL	t_{low}	t_{high}	T_{I2C} $(=t_{low}+t_{high})$	Baud rate $(1/T_{I2C})$
0 (Standard)	X	50%	$CCR \times T_{PCLK}$	$CCR \times T_{PCLK}$	$2 \times CCR \times T_{PCLK}$	$F_{PCLK}/(2 \times CCR)$
1 (Fast)	0	33.3%	$2 \times CCR \times T_{PCLK}$	$CCR \times T_{PCLK}$	$3 \times CCR \times T_{PCLK}$	$F_{PCLK}/(3 \times CCR)$
1 (Fast)	1	36%	$16 \times CCR \times T_{PCLK}$	$9 \times CCR \times T_{PCLK}$	$25 \times CCR \times T_{PCLK}$	$F_{PCLK}/(25 \times CCR)$

Table 16-3: Duty Cycle and Baud Rate for I2C

I2C_TRISE (Trise)

In I2C protocol, the transistors of master devices pull down the lines and the pull-up resistors pull the lines up. Since the internal resistance of transistors are much less than the pull-up resistors, the fall time is much less than the rise time. Using the I2C_TRISE, we mention the amount of time that the rise time might take. For Standard mode, I2C_TRISE is usually set to (PCLK/1M) + 1. For example, when the peripheral clock (PCLK) is 32MHz, we should set it to 33 in standard mode (32 + 1 = 33). In Fast mode, I2C_TRISE should be set to 0.3 × (PCLK/1M) + 1. For example, if PCLK is 40MHz, the I2C_TRISE should be set to (0.3 × 40) + 1 = 13.

Example 16-6

Assume the Peripheral Clock (PCLK1) frequency is 36MHz. Find the values for the I2C_CR2, I2C_CCR, and I2C_TRISE registers if we want I2C clock of (a) 100Kbps, (b) 400Kbps. Disable the interrupts.

Solution:

PCLK1 is 36MHz. So, FREQ should be set to 36 (100000 in binary) in both (a) and (b).

I2C_CR2	Res.	LAST	DMAEN	ITBUFEN	ITEVTEN	ITERREN	Res.	FREQ
	000	0	0	0	0	0	00	100000

I2C_CR2 = 0x0020

(a) 100Kbps is standard. baud rate = PCLK/(2×CCR) ➔ 100K = 36M/(2×CCR) ➔ CCR = 180. So, F/S = 0 and CCR = 000010110100. DUTY can be 0 or 1.

I2C_CCR	F/S	DUTY	Res.	CCR
	0	X	00	0000 1011 0100

I2C_CCR = 0x00B4
I2C_TRISE = (PCLK/1M) + 1 = 36 + 1 = 37.

(b) According to Table 16-2, 400Kbps is fast. If we use DUTY = 0 (33%), we have:
Baud rate = PCLK/(3×CCR) ➔ 400K = 36M/(3×CCR) ➔ CCR = 30 (000000011110 in binary)

I2C_CCR	F/S	DUTY	Res.	CCR
	1	0	00	0000 0001 1110

I2C_CCR = 0x0801E
I2C_TRISE = [0.3 × (PCLK/1M)] + 1 = [0.3 × 36] + 1 = 11.

I2C_OAR1 (Own Address Register 1)

When the STM32 device is designated as a slave, it needs to have a calling address so that it can be addressed by the master by its slave address. We use I2Cx_OAR1 (I2C Own Address 1 Register) to hold the address as the slave device. When the device is in slave mode, it compares the addresses with the contents of I2C_OAR1 to see if it is addressed or not. Notice, the addresses in I2C are only 7 bits (maximum of 127 devices). In the I2Cx_OAR1 register, the D7-D1 bits are used for the slave address and the LSB of D0 is unused and is 0. (The STM32 has the option of 10-bit address, as well. See the STM32 reference manual for more information. The STM32F10x I2C modules can have 2 slave addresses, as well. For more information, see I2C_OAR2 in the reference manual.)

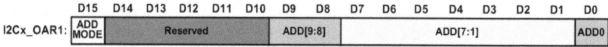

Figure 16-14: I2C_OAR1 (Own Address Register 1)

I2C_DR (Data Register)

See Figure 16-15. I2C_DR is an 8-bit register. We place address and data in I2Cx_DR (I2C Data Register) for transmission. In receive mode, we read I2C_DR to get the received data.

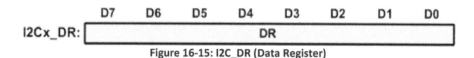

Figure 16-15: I2C_DR (Data Register)

I2C_SR1 (Status Register 1) and I2C_SR2 (Status Register 2)

See Figures 16-16 and 16-17. The registers I2C_SR1 (Status Register 1) and I2C_SR2 (Status Register 2) show the status of I2C. Some of the most useful flags of the registers are as follow:

- **SB:** in master mode, the flag sets when you successfully send a start condition.
- **ADDR:** If STM32 is in master mode, the flag rises when the address is successfully sent. In slave mode, the flag rises when the address of the device is mentioned on the bus.
- **RxNE:** When the flag is set, we should read the contents of data register (I2C_DR).
- **TxE:** When the flag is set, the I2C hardware is waiting for us to write into the data register (I2C_DR).
- **STOPF:** In Slave mode, if a Stop condition is detected on the bus, the flag sets.
- **AF (Acknowledge Failed):** If none of the devices answers us with ACK, the line remains high and the AF flag sets.
- **ARLO (Arbitration Lost):** In a multi-master bus, the flag is set if the device loses arbitration.
- **TRA (Transmitter/Receiver):** In slave mode, we check the flag to see if we should send data or receive it.
- **MSL (Master/Slave):** The flag shows if the microcontroller is in master mode or slave mode.
- **Busy (Bus busy):** The flag shows if the bus is in use or not. When a communication is in progress, the flag is set. Before making a start condition, you should make sure that the bus is not busy. Otherwise, a bus collision can occur.

	D15	D14	D13	D12	D11	D10	D9	D8	D7	D6	D5	D4	D3	D2	D1	D0
I2Cx_SR1:	SMB ALERT	TIME OUT	Res.	PEC ERR	OVR	AF	ARLO	BERR	TxE	RxNE	Res.	STOPF	ADD10	BTF	ADDR	SB

Field	Bit	Descriptions
SMBALERT	15	SMB Alert (Not used in I2C)
TIMEOUT	14	Time out (0: No timeout, 1: SCL remained LOW for 25ms)
PECERR	12	PEC Error in reception
OVR	11	Overrun/Underrun (0: No overrun/underrun, 1: overrun or underrun) The flag might rise when NOSTRETCH=1. See the reference manual.
AF	10	Acknowledge Failure (0: No Acknowledge failure, 1: Acknowledge failure)
ARLO	9	Arbitration lost in master mode (0: no arbitration lost, 1: arbitration lost)
BERR	8	Bus Error (0: no bus error, 1: bus error) As discussed earlier, when SCL is high, SDA should not change. Otherwise, it is considered as STOP or START conditions. While transferring data, if the value of SDA changes when SCL is high, the BERR flag sets.
TxE	7	Transmit Empty (0: Transmit not empty, 1: Empty) The flag is set by hardware when it is waiting for us to write data to the I2C_DR register.
RxNE	6	Receive Not Empty The flag is set when a new byte of data is in I2C_DR waiting to be read.
STOPF	4	Stop detection in slave mode (0: No Stop condition, 1: Stop condition)
ADD10	3	10-bit header sent (Used in 10-bit address mode)
BTF	2	Byte transfer finished (0: transfer not done, 1: transfer successfully finished)
ADDR	1	Address sent (master mode)/matched (slave mode) In slave mode, if the received address matches with I2C_OAR, the flag sets. In master mode, when the address is sent, the flag sets. To clear the flag, read I2C_SR1 and then I2C_SR2.
SB	0	Start bit (0: No start generated, 1: start condition generated) In master mode, the flag sets as soon as a Start condition is generated. To clear the flag, read I2C_SR1 and then write to I2C_DR.

Figure 16-16: I2C_SR1 (Status Register 1)

	D15	D14	D13	D12	D11	D10	D9	D8	D7	D6	D5	D4	D3	D2	D1	D0
I2Cx_SR2:					PEC				DUALF	SMB HOST	SMBDE FAULT	GEN CALL	Res.	TRA	BUSY	MSL

Field	Bit	Descriptions
PEC	15-8	Packet Error Checking (For more information, see the manual.)
DUALF	7	Dual Flag (It is used when dual address is enabled. See the manual.)
SMBHOST	6	Used in SMBus. (For more information, see the manual.)
SMBDEFAULT	5	Used in SMBus. (For more information, see the manual.)
GENCALL	4	General call detected (0: no general call, 1: General call address received)
TRA	2	Transmitter/receiver (0: receiver, 1: transmitter) In slave mode, the flag shows if the device is in receiver or transmitter mode. The hardware sets the flag according to the R/W signal.
BUSY	1	Bus Busy (0: no communication on the bus, 1: communication on the bus)

		The flag indicates that a communication is in progress. Hardware sets the flag when SCL or SDA become low. The flag clears when a stop condition is detected.
MSL	0	Master/Slave (0: Slave, 1: Master) The hardware sets the flag when the I2C module is in master mode. The flag is set when a stop condition or arbitration lost is detected.

Figure 16-17: I2C_SR2 (Status Register 2)

Configuring GPIO for I2C

Most of the digital output pins are configured as push-pull outputs. This configuration allows for faster transition when the output is switching from high to low or from low to high. The problem with a push-pull output is when more than one output is connected together and one output is high the other is low, the high outputs push the current out and the low outputs pull in the current. A large amount of current could flow between the outputs and damages the circuit.

One common solution to allow multiple outputs connected together is to use the open-drain output (open-drain for CMOS devices or open-collector for TTL devices). In this output configuration, the output pin is connected to the drain of the output transistor while the source of that transistor is grounded. When the transistor is on, the output pin is grounded and when the transistor is off, the output pin (the drain) is open. The open-drain outputs may be connected together. A pull-up resistor is added so that the signal is high when none of the outputs is active. When any one of the outputs is active, the signal is low. It forms a "wired-AND" logic and is exactly what is required for the I2C bus. Figure 16-1 showed the physical connection of the I2C buses. The I2C pins of the STM32 must be configured as open-drain output when the pins are assigned to an I2C module.

See Figure 8-3. When CNF=11, the pin is configured as alternate function Open-drain output.

Example 16-7

Configure PB6 and PB7 as SCL1 and SDA1.

Solution:

To make the pins output, the MOD bits must have values other than 00. We set the MODE bits to 11. The I2C pins must be configured as alternate function open-drain output. So, CNF bits must be 11. The values of CNF and MOD make the number 1111 in binary which is 0xF in hex. SCL1 and SDA1 are pins 6 and 7 of port B. So,

CRL = 0xFF444444; /*configuring pins 6 and 7 as I2C pins */

Transferring data using I2C

Configuring for I2C

We need to take the following steps to configure the I2C:

1. Enable the clock to I2C module and the GPIO using APB1ENR and APB2ENR,
2. Configure I2C pins as alternate function open-drain output,
3. Initialize the FREQ field of CR2 with the PCLK frequency,
4. Initialize CCR to make proper baud rate in master mode,
5. Initialize the TRISE register,
6. Enable the I2C module by setting the PE bit of CCR1.

Sending data in master mode

To send data in master mode we should do the followings:

1. Check the busy flag of SR2 to make sure the bus is not busy.
2. Set the START bit of CR1 to make a start condition.
3. Monitor the SB bit of SR1 until the start condition is generated.
4. Put the slave address in the data register (DR). Bits 1 to 7 should contain the slave address and the bit 0 is R/W. To send data the R/W needs to be 0.
5. Monitor the status registers. If the address is sent successfully the ADDR flag sets and you can continue the progress. If the ARLO (Arbitration Lost) is set, you should wait until the bus becomes free and you should repeat steps 1 to 5.
6. Load the data register with the data to be sent.
7. Monitor the TxE flag. The flag is set when an ACK is received.
8. Repeat steps 6 and 7 if you have more bytes to send. Otherwise, set the STOP bit of CR1 to make a stop condition. See Program 16-1.

Receiving data in master mode

Receiving data is similar to sending. To receive data in master mode we should do the followings:

1. Check the busy flag of SR2 to make sure the bus is not busy.
2. Set the START bit of CR1 to make a start condition.
3. Monitor the SB bit of SR1 until the start condition is generated.
4. Put the slave address in the data register (DR). Bits 1 to 7 should contain the slave address and the bit 0 is R/W. To receive data, the R/W needs to be 1.
5. Monitor the status registers. If the address is sent successfully the ADDR flag sets and you can continue the progress. If the ARLO (Arbitration Lost) is set, you should wait until the bus becomes free and you should repeat steps 1 to 5.
6. If you want to send an ACK in response, set the ACK bit of CR1.
7. Monitor the RxNE flag. The flag is set when a byte is received. Then, read the data register to get the received byte.
8. Repeat steps 6 and 7 if you want to receive more bytes. Otherwise, set the STOP bit of CR1 to make a stop condition.

```c
#include <stm32f10x.h>

void i2c_init(void);
void i2c_waitForReady(void);
void i2c_sendStart(void);
uint8_t i2c_sendAddrForWrite(uint8_t addr);
uint8_t i2c_sendData(uint8_t data);
void i2c_sendStop(void);

int main()
{
   i2c_init();

   do{
      i2c_waitForReady();
      i2c_sendStart();
   }while(i2c_sendAddrForWrite(0x68) != 0);

   i2c_sendData(0x0E);
   i2c_sendData(0);
   i2c_sendStop();

   while(1);
}

void i2c_init()
{
   RCC->APB2ENR |= (0xFC); /* enable clocks for GPIOs */
   RCC->APB1ENR |= (1<<21); /* enable clock for I2C1 */
   GPIOB->CRL |= 0xFF000000; /* configure PA6 and PA7 as alt. func. open drain */

   I2C1->CR2 = 0x0020;
   I2C1->CCR = 0x00B4;
   I2C1->TRISE = 37;
   I2C1->CR1 = 1;    /* PE = 1 */
}

void i2c_waitForReady()
{
   while((I2C1->SR2&(1<<1)) != 0); /* check bus busy */
}

void i2c_sendStart()
{
   I2C1->CR1 |= (1<<8); /* start */
   while((I2C1->SR1&(1<<0)) == 0); /* wait for SB */
   int stat = I2C1->SR2;
}

void i2c_sendStop()
{
   I2C1->CR1 |= (1<<9); /* stop */
   while((I2C1->SR2&(1<<0)) != 0); /* wait for becoming slave */
}
```

```
uint8_t i2c_sendAddr(uint8_t addr)
{
   I2C1->DR = addr;
   int stat;
   do{
      stat = I2C1->SR1;
      if((stat&(1<<9)) != 0) /* arbitration lost */
         return 1;

      if((stat&(1<<1)) != 0) /* address sent */
      {
         stat = I2C1->SR2; /* read SR2 to clear ADDR flag */
         return 0;
      }
   }while(1);
}

uint8_t i2c_sendAddrForRead(uint8_t addr)
{
   return i2c_sendAddr((addr<<1) + 1); /* addr+Read(1) */
}

uint8_t i2c_sendAddrForWrite(uint8_t addr)
{
   return i2c_sendAddr((addr<<1)); /* addr+Write(0) */
}

uint8_t i2c_sendData(uint8_t data)
{
   I2C1->DR = data;
   int stat1;
   do{
      stat1 = I2C1->SR1;
      if((stat1&(1<<7)) != 0) /* TxE = 1 */
         return 0;
   } while(1);
}

uint8_t i2c_readData(uint8_t ack)
{
   if(ack!= 0)
      I2C1->CR1 |= 1<<10;
   else
      I2C1->CR1 &= ~(1<<10);
   while((I2C1->SR1&(1<<6)) == 0); /* waiting for RxNE */
   return I2C1->DR;
}
```

Review Questions

1. True or false. The I2C module in STM32F10x chips support Standard and Fast speeds.
2. True or false. The I2Cx_CR1 is used to enable the I2Cx module.
3. True or false. There is no CS (chip select) pin in I2C.
4. In STM32F10x, which register is used to configure the baud rate?

Section 16.3: DS3231 RTC Interfacing and Programming

The real-time clock (RTC) is a widely used device that provides accurate time and date information for many applications. Many systems such as the PC come with such a chip on the motherboard. The RTC chip in the PC provides the time components of hour, minute, and second, in addition to year, month, and day. Many RTC chips use an external battery, which keeps the time and date even when the power of the system is off. Although some microcontrollers come with the RTC already embedded into the chip, we have to interface the vast majority of them to an external RTC chip. The DS3231 is a serial RTC with an I2C bus. In this section, we interface and program the DS3231 RTC. According to the DS3231 data sheet from Maxim, the clock/calendar provides seconds, minutes, hours, day, date, month, and year information. The end of the month date is automatically adjusted for months with fewer than 31 days, including corrections for leap year. The clock operates in either the 24-hour or 12-hour format with AM/PM indicator. The DS3231 has a built-in power-sense circuit that detects power failures and automatically switches to the battery supply. The DS3231 has an internal crystal and an internal temperature sensor, as well. The DS3231 does not support the Daylight Savings Time option. Next, we describe the pins of the DS3231. See Figure 16-18.

Vcc

Vcc is used as the primary voltage supply to the chip. The voltage source can be between 2.3 V to 5.5 V. When Vcc is above 2.3 V, the DS3231 starts working and keeps the time. But the I2C interface is disabled unless the Vcc is above 2.5 V.

V_BAT

V_BAT can be connected to an external battery, thereby providing the power source to the chip when the external supply voltage is not available.

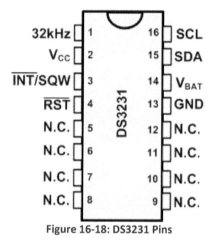

Figure 16-18: DS3231 Pins

GND

Pin 13 is the ground.

SDA (Serial Data) and SCL (Serial Clock)

SDA and SCL pins must be connected to the SDA and SCL line of the I2C bus, respectively.

SWQ/INT

Pin 3 is an output pin providing 1 Hz, 1kHz, 4 kHz, or 8 kHz frequency if enabled. This pin needs an external pull-up resistor to generate the frequency because it is open drain. If you do not want to use this pin you can omit the external pull-up resistor. The pin can be used as the output for INT, as well. The DS3231 has two Alarms: Alarm 1 and Alarm 2. If any of the alarms is enabled, the INT pin is asserted when the current time and date matches the values of the Alarm registers.

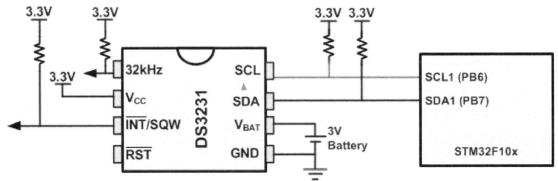

Figure 16-19: DS3231 Connections

Address map of the DS3231

The DS3231 has a total of 18 bytes of RAM space with addresses 00–12H. The first seven locations, 00–06, are set aside for RTC values of time and date. Locations 07H-0DH are set aside for Alarm 1 and Alarm 2 registers. The next 3 bytes are used for control and status registers. Locations 11H and 12H contain the temperature with accuracy of 3 degrees of centigrade. Table 16-4 shows the address map of the DS3231. Next, we study the control register, and time and date access in DS3231.

Address	Bit7	Bit6	Bit5	Bit4	Bit3	Bit2	Bit1	Bit0	Function	Range
00H	0	10 Seconds			Seconds				Seconds	00-59
01H	0	10 Minutes			Minutes				Minutes	00-59
02H	0	12/24	PM/AM 20 hour	10hour	Hours				Hours	1-12+AM/PM 0-23
03H	0	0	0	0	0	Day			Day	1-7
04H	0	0	10 Date		Date				Date	01-31
05H	Century	0	0	10Month	Month				Month Century	1-12+Century
06H	10 Year				Year				Year	00-99
07H	A1M1	10 Seconds			Seconds				Alarm 1 Seconds	00-59
08H	A1M2	10 Minutes			Minutes				Alarm 1 Minutes	00-59
09H	A1M3	12/24	AM/PM 20 Hour	10 Hour	Hour				Alarm 1 Hours	1-12 00-23
0AH	A1M4	DY/DT	10 Date		Day				Alarm 1 Day	1-7
					Date				Alarm 1 Date	01-31
0BH	A2M2	10 Minutes			Minutes				Alarm 2 Minutes	00-59
0CH	A2M3	12/24	AM/PM 10 Hour	10 Hour	Hour				Alarm 2 Hours	1-12 00-23
0DH	A2M4	DY/DT	10 Date		Day				Alarm 2 Day	1-7
					Date				Alarm 2 Date	01-31
0EH	EOSC#	BBSQW	CONV	RS2	RS1	INTCN	A2IE	A1IE	Control	-
0FH	OSF	0	0	0	EN32kHz	BSY	A2F	A1F	Control/Status	-
10H	SIGN	DATA	DATA	DATA	DATA	DATA	DATA	DATA	Aging Offset	-
11H	SIGN	DATA	DATA	DATA	DATA	DATA	DATA	DATA	MSB of Temp	-
12H	DATA	DATA	0	0	0	0	0	0	LSB of Temp	-

Table 16-4: DS3231 Address Map

Time and date address locations and modes

The byte addresses 0–6 are set aside for the time and date, as shown in Table 16-4. The DS3231 provides date and time in BCD format. Notice the data range for the hour mode. We can select 12-hour or 24-hour mode with bit 6 of Hours register at location 02. When bit 6 is 1, the 12-hour mode is selected, and bit 6 = 0 provides us the 24-hour mode. In the 12-hour mode, bit 5 indicates whether it is AM or PM. If bit 5 = 0, it is AM; and if bit 5 = 1, it is PM. See Example 16-8.

Example 16-8

What value should be placed at location 02 to set the hour to: (a) 21, (b) 11AM, (c) 12 PM.

Solution:

(a) For 24-hour mode, we have D6 = 0. Therefore, we place 0010 0001 (or 0x21) at location 02, which is 21 in BCD.
(b) For 12-hour mode, we have D6 = 1. Also, we have D5 = 0 for AM. Therefore, we place 0101 0001 at location 02, which is 51 in BCD.
(c) For 12-hour mode, we have D6 = 1. Also, we have D5 = 1 for PM. Therefore, we place 0111 0010 at location 02, which is 72 in BCD.

The DS3231 control register

As shown in Table 16-4, the control register has an address of 0EH. In the DS3231 control register, the bits control the function of the SQW/INT pin. Figure 16-20 shows the simplified diagram for SQW/INT pin.

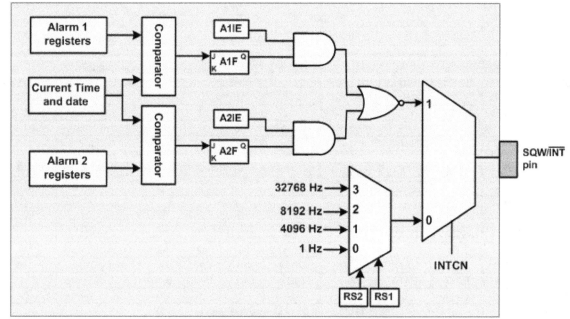

Figure 16-20: Simplified Structure of SQW/INT Pin

The SQW/INT pin can be used as a square wave generator or an interrupt generator. When the INTCN bit of control register is 0, the pin works as a wave generator. Using the RS2 and RS1 bits, the frequency of the generated wave is chosen. RS2-RS1 (rate select) bits select the output frequency of the generated wave according to Table 16-5.

RS2	RS1	Output Frequency
0	0	1 Hz
0	1	1.024 kHz
1	0	4.096 kHz
1	1	8.192 kHz

Table 16-5: RS bits

Example 16-9

What value should be placed at location 0x0E to generate a 1Hz wave on the SQW pin?

Solution:

To generate a 1Hz wave RS2 and RS1 need to be 00 and the interrupt bits must be 0 (disabled).

EOSC	BBSQW	CONV	RS2	RS1	INTCN	A2IE	A1IE
0	0	0	0	0	0	0	0

So, if we write 0 to location 0x0E the DS3231 generates a 1Hz square wave on the SQW pin.

When INTCN = 1, the SQW/INTB works as an interrupt generator. Locations 07H-0DH of DS3231 memory are related to Alarm 1 and 2. The contents of the Alarm 1 registers (locations 07H-0A) are compared with the values current time and date (locations 00H-06H). When the current date and time matches the alarm 1 values, the A1F flag of status register (location 0FH) goes high. If the A1IE (Alarm1 Interrupt Enable) bit of the control register is set, the INT becomes 0. The pin remains 0 until the A1F flag is cleared by software. To clear the A1F flag, write 0 into it. In the same way, the contents of the Alarm 1 registers (locations 0BH-0D) are compared with the current time and date. See Figure 16-20.

It can make an interrupt every minute, hour, day, or date. The bit 7 of alarm registers, are mask registers. If it is 0, the value of the register is compared with the timekeeping registers; otherwise, it is masked. Table 16-6 shows how to make interrupts every minute, hour, day, or date.

DY/DT	A1M4	A1M3	A1M2	A1M1	ALARM RATE
X	1	1	1	1	Alarm once per second
X	1	1	1	0	Alarm once per minute (when seconds match)
X	1	1	0	0	Alarm when minutes and seconds match
X	1	0	0	0	Alarm when hours, minutes, and seconds match
0	0	0	0	0	Alarm when date, hours, minutes, and seconds match
1	0	0	0	0	Alarm when day, hours, minutes, and seconds match

Table 16-6: Alarm 1 Register Mask Bits

The bit 7 of the control register is EOSC (Enable Oscillator) bit. This bit is active low. If it is 0, the oscillator works.

Register pointer

In DS3231, there is a register pointer that specifies the byte that will be accessed in the next read or write command. The first read or write operation sets the value of the pointer. After each read or write operation, the content of the register pointer is automatically incremented to point to the next location. This is useful in multi-byte read or write. When it points to location 0x12 (the last location), in the next read/write it rolls over to 0.

Writing to DS3231

To set the value of the register pointer and write one or more bytes of data to DS3231, you can use the following steps:

1. To access the DS3231 for a write operation, after sending a START condition, you should transmit the address of DS3231 (0x68) followed by 0 to indicate a write operation.
2. The first byte of data in the write operation will set the register pointer. For example, if you want to write to the control register you should send 0x07.
3. Check the acknowledge bit to be sure that DS3231 responded.
4. If you want to write one or more bytes of data, you should transmit them one byte at a time and check the acknowledge bit at the end of each byte sent. Remember that the register pointer is automatically incremented and you can simply transmit bytes of data to consecutive locations in a multi-byte burst write.
5. Transmit a STOP bit condition.

In the last section, we wrote a program that sends 0x0E and 0 to slave address 0x68. If you connect the DS3231 to your microcontroller, the program will set the register pointer to location 0x0E and writes 0 to location 0x0E. This configures the DS3231 to generates a 1Hz wave on the SQW pin, as discussed in Example 16-9.

Reading from DS3231

Notice that before reading a byte, you should load the address of the byte to the register pointer by doing a write operation as mentioned before.

To read one or more bytes of data from the DS3231 you should do the following steps:

1. To access the DS3231 for a read operation, you need to set the register pointer first. After sending a START condition, you should transmit the address of DS3231 (1101 000) followed by 0 to indicate a write operation (writing the register pointer).
2. Check the acknowledge bit to be sure that DS3231 responded.
3. The byte of data in the write operation will set the register pointer. For example, if you want to read from the control register you should send 0x07. Check the acknowledge bit to be sure that DS3231 responded.

4. Now you need to change the bus direction from a transmit to receive. Send a START condition (a REPEATED START), then transmit the address of DS3231 (0x68) followed by 1 to indicate a read operation. Check the acknowledge bit to be sure that DS3231 responded.
5. You can read one or more bytes of data. Remember that the register pointer indicates which location will be read. If you want to read more bytes, set the ACK bit in the CR1 register to acknowledge the DS3231. Also notice that the register pointer is automatically incremented and you can simply receive consecutive bytes of data in a multi-byte burst read.
6. Before reading the last byte, clear the ACK bit in the CR1 register. The last byte read will have a NACK to signal the DS3231 that the burst read is complete.
7. Transmit a STOP bit condition.

Setting the Time and Date of DS3231

Program 9-2 shows how to set the date to Sunday September 15th, 2019 and initializes the clock at 19:14:35 using the 24-hour clock mode. It uses multi-byte burst mode for writing to the DS3231. As you can see in the program, to access the location of second, you should write 0x00 into the register pointer and then you can use multi-byte burst write to write values of second, minute, …, month and year in the consecutive locations.

Program 16-2: Setting the time and date of DS3231 using burst write

```
/* The program sets time and date to 15/9/2019 19:14:35 */
#include <stm32f10x.h>

void i2c_init(void);
void i2c_waitForReady(void);
void i2c_sendStart(void);
uint8_t i2c_sendAddrForWrite(uint8_t addr);
uint8_t i2c_sendData(uint8_t data);
void i2c_sendStop(void);

int main()
{
  i2c_init();

  do{
     i2c_waitForReady();   /* wait while the bus is busy */
     i2c_sendStart();      /* generate a start condition */
  }while(i2c_sendAddrForWrite(0x68) != 0);    /* send slave addr. 0x68 for write.
repeat from beginning if arbitration lost */

  i2c_sendData(0x0); /* set addr. pointer to 0 */
  i2c_sendData(0x35); /* second */
  i2c_sendData(0x14); /* min */
  i2c_sendData(0x19); /* hour */
  i2c_sendData(0x07); /* day of week */
  i2c_sendData(0x15); /* day of month */
  i2c_sendData(0x09); /* month */
  i2c_sendData(0x19); /* year */

  i2c_sendStop();   /* generate a stop condition */
```

```
      while(1);
}
/* Copy all the I2C functions from Program 16-1 to here */
```

Reading the date and time of DS3231

Program 16-3 reads the date and time from DS3231 using multi-byte burst mode and sends to the computer using USART1. As you can see in the program, the register pointer is set to 0 and then you can use multi-byte burst read to read the values of second, minute, hour, day, date, month and year in the consecutive locations.

Program 16-3: Reading date and time of DS3231

```c
#include <stm32f10x.h>
#include <stdio.h>

void i2c_init(void);
void i2c_waitForReady(void);
void i2c_sendStart(void);
void i2c_sendStop(void);
uint8_t i2c_sendAddrForRead(uint8_t addr);
uint8_t i2c_sendAddrForWrite(uint8_t addr);

void getTime(uint8_t *year, uint8_t *month, uint8_t *day, uint8_t *hour, uint8_t *min,
uint8_t *sec);

void usart1_init(void);
void usart1_sendByte(unsigned char c);
void usart1_sendStr(char *str);

void delay_ms(uint16_t t);

int main()
{
   char str[30];
   uint8_t year, month, day, hour, min, sec;

   RCC->APB2ENR |= (0xFC)|(1<<14); /* enable clocks for GPIOs and usart1 */

   i2c_init();
   usart1_init();

   usart1_sendStr("\n\rDate and Time");

   while(1)
   {
      getTime(&year, &month, &day, &hour,&min,&sec);
      sprintf(str,"\n\r%d/%d/%d %d:%d:%d",day,month,year,hour,min,sec);
      usart1_sendStr(str);
      delay_ms(1000);
   }
}

/* The function gets a BCD number and converts it to binary */
uint8_t bcd2int(uint8_t n)
```

436

```
{
   return ((n&0xF0)>>4)*10 + (n&0x0F);
}

void getTime(uint8_t *year, uint8_t *month, uint8_t *day, uint8_t *hour, uint8_t *min,
uint8_t *sec)
{
   do{
   i2c_waitForReady(); /* wait while the bus is busy */
   i2c_sendStart(); /* generate start condition */
   }while(i2c_sendAddrForWrite(0x68) != 0); /* send addr. 0x68 for write; repeat if
arbitration lost */

   i2c_sendData(0x00); /* set register pointer to 0 */
   i2c_sendStop();          /* generate stop condition */

   do{
   i2c_waitForReady();       /* wait while the bus is busy */
   i2c_sendStart(); /* generate start condition */
   }while(i2c_sendAddrForRead(0x68) != 0); /* send addr. 0x68 for read; repeat if
arbitration lost */

   *sec = bcd2int(i2c_readData(0x01)); /* read sec */
   *min = bcd2int(i2c_readData(0x01));   /* read min */
   *hour = bcd2int(i2c_readData(0x01));   /* read hour */
   i2c_readData(0x01);       /* read day of week */
   *day = bcd2int(i2c_readData(0x01));   /* read day of month */
   *month = bcd2int(i2c_readData(0x01)); /* read month */
   *year = bcd2int(i2c_readData(0x00)); /* read year (the last read) */

   i2c_sendStop();          /* generate stop */
}

void delay_ms(uint16_t t)
{
   volatile unsigned long l = 0;
   for(uint16_t i = 0; i < t; i++)
      for(l = 0; l < 6000; l++)
      {
      }
}

/* Copy the I2C functions from Program 16-1 to here. */
/* Copy the USART1 functions from Program 11-1 to here. */
```

Review Questions

1. How many bytes of RAM in the DS3231 are set aside for the clock and date?

   (a) 7 bytes   (b) 8 bytes   (c) 56 bytes   (d) 64 bytes

2. Which pin of the DS3231 is used for clock in I2C connection?

3. Which pins of the DS3231 is used to provide main power?

4. Which pins of the DS3231 does battery connect to?

5. What is the address location for the control register?

   (a) 07H   (b) 08H   (c) 0x0E   (d) 64H

Problems

Section 16-1

1. True or false. The I2C bus needs an external clock.
2. True or false. The SDA pin is internally pulled up.
3. True or false. The I2C bus needs two wires to transfer data.
4. True or false. The SDA line is output for the master device.
5. True or false. When a device is used as a slave, the SCL is an input pin.
6. True or false. In I2C, the data frame is 8 bits long.
7. True or false. In I2C devices, each bit of information (data, address, ACK/NACK) is transferred with a single clock pulse.
8. True or false. In I2C devices, the 8-bit data is followed by an ACK/NACK.
9. In terms of data pins, what is the difference between the I2C and UART connections?
10. How does the I2C protocol distinguish between the read and write cycles?

Section 16-2

11. Which register is used to generate START or STOP conditions in the STM32F10x?
12. Which register is used to specify the clock of the I2C?
13. Which register is used to enable interrupts of I2C?
14. Write a program to read a byte from a slave with address 0110 100 (0x34) and write the byte to a slave with address 0110 101 (0x35).
15. The DS3231 DIP package is a(n) _____-pin package.
16. True or false. The DS3231 needs an external battery.
17. True or false. The DS3231 needs an external crystal oscillator.
18. What is the maximum year that the DS3231 can provide?
19. Describe the function of the SQW pin.
20. SQW is an _____ (input, output) pin.
21. The frequency of SQW pin is controlled by _____ and _____ bits.
22. DS3231 has a total of _____ bytes of RAM locations.
23. When does the DS3231 switch to a battery energy source?
24. What are the addresses assigned to time (clock)?
25. What are the addresses assigned to the calendar?
26. Which bit is used to select between 12-hour and 24-hour modes?
27. At what memory location does the DS3231 store the year?
28. What is the address of the last location of RAM for the DS3231?
29. True or false. The DS3231 provides date in BCD format.
30. Write a program to set the time to 9:15:05 PM.
31. Write a program to set the time to 22:47:19.
32. Write a program to set the date to May 14, 2019.

Answers to Review Questions

Section 16-1
1. True
2. 9, the 9th bit is for acknowledge
3. False, START and STOP conditions are generated when the SCL is high.
4. Clock stretching.
5. False, the master who won the arbitration will continue.

Section 16-2
1. True
2. True
3. True
4. I2C_CCR

Section 16-3
1. (a)
2. SCL
3. VCC and GND
4. VBAT and GND

Chapter 17: SPI Protocol and Devices

The SPI (serial peripheral interface) is a bus interface incorporated in many devices such as ADC and EEPROM. In Section 17.1 we will examine the signals of the SPI bus and show how the read and write operations in the SPI work. Section 17.2 examines the STM32 SPI registers. In Section 17.3 we show MAX7219 7-segment driver interfacing to the microcontroller using SPI bus.

Section 17.1: SPI Bus Protocol

The SPI bus was originally started by Motorola (now NXP), but in recent years has become a widely used by many semiconductor chip companies. SPI devices use only 2 pins for data transfer, called SDI (Din) and SDO (Dout), instead of the 8 or more pins used in traditional buses. This reduction of data pins reduces the package size and power consumption drastically, making them ideal for many applications in which space is a major concern. The SPI bus has the SCLK (serial clock) pin to synchronize the data transfer between two chips. The last pin of the SPI bus is CE (chip enable), which is used to initiate and terminate the data transfer. These four pins, SDI, SDO, SCLK, and CE, make the SPI a 4-wire interface. See Figure 17-1.

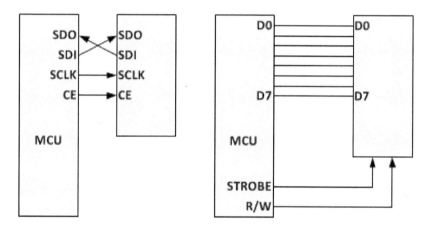

Figure 17-1: SPI Bus vs. Traditional Parallel Bus Connection to Microcontroller

In many chips, the SDI, SDO, SCLK, and CE signals are alternatively named as MOSI, MISO, SCK, and SS as shown in Figure 17-2 (compare with Figure 17-1). There is also a widely used standard called a 3-wire interface bus. In a 3-wire interface bus, we have SCLK and CE, and only a single pin for data transfer. The SPI 4-wire bus can become a 3-wire interface when the SDI and SDO data pins are tied together. However, there are some major differences between the SPI and 3-wire devices in the data transfer protocol. For that reason, a device must support the 3-wire protocol internally in order to be used as a 3-wire device. Many devices support both SPI and 3-wire protocols.

How SPI works

SPI consists of two shift registers, one in master and the other in the slave side. Also, there is a clock generator in the master side that generates the clock for the shift registers.

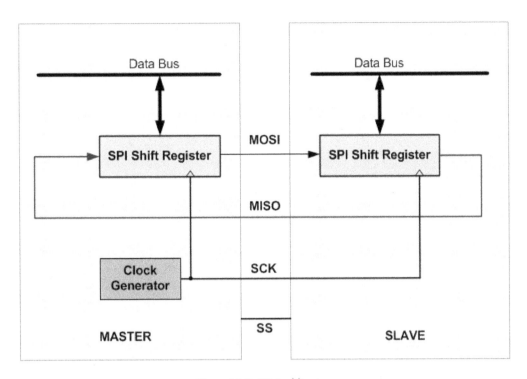

Figure 17-2: SPI Architecture

As you can see in Figure 17-2, serial-out pin of the master shift register is connected to the serial-in pin of the slave shift register by MOSI (Master Out Slave In) and the serial-in pin of the master shift register is connected to the serial-out pin of the slave shift register by MISO (Master In Slave Out). The master clock generator provides clock to shift register in both master and slave shift registers. The clock input of the shift registers can be falling- or rising-edge triggered. This will be discussed shortly.

In SPI, the shift registers are 8 bits long. It means that after 8 clock pulses, the contents of the two shift registers are interchanged. When the master wants to send a byte of data, it places the byte in its shift register and generates 8 clock pulses. After 8 clock pulses, the byte is transmitted to the slave shift register. When the master wants to receive a byte of data, the slave side should place the byte in its shift register and after 8 clock pulses the data will be received by the master shift register. It must be noted that SPI is full duplex meaning that it sends and receives data at the same time.

The SS (Slave Select) pin

The SS pin is in fact the Chip Enable (CE) pin for the slave device. Slave ignores any changes to other pins, as long as the SS pin is high. When the master device wants to communicate with a slave, it pulls down the SS pin of the slave device and the SPI interface of slave device becomes enabled. To enable the SPI interface of the slave device permanently, we can connect the SS pin of slave to ground.

The SS pin is used when there are more than one slaves. In the case, the SS pins of slaves are connected to different I/O pins of the master device. See Figure 17-3. When master wants to communicate with a slave, it pulls down the SS pin of the slave device and leaves the SS pins of the other slaves high. So, the other slaves ignore the transmissions.

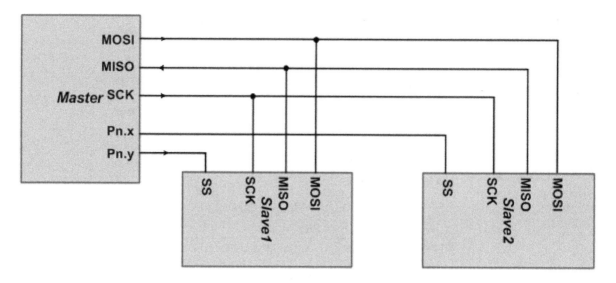

Figure 17-3: a Master with more than one Slaves

SS can be used in multi-master communications, as well. Consider the 2 devices in Figure 17-2. The devices leave the SS pin high as long as they do not transfer data. When a device needs to transfer, it makes the SS pin low. So, the other device goes to slave mode and gets ready to transfer data.

Steps for sending and receiving as a master

The following steps are used to send and receive data in SPI devices, as a master:

1. Make the SS pin of the slave low (SS = 0).
2. Send a byte of data. The data is shifted out, one bit at a time, with each edge of SCLK and at the same time a bit of data will be received. Repeat sending and receiving data as many times as you need.
3. Make SS = 1.

Clock polarity and phase in SPI device

In SPI communication, both master and slave use the same clock. The master must choose a clock rate that can be handled by the slave. If the master is driving the clock faster than the slave can handle, the transmission will fail. The master and slave(s) must agree on the clock polarity and phase with respect to the data. NXP names these two options as CPOL (clock polarity) and CPHA (clock phase), respectively, and most companies have adopted that convention. CPOL determines the idle state of the clock. When CPOL= 0 the idle value of the clock is zero while at CPOL=1 the idle value of the clock is one. CPHA determines when to sample the data. CPHA=0 means data should be sampled on the leading (first) clock edge, while CPHA=1 means data should be sampled on the trailing (second) clock edge. Notice that if the idle value of the clock is zero the leading (first) clock edge is a rising edge but if the idle value of the clock is one, the leading (first) clock edge is a falling edge. See Table 17-1 and Figure 17-4.

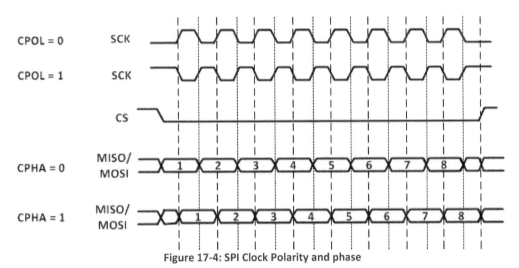

Figure 17-4: SPI Clock Polarity and phase

CPOL	CPHA	Data Read and change time	SPI Mode
0	0	read on rising edge, changed on a falling edge	0
0	1	read on falling edge, changed on a rising edge	1
1	0	read on falling edge, changed on a rising edge	2
1	1	read on rising edge, changed on a falling edge	3

Table 17-1: SPI Clock Polarity and phase

Review Questions

1. True or false. SPI is an Asynchronous protocol.
2. True or false. In the SPI protocol, the clock is always generated by the master device.

Section 17.2: SPI programming in STM32

The STM32 chips come with on-chip SPI modules. The medium density STM32F103 chips have 2 SPI modules while the high-density chips have 3 SPI modules. The SPI modules of STM32 support I2S protocol, as well. In this chapter, we concentrate on the registers which are used in SPI protocol. The following table shows some of the registers in SPI modules with their addresses.

Register name	Offset	Description
SPI_CR1	0x00	SPI Control Register 1
SPI_CR2	0x04	SPI Control Register 2
SPI_SR	0x08	SPI Status Register
SPI_DR	0x0C	SPI Data Register

Table 17-2: Some of the STM32 SPI Registers

SPI_CR1 (Control Register 1)

The SPIx_CR1 (SPI Control Register 1) sets SPI configuration. Figure 17-5 shows the bits of SPIx_CR1. We must use SPI_CR1 register to select the SPI mode operation of the STM32. Notice that the SPE bit in the SPIx_CR1 register must be set to HIGH to allow the use of the STM32 pins for SPI data bus protocol. We choose the SPI Master mode by using the MSTR bit of SPIx_CR1 register. The CPOL bit is used for selecting an inverted or non-inverted SPI clock. In the active-HIGH (non-inverted) SCK, it is low in the idle state. We use the CPHA bit in the SPIx_CR1 register to select the rising or falling edge of the SCK for sampling of data. Using the LSBFIRST, we have the option of sending out the LSB or the MSB first. We must

443

make sure that the SPI slave device and the master have the same SCK polarity and phase and send the same bit first (CPOL, CPHA, and LSBFIRST bits).

	D15	D14	D13	D12	D11	D10	D9	D8	D7	D6	D5	D4	D3	D2	D1	D0
SPI_CR1:	BIDI MODE	BIDI OE	CRC EN	CRC NEXT	DFF	RX ONLY	SSM	SSI	LSB FIRST	SPE		BR		MSTR	CPOL	CPHA

Field	Bit	Descriptions
BIDIMODE	15	Bidirectional data mode enable (0: 2-line unidirectional, 1: 1-line bidirectional)
BIDIOE	14	When BIDIMODE=1 (1-line bidirectional), the BIDIOE bit selects the direction of transfer (0: receive, 1: transmit)
CRCEN	13	Hardware CRC calculation enable (0: disabled, 1: enabled)
CRCNEXT	12	CRC transfer next (0: data transfer, 1: next transfer is CRC)
DFF	11	Data Frame format (0: 8-bit data frame, 1: 16-bit data frame)
RXONLY	10	Receive only (0: Both transmit & receive (Full-duplex), 1: Receive only)
SSM	9	Software Slave Management (0: NSS pin, 1: SSI bit) If the bit is set, the SSI bit manages the communication instead of the NSS pin.
SSI	8	Internal Slave Select
LSBFIRST	7	LSB First Enable 1 = Data is transferred least significant bit first. 0 = Data is transferred most significant bit first.
SPE	6	SPI System Enable bit 1 = Enables SPI port and configures pins as serial port pins 0 = Disables SPI port and configures these pins as I/O ports
BR	5-3	Baud rate control <table><tr><th>BR</th><th>0</th><th>1</th><th>2</th><th>3</th><th>4</th><th>5</th><th>6</th><th>7</th></tr><tr><th>Speed</th><td>PCLK/2</td><td>PCLK/4</td><td>PCLK/8</td><td>PCLK/16</td><td>PCLK/32</td><td>PCLK/64</td><td>PCLK/128</td><td>PCLK/256</td></tr></table>
MSTR	2	SPI Master/Slave mode Select bit. This bit selects master or slave mode. 1 = SPI in master mode 0 = SPI in slave mode
CPOL	1	SPI Clock Polarity bit 1 = Active-LOW clocks selected. In idle state SCK is high. 0 = Active-HIGH clocks selected. In idle state SCK is low.
CPHA	0	SPI Clock Phase bit 1 = Sampling of data occurs at even edges of the SCK clock. 0 = Sampling of data occurs at odd edges of the SCK clock.

Figure 17-5: SPI_CR1 (SPI Control Register 1)

NSS pin and the SSM bit

As we mentioned earlier, the Slave Select (SS) pin of slave device can be connected to ground and the SS pin of masters should be left high. In STM32, the SS pin is named as NSS. See Figure 17-6. The SSM bit of SPI_CR1 selects between the NSS pin and the SSI bit of SPI_CR1. You can set the SSM bit to one and make the SSI bit 0 or 1, instead of connecting the NSS pin to ground or VCC. In the case, you can freely use the NSS pin as a GPIO pin or for other purposes.

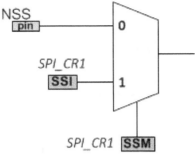

Figure 17-6: SSM Bit of the SPI_CR1

Baud rate

The BR bits of the SPI_CR1 register are used to set the SCK clock speed for the SPI module in master mode. See the Figure 17-7. See the Example 17-1.

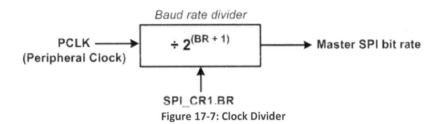

Figure 17-7: Clock Divider

Example 17-1

Assuming that the PCLK2 (peripheral clock for APB2) is 72MHz, (a) find the SPI_CR1 value to initialize the SPI device as a master device, with bit rate of 1.125MHz, with active-high clock, sampling on rising edge, and 8-bit data MSB first. Make the NSS pin free, (b) repeat for slave mode, (b) repeat for a 16-bit data and master mode.

Solution:

(a) 72MHz / 2.25MHz = 64 = 2^6 ➜ BR = 5

SPI_CR1	BIDIMODE	BIDIOE	CRCEN	CRCNEXT	DFF	RXONLY	SSM	SSI	LSBFIRST	SPE	BR	MSTR	CPOL	CPHA
	0	0	0	0	0	0	1	1	0	1	101	1	0	0

SPI_CR1 = 0x036C

(b) For slave mode, the MSTR bit becomes 0. In slave mode, there is no need to set the BR bits as well since the master device generates the clock and BR has no effect in slave mode.

SPI_CR1	BIDIMODE	BIDIOE	CRCEN	CRCNEXT	DFF	RXONLY	SSM	SSI	LSBFIRST	SPE	BR	MSTR	CPOL	CPHA
	0	0	0	0	0	0	1	1	0	1	xxx	0	0	0

(c)

SPI_CR1	BIDIMODE	BIDIOE	CRCEN	CRCNEXT	DFF	RXONLY	SSM	SSI	LSBFIRST	SPE	BR	MSTR	CPOL	CPHA
	0	0	0	0	1	0	1	1	0	1	101	1	0	0

SPI_CR1 = 0x0B6C

The MSTR bit and the roles of pins

The MSTR bit of the SPI_CR1 register chooses between master and slave modes. See Figure 17-7. The roles of the SPI pins change when the mode changes. In master mode, MOSI (Master Out Slave In) sends data and the SCK provides the clock for the slave and MISO (Master In Slave Out) receives data. In slave mode, MOSI (Master Out Slave In) and SCK pins are inputs and MISO (Master In Slave Out) is an output pin.

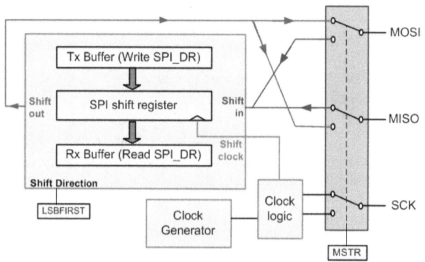

Figure 17-8: SPI Shift Register and the SPI Pins

SPI_DR (Data Register)

SPI_DR is similar to the data register (USART_DR) in USARTs. SPI_DR is made of two separate registers: a send buffer and a receive buffer. See Figure 17-8. When we write to the SPI_DR, the value is loaded to the send buffer. We read the receive buffer when we read the SPI_DR.

The STM32 SPI supports 8-bit and 16-bit data. We can choose the length of transferring data using the DFF (Data Frame Format) bit of SPI_CR1. When it is in 8-bit mode, only the low byte of the SPI_DR is used.

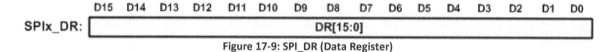

Figure 17-9: SPI_DR (Data Register)

SPI_SR (Status Register)

SPI_SR shows the status of the SPI module See Figure 17-10 and Table 17-3. The RXNE (Receive Not Empty) is set when a new data is received and the TXE (Transmit Empty) is set when the transmit buffer is ready to send another data. On reset, the TXE flag is set since the send buffer is empty and ready to send.

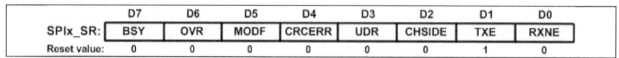

Figure 17-10: SPI_SR (Status Register)

Field	Bit	Description
BSY	7	Busy flag (0: not busy, 1: busy)
		The bit is set when the SPI module is transferring data or TX buffer is not empty. The flag is set and cleared by hardware.
OVR	6	Overrun (0: no overrun, 1: overrun occurred)
		The flag sets when a new data is received and the previous data is not read.
MODF	5	Mode fault (0: no mode fault occurred, 1: mode fault occurred)
		If the STM32 is in master mode and the NSS pin is pulled down with another device, STM32 goes to slave mode and the flag is set.
CRCERR	4	CRC Error (For more information, see the user manual.)
UDR	3	Underrun (0: no underrun, 1: underrun occurred)
		The flag sets if the STM32 is in slave mode and a SCLK clock is appeared while we have not loaded data to the data register (SPI_DR).
CHSIDE	2	Channel side (It is not used in SPI mode.)
TXE	1	Transmit buffer empty (0: not empty, 1: empty)
		The flag is set if the SPI transmit buffer is empty and it is ready to send another data.
RXNE	0	Receive buffer not empty (0: empty, 1: not empty)
		The flag is set if a new data is received.

Table 17-3: SPI_SR (Status Register)

Sending and receiving data as a master device

To transfer data as a master, you should do the followings:

1. Enable the clocks for SPI and GPIO. The clock enable for SPI1 is bit 12 of APB2ENR. Bits 14 and 15 of APB1ENR are the clock enable bits for SPI2 and SPI3, respectively.
2. Initialize MOSI and SCK as alternate function output push-pull (CNFx = 10) and make MISO an input pin.
3. Initialize SPI_CR1 with proper value:
   a. The clock polarity and phase must be the same as the slave device
   b. Set the MSTR bit.
   c. Set both SSI and SSM bits to 1 to free the NSS pin.
   d. Consider the maximum speed that the slave device supports and set the BR bits to proper values.
   e. In most devices, the MSB bit is sent first. But if your slave device sends the LSB bit, set the LSBFIRST.
   f. Set the SPE bit to enable the SPI module.
   g. If your slave device sends 16-bit data, set the DFF bit. Otherwise, clear it.
4. Consider a GPIO pin for the SS pin of each slave device, as shown in Figure 17-3. Initialize the GPIO pins as outputs and make them high. If there is only one slave, you can connect its SS pin to ground as discussed in Section 17-1.
5. Make the SS pin of the desired slave low.
6. Load SP_DR to send data.
7. Monitor the SPI_SR register until the TXE is set (or RXNE is set).

8. Read SP_DR to get the received data.
9. Repeat steps 6 to 8 until all data are transferred.
10. Make the SS pin of the slave device high.

See Program 17-1. It sends characters 'A' to 'Z' via SPI using the spi1_transfer() function. See spi1_init() and spi1_transfer() functions.

Program 17-1: Sending 'A' to 'Z' characters via SPI

```c
#include <stm32f10x.h>

void spi1_init(void);
uint8_t spi1_transfer(uint8_t d);

int main()
{
   RCC->APB2ENR |= 0xFC;   /* enable clocks for GPIO */

   spi1_init(); /* initialize the SPI module */

   /*--- make the SS pin of the slave low if needed ---*/

   for(char c = 'A'; c <= 'Z'; c++) /* send characters 'A' to 'Z' */
      spi1_transfer(c); /* send c through SPI */

   /*--- make the SS pin high ---*/

   while(1)
   {
   }
}

/* The function initializes the SPI module */
void spi1_init()
{
   RCC->APB2ENR |= 0xFC|(1<<12);  /* enable clocks for GPIO and SPI1 */
   GPIOA->CRL = 0xB4B44444;       /* MOSI (PA7) and SCK(PA5): alt. func. out, MISO
(PA6): input */
   SPI1->CR1 = 0x35C;        /* SPE = 1, BR = 3, FFD = 0, SSI and SSM = 1 */
}

/* The function sends a byte of data via SPI */
/* argument d: the byte to be sent */
/* return value: the received data */
uint8_t spi1_transfer(uint8_t d)
{
   SPI1->DR = d; /* send the contents of d */
   while((SPI1->SR&(1<<0)) == 0); /* wait until RXNE is set */
   return SPI1->DR;  /* return the received data */
}
```

If you do not have access to SPI probe, you can connect the MOSI and MISO pins together. Then, the microcontroller receives the data that it sends. In Program 17-2, we send characters 'A' to 'Z' through

SPI. In the program, when we receive data via SPI, we send it to the PC through USART. So, if you connect the MOSI and MISO pins together, the characters 'A' to 'Z' are sent back to the microcontroller, and the microcontroller sends them to the PC. So, you will receive letters 'A' to 'Z' on your PC.

Program 17-2: Sending 'A' to 'Z' via SPI. Forwarding the received data from SPI to the PC

```c
#include <stm32f10x.h>
#include <stdio.h>

void usart1_init(void);
void usart1_sendByte(unsigned char c);
void usart1_sendInt(unsigned int i);

void spi1_init(void);
uint8_t spi1_transfer(uint8_t d);

int main()
{
   RCC->APB2ENR |= 0xFC;    /* enable clocks for GPIO */
   RCC->APB2ENR |= (1<<14);       /* enable clock for usart1 */

   spi1_init();
   usart1_init();

   /*--- make the SS pin of the slave low if needed ---*/

   for(char c = 'A'; c <= 'Z'; c++)
     usart1_sendByte(spi1_transfer(c));

   /*--- make the SS pin high ---*/

   while(1)
   {
   }
}

/* Copy spi1_init and spi1_transfer from Program 17-1 to here. */
/* Copy usart1_init and usart1_sendByte from Program 11-1 */
```

Transferring data as a slave device (Case Study)

Initializing the STM32 as a slave device

To initialize STM32 as a slave device, you should do the followings:

1. Enable the clocks for SPI and GPIO. The clock enable for SPI1 is bit 12 of APB2ENR. Bits 14 and 15 of APB1ENR are the clock enable bits for SPI2 and SPI3, respectively.
2. Initialize MOSI and SCK as input pins and set MISO as alternate function output push-pull.
3. Initialize SPI_CR1 with proper value:
   a. The clock polarity and phase must be the same as the master device

b. If you want to use NSS as the SS (Slave Select) pin, clear the SSM bit of the SPI_CR, and initialize the NSS pin as input. Otherwise, set the SSM bit to 1 and clear the SSI bit.
c. Clear the MSTR bit.
d. In most devices, the MSB bit is sent first. But if your master device sends the LSB bit, set the LSBFIRST.
e. Set the SPE bit to enable the SPI module.
f. If your master device sends 16-bit data, set the DFF bit. Otherwise, clear it.

Sending data as a slave device

To send data you load SPI_DR. But since you are in slave mode, the data will not be sent unless the master asserts the SS pin and sends clocks through SCK. To check if your previous data is sent or not, you can check the TXE bit of the SPI_SR register.

Receiving data as a slave device

When a new data is received, the RXNE flag is set. Then read the value of SPI_DR to get the received data.

Review Questions
1. True or false. STM32 does not support SPI protocol.
2. True or false. In STM32F10x chips, the SPI module can transfer 8-bit and 16-bit data.
3. In STM32, which register is used to choose the clock polarity?
4. In STM32, which register is used to set the SPI baud rate?
5. How can we configure the device as master?

Section 17.3: MAX7219/MAX7221 SPI 7-Segment Driver

Chapter 2 examined 7-seg concepts. To save pins we can use MAX7219/21 chip. In this section we show an SPI-based 7-seg driver and its interfacing to STM32. MAX7219 is an SPI serial 7-segment driver from Maxim Corporation. It can support up to 8-digit seven-segment display.

There are two types of 7-segments, common anode and common cathode. The MAX7219 supports common cathode only. See Figure 17-11.

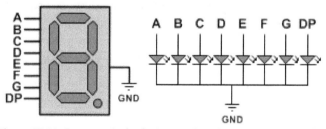

Figure 17-11: Common Cathode Connections in a 7-Segment Display

In many applications you need to connect two or more 7-segment LEDs to a microcontroller. For example, if you want to connect four 7-segment LEDs directly to a microcontroller you need 4 × 8 = 32 pins. This is costly. The MAX7219 IC is an ideal chip for such applications since it supports up to eight 7-segment LEDs. We can connect the MAX7219 to the microcontroller using SPI protocol and control up to eight 7-segment LEDs. The MAX7219 contains an internal decoder that can be used to convert binary numbers to 7-segment codes. It activates the digits one at a time. That means the CPU does not need to refresh the 7-segment LEDs. All you need to do is to send a binary number to the MAX7219, and the chip decodes the binary data and displays the number. The device includes analog and digital brightness control, an 8×8 static RAM that stores each digit, and a test mode that forces all LEDs on. Next, we will show how to interface an MAX7219 to the STM32 and program it using SPI protocol.

MAX7219 pins and connections

The MAX7219 is a 24-pin DIP chip. It can be directly connected to the microcontroller and control up to eight 7-segment LEDs. A resistor or a potentiometer is the only external component that you need. Next, we will discuss the pins of the MAX7219. See Figure 17-12.

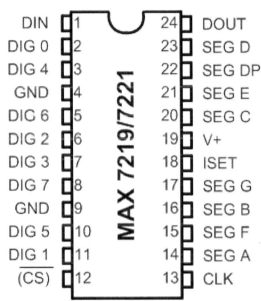

Figure 17-12: MAX7219 and MAX7221

GND

Pin 4 and pin 9 are the ground. Notice that both of the ground pins should be connected to system ground and you cannot leave any of them unconnected.

VCC

Pin 19 is the VCC and should be connected to the +5 V power supply. Notice that this pin also supplies the power to drive the 7-segments and the connecting wire to this pin should be able to handle 100–300 mA.

ISET

Pin 18 is ISET and sets the maximum segment current. This pin should be connected to VCC through a resistor. A 10 kΩ resistor can be connected to this pin. If you want to manually control the segments' light intensity, you can replace the resistor with a 50K potentiometer. For more details about how to calculate the value of the resistor you can look at the datasheet of the chip.

CS

Pin 12 is the chip select pin and should be connected to the SS (slave select) pin of the microcontroller. Serial data is loaded into the chip while CS is low, and the last 16 bits of the serial data are latched on the rising edge of CS.

DIN

Pin 1 is the serial data input and should be connected to the MOSI pin of the microcontroller. On CLK's rising edge, data on this pin is loaded into the internal shift register. Notice that the MAX7219 reads the bit on rising edge.

CLK

Pin 13 is the serial clock input and should be connected to the SCK pin of the microcontroller. On MAX7219 the clock input is inactive when CS is high.

DOUT

Pin 24 is the serial data output and is used to connect more than one MAX7219 to a single SPI bus.

DIG0–DIG7

The DIG pins are the 7-segment selector pins and should be connected to the 7-segments' common cathode pin. The MAX7219 chip can control up to eight 7-segment LEDs. These eight 7-segment displays are designated as DIG0 to DIG7.

SEGA–SEGG and DP

These pins select each segment and should be connected to segments of each 7-segment accordingly. Figure 17-13 shows the connection for two 7-segments. You can connect up to eight 7-segments to MAX7219.

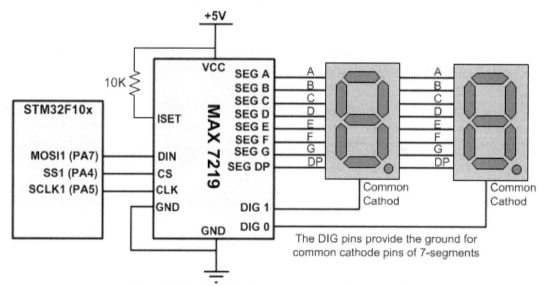

Figure 17-13: MAX7219 Connections to the Microcontroller

MAX7219 data packet format

In MAX7219, data packets are 16 bits long (two bytes). You should first make CS low before transmitting; then you transmit two bytes of data and terminate the transmission by making CS high.

The first byte (MSBs) of each packet contains the command control bits, and the second byte is the data to be displayed. See Figure 17-14.

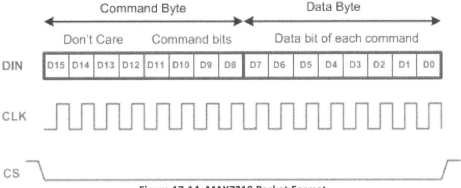

Figure 17-14: MAX7219 Packet Format

The upper four bits (D15–D12) of the command byte are "don't cares" and the lower four bits (D11–D8) are used to identify the meaning of the data byte that follows. The second byte (D7–D0) of the two-byte packet is called the data byte and is the actual data to be displayed or control the 7-segment driver. Table 17-4 shows the binary and hex values of each command. Next, we will discuss the commands in more detail.

Command	D15-12	D11	D10	D9	D8	Hex Code
No operation	X	0	0	0	0	X0
Set value of digit 0	X	0	0	0	1	X1
Set value of digit 1	X	0	0	1	0	X2
Set value of digit 2	X	0	0	1	1	X3
Set value of digit 3	X	0	1	0	0	X4
Set value of digit 4	X	0	1	0	1	X5
Set value of digit 5	X	0	1	1	0	X6
Set value of digit 6	X	0	1	1	1	X7
Set value of digit 7	X	1	0	0	0	X8
Set decoding mode	X	1	0	0	1	X9
Set intensity of light	X	1	0	1	0	XA
Set scan limit	X	1	0	1	1	XB
Turn on/off	X	1	1	0	0	XC
Display test	X	1	1	1	1	XF
Notes:						
X means don't care.						
Digits are designated as 0-7 to drive total of eight 7-segment LEDs.						

Table 17-4: List of Commands in MAX7221/MAX7219

Set value of digit 0–digit 7 (commands X1–X8)

These commands set what is to be displayed on each 7-segment. You can either send a binary number to the chip decoder and let it turn on/off the segments accordingly, or you may decide to turn on/off each segment of the 7-segment by yourself. The first way is useful when you do not want to deal with converting a binary number to 7-segment codes. The second way is useful when you want to show a character or any other thing that is not predefined. For example, if you want to show letter 'U', you should use the second way and turn on/off segments yourself. Next, you will see how to enable or bypass the decoder for each 7-segment.

Set decoding mode (command X9)

This command lets you enable or bypass the binary to 7-segment decoding function for each 7-segment digit. Each bit in the data byte (second byte) is assigned to one digit of 7-segment. D0 is assigned to Digit 0, D1 is assigned to Digit 1, and so on. If you want to enable the decoding function for a digit you should set to one the bit assigned to that digit, and if you want to disable the decoding function you should clear the bit for that digit. Figure 17-15 shows the structure of the set decoding mode command. See Examples 17-2 and 17-3.

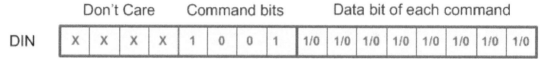

Figure 17-15: Set Decoding Mode Command Format

Example 17-2

What sequence of bytes should be sent to the MAX7219 in order to enable the decoding function for digit 0 and digit 2, and disable the decoding function for other digits?

Solution:

The first byte should be xxxx 1001 (X9 hex) to execute the "Set decoding mode" command, and the second byte (argument of the command) should be 0000 0101 to enable the decoding function for digit 0 and digit 2.

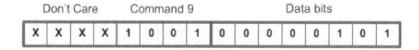

Example 17-3

After running Example 17-2, what sequence of numbers should be sent to the MAX7219 in order to write 5 on digit 2?

Solution:

The first byte should be xxxx 0011 (X3 hex) to execute the "Set value of digit 2" command, and the second byte (argument of the command) should be 0000 0101 (05 hex) to write 5 on digit 2. Notice that the decoding function for digit 2 has been enabled before.

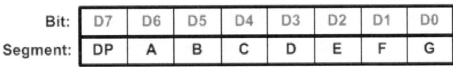

If you want to turn on/off each segment by yourself to display a specific letter on a 7-segment, you should bypass the decoding function and then use the "Set value of digit x" command to turn on/off each bit of a segment. As you see in Figure 17-18, each bit of the data bits is assigned to a segment of the 7-segment. For example, D0 is assigned to the G segment, D1 is assigned to the F segment, and so on. If you want to turn on a segment, you should write one to the corresponding bit, and if you want to turn off a segment, you should write zero to its bit. Figure 17-16 shows the bit assigned to each segment. See Example 17-4.

Bit:	D7	D6	D5	D4	D3	D2	D1	D0
Segment:	DP	A	B	C	D	E	F	G

Figure 17-16: Bits Assigned to Segments

Example 17-4

After running Example 17-4, what sequence of numbers should be sent to the MAX7219 in order to write U on digit 1?

Solution:

The decoding function for digit 1 has been disabled before in Example 17-4, and we have to turn on/off each segment manually. As you see in the figure, segments B, C, D, E, and F should be turned on. To turn on these segments of digit 1, we should send the first byte xxxx 0010 (X2 hex) to execute the "Set value of digit 1" command and then we should send 0011 1110 (3E hex) to write U on digit 1. Notice that the decoding function for digit 1 has been enabled before. The figure below shows the bits.

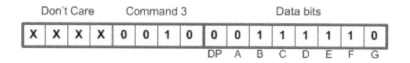

Don't Care				Command 3				Data bits							
X	X	X	X	0	0	1	0	0	0	1	1	1	1	1	0
								DP	A	B	C	D	E	F	G

Set Intensity of Light (command XA)

This command sets the light intensity of the segments. The intensity can be any value between 0 and 16 (0F hex). 0 is the minimum value of intensity, and 16 is the maximum value of intensity. Notice that 0 does not mean off but it is the minimum intensity. As we mentioned before, you can also change the light intensity of segments by changing the resistor that connects the ISET pin to VCC.

Set Scan Limit (command XB)

The number of 7-segments that are connected to the chip can vary from 1 to 8. This command sets the index of last digit to be scanned. For example, to scan digits 0 to 5, send the XB command together with 5 as data. Similarly, to scan digits 0 and 1, send the command together with 1.

You can also use the command if you want to make the left most digits off. For example, sending the command together with 3, makes the chip scan digits 0 to 3 leaving the other digits off.

Turn On/ Off (command XC)

This command turns the display on or off. 1 (01 hex) turns the display on, while 0 (00 hex) turns off the display. This command is useful when you want to reduce the power consumption of your device.

Display Test (command XF)

This command is used to test the display. If you send 1 (01 hex) after sending the display test command to the chip, it enters display-test mode and turns on all segments. This lets you check to see if all segments work properly. When you want to return to normal operation mode, you should execute the command but send 0 (00 hex) as data to the chip.

MAX7219 programming in the STM32

To program MAX7219 in the STM32 you should do the following steps. Notice that step 4 is optional and can be ignored:

1. Initialize the SPI to operate in master mode so that data is stable on rising edge and changes on falling edge.
2. Enable or disable decoding mode by executing command 9 (x9 hex).
3. Set the scan limit.
4. Set the intensity of light (optional).
5. Disable test mode
6. Turn on the display.
7. Set the values of each digit.

See Programs 17-2 and 17-3. Program 17-2 shows how to display 49 on the 7-segment display of Figure 17-15 using the decoding function.

```c
#include <stm32f10x.h>

void spi1_init(void);
uint8_t spi1_transfer(uint8_t d);

void max7219_send(uint8_t cmd, uint8_t data);

int main()
{
   RCC->APB2ENR |= 0xFC;    /* enable clocks for GPIO */

   spi1_init(); /* initialize the SPI module */

   max7219_send(0x09, 0xFF);      /* enable decoding for all digits */
   max7219_send(0x0B, 1);         /* 2 (1+1) digits */
   max7219_send(0x0C, 0x01);      /* turn on */

   max7219_send(0x01, 9);  /* show 9 on digit 1 */
   max7219_send(0x02, 4);  /* show 4 on digit 2 */

   while(1)
   {
   }
}

/* The function initializes the SPI module */
void spi1_init()
{
   RCC->APB2ENR |= 0xFC|(1<<12);  /* enable clocks for GPIO and SPI1 */
   GPIOA->CRL = 0xB4B34444;       /* MOSI (PA7) and SCK(PA5): alt. func. out, PA4:
output for CS */
   SPI1->CR1 = 0x35C;       /* SPE = 1, BR = 3, FFD = 0, SSI and SSM = 1 */
}

/* The function sends a byte of data through SPI */
/* argument d: the byte to be sent */
/* return value: the received data */
uint8_t spi1_transfer(uint8_t d)
{
   SPI1->DR = d; /* send the contents of d */
   while((SPI1->SR&(1<<0)) == 0); /* wait until RXNE is set */
   return SPI1->DR;  /* return the received data */
}

void max7219_send(uint8_t cmd, uint8_t data)
{
   GPIOA->BRR = (1<<4);     /* Enable Chip Select */
   spi1_transfer(cmd);
   spi1_transfer(data);
   GPIOA->BSRR = (1<<4);    /* Disable Chip Select */
}
```

Program 17-4 shows how to display "2U" on the 7-segment of Figure 17-13 without using the decoding function for letter 'U'.

Program 17-4: Displaying 2U on the 7-segment

```
#include <stm32f10x.h>

void spi1_init(void);
uint8_t spi1_transfer(uint8_t d);

void max7219_send(uint8_t cmd, uint8_t data);

int main()
{
  RCC->APB2ENR |= 0xFC;   /* enable clocks for GPIO */

  spi1_init(); /* initialize the SPI module */

  max7219_send(0x09, 0x02);      /* enable decoding for digit2 and disable for digit1
*/
  max7219_send(0x0B, 1);         /* 2 (1+1) digits */
  max7219_send(0x0C, 0x01);      /* turn on */

  max7219_send(0x01, 0x3E);      /* show U on digit 1 */
  max7219_send(0x02, 2);   /* show 2 on digit 2 */

  while(1)
  {
  }
}

/* Copy spi1_init, spi1_transfer, and max7219_send from Program 17-3 */
```

Review Questions
1. How many 7-segments can be controlled by MAX7219?
2. What would happen if you do not set the scan limit?
3. True or False. If you want to show P on a 7-segment you can use the decoding function.
4. Which segments should be on to display P on a 7-segment?
5. What is the recommended value of the ISET resistor?

Problems

Section 17.1
1. True or false. The SPI bus needs an external clock.
2. True or false. In SPI, SS is active-LOW.
3. True or false. The SPI bus has a single Din pin.
4. True or false. The SPI bus has multiple Dout pins.
5. True or false. When the SPI device is used as a slave, the SCLK is an input pin.
6. True or false. In SPI devices, each bit of data is transferred with a single clock pulse.

Section 17.2

7. How do we set the SPI to slave mode?
8. True or false. The NSS pin must be taken low externally in order to enable the SPI module in slave mode.
9. Find the value for BR to set the SPI speed to 1MHz. Assume PCLK is 32MHz.
10. Find the value for BR to set the SPI speed to 8MHz. Assume PCLK is 64MHz.

Section 17.3

11. The MAX7219 DIP package is a(n) _____-pin package.
12. Which pin is assigned as Vcc?
13. How much is the maximum current of the Vcc pin?
14. True or false. The MAX7219 has a pin for controlling the intensity of light of the segments.
15. What is the recommended resistor value for light intensity?
16. How many 7-segments can be interfaced by a single MAX7219?
17. What is the first byte in a 16-bit packet in the MAX7219?
18. What is the second byte in a 16-bit packet in the MAX7219?
19. True or false. The decoding function should be enabled to write L on a 7-segment.

Answers to Review Questions

Section 17.1

1. False
2. True

Section 17.2

1. False
2. True
3. SPI_CR1
4. SPI_CR1
5. We set the MSTR bit of the SPI_CR1 register

Section 17.3

1. 8
2. The scan limit would be 0 and nothing would be shown on the 7-segment.
3. False
4. A, B, E, F, G
5. 10 kΩ

Chapter 18: Programming Graphic LCD

Chapter 9 used the character LCD. In this chapter, we examine the graphic LCDs and show some programming examples, although an entire book can be dedicated to graphic LCD and its programming. Section 18.1 covers some basic concepts of graphic LCDs. In Section 18.2, you learn to make fonts to display texts on the graphic LCDs. Then, the PCD8544 graphic LCD is discussed.

Section 18.1: Graphic LCDs

The screen of graphic LCDs is made of pixels. The pictures and the texts are created using pixels and the programmers have control over each and every individual pixel. See Figures 18-1 and 18-2.

Figure 18-1: A picture on a Mono-color LCD

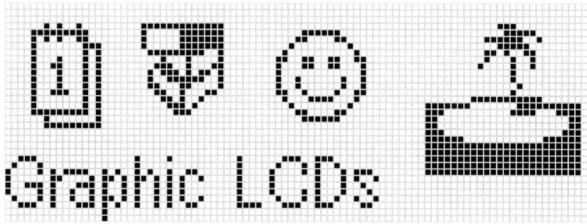

Figure 18-2: A Zoomed Picture on a Mono-color LCD

The graphic LCDs can be mono-colored (monochorme) or colored. In mono-colored LCDs each pixel can be on or off or different shades of gray; in contrast in colored LCDs each pixel can have different colors. In fact, the colored pixels can display red, green, and blue; using the 3 primary color lights they make different colors.

Some LCD Characteristics

Resolution

The total number of pixels (dots) per screen is a major factor in assessing an LCD and is shown below:

$$\text{Resolution} = \text{Pixels per line} \times \text{number of lines}$$

For example, when the resolution of an LCD is 720 × 350, there are 720 pixels per line and 350 lines per screen, giving a total of 252,000 pixels. The total number of pixels per screen is determined by the size of the pixel and how far apart the pixels are spaced. For this reason, one must look at what is called the *dot pitch* in LCD specifications.

Dot pitch

Dot pitch is the distance between adjacent pixels (dots) and is given in millimeters. For example, a dot pitch of 0.31 means that the distance between pixels is 0.31 mm. Consequently, the smaller the size of the pixel itself and the smaller the space between them, the higher the total number of pixels and the better the resolution. Dot pitch varies from 0.6 inch in some low-resolution LCDs to 0.2 inch in higher-resolution LCDs. Figure 18-3 shows Dot Pitch and Dot Size parameters.

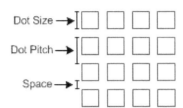

Figure 18-3: Dot Pitch and Dot Size

The specifications of a sample mono-colored LCD are shown in Figure 18-4.

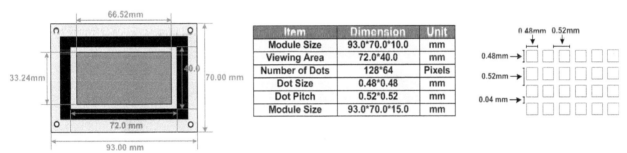

Item	Dimension	Unit
Module Size	93.0*70.0*10.0	mm
Viewing Area	72.0*40.0	mm
Number of Dots	128*64	Pixels
Dot Size	0.48*0.48	mm
Dot Pitch	0.52*0.52	mm
Module Size	93.0*70.0*15.0	mm

Figure 18-4: Mechanical specifications of a GDM12864 128x64 LCD

In some LCD specifications, it is given in terms of the number of dots per square inch, which is the same way it is given for laser printers, for example, 300 DPI (dots per inch).

Dot pitch and LCD size

LCDs, like televisions, are advertised according to their diagonal size. For example, a 14-inch monitor means that its diagonal measurement is 14 inches. There is a relation between the number of horizontal and vertical pixels, the dot pitch, and the diagonal size of the image on the screen. The diagonal size of the image must always be less than the LCD's diagonal size. The following simple equation can be used to relate these three factors to the diagonal measurement. It is derived from the Pythagorean Theorem:

$$\text{(image diagonal size)}^2 = \text{(number of horizontal pixels} \times \text{dot pitch)}^2$$
$$+ \text{(number of vertical pixels} \times \text{dot pitch)}^2$$

Since the dot pitch is in millimeters, the size given by the equation above would be in mm, so it must be multiplied by 0.039 to get the size of the monitor in inches. See Example 18-1.

Example 18-1

A manufacturer has advertised a 14-inch monitor of 1024 × 768 resolution with a dot pitch of 0.28. Calculate the diagonal size of the image on the screen. It must be less than 14 inches.

Solution:

The calculation is as follows:
(image diagonal size)^2 = (number of horizontal pixels × dot pitch)^2 + (number of vertical pixels × dot pitch)^2
(diagonal size)^2 = (1024 × 0.28 mm)^2 + (768 × 0.28 mm)^2 = 358.4 mm
diagonal size (inches) = 358.4 mm × 0.039 inch per mm = 13.98 inches
In the LCD the diagonal size of the image area is 13.98 inches while the diagonal size of the viewing area is 14 inches.

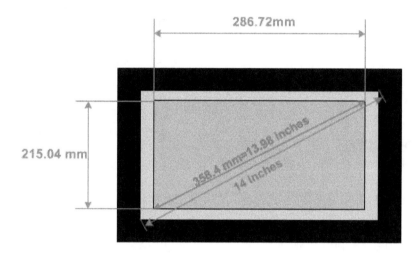

Displaying on the graphic LCDs

To display a picture on the screen, a distinct color must be shown on each pixel of the LCD. To do so, there is a display memory (frame buffer) that retrieves the attributes (colors) of the entire pixels of the screen and there is an LCD controller which displays the contents of the frame buffer memory on the LCD. See Figure 18-5.

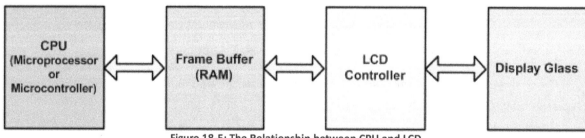

Figure 18-5: The Relationship between CPU and LCD

Graphic LCDs might come with or without frame buffer and the LCD controller. In cases that the LCD does not have frame buffer memory or controller they must be provided externally. Some new microcontrollers have the LCD controllers internally which can directly drive the LCDs. To display a picture on the screen the microcontroller writes it to the frame buffer memory.

Since the attributes (colors) of the entire pixels are stored in the frame buffer memory, the higher the number of pixels and colors options, the larger the amount of memory is needed to store them. In other words, the memory requirement goes up as the resolution and the number of supported colors go up. The number of colors displayed at one time is always 2^n where n is the number of bits set aside for the color. For example, when 4 bits are assigned for the color of the pixel, this allows 16 combinations of colors to be displayed at one time because $2^4 = 16$. The number of bits used for a pixel color is called color depth or bits per pixel (BPP). See Table 18-1.

BPP	Colors
1	on or off (monochrome)
2	4
4	16
8	256
16	65,536
24	16,777,216

Table 18-1: BPP (bit per pixel) vs. color

In Table 18-1, notice that in a monochrome LCD a single bit is assigned for the color of the pixel and it is for "on" or "off".

Mixing RGB (Red, Green, Blue) colors

We can get other colors by mixing the three primary colors of Red, Green, and Blue. The intensity (proportion) of the colors mixed can also affect the color we get. In many high-end graphics systems, an 8-bit value is used to represent the intensity. Its value can be between 0 and 255 (0 to 0xFF) representing high intensity (255) and zero intensity. See Table 18-2. Using three primary colors and intensity, we can make many colors we want. See Figure 18-6.

I	R	G	B	Color
0	0	0	0	Black
0	0	0	1	Blue
0	0	1	0	Green
0	0	1	1	Cyan
0	1	0	0	Red
0	1	0	1	Magenta
0	1	1	0	Brown
0	1	1	1	Light Gray
1	0	0	0	Dark Gray
1	0	0	1	Light blue
1	0	1	0	Light green
1	0	1	1	Light cyan
1	1	0	0	Light red
1	1	0	1	Light Magenta
1	1	1	0	Yellow
1	1	1	1	White

Table 18-2: The 16 Possible Colors

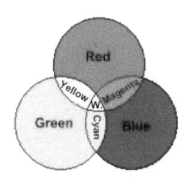

Figure 18-6: Making New Light Colors by Mixing the 3 Primary Light Colors

Example 18-2

In a certain graphic LCD, a maximum of 256 colors can be displayed at one time. How many bits are set aside for the color of the pixels?

Solution:

To display 256 colors at once, we must have 8 bits set for color since $2^8 = 256$.

LCD Buffer memory size and color

In discussing the graphics, we need to clarify the relationship between pixel resolution, the number of colors supported, and the amount of frame buffer RAM needed to store them. There are two facts associated with every pixel on the screen:

1. The location of the pixel
2. Its attributes: color and intensity

These two facts must be stored in the frame buffer RAM. The higher the number of pixels and colors options, the larger the amount of memory that is needed to store them. In other words, the memory requirement goes up as the resolution and the number of colors supported goes up. As we just mentioned, the number of colors displayed at one time is always 2^n where n is the number of bits set aside for the color. For example, when 4 bits are assigned for the color of the pixel, this allows 16 combinations of colors to be displayed at one time because $2^4 = 16$. The commonly used graphics resolutions are 176 x 144 (QCIF), 352x288 (CIF), 320x240 (QVGA), 480x272 (WQVGA), 640x480 (VGA) and 800x480 (WVGA). You may find the definitions of these abbreviations on the Internet.

We use the following formula to calculate the minimum frame buffer memory requirement for a graphic LCD:

$$Buffer\ memory\ size\ (in\ byte) = \frac{Horizontal\ Pixels\ \times Vertical\ Pixels\ \times BPP}{8}$$

Example 18-3 shows how to calculate the memory need for various resolutions and color depth.

Example 18-3

Find the frame buffer RAM needed for (a) 176x144 with 4 BPP and (b) 640x480 resolution with 256 colors.

Solution:

(a) For this resolution, there are a total of 25,344 pixels (176 columns × 144 rows = 25,344). With 4 bits for the color of each pixel, we need total of (25,344 × 4)/8= 16,672 bytes of frame buffer RAM. These 4 bits give rise to 16 colors.

(b) For this resolution, there are a total of 640 × 480=307200 pixels. With 256 colors, we need 8 bits for color of each pixel. Now, total of (640 × 480 × 8) / 8 = 307200 bytes of frame buffer RAM needed.

In VGA, 640 x 480 resolution with support for 256 colors displayed at one time requires a minimum of 640 × 480 × 8 = 2,457,600 bits =307,200 bytes of memory, but due to the memory organization used, the amount of memory used is higher.

Storing pixels in the memory of mono-color LCDs

In mono-colored LCDs each pixel can be on or off. Therefore, 1 bit can preserve the state of 1 pixel and a byte preserves 8 adjacent pixels. In some LCDs, e.g. GDM12864A and PCD8544, pixels are stored vertically in the bytes, as shown in Figure 18-7, while in some other LCDs, e.g. T6963, the pixels are stored horizontally.

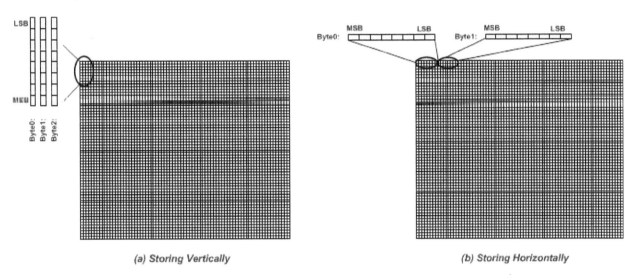

(a) Storing Vertically   (b) Storing Horizontally

Figure 18-7: Storing Data in the LCD Memory of Mono-colored LCDs

Review Questions
1. As the number of pixels goes up, the size of display memory _____ (increases, decreases).
2. If a total of 24 bits is set aside for color, how many colors are available?
3. Calculate the total video memory needed for 1024 × 768 resolution with 16 colors displayed at the same time.
4. With BPP of 16, we get _____colors.

Section 18.2: Displaying Texts on Graphic LCDs

As shown in Figure 18-8 each character can be made by putting pixels next to each other.

To display characters on the screen, we must have the pixel patterns of the entire characters. Whenever we want to display a character on the screen, we copy its pixel pattern into the display memory. See Figure 18-9.

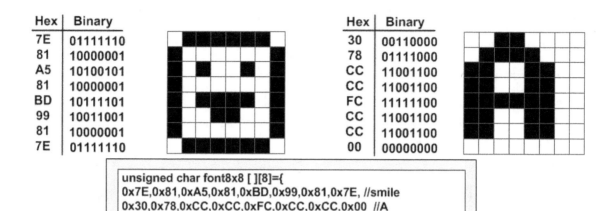

Hex	Binary
7E	01111110
81	10000001
A5	10100101
81	10000001
BD	10111101
99	10011001
81	10000001
7E	01111110

Hex	Binary
30	00110000
78	01111000
CC	11001100
CC	11001100
FC	11111100
CC	11001100
CC	11001100
00	00000000

```
unsigned char font8x8 [ ][8]={
0x7E,0x81,0xA5,0x81,0xBD,0x99,0x81,0x7E, //smile
0x30,0x78,0xCC,0xCC,0xFC,0xCC,0xCC,0x00  //A
};
```

Figure 18-8: Pixel Patterns of Characters Happy Face and Letter A

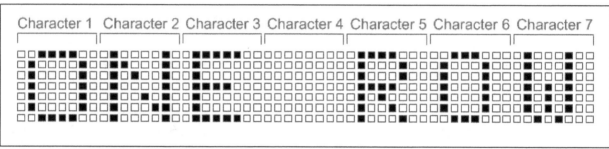

Figure 18-9: A Sample Text

The pixel patterns are stored in an array in the same way that they should be stored in the LCD memory. This means that for horizontal LCDs the bits are stored horizontally and for vertical LCDs the pixels are stored vertically. Figure 18-8 shows the way patterns are stored for horizontal LCDs. In Figure 18-10 the same patterns are stored for vertical LCDs.

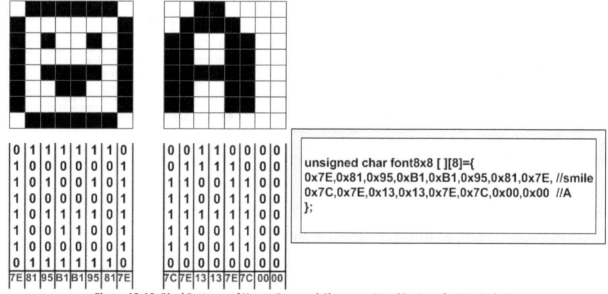

0	1	1	1	1	1	1	0
1	0	0	0	0	0	0	1
1	0	1	0	0	1	0	1
1	0	0	0	0	0	0	1
1	0	1	1	1	1	0	1
1	0	0	1	1	0	0	1
1	0	0	0	0	0	0	1
0	1	1	1	1	1	1	0
7E	81	95	B1	B1	95	81	7E

0	0	1	1	0	0	0	0
0	1	1	1	1	0	0	0
1	1	0	0	1	1	0	0
1	1	0	0	1	1	0	0
1	1	1	1	1	1	0	0
1	1	0	0	1	1	0	0
1	1	0	0	1	1	0	0
0	0	0	0	0	0	0	0
7C	7E	13	13	7E	7C	00	00

```
unsigned char font8x8 [ ][8]={
0x7E,0x81,0x95,0xB1,0xB1,0x95,0x81,0x7E, //smile
0x7C,0x7E,0x13,0x13,0x7E,0x7C,0x00,0x00  //A
};
```

Figure 18-10: Pixel Patterns of Happy Face and Character A and its Font for Vertical LCD

To get better-looking characters, the font resolution must be increased, which translates to more pixels horizontally and vertically. See Figure 18-11.

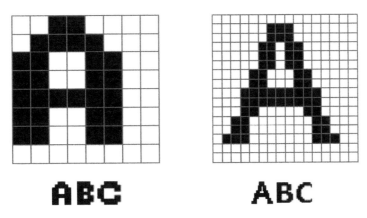

(a) a small font displayed
on a low resolution LCD

(b) a bigger font displayed
on a high resolution LCD

Figure 18-11: A Bigger Font vs. a Smaller Font

PCD8544 LCD

Next, we use PCD8544 Graphic LCD. PCD8544 (also known as Nokia5110) is a 48×84 monochrome graphic LCD. The LCD has the following pins:

VCC (VDD) and GND (VSS): The pins provide power to the LCD.

CE (SS), DIN (MOSI), and SCLK: Using the pins the LCD is connected to the microcontroller using the SPI protocol. The pins are used to send command and data to the LCD.

D/C# (Data/Command#): If D/C is high, the sending value is considered as data. Otherwise, the value is considered as command.

Reset#: The reset pin, initializes the LCD and it is active low.

BL (Backlight): The LCDs have some backlight LEDs. Connect the BL pin to VCC to turn on the backlight.

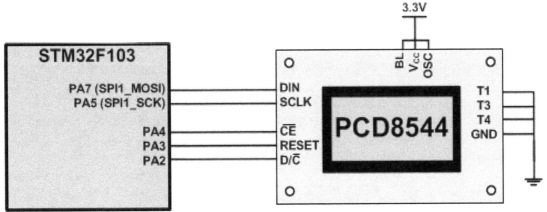

Figure 18-12: The PCD8544 LCD connection to the STM32F10x

The connection between the PCD8544 LCD and the microcontroller is shown in Figure 18-12. In most LCDs, the pins T1, T2, T3, and T4 are grounded internally and you do not have access to them. If your LCD provide access to the pins, ground them as shown in the picture. In most LCDs, you do not have access to the OSC (Oscillator) pins, neither. But if it is provided to you, connect it to 3.3V.

DDRAM (Display Data RAM) and the Address Counter

The LCD has a 512-byte SRAM memory which stores the contents of display. See Figures 18-13 and 18-14. In the RAM, the pixels are grouped vertically. For example, the top-left corner pixel of the LCD is stored in the least significant bit of the first byte.

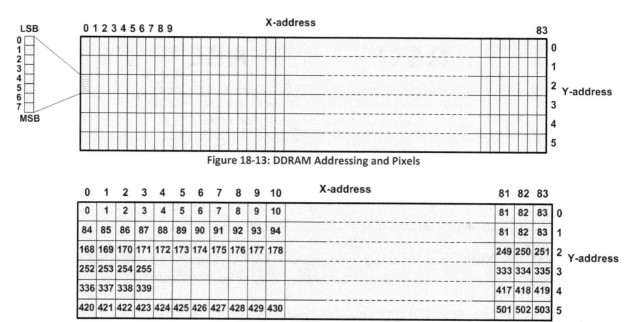

Figure 18-13: DDRAM Addressing and Pixels

Figure 18-14: DDRAM Addresses

If the D/C pin is high, whenever you send data via SPI, they are stored in the DDRAM. The LCD has an address counter which can point to different locations of the DDRAM. To draw on the LCD, you move the address counter using commands and then you write to the memory by sending data. When you send a byte of data, the address counter automatically increments and you can write to the next byte.

PCD8544 Commands

PCD8544 has two groups of command: basic and extended. The extended commands are used to initialize the LCD while the basic instructions are used for drawing. The function set command is available both in basic and extended modes and you can use it to switch between basic and extended modes.

Function Set

See Figure 18-15 and Table 18-3. The H bit of the command, chooses between basic and extended modes. Sending the command 0x21 switches the command mode to extended while sending 0x20 switches the mode to basic.

D/C	D7	D6	D5	D4	D3	D2	D1	D0
0	0	0	1	0	0	PD	V	H

Figure 18-15: Function Set

Bit	Usage	Description
H	Extended/Basic	0: Basic, 1: Extended
V	Vertical/Horizontal	The bit chooses the direction of moving the address counting when auto increments. For common uses it is 0. (0: Horizontal, 1: Vertical)
PD	Power Down/Active	0: Active, 1: Power down

Table 18-3: The Bits of Function-Set Command

Extended commands

Figure 18-16 lists the extended commands. The commands are sent to initialize the LCD. For more information, see the PCD8544 datasheet.

Instruction	D/C	D7	D6	D5	D4	D3	D2	D1	D0
Set Temperature Coefficient	0	0	0	0	0	0	1	TC1	TC0
Set Bias	0	0	0	0	1	0	BS2	BS1	BS0
Set VOP	0	1	VOP6	VOP5	VOP4	VOP3	VOP2	VOP1	VOP0

Figure 18-16: Extended Commands (H = 1)

Basic commands

See Figure 18-17. The Set X and Set Y commands are used to move the address counter. See Figure 18-13. Set X can give values between 0 to 83 as argument and moves the address counter to columns 0 to 83. Set Y gives values between 0 to 5 and moves the address counter in vertical direction. Using the Display command, you can inverse the whole display. For normal display mode, send a 0x0C command.

Instruction	D/C	D7	D6	D5	D4	D3	D2	D1	D0
Display Configuration	0	0	0	0	0	1	D	0	E
Set Y address of RAM	0	0	1	0	0	0	Y2	Y1	Y0
Set X address of RAM	0	1	X6	X5	X4	X3	X2	X1	X0

Figure 18-17: Basic Commands (H = 0)

D	E	Display Mode
0	0	Display blank
0	1	All display segments on
1	0	Normal mode
1	1	Inverse video mode

Table 18-4: Display Configuration Modes

See Program 18-1. The glcd_init function uses extended commands to initialize the LCD and the glcd_setCursor function uses basic commands to move cursor. A lookup table of the pixel patterns of the characters is made using an array. The glcd_putchar function accesses the lookup array to display characters on the LCD.

Program 18-1: Displaying a text on the PCD8544 GLCD

```
/* The program shows HELLO on the PCD8544 (Nokia5110) LCD */
#include <stm32f10x.h>

#define GLCD_CS     4
#define GLCD_RESET  3
```

469

```c
#define GLCD_DC      2

void delay_ms(uint16_t t);

uint8_t spi1_transfer(uint8_t d);
void glcd_init(void);
void glcd_setCursor(uint8_t x, uint8_t y);
void glcd_putchar(char c);
void glcd_clear(void);

int main()
{
   glcd_init(); /* initialize the lcd */
   glcd_clear();

   glcd_setCursor(20, 2); /* move cursor to 20, 2 */

   glcd_putchar('H');
   glcd_putchar('E');
   glcd_putchar('L');
   glcd_putchar('L');
   glcd_putchar('O');

   while(1)
   {
   }
}

/* The function sends a byte of data through SPI */
/* argument d: the byte to be sent */
uint8_t spi1_transfer(uint8_t d)
{
   GPIOA->BRR = (1 << GLCD_CS);

   SPI1->DR = d; /* send the contents of d */
   while((SPI1->SR&(1<<0)) == 0); /* wait until RXNE is set */

   GPIOA->BSRR = (1 << GLCD_CS);
   return SPI1->DR; /* return the received data */
}

void glcd_cmd(uint8_t cmd)
{
   GPIOA->BRR = (1 << GLCD_DC);    /* DC = 0 (command) */
   spi1_transfer(cmd);
}

void glcd_data(uint8_t data)
{
   GPIOA->BSRR = (1 << GLCD_DC);   /* DC = 1 (data) */
   spi1_transfer(data);
}

void glcd_init()
{
   RCC->APB2ENR |= 0xFC|(1<<12);  /* enable clocks for GPIO and SPI1 */
```

```
    GPIOA->CRL = 0xB4B33344; /* MOSI (PA7) and SCK(PA5): alt. func. out, PA4: output
for CS */
    SPI1->CR1 = 0x35C;          /* SPE = 1, BR = 3, FFD = 0, SSI and SSM = 1 */
    GPIOA->BSRR = (1<<GLCD_CS)|(1<<GLCD_RESET); /* make CS and RESET pins high */

    delay_ms(10);
    GPIOA->BRR = (1<<GLCD_RESET);  /* reset the LCD (RESET = 0) */
    delay_ms(70);
    GPIOA->BSRR = (1<<GLCD_RESET); /* release the RESET pin */

    glcd_cmd(0x21);   /* switch to extended command mode */
    glcd_cmd(0x06);   /* set temp. coefficient 2 */
    glcd_cmd(0x13);   /* set LCD bias mode 1:48 */
    glcd_cmd(0xC2); /* set VOP to 7V (VOP = 66) */
    glcd_cmd(0x20); /* switch to basic mode */
    glcd_cmd(0x0C);   /* set lcd display to normal mode */
}

void glcd_setCursor(uint8_t x, uint8_t y)
{
    glcd_cmd(0x80 | x);      /* set x */
    glcd_cmd(0x40 | y);      /* set y (bank) */
}

void glcd_clear(void)
{
    uint16_t i;

    glcd_setCursor(0, 0);

    for (i = 0 ; i < 504 ; i++)
        glcd_data(0x00);
}

/* sample font table */
const char font_table[][5] = {
    { 0x7e, 0x11, 0x11, 0x11, 0x7e }, /* A */
    { 0x7f, 0x49, 0x49, 0x49, 0x36 }, /* B */
    { 0x3e, 0x41, 0x41, 0x41, 0x22 }, /* C */
    { 0x7f, 0x41, 0x41, 0x22, 0x1c }, /* D */
    { 0x7f, 0x49, 0x49, 0x49, 0x41 }, /* E */
    { 0x7f, 0x09, 0x09, 0x09, 0x01 }, /* F */
    { 0x3e, 0x41, 0x49, 0x49, 0x7a }, /* G */
    { 0x7f, 0x08, 0x08, 0x08, 0x7f }, /* H */
    { 0x00, 0x41, 0x7f, 0x41, 0x00 }, /* I */
    { 0x20, 0x40, 0x41, 0x3f, 0x01 }, /* J */
    { 0x7f, 0x08, 0x14, 0x22, 0x41 }, /* K */
    { 0x7f, 0x40, 0x40, 0x40, 0x40 }, /* L */
    { 0x7f, 0x02, 0x0c, 0x02, 0x7f }, /* M */
    { 0x7f, 0x04, 0x08, 0x10, 0x7f }, /* N */
    { 0x3e, 0x41, 0x41, 0x41, 0x3e }, /* O */
    { 0x7f, 0x09, 0x09, 0x09, 0x06 }, /* P */
    { 0x3e, 0x41, 0x51, 0x21, 0x5e }, /* Q */
    { 0x7f, 0x09, 0x19, 0x29, 0x46 }, /* R */
    { 0x46, 0x49, 0x49, 0x49, 0x31 }, /* S */
    { 0x01, 0x01, 0x7f, 0x01, 0x01 }, /* T */
    { 0x3f, 0x40, 0x40, 0x40, 0x3f }, /* U */
```

```
   { 0x1f, 0x20, 0x40, 0x20, 0x1f }, /* V */
   { 0x3f, 0x40, 0x38, 0x40, 0x3f }, /* W */
   { 0x63, 0x14, 0x08, 0x14, 0x63 }, /* X */
   { 0x07, 0x08, 0x70, 0x08, 0x07 }, /* Y */
   { 0x61, 0x51, 0x49, 0x45, 0x43 }, /* Z */
};

void glcd_putchar(char c)
{
   uint8_t i;

   if((c < 'A')||(c > 'Z'))
      return;

   for (i = 0; i < 5; i++)
      glcd_data(font_table[c-'A'][i]);

   glcd_data(0); /* an empty column between chars */
}

void delay_ms(uint16_t t)
{
   volatile unsigned long l = 0;
   for(uint16_t i = 0; i < t; i++)
      for(l = 0; l < 6000; l++)
      {
      }
}
```

Review Questions

1. True or false. The same font can be used for vertical and horizontal LCDs.
2. True or false. To display a character on the LCD, its pixel pattern should be copied onto the LCD display memory.

Problems

Section 18-1

1. What is the difference between monochrome and colored LCDs?
2. True or False. The resolution of two LCDs are the same. One of them is 15-inch and the other one is 14-inch. The 15-inch LCD has a bigger dot pitch.
3. What does an LCD controller do?
4. Find the frame buffer RAM needed for a monochrome 640x480 LCD.

Section 18-2

5. True or False. PCD8544 is a monochrome LCD.
6. What is the resolution for PCD8544?
7. True or False. The role of D/C pin in the PCD8544 is similar to the RS pin in character LCDs.
8. True or False. The font which is used in Program 18-1, is a vertical font.

Answers to Review Questions

Section 18-1
1. increases
2. 2^{24} = 16.7 million
3. $1024 \times 768 \times 4$ = 3,145,728 bits = 384K bytes, but it uses 512 KB due to bit planes.
4. 2^{16} = 65,536

Section 18-2
1. False
2. True

Chapter 19: Direct Memory Access (DMA)

Section 19.1: Introduction to DMA

Sometimes it is needed to copy a huge amount of data. For example, to display a picture on the LCD, the data of the picture should be copied to the graphic memory. In C, you can use two pointers, a counter, and a for to copy data. See Program 19-1.

Program 19-1

```c
#include <stm32f10x.h>

void copyContents(char *destP, char *srcP, uint16_t count)
{
   while(count > 0)
   {
      *destP = *srcP;
      destP++;
      srcP++;
      count--;
   }
}

int main()
{
   char source[10]="ABCDEDFHI";
   char dest[10];

   copyContents(dest, source, 10);

   while(1);
}
```

It needs two memory cycles to copy a byte of data (one read and a write). But in the above program it takes around 12 machine cycles to copy a byte since the CPU has to fetch, decode, and execute the instructions. It can become too time consuming specially when a huge amount of data should be copied. This also makes the CPU busy and waists its time. See Example 19-1.

Example 19-1

A 640x480 picture is stored in the flash memory. Assuming that BPP is 24-bit (3 bytes are used to store each pixel of picture) and the CPU crystal is 72MHz, how long does it take to copy the picture from flash to memory using Program 19-1?

Solution:

Copying each byte takes = 12 machine cycle $\times \dfrac{1}{72M} = 166ns$

Number of bytes = 640 × 480 × 3 = 921600

921600 × 166 ns = 152985600 ns = 0.152985600 seconds

To copy data, we can make a simple hardware using 4 registers and a control circuit, as shown in Figure 19-1.

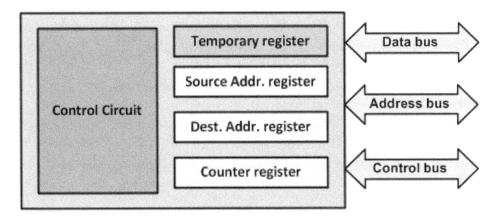

Figure 19-1: A Simplified DMA

The CPU initializes the source address register and the destination address register with the source and destination addresses and the counter register is loaded with the number of bytes to be copied. Then, the CPU waits until the data is copied.

The hardware puts the value of the source register on the address bus and reads from the source location of memory into the temporary register; then, it puts the value of the temporary register on the data bus and the value of destination address register on the address bus and enables the WR (write) signal. So, the data is copied into the destination address. Then, the source and destination registers are incremented and the counter register decrements and the process is repeated as long as the counter is not zero. See Figure 19-2.

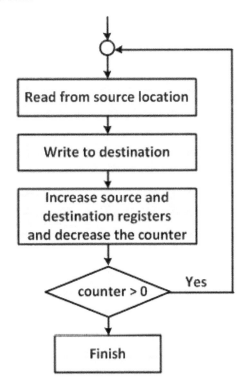

Figure 19-2: Copying Steps

DMA controllers work similar to the above hardware. But they are more advanced. In real DMA controllers:

- There are more than 1 channel. Each channel can be programmed to copy data from a separate source address to the destination address.

475

- In Arm Cortex microcontrollers, including the STM32, DMA controllers have separate buses. So, CPU and DMA can access to devices at the same time, as long as they access different devices.

- Sometimes it is needed to transfer data between memory and a peripheral register. For example, you might want to send 1KB of data through the USART or the SPI. So, the DMA controller should transfer data when the peripheral is ready. The data registers for the peripherals have a fixed address, as well. So, in DMA controllers we can choose if the registers should increase after each transfer or not.

Review Questions
5. True or false. DMA does not need any initialization.
6. True or false. Each channel of DMA copies data from a source to the destination.
7. True or false. In Cortex-M, there is a separate bus for each DMA controller.
8. True or false. DMA can only copy data from peripherals to memory.

Section 19-2: DMA in STM32F10x

In STM32F10x, there are two DMA controllers. DMA1 has 7 channels and DMA2 has 5 channels. See Figure 19-3.

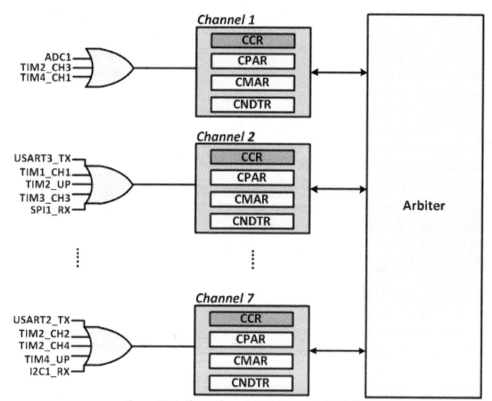

Figure 19-3: Channels and DMA Requests in DMA1

Memory-to-Memory vs. Peripheral-Memory mode

Each channel can be used to transfer data between two parts of memory or between memory and a peripheral.

476

If the channel is configured to the memory-to-memory mode, when the software enables the channel, it begins copying data from source to destination. The channel stops copying when the counter becomes zero.

If the channel is configured to peripheral-memory mode, before each transfer, the channel waits until the peripheral sends a DMA request signal. Then, the channel transfers data. For example, if channel4 is configured for USART1 transmit, USART1 makes a DMA request whenever it is ready to transmit data. Then, the channel transfers data and loads the USART1 data register with the data to be sent. The USART1 makes the next DMA request when it is ready to send the next byte and the DMA channel sends the next byte to the USART1 data register.

As shown in Figure 19-3 and Tables 19-1 and 19-2, each channel is dedicated to some peripherals. At any given time, we should use each channel for just one peripheral (not more).

Peripherals	Channel 1	Channel 2	Channel 3	Channel 4	Channel 5	Channel 6	Channel 7
ADC1	ADC1	-	-	-	-	-	-
SPI/I^{2}S	-	SPI1_RX	SPI1_TX	SPI2/I2S2_RX	SPI2/I2S2_TX	-	-
USART	-	USART3_TX	USART3_RX	USART1_TX	USART1_RX	USART2_RX	USART2_TX
I^{2}C	-	-	-	I2C2_TX	I2C2_RX	I2C1_TX	I2C1_RX
TIM1	-	TIM1_CH1	-	TIM1_CH4 TIM1_TRIG TIM1_COM	TIM1_UP	TIM1_CH3	-
TIM2	TIM2_CH3	TIM2_UP	-	-	TIM2_CH1	-	TIM2_CH2 TIM2_CH4
TIM3	-	TIM3_CH3	TIM3_CH4 TIM3_UP	-	-	TIM3_CH1 TIM3_TRIG	-
TIM4	TIM4_CH1	-	-	TIM4_CH2	TIM4_CH3	-	TIM4_UP

Table 19-1: DMA1 Peripheral Requests for Channels

Peripherals	Channel 1	Channel 2	Channel 3	Channel 4	Channel 5
ADC3[1]	-	-	-	-	ADC3
SPI/I2S3	SPI/I2S3_RX	SPI/I2S3_TX	-	-	-
UART4	-	-	UART4_RX	-	UART4_TX
SDIO[1]	-	-	-	SDIO	-
TIM5	TIM5_CH4 TIM5_TRIG	TIM5_CH3 TIM5_UP	-	TIM5_CH2	TIM5_CH1
TIM6/ DAC_Channel1	-	-	TIM6_UP/ DAC_Channel1	-	-
TIM7	-	-	-	TIM7_UP/ DAC_Channel2	-
TIM8	TIM8_CH3 TIM8_UP	TIM8_CH4 TIM8_TRIG TIM8_COM	TIM8_CH1	-	TIM8_CH2

Table 19-2: DMA1 Peripheral Requests for Channels

Arbiter and the priority bits

For each channel there are priority bits. The bits can be used to configure the priorities for channels. If two channels want to transfer data at the same time, the channel with higher priority transfers first and then the other channel transfers data. When the priority of two channels are configured to the same value, the channel with lower channel number will transfer first. For example, if both channel 2 and 5 want to transfer data and they have the same priority, channel 2 transfers data first and then channel 5 is allowed to transfer data.

The DMA registers

CPAR (Channel Peripheral Address Register) and CMAR (Channel Memory Address Register)

CPAR and CMAR are the source and destination address registers. There is no difference between CPAR and CMAR. They are 32-bit registers and they can point to any locations in the memory or the peripherals. In DMA, CPAR and CMAR play the same role as the destP and srcP pointers in Program 19-1. The registers are initialized by software with the address of source and destination. We must not write to the registers when the channel is enabled. See Figures 19-4 and 19-5.

D31	D30	......	D3	D2	D1	D0

DMAn_CPARx: | PA |

DMAn_Channelx->CPAR

Figure 19-4: CPAR (Channel Peripheral Address Register)

D31	D30	......	D3	D2	D1	D0

DMAn_CMARx: | MA |

DMAn_Channelx->CMAR

Figure 19-5: CMAR (Channel Memory Address Register)

CNDTR (Channel Number of Data Register)

The CNDTR is a 16-bit register. It is initialized with number of data to be copied. When the channel is enabled, the register becomes read only and its value shows the remained number of data to be copied. See Figure 19-6.

D15	D14	......	D3	D2	D1	D0

DMAn_CNDTRx: | NDT |

DMAn_Channelx->CNDTR

Figure 19-6: CNDTR (Channel Number of Data) Register

CCR (Channel Configuration Register)

There is a CCR register for each channel. It is used to configure the channel. See Figures 19-3 and 19-7.

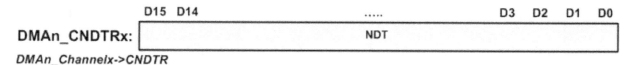

D14	D13	D12	D11	D10	D9	D8	D7	D6	D5	D4	D3	D2	D1	D0
MEM2 MEM	PL		MSIZE		PSIZE		MINC	PINC	CIRC	DIR	TEIE	HTIE	TCIE	EN

DMAn_CCRx:

DMAn_Channelx->CCR

478

Field	Bit	Description
MEM2MEM	14	Memory to memory (0: Peripheral-memory, 1: Memory to memory) If the bit is set, the DMA channel copies data from source to destination until the counter becomes zero and transfer finishes. If the bit is zero, before each transfer, the channel waits to receive a DMA request from the peripherals and then it makes a transfer.
PL	13-12	Priority Level (00: Low, 01: Medium, 10: High, 11: very high) The bits are used to set the priority of the channel.
MSIZE	11-10	Memory size (00: 8-bit, 01: 16-bit, 10: 32-bit, 11: reserved)
PSIZE	9-8	Peripheral size (00: 8-bit, 01: 16-bit, 10: 32-bit, 11: reserved)
MINC	7	Memory increment mode (0: fixed address, 1: increment) If the MINC bit is 1, after each transfer the CMAR register is incremented. Otherwise, the value of the CMAR remains fixed.
PINC	6	Peripheral increment mode (0: fixed, 1: increment) If PINC is 1, after each transfer the CPAR register is incremented. Otherwise, the value of the CPAR remains fixed.
CIRC	5	Circular mode (0: disabled, 1: enabled) If the bit is 1, when the counter becomes zero, the CPAR, CMAR, and CNDTR are reloaded with their initial values and the channels transfers data from beginning. Otherwise, when the counter becomes zero, the channel stops.
DIR	4	Direction (0: copy from CPAR to CMAR, 1: copy from CMAR to CPAR) If DIR=0, CPAR is considered as the source address, and CMAR as the destination address and the channel reads data from CPAR and copies to CMAR. Otherwise, CPAR is considered as the destination address and CMAR as the source address and the channel reads from the address pointed to by CMAR and copies to the address pointed to by CPAR.
TEIE	3	Transfer Error Interrupt Enable (0: disabled, 1: enabled)
HTIE	2	Half Transfer Interrupt Enable (0: disabled, 1: enabled)
TCIE	1	Transfer Complete Interrupt Enable (0: disabled, 1: enabled)
EN	0	Channel Enable (0: disabled, 1: enabled)

Figure 19-7: CCR (Channel Configuration Register)

Memory size and Peripheral size

MSIZE is used to set the size of data that is pointed to by the CMAR register. If MSIZE is set to 00 (8-bit mode) each data which is pointed to by CMAR is considered as an 8-bit data. So, when the channel accesses the CMAR, it reads/writes a byte of data and after each transfer if MINC=1, the CMAR is incremented by one.

When MSIZE is 01 (16-bit), the channel considers the data which is pointed to by CMAR as 16-bit. So, on each transfer the channel reads/writes 2 bytes of data from/to the location which is pointed to by CMAR, and if MINC is 1, after each transfer CMAR is increased by 2. When MSIZE is 01, the channel considers data to be aligned 16-bit, and considers the least significant bit of CMAR as 0, even if it is initialized to 1.

When the MSIZE bits are set to 10, the size of data is considered 32-bit. So, CMAR is incremented by 4 when the MINC bit is set. See Table 19-3.

MSIZE	Mode	Read/write size	CMAR Increment	Alignment
00	8-bit	8-bit	+1	Byte
01	16-bit	16-bit	+2	Half-word
10	32-bit	32-bit	+4	word

Table 19-3: MSIZE

Similarly, PSIZE is used to set the data size which is pointed to by CPAR. In fact, if we consider CMAR and CPAR as two pointers, the MSIZE and PSIZE bits choose the type of the pointers. When PSIZE and MSIZE are 00, the pointers are considered as char*. When PSIZE and MSIZE are 01, they are short*, and whenever PSIZE and MSIZE are 10, CMAR and CPAR are considered as long*.

Circular mode

If CIRC=0, the channel copies data from source to destination until the counter becomes zero. But if we set CIRC to 1, the channel works in circular mode. In the circular mode, when the counter reaches zero, the registers CMAR, CPAR, and CNDTR will be reloaded with their initial values and the channel begins copying data from the beginning.

Interrupts

There are 3 interrupt sources for each DMA channel: Transfer Error, Half-transfer complete, and Transfer complete. Each of the interrupts can be enabled/disabled separately using TEIE, HTIE, and TCIE.

- Transfer Error: While transferring data, if the channel tries to read or write from an invalid address (reserved address), the interrupt is generated. In the case, the channel becomes disabled, automatically. The interrupt can be enabled/disabled using TEIE.

- Half-transfer: When the channel transfers half of data, the interrupt can be generated. The HTIE bit can be used to enable/disable the interrupt.

- Transfer-complete: The interrupt is generated when the transfer is finished and the counter becomes zero. The interrupt can be enabled/disabled using the TCIE bit.

For each channel, there is an interrupt vector. So, if we enable more than 1 interrupt, we should check the status register (DMA_ISR) to find the interrupt source.

DMAn_ISR (Interrupt Status Register)

It is a read-only register that shows the status of all channels of a DMA. See Figure 19-8.

	D31	D30	D29	D28	D27	D26	D25	D24	D23	D22	D21	D20	D19	D18	D17	D16
DMAn_ISR: *DMAn->ISR*	Reserved				TEIF7	HTIF7	TCIF7	GIF7	TEIF6	HTIF6	TCIF6	GIF6	TEIF5	HTIF5	TCIF5	GIF5
	D15	D14	D13	D12	D11	D10	D9	D8	D7	D6	D5	D4	D3	D2	D1	D0
	TEIF4	HTIF4	TCIF4	GIF4	TEIF3	HTIF3	TCIF3	GIF3	TEIF2	HTIF2	TCIF2	GIF2	TEIF1	HTIF1	TCIF1	GIF1

Field	Description
TEIFx	Channel x Transfer-Error flag: the bit is set when there is an error.
HTIFx	Channel x Half-transfer complete flag: It is set when half of data is transferred.
TCIFx	Channel x Transfer-Complete flag: It is set when the transfer is completed.
GIFx	Channel x Global flag: It is set when any flags of the channel is set.

Figure 19-8: DMA_ISR (DMA Status Register)

DMAn_IFCR (Interrupt Flag Clear Register)

The IFCR register is used to clear the flags of the ISR register. Writing 1 to each bit of IFCR, clears the corresponding flag in the DMAn_ISR register. See Figure 19-9. For example, the following line of code clears the transmit complete flag of channel 2 (TCIF2):

```
DMA1->IFCR = (1<<5); /* clear TCIF2 (bit 5 of DMA_ISR) */
```

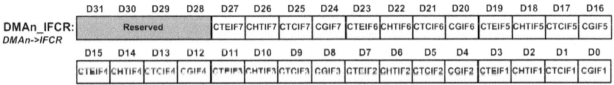

Figure 19-9: DMA_IFCR (Interrupt Flag Clear Register)

Writing programs using DMA

To program a DMA channel, the following sequence should be followed:

- Initialize CPAR and CMAR with source and destination addresses.

- Initialize CNDTR with the number of data to be copied. The value of CNDTR is decremented after each transfer and the channel stops transferring when the CNDTR becomes zero.

- Using the CCR register configure the channel priority, transfer direction, data size, memory and peripheral increment mode, circular mode, and the interrupts.

- Enable the channel by setting the Enable bit of the CCR register.

Copying memory to memory

See Program 19-2. It copies the source array to the dest array using channel1 of DMA1. In the program, the CPAR is initialized with the source address and the CMAR with the destination address. So, the channel should copy from CPAR to CMAR and the DIR bit of the CCR is cleared. Since the type of transferring data is char, both PSIZE and MSIZE are set to 00.

In the program, when the channel is enabled, the channel begins copying data from source to dest. Since the CNDTR register is set to 10, the channel copies 10 bytes of data from source to destination and then stops. When the transfer is finished the TCIF flag is set and the channel interrupt is invoked. In the interrupt routine, we set the transmitComplete to 1 to aware the main program that the copy is finished. Write the program in Keil and single step in debug mode. Watch the values of dest and source.

Program 19-2: Copying data from memory to memory (Program 19-1 using DMA)

```c
#include <stm32f10x.h>

volatile char transmitComplete = 0;

int main()
{
    char source[10]="ABCDEFGHI";
    char dest[10];

    RCC->AHBENR = (1<<0); /* DMA clock enable (RCC_AHBENR_DMA1EN) */

    DMA1_Channel1->CPAR = (uint32_t) source; /* CPAR = the addr. of source array */
    DMA1_Channel1->CMAR = (uint32_t) dest; /* CMAR = the addr. of dest array */
    DMA1_Channel1->CNDTR = 10;      /* number of bytes to be copied = 10 */
    DMA1_Channel1->CCR = (1<<14)|(1<<7)|(1<<6)|(1<<1); /* mem2mem=1, mem inc.=1, per.
inc.=1, TCIE=1 */
    DMA1_Channel1->CCR |= 1; /* enable the channel */
    NVIC_EnableIRQ(DMA1_Channel1_IRQn); /* enable the interrupt for channel 1 */

    while(transmitComplete == 0); /* wait until the transmission is complete */

    while(1);
}

void DMA1_Channel1_IRQHandler(void)
{
    if((DMA1->ISR&(1<<1)) != 0) /* is transmit complete */
    {
        DMA1->IFCR = 1<<1; /* clear the TCIF flag */
        transmitComplete = 1; /* set the flag */
    }
}
```

Program 19-3 is similar to Program 19-2. But it copies an array of long integer. So, the PSIZE and MSIZE bits are set to 10.

Program 19-3: Copying an array of long int using DMA

```c
#include <stm32f10x.h>

volatile char transmitComplete = 0;

int main()
{
    long source[10]=
{0x11111111,0x22222222,0x33333333,0x44444444,0x55555555,0x66666666,0x77777777,0x88888888,0x99999999,0xaaaaaaaa};
    long dest[10];
```

```
RCC->AHBENR = (1<<0); /* DMA clock enable (RCC_AHBENR_DMA1EN) */

DMA1_Channel1->CPAR = (uint32_t) source; /* CPAR = the addr. of source array */
DMA1_Channel1->CMAR = (uint32_t) dest; /* CMAR = the addr. of dest array */
DMA1_Channel1->CNDTR = 10;        /* number of bytes to be copied = 10 */
DMA1_Channel1->CCR = (1<<14)|(1<<11)|(1<<9)|(1<<7)|(1<<6)|(1<<1);       /*
mem2mem=1,MSIZE=10 (32-bit), PSIZE=10 (32-bit), mem inc.=1, per. inc.=1, TCIE=1 */
DMA1_Channel1->CCR |= 1; /* enable the channel */
NVIC_EnableIRQ(DMA1_Channel1_IRQn); /* enable the interrupt for channel 1 */

while(transmitComplete == 0); /* wait until the transmission is complete */

while(1);
}

void DMA1_Channel1_IRQHandler(void)
{
   if((DMA1->ISR&(1<<1)) != 0) /* is transmit complete */
   {
      DMA1->IFCR = 1<<1; /* clear the TCIF flag */
      transmitComplete = 1; /* set the flag */
   }
}
```

Sending data via USART using DMA

To use DMA for peripherals, we should enable the DMA for the peripheral and disable the MEM2MEM bit of the CCR register. We should also check Tables 19-1 and 19-2 to find the related channel. See Figure 19-10. The DMAT bit of the CR3 register enables the DMA for USART transmit.

	D31	....	D11	D10	D9	D8	D7	D6	D5	D4	D3	D2	D1	D0
USART_CR3: *USART->CR3*	Reserved			CTSIE	CTSSE	RTSE	DMAT	DMAR	SCEN	NACK	HDSEL	IRLP	IREN	EIE

Field	Bit	Description
DMAT	7	DMA enable Transmitter 0: Disable the DMA for USART transmit 1: Enable the DMA for USART transmit
DMAR	6	DMA enable receiver 0: the DMA for USART receive is disabled 1: the DMA for USART receive is enabled

Figure 19-10: USART_CR3 (Control Register 3)

According to Table 19-1, the USART1_TX DMA request is related to channel 4. So, we should initialize channel 4 to use the DMA for USART1_TX.

In Program 19-4, the contents of ourMsg are sent via USART1 using DMA. When the USART1_TX is ready, it sends a DMA request to channel 4. Then, channel 4 transfers a byte of data from ourMsg to the USART1_DR and increases the CMAR register and decreases the CNDTR. When the byte of data is transmitted via USART, USART1_TX sends another DMA request to channel 4, and channel 4 sends another byte of data to USART1_DR. The process repeats until 12 bytes of data are transferred by DMA and the CNDTR becomes zero.

Program 19-4: Sending data via USART1 using DMA

```c
#include <stm32f10x.h>

int main()
{
    char ourMsg[]="ABCDEFGHIJKL";

    RCC->AHBENR = (1<<0); /* DMA1 clock enable */
    RCC->APB2ENR |= 0xFC | (1<<14); //enable GPIO clocks

    //USART1 init.
    GPIOA->ODR |= (1<<10);  /* pull-up PA10 */
    GPIOA->CRH = 0x444448B4; /* RX1=input with pull-up, TX1=alt. func. output */
    USART1->BRR = 7500;      /* 72MHz/9600bps = 7500 */
    USART1->CR1 = 0x200C; /* enable usart transmit and receive */
    USART1->CR3 = (1<<7); /* DMA Trans. enable */

    DMA1_Channel4->CPAR = (uint32_t) &USART1->DR; /* to USART1->DR */
    DMA1_Channel4->CMAR = (uint32_t) ourMsg; /* from ourMsg array in mem. */
    DMA1_Channel4->CNDTR = 12;      /* copy 12 bytes */
    DMA1_Channel4->CCR = (1<<7)|(1<<4);   /* mem inc., read from mem */
    DMA1_Channel4->CCR |= 1; /* enable channel 4 */

    while(1)
    {
    }
}
```

Review Questions
1. True or false. CPAR can only be used to point to peripheral registers.
2. True or false. CNDTR is initialized with the number of data to be copied.
3. To use a channel, which registers must be initialized by software?
4. True or false. If both PSIZE and MSIZE are 01, two bytes of data are copied in each step.
5. True or false. The priorities of channels are fixed.

Problems

Section 19-1
1. True or false. DMA can copy data faster than CPU.
2. True or false. DMAs copy just one byte of data and then they must be reinitialized by software.
3. True or false. Each DMA has only a channel.
4. True or false. DMA can copy data just from flash memory to RAM.

Section 19-2

5. True or false. CMAR cannot address the peripheral registers.
6. True or false. Reading the CNDTR register while the channel is enabled gives us the remained number of bytes to be copied.
7. What should we load to PSIZE if we want to use CPAR to address 16-bit registers?
8. Find the values for PSIZE and MSIZE if we want to copy 32-bit data.
9. True or false. When a channel tries to access a reserved address, a transfer error occurs.
10. Write a program that sends "Hello world!" via USART2 using DMA.

Answers to Review Questions

Section 19-1

1. False
2. True
3. True
4. False

Section 19-2

1. False
2. True
3. CMAR, CPAR, CNDTR, and CCR
4. True
5. False

Appendix A: ARM Cortex-M3 Instruction Description

Section A.1: List of ARM Cortex-M3 Instructions

ADC	Add with Carry
ADD	Add
ADR	Load PC-Relative Address
AND	Logical AND
ASR	Arithmetic Shift right
B	Branch (unconditional jump)
Bxx	Branch Conditional
BFC	Bit Field Clear
BFI	Bit Field Insert
BIC	Bit Clear
BKPT	Breakpoint
BL	Branch with Link (this is Call instruction)
BLX	Branch Indirect with Link
BX	Branch Indirect (BX LR is used for Return)
CBNZ	Compare and Branch on Non-Zero
CBZ	Compare and Branch on Zero
CDP	Coprocessor Data processing
CLREX	Clear Exclusive
CLZ	Count Leading Zero
CMN	Compare Negative
CMP	Compare
CPSID	Change processor ID and Disable Interrupt
CPSIE	Change Processor State and Enable Interrupt
DMB	Data Memory Barrier
DSB	Data Synchronization Barrier
EOR	Exclusive OR
ISB	Instruction Synchronization Barrier
IT	If-Then Condition Block

LDC   Load Coprocessor

LDM   Load Multiple registers

LDMDB   Load Multiple registers and Decrement Before each access

LDMEA   Load Multiple registers from Empty Ascending

LDMFD   Load Multiple registers Full Descending

LDMIA   Load Multiple registers and Increment after each Access

LDR   Load Register

LDR Rx, =Value   Load Register with 32-bit value

LDRB   Load Register Byte

LDRH   Load Register Halfword

LDRSB   Load Register signed Byte

LDRSH   Load Register Signed Halfword

LDRT   Load Register with Translation

LSL   Logical Shift Left

LSR   Logical Shift Right

MCR   Move to Coprocessor from ARM Register

MLA   Multiply Accumulate

MLS   Multiply and Subtract

MOV   Move (ARM7)

MOV   Move (ARM Cortex)

MOVT   Move Top

MOVW   Move 16-bit constant

MRC   Move to ARM Register from Coprocessor

MRS   Move to general Register from Special register

MSR   Move to Special register from general Register

MUL   Unsigned Multiplication

MVN   Move Negative

NOP   No Operation

ORN   Logical OR Not

ORR   Logical OR

POP   POP register from Stack

PUSH   PUSH register onto stack

RBIT	Reverse Bits
REV	Reverse byte order in a word
RV16	Reverse byte order in 16-bit
REVSH	Reverse byte order in bottom halfword and sign extend
ROR	Rotate Right
RRX	Rotate Right with extend
RSB	Reverse Subtract
SBC	Subtract with Carry (Borrow)
SBFX	Sign Bit Field extract
SDIV	Signed Divide
SEV	Send Event
SMLAL	Signed Multiply Accumulate Long
SMULL	Signed Multiply Long
SSAT	Sign Saturate
STM	Store Multiple
STMDB	Store Multiple register and Decrement Before
STMEA	Store Multiple register Empty Ascending
STMIA	Store Multiple register Empty Ascending
STMFD	Store Multiple register Full Descending
STR	Store Register
STRB	Store Register Byte
STRD	Store Register Double (two words)
STRH	Store Register Halfword
STRT	Store Register
SUB	Subtract
SUBS	Subtract
SVC	supervisor Call (Software Interrupt)
SXTB	Sign Extend byte
SXTH	Sign Extend Halfword
TBB	Table Branch Byte
TBH	Table Branch halfword
TEQ	Test Equivalence

TST	Test
UBFX	Unsigned Bit filed extract
UDIV	Unsigned Divide
UMLAL	Unsigned Multiply with Accumulate
UMULL	Unsigned Multiply Long
UXBT	Zero extend a byte
UXTH	Zero extend halfword
WFE	Wait for event
WFI	Wait for interrupt

Section A.2: ARM Instruction Description

ADC   Add with Carry

Flags: Unaffected.

Format: ADC Rd, Rn, Op2   ; Rd = Rn + Op2 + C

Function: If C = 1 prior to this instruction, then after execution of this instruction, Op2 is added to Rn plus 1 and the result is placed in Rd. If C = 0, Op2 is added to Rn plus 0. Used widely in multiword additions. After the execution the flags are not updated. The ADCS instruction updates the flags.

Example 1:
```
    LDR   R0,  =0xFFFFFFFB         ;   R0=0xFFFFFFFB
    LDR   R1,  =0xFFFFFFFF         ;   R1=0xFFFFFFFF
    MOV   R2,  #3                  ; R2=3
    MOV   R3,  #4                  ; R3=4
    ADDS  R4,  R0,  R1             ; R4=R0+R1,  C=1
    ADC   R5,  R2,  R3             ; R5=R2+R3+C=R2+R3+1
```

ADD   ADD

Flags: Unaffected

Format: ADD Rd, Rn, Op2   ; Rd = Rn + Op2

Function: Adds source operands together and places the result in destination. This will not update the flags. To update the flags, we must use ADDS.

Example 1:
```
    LDR   R0,  =0xFFFFFFFF         ;  R0=0xFFFFFFFB
    MOV   R1,  #0x5               ;  R1=0x5
    ADD   R2,  R0,  R1
      ; R2=R0+R1=0xFFFFFFFB+0x5=00000000
      ; flags unchanged
```

Example 2:

```
LDR   R0, =0xFFFFFFFF              ; R0=0xFFFFFFFF
ADD   R2, R0, #0xF1
   ; R2=R0+0xF1=R1=0xFFFFFFFF+0xF1=000000F0
   ; flags unchanged
```

ADR   Load PC-Relative Address

Flags: Unaffected:

Format: ADR Rd, label   ; Rd= address of label

Function: This allows loading into Rd register an address relative to the current PC (program counter). The label target address must be within the -4,095 to +4,096 bytes from the address in PC register. That is no farther than 1024 instructions in either direction of backward or forward.

Example:

```
      ADR   R3, MyMessage
HERE     B  HERE
MyMessage  DCB   "Hello"
```

AND   Logical AND

Flags: Unaffected

Format: AND Rd, Rn, Op2   ; Rd= Rn ANDed Op2

Function: Performs logical AND on the operands, bit by bit, storing the result in the destination. This will not update the flags. To update the flags, we must use ANDS. Notice that C flag is updated during calculation of Op2 when LSR or LSL are used.

Inputs		Output
X	Y	X AND Y
0	0	0
0	1	0
1	0	0
1	1	1

Example 1:

```
MOV   R0, #0x39   ; R0=0x39
MOV   R1, #0x0F   ; R1=0x0F
AND   R2, R1, R0  ; R2=09
         ; 39  0011 1001
         ; 0F  0000 1111
         ; --  ---------
         ; 09  0000 1001  Flags unchanged
```

Example 2:

```
MOV   R0, #0x37   ; R0=0x37
AND   R1, R0, #0x0F      ; R1 = R0 ANDed 0x0F = 07
         ; 37  0011 0111
         ; 0F  0000 1111
         ; --  ---------
```

```
                    ; 07 0000 0111  Flags unchanged
```

ASR   Arithmetic Shift right

Flags: Unaffected. Except C

Format: ASR Rd, Rm, Rn

Function: As each bit of Rm register is shifted right, the LSB is removed and the empty bits filled with the sign bit (MSB). The number of bits to be shifted right is given by Rn and the result is placed in Rd register. The flags are unchanged. To update the flags, use ASRS instruction.

Example 1:
```
    LDR   R2, =0xFFFFFF82
    ASR   R0, R2, #6  ; R0=R2 is shifted right 6 times
          ; now, R0 = 0xFFFFFFFE
```

Example 2:
```
    LDR   R0, =0x2000FF18
    MOV   R1, #12
    ASR   R2, R0, R1  ; R2=R0 is shifted right R1 number of times.
          ; now, R2 = 0x0002000F
```

Example 3:
```
    LDR   R0, =0x0000FF18
    MOV   R1, #16
    ASR   R2, R0, R1  ; R2=R0 is shifted right R1 number of times
          ; now, R2 = 0x00000000
```
ASR arithmetic shift is used for signed number shifting. ASR essentially divides Rm by a power of 2 for each bit shift.

B   Branch (unconditional jump)

Flags: Unchanged.

Format: B target   ; jump to target address

Function: This instruction is used to transfer control unconditionally to a new address. The difference between B and BL is that the BL instruction saves the address of the next instruction to LR (the link register, R14). For ARM7, the target address is calculated by (a) shifting the 24-bit signed (2's comp) offset left two bits, (b) sign-extend the result to 32-bit, and (c) add it to contents of PC (program counter). This means the target address could be within the −32M bytes to +32M bytes of address space from the current program counter. For ARM Cortex M3. the target address must be within −16MB to +16 MB address space from current instruction.

Bxx   Branch Conditional

Flags: Unaffected.

Format: Bxx target ; jump to target upon condition

Function: Used to jump to a target address if certain conditions are met. In ARM7, the target address cannot be more than −32MB to +32MB bytes away. For ARM Cortex M3. the target address must be within −16MB to +16 MB address space from current instruction. The conditions are indicated by the flag register. The conditions that determine whether the jump takes place can be categorized into three groups:

1. flag values,
2. the comparison of unsigned numbers, and
3. the comparison of signed numbers.

Each is explained next.

1. "B condition" where the condition refers to flag values. The status of each bit of the flag register has been decided by execution of instructions prior to the jump. The following "B condition" instructions check if a certain flag bit is raised or not.

Instruction		Condition
BCS	Branch if Carry Set	jump if C=1
BCC	Branch if Carry Clear	jump if C=0
BEQ	Branch if Equal	jump if Z=1
BNE	Branch if Not Equal	jump if Z=0
BMI	Branch if Minus/Negative	jump if N=1
BPL	Branch if Plus/Positive	jump if N=0
BVS	Branch if Overflow	jump if V=1
BVC	Branch if No overflow	jump if V=0

2. "B condition" where the condition refers to the comparison of unsigned numbers. After a compare (CMP Rn, Op2) instruction is executed, C and Z indicate the result of the comparison, as follows:

	C	Z
Rn > Op2	1	0
Rn = Op2	1	1
Rn < Op2	0	0

Since the operands compared are viewed as unsigned numbers, the following "B condition" instructions are used.

Instruction		Condition
BHI	Branch if Higher	jump if C=1 and Z=0
BEQ	Branch if Equal	jump if C=1 and Z=1
BLS	Branch if Lower or same	jump if C=0 or Z=1

In reality, the "CMP Rn, Op2" is a subtract instruction (Rn-Op2). After the subtraction the result is discarded and flags are changed according to the result. Notice in ARM the subtract affects the C flag setting differently from the x86 and other CPUs. See the SUB instruction.

3. "B condition" where the condition refers to the comparison of signed numbers. In the case of the signed number comparison, although the same instruction, "CMP Rn, Op2", is used, the flags used to check the result are as follows:

Rn > Op2	V=N or Z=0
Rn = Op2	Z=1
Rn < Op2	V inverse of N

Consequently, the "B condition" instructions used are different. They are as follows:

Instruction		
BGE	Branch Greater or Equal	jump if N=1 and V=1 or N=0 and V=0 (V=N)
BLT	Branch Less than	jump if N=1 and V=0 or N=0 and V=1 (N not equal to V)
BGT	Branch Greater than	jump if Z=0 and either N=1 and V=1 or N=0 and V=0 (N=V)
BLE	Branch Less or Equal	jump if Z=1 or N=1 and V=0. Or N=0 and V=1 (Z=1 or N not equal to V)
BEQ	Branch if Equal	jump if Z = 1

All "B condition" instructions are short jumps, meaning that the target address cannot be more than -32M bytes backward or +32M bytes forward from the PC of the instruction following the jump. In ARM Cortex M3 it is 16MB in each direction. What happens if a programmer needs to use a "B condition" to go to a target address beyond the -32MB to +32MB range? The solution is to use the "BX condition, Rm" since Rm can be 32-bit address and covers the entire 4GB address space of the ARM. This is shown next.

```
        LDR   R4, =MYTARGET
        ADDS  R1, R2, R3
        BXEQ  R4 ; branch to address held by R4 if Z=1
MYTRGT  SUBS  R7, #4
        NOP
        NOP
        ....
```

	C	Z	N	V
Rn > Op2	0	0	0	N
Rn = Op2	0	1	0	N
Rn < Op2	1	0	1	Inverse of N

BFC   Bit Field Clear

Flags: Unaffected.

Format: BFC Rd, #LSB, #Width

Function: Clears selected bits of Rd. The start location of the Rd bit is indicated by #LSB and must be in the range of 0–31. How many bits should be cleared is indicated by #Width and must be in the range of 1–32.

Example 1:
```
LDR   R1, =0xFFFFFFFF   ; R1=0xFFFFFFFF
BFC   R1, #2, #14       ; now R1=0xFFFF0003
```

Example 2:
```
LDR   R2, =0x999999999   ; R2=0x99999999
BFC   R2, #8, #24        ; now R2=0x00000099
```

BFI   Bit Field Insert

Flags: Unaffected.

Format: BFI Rd, Rn, #LSB, #Width

Function: Selected bits of Rn are copied to Rd. The start location of the Rd bit is indicated by #LSB and must be in the range of 0 – 31. How many bits should be copied is indicated by #Width and must be in the range of 1–32. The start bit location of Rn is always bit 0 (D0).

Example:
```
LDR   R1, =0xABCDABCD   ; R1=0xABCDABCD
LDR   R2, =0x12345678   ; R2=0x12345678
BFI   R1, R2, #4, #8    ; now R1=0xABCDA78D
```

BIC   Bit Clear

Flags: Unaffected.

Format: BIC Rd, Rn, Op2 ; Rd=Rn ANDed with NOT of Op2

Function: Selected bits of Rn are cleared and placed in Rd. The Op2 provides the bits selection. If the selected bits in Op2 are high, then corresponding bits in Rn are cleared and the result is placed in Rd. If the selected bits in Op2 are low the corresponding bits in Rn are left unchanged and the result is placed in Rd. In reality, the BIC performs the AND operation on the bits of Rn with the complement of the bits in Op2. The BIC will not update the flags. To update the flags, we must use BICS.

Inputs		Output
X	Y	X AND (NOT Y)
0	0	0
0	1	0
1	0	1
1	1	0

Example:
```
LDR   R1, =0xFFFFFF00   ; R1=0xFFFFFF00
LDR   R2, =0x99999999   ; R2=0x9999999
```

```
   BIC   R3, R2, R1          ; now R3=0x00000099
```

BKPT   Breakpoint

Flags: Unaffected.

Format: BKPT #imme_value

Function: used by compiler to insert breakpoint into programs. Upon execution of the BKPT instruction the program enters the Debug mode. See your ARM compiler for more information

BL   Branch with Link (this is Call instruction)

Flags: Unchanged.

Format: BL Subroutine_Addr   ; transfer control to a subroutine

Function: Transfers control to a subroutine. This instruction saves the address of the instruction after the BL in R14 (link register). At the end of the subroutine the control to the instruction after the BL is achieved by copying the LR (R14) register to PC. In ARM7, the target address cannot be more than −32MB to +32MB bytes away. For ARM Cortex M3, the target address must be within −16MB to +16 MB address space from current instruction.

Example:
```
  LDR   R7, =20000000
  BL DELAY   ; Call subroutine MY_DELAY
  ADD   R3, #4      ; address of this instruction is saved in R14
  . . .
  . . .
DELAY SUBS R7, #4
  NOP
  NOP
  MOV   PC, R14      ; Return, could have used "BX LR" instruction
```

BLX   Branch Indirect with Link

Flags: Unaffected.

Format: BLX Rm   ; transfer control to a subroutine whose

                            ; address is given by Rm

Function: Transfers control to a subroutine whose address is given by the Rm register. This instruction saves the address of the instruction after the BL in R14 (link register). At the end of the subroutine the control to the instruction after the BL is achieved by copying the LR (R14) register to PC. One can use "BX LR" as return instruction. Notice the difference between this instruction and "BL Target_Addr" instruction. In the "BL Target_Addr" instruction the target address of the subroutine is given right there. However, in the "BLX Rm" instruction, the target address of the subroutine is held by register Rm.

Example:
```
  ADR   R2, DELAY
```

```
   BLX   R2    ; Call subroutine pointed to by R2
   ADD   R3, #4  ; address of this instruction is saved in R14
   ...
   ...
DELAY SUBS  R1, #4
   NOP
   NOP
   BX LR   ; return
```

BX Branch Indirect (BX LR is used for Return)

Flags: Unchanged.

Format: BX Rm   ; BX LR is used for Return from a subroutine

Function: The most widely usage of this instruction is in the form of "BX LR" for the purpose of return instruction at the end of subroutine.

Example:
```
   LDR   R1, =20000000
   BL DELAY   ; Call subroutine MY_DELAY
   ADD   R3, #4  ; address of this instr. is saved in R14
   ......
DELAY SUBS  R1, #4
   NOP
   NOP
   BX    LR   ; return to caller
```

CBNZ Compare and Branch on Non-Zero

Flags: Unchanged.

Format: CBNZ Rn, Target

Function: Transfers control to the target location if Rn is not equal to zero. The Rn must be in the range of R0–R7 and target address cannot be farther than 130 bytes away from the instruction. This instruction compares the Rn with zero and jumps only if Rn is not zero. The comparison has no effect on flags. This can be used for loops in which the body of the loop is no more than 20 instructions.

Example 1:
```
   MOV   R1, #10 ; R1=10
L1 NOP
   NOP
   NOP
   SUB   R1, R1, #1  ; R1=R1-1
   CBNZ  R1, L1
```

CBZ Compare and Branch on Zero

Flags: Unaffected.

Format: CBZ Rn, Target

Function: Transfers control to the target location if Rn is zero. The Rn must be in the range of R0–R7 and target address cannot be farther than 130 bytes away from the instruction. This instruction compares the Rn with zero and jumps only if Rn is zero. The comparison has no effect on flags. This can be used to test a register value after reading a port.

Example 1:
```
   LDR  R0, =MYPORT_ADR    ; R0 = MYPORT address
HERE LDR  R2, [R0]         ; read from MYPORT
   CBZ  R2, HERE           ; keep reading MYPORT until it is zero
```

CDP   Coprocessor Data processing
See ARM Cortex-M Manual.

CLREX   Clear Exclusive
See ARM Cortex-M Manual.

CLZ   Count Leading Zero
Flags: Unchanged.

Format: CLZ Rd, Rn

Function: Scans the Rn register contents from most significant bit (D31) toward least significant bit (D0) until it find the first HIGH. The number of binary zero bits before it encounters the first binary HIGH is placed in Rd.

Example:
```
   LDR  R3, =0x01FFFFFF
   CLZ  R1, R3     ; R1=7 since there are 7 zeros before the first binary 1
```

CMN   Compare Negative
Flags: Affected: V, N, Z, C.

Format: CMN Rn, Op2   ; sets flags as if "Rn + Op2"

   ; Notice, the Rn -(-Op2)=Rn+Op2

Function: Compares Rn register value with the negative of Op2 value. This is done by Rn - (negative of Op2) which is Rn - (-Op2) = Rn + Op2. The Rn and Op2 operands are not altered. In other words, the CMN adds the Op2 to Rn (Rn+Op2) and sets the flags accordingly. This is the same as ADDS instruction except the operands are unchanged and the result is discarded. See Bxx instruction for possible cases of comparison.

CMP   Compare
Flags: Affected: V, N, Z, C.

Format: CMP Rn, Op2   ; sets flags as if "Rn-Op2"

Function: Compares two operands. The operands are not altered. Performs comparison by subtracting the Op2 operand from the Rn and updates flags as if SUBS were performed. As we can see in SUBS, the CMP perform the operation of Rn + 2's comp of Op2 and sets the flags according to the result. See Bxx instruction for possible cases of comparison.

CPSID   Change processor ID and Disable Interrupt
Flags: Unaffected

Format: CPSID iflag   ; iflag is i in PRIMASK or f in FAULTMASK

Function: Used for disabling the interrupt flags in PRIMASK or FAULTMASK registers. See ARM Cortex manual.

CPSIE   Change Processor State and Enable Interrupt
Flags: Unaffected

Format: CPSIE iflag   ; iflag is i in PRIMASK or f in FAULTMASK

Function: Used for enabling the interrupt flags in PRIMSK or FAULTMASK registers. See ARM Cortex manual.

DMB   Data Memory Barrier
Flags: Unaffected

Format: DMB

Function: It makes sure that all the explicit memory accesses prior to DMB instruction are completed before the explicit memory accesses after the DMB. See ARM Cortex manual.

DSB   Data Synchronization Barrier
Flags: Unaffected

Format: DSB

Function: It makes sure that all the explicit memory accesses prior to DSB instruction are completed before the DSB instruction is executed. See ARM Cortex manual.

EOR   Exclusive OR
Flags: Unaffected

Format: EOR Rd, Rn, Op2

Function: Performs logical Ex-OR on the Rn and Op2 operands, bit by bit, storing the result in the Rd. This will not update the flags. Use EORS instruction to updates the flags.

Inputs		Output
X	Y	X EOR Y
0	0	0
0	1	1
1	0	1
1	1	0

Example 1:
```
MOV   R0, #0xAA    ; R0=0xAA
EOR   R2, R0, #0xFF      ; now, R2=0x55
         ; AA  1010 1010
         ; FF  1111 1111
         ; --  ---------
         ; 55  0101 0101  flags unchanged
```

Example 2:
```
LDR   R0, =0xAAAAAAAA    ; R0=0xAAAAAAAA
LDR   R1, =0x55555555    ; R1=0x55555555
EOR   R2, R1, R0         ; R2=0xFFFFFFFF
           ; AA   1010 1010
           ; 55   0101 0101
           ; --   ---------
           ; FF   1111 1111    flags unchanged
```
The "EOR Rd, Rx, Rx" can be used to clear Rd.

Example 3:
```
MOV   R1, #0x55
EOR   R2, R1, R1  ; R2=0
         ; 55 0101 0101
         ; 55 0101 0101
         ; -- ---------
         ; 00 0000 0000  flags unchanged
```
To complement the bits of Rn, EX-OR it with 0xFF.

Example 4:
```
LDR   R0, =0xAAAAAAAA    ; R0=0xAAAAAAAA
LDR   R1, =0xFFFFFFFF    ; R1=0xFFFFFFFF
EOR   R2, R1, R0         ; R2=0x55555555
           ; AA   1010 1010
           ; FF   1111 1111
           ; --   ---------
           ; 55   0101 0101    flags unchanged
```

ISB   Instruction Synchronization Barrier

Flags: Unaffected.

Format: ISB

Function: It flushes the pipeline to make sure the instructions executed right after the ISB instruction are fetched fresh from the cache or memory.

IT   If-Then Condition Block
Flags: Unaffected

Format: See ARM manual

Function: It allows the execution of up to four instructions after the IT to be conditional.

LDC   Load Coprocessor
See the ARM Manual

LDM   Load Multiple registers
Flags: Unaffected.

Format: LDM Rn, {Rx, Ry, ...}

Function: Loads into registers from consecutive memory locations. The starting address of memory location is given by Rn register. The destination registers separated by comma and placed in braces. In the ARM Cortex, the stack is descending meaning that as information is pushed onto stack the stack pointer is decremented. This IA (Increment the address after each Access) is the default for loading (Poping). This instruction is widely used for Poping (loading) multiple words from descending stack into CPU registers.

Example:
```
; Assume the following memory locations with the contents:
; 12000=(46)
; 12001=(10)
; 12002=(38)
; 12003=(82)
; 12004=(56)
; 12005=(50)
; 12006=(58)
; 12007=(15)
; 12008=(63)
; 12009=(60)
; 1200A=(68)
; 1200B=(39)
; 1200C=(79)
; 1200D=(70)
; 1200E=(75)
; 1200F=(92)

LDR  R7, =0x12000
LDM  R7, {R0, R2, R4}
; now, R0=0x82381046, R2=0x15585056, ...
; the contents of memory locations 0x12000-0x12003 are
; moved to register R0, and the contents of memory
; locations 0x12004-0x12007 are moved to register
; R2, and so on. Therefore, we have R0=0x82381046,
; R2=0x15585056, and R4=0x39686063.
```

LDMDB Load Multiple registers and Decrement Before each access
Flags: Unaffected.

Format: LDMDB Rn, {Rx, Ry, ...}

Function: This is the same as LDMEA (load multiple registers from Empty Ascending) used for cases in which the stack is ascending. See LDMEA instruction.

LDMEA Load Multiple registers from Empty Ascending
Flags: Unaffected.

Format: LDMEA Rn, {Rx, Ry, ...}

Function: Loads into registers from consecutive memory locations. The starting address of memory location is given by Rn register. The destination registers separated by comma and placed in braces. In the ARM Cortex, the default for stack is descending meaning that as information are pushed onto stack the stack pointer is decremented. The IA (Increment the address after each Access) is the default. If we change the default of descending stack to ascending stack, then we have to use the EA (Empty Ascending). The ascending stack means as information are pushed onto stack the stack pointer is incremented. The LDMEA is used for Popping (loading) multiple words from ascending stack into CPU registers.

LDMFD Load Multiple registers Full Descending
Flags: Unaffected.

Format: LDMFD Rn, {Rx, Ry, ...}

Function: This is the same as LDM and LDMIA.

LDMIA Load Multiple registers and Increment after each Access
Flags: Unaffected.

Format: LDM Rn, {Rx, Ry, ...}

Function: This is the same as the LDM instructions. In the ARM Cortex, the stack is descending meaning that as information are pushed onto stack the stack pointer is decremented. This IA (Increment the address after each Access) is the default. We use this for Popping (loading) multiple words from descending stack into CPU registers.

LDR   Load Register
Flags: Unaffected.

Format: LDR Rd, [Rx]   ; load into Rd a word from memory location pointed to be Rx

Function: Loads into destination register the contents of four memory locations. The [Rx] points to address of memory location. This is widely used to load 32-bit data from memory into Rd register of the ARM since in the "MOV Rd, #immediate_value" the immediate value cannot be larger than 0xFF.

Example:

```
; Assume the following memory locations with the contents:
; 12000=(46)
; 12001=(10)
; 12002=(38)
; 12003=(82)
LDR R0, =0x12000
LDR R1, [R0]
; now, R1=82381046.
```

LDR Rx, =Value   Load Register with 32-bit value

Flags: Unaffected.

Format: LDR Rd, =32_bit_value   ; load Rd with 32-bit value

Function: Loads into destination register a 32-bit immediate value. This is widely used to load 32-bit immediate value into Rd register of the ARM since in the "MOV Rd, #immediate_value" the immediate value cannot be larger than 0xFF.

Example:

```
LDR   R0, =0x1200000     ; R0=0x1200000
LDR   R1, =0x2FFFF       ; R1=0x2FFFF
LDR   R0, =0xFFFFFFFF    ; R0=0xFFFFFFFF
LDR   R1, =200000000     ; R1=200000000
```

LDRB   Load Register Byte

Flags: Unaffected.

Format: LDRB Rd, [Rx]   ; load into Rd a byte from memory location pointed to be Rx

Function: Loads into destination register the contents of a single memory location indicated by Rx.

Example:

```
; Assume the following memory locations with the contents:
; 12000=(46)
; 12001=(10)
; 12002=(38)
; 12003=(82)
LDR   R0, =0x12000
LDRB R1, [R0]
; now, R0=00000046
```

LDRH   Load Register Halfword

Flags: Unaffected.

Format: LDRH Rd, [Rx]   ; load into Rd a 2-byte from memory location pointed to be Rx

Function: Loads into destination register the contents of the two consecutive memory locations (halfword) indicated by Rx.

Example:

```
; Assume the following memory locations with the contents:
```

502

```
; 12000=(46)
; 12001=(10)
; 12002=(38)
; 12003=(82)
LDR  R0, =0x12000
LDRH R1, [R0]
; now, R0=00001046
```

LDRSB   Load Register signed Byte

Flags: Unaffected.

Format: LDRSB Rd, [Rx]

Function: Loads into Rd register a byte from memory location pointed to by Rx and sign-extends the byte to 32-bit word. That means the sign (D7) of the byte is copied to all the upper 24 bits of the Rd register.

Example 1:
```
; Assume the following memory locations with the contents:
; 12000=(85)
; 12001=(10)
; 12002=(38)
; 12003=(82)
LDR  R0, =0x12000
LDRB R1, [R0]    ; now  R1=FFFFFF85 because MSB of 85 is 1
```

Example 2:
```
; Assume the following memory locations with the contents:
; 12000=(15)
; 12001=(20)
; 12002=(3F)
; 12003=(82)
LDR  R0, =0x12000
LDRB R1, [R0]    ; now, R1-00000015 because MSB of 15 is 0
```

LDRSH   Load Register Signed Halfword

Flags: Unaffected.

Format: LDRSH Rd, [Rx]

Function: Loads Into Rd register a half-word (2-byte) from memory location pointed to by Rx and sign-extends it to 32-bit word. That means the sign (D15) of the 16-bit operand is copied to all the upper 16 bits of the Rd register.

Example 1:
```
; Assume the following memory locations with the contents:
; 12000=(46)
; 12001=(F3)
; 12002=(38)
; 12003=(82)
LDR  R0, =0x12000
LDRB R1, [R0]    ; now, R0=FFFFF346 because MSB of F3 is 1
```

Example 2:
```
; Assume the following memory locations with the contents:
; 12000=(4F)
; 12001=(23)
; 12002=(18)
; 12003=(B2)
LDR  R0, =0x12000
LDRB R1, [R0]    ; now, R1=0000234F because MSB of 23 is 0
```

LDRT   Load Register with Translation
Flags: Unaffected

Format: LDRT Rd, [Rx]

Function: Loads into Rd register a byte from memory location pointed to by Rx and zero-extends the byte to 32-bit word. That means a zero is copied to all the upper 24 bits of the Rd register. Used for unprivileged memory access.

Example:
```
; Assume the following memory locations with the contents:
; 12000=(46)
; 12001=(10)
; 12002=(38)
; 12003=(82)
LDR  R0, =0x12000
LDRB R1, [R0]    ; now, R1=00000046
```

LSL   Logical Shift Left
Flags: Unaffected.

Format: LSL Rd, Rm, Rn

Function: As each bit of Rm register is shifted left, the MSB is removed and the empty bits are filled with zeros. The number of bits to be shifted left is given by Rn and the result is placed in Rd register. The LSL does not update the flags.

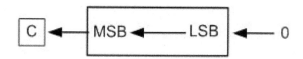

Example 1:
```
LDR  R2, =0x00000010
LSL  R0, R2, #8  ; R0=R2 is shifted left 8 times
        ; now, R0= 0x00001000, flags not changed
```

Example 2:
```
LDR  R0, =0x00000018
MOV  R1, #12
LSL  R2, R0, R1  ; R2=R0 is shifted left R1 number of times
        ; now, R2= 0x000018000, flags not changed
```

Example 3:
```
LDR   R0, =0x0000FF18
MOV   R1, #16
LSL   R2, R0, R1  ; R2=R0 is shifted left R1 number of times
        ; now, R2= 0xFF180000, flags not changed
```
The logical shift left used for unsigned number shifting. LSL essentially multiplies Rm by a power of 2 for each bit shift.

LSR   Logical Shift Right

Flags: Unaffected.

Format: LSR Rd, Rm, Rn

Function: As each bit of Rm register is shifted right, the LSB is removed and the empty bits are filled with zeros. The number of bits to be shifted left is given by Rn and the result is placed in Rd register. The LSR does not update the flags.

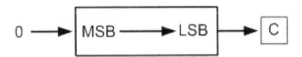

Example 1:
```
LDR   R2, =0x00001000
LSR   R0, R2, #8  ; R0=R2 is shifted right 8 times
        ; now, R0= 0x00000010, C=0
```

Example 2:
```
LDR   R0, =0x000018000
MOV   R1, #12
LSR   R2, R0, R1  ; R2=R0 is shifted right R1 number of times
        ; now, R2= 0x00000018, C=0
```

Example 3:
```
LDR   R0, =0x7F180000
MOV   R1, #16
LSR   R2, R0, R1  ; R2=R0 is shifted right R1 number of times
        ; now, R2=0x00007F18, C=0
```
The logical shift right used for shifting unsigned numbers. LSR essentially divides Rm by a power of 2 for each bit shift.

MCR   Move to Coprocessor from ARM Register
See ARM Manual.

MLA   Multiply Accumulate
Flags: Unaffected

Format: MLA Rd, Rs1, Rs2, Rs3 ; Rd= (Rs1 × Rs2) + Rs3

Function: Multiplies an unsigned word held by Rs1 by an unsigned word in Rs2 and the result is added to Rs3 and placed in Rd.

Example:
```
MOV   R0, #0x20   ; R0=0x20
MOV   R1, #0x50   ; R1=0x50
MOV   R2, #0x10   ; R2=0x10
MLA   R4, R0, R1, R2    ; now R4= (0x20 × 0x50)+10= 0xA10
```

MLS   Multiply and Subtract
Flags: Unaffected

Format: MLS Rd, Rm, Rs, Rn ; Rd= Rn -(Rs × Rm)

Function: Multiplies an unsigned word held by Rm by an unsigned word in Rs and the result is subtracted from Rn and placed in Rd.

Example:
```
MOV   R0, #0x20   ; R0=0x20
MOV   R1, #0x50   ; R1=0x50
LDR   R2, =0x1000 ; R2=0x1000
MLS   R4, R0, R1, R2    ; now R4= 0x1000-(0x20×0x50)=0x600
```

MOV   Move   (ARM7)
Flags: Unaffected.

Format: MOV Rd, #imm_value   ; Rd=imm_Value < 0x200

Function: Load the Rd register with an immediate value. The immediate value cannot be larger than 0xFF (0–255). After the execution the flags are not updated. The MOVS instruction updates the flags.

Example 1:
```
MOV   R0, #0x25   ; R0=0x25
MOV   R1, #0x5F   ; R1=0x5F
```
To load the ARM register with value larger than 0xFF we must use the "LDR Rd, = 32_bit_data." For example, we can use LDR R2, =0xFFFFFFFF.

Example 2:
```
LDR    R0, =0x2000000          ; R0=0x2000000
```

MOVT   Move Top
Flags: Unaffected.

Format: MOVT Rd, #imm_value ; imm_value < 0x10000

Function: Loads the upper 16-bit of Rd register with an immediate value. The immediate value cannot be larger than 0xFFFF (0–65535). The lower 16-bit of the Rd register remains unchanged.

Example:
```
LDR   R0, =0x25579934    ; R0=0x25579934
MOVT  R0, #0xAAAA        ; R0=0xAAAA9934
```

MOVW   Move 16-bit constant

Flags: Unaffected.

Format: MOVW Rd, #imm_value   ; imm_value < 0x10000

Function: Load the Rd register with an immediate value. The immediate value cannot be larger than 0xFFFF (0–65535).

Example:
```
MOVW R1, #0x5555 ; R1=0x5555
```
To load the ARM register with value larger than 0xFFFF we must use the "LDR Rd, = 32_bit_data." For example, we can use LDR R2, =0xFFFFFFFF.

MRC   Move to ARM Register from Coprocessor
See ARM manual

MRS   Move to general Register from Special register

Flags: Unaffected.

Format: MRS Rd, special_reg   ; copy special_reg to Rd

Function: Copies the contents of a special function register to a general-purpose register. This instruction along with the MSR is widely used to modify the special function registers such as CONTROL, PRIMASK, and ISPR. This is the only way we can access the special function registers.

Example:
```
MRS   R1, CONTROL ; R1=CONTROL
AND   R1, #0x00   ; mask the lower 8 bits
MSR   CONTROL, R1
```

MSR   Move to Special register from general Register

Flags: Unaffected.

Format: MSR special_reg, Rn   ; copy special_reg to Rn

Function: Copies the contents of a general-purpose register to special function register. This instruction along with the MRS is widely used to modify the contents of special function registers such as CONTROL, PRIMASK, and ISPR. This is the only way we can access the special function registers.

Example:
```
MRS  R1, CONTROL ; R1=CONTROL
AND  R1, #0x00   ; mask the lower 8 bits
MSR  CONTROL, R1 ; mask the lower 8 bits of CONTROL reg.
```

MUL   Unsigned Multiplication

Flags: Affected: N, Z, Unaffected: C, V

Format: MUL Rd, Rn, Rm   ; Rd = Rn × Rm

Function: Multiplies a word in register Rn by a word in register Rm and places the result in Rd.

Example 1:
```
MOV   R0, #100    ; R0=100
MOV   R1, #200    ; R1=200
MUL   R3, R0, R1  ; R3 = R0 x R1 = 100 x 200 =20000
```

Example 2:
```
LDR   R0, =10000  ; R0=10000
LDR   R1, =20000  ; R1=20000
MUL   R3, R0, R1  ; R3 = R0 x R1= 10000 x 20000 = 200000000
```

MVN   Move Negative
Flags: Unaffected.

Format: MVN Rd, Op2   ; Rd = 1's comp. of Op2

Function: Places in Rd the negation (the 1's complement) of Op2. Each bit of Op2 is inverted (logical NOT) and placed in Rd while flags remain unchanged.

Example 1:
```
MOV   R0, #0xAA   ; R0=0xAA
MVN   R2, R0  ; now, R2=0xFFFFFF55
```

Example 2:
```
LDR   R0, =0xAAAAAAAA    ; R0=0xAAAAAAAA
MVN   R1, R0      ; R1=0x55555555
```

Example 3:
```
MVN   R0, #0x0F   ; R0=0xFFFFFFF0
```

Example 4:
```
MVN   R2, #0x0   ; R0=0xFFFFFFFF widely used to load Rx with all 1s
```

NOP   No Operation
Flags: Unaffected.

Format: NOP

Function: Performs no operation. Sometimes used for timing delays to waste clock cycles. Updates PC (program counter) to point to next instruction following NOP. In some ARM CPUs, the pipeline removes the NOP before it reaches the execution stage.

ORN   Logical OR Not
Flags: Unaffected.

Format: ORN Rd, Rn, Op2   ; Rd = Rn ORed with 1's comp of Op2

Function: Performs the OR operation on the bits of Rn with the complement of the bits in Op2. The ORN will not update the flags. To update the flags, we must use ORNS.

Inputs		Output
A	B	A OR (NOT B)
0	0	1
0	1	0
1	0	1
1	1	1

Example 1:
```
LDR   R1, =0xFFFFFF00      ; R1=0xFFFFFF00
LDR   R2, =0x99999999      ; R2=0x9999999
ORN   R3, R2, R1           ; now R3=0x999999FF
```

Example 2:
```
MOV   R1, #0               ; R1=0
LDR   R0, =0xFFFFFFFF           ; R0=0xFFFFFFFF
ORN   R2, R1, R0           ; now, R2=0x0
```

ORR   Logical OR

Flags: Unaffected

Format: ORR Rd, Rn, Op2   ; Rd= Rn ORed Op2

Function: Performs logical OR on the bits of Rn and Op2, and places the result in Rd. Often used to turn a bit on. ORR will not update the flags.

Example 1:
```
MOV   R0, #0xAA   ; R0=0xAA
ORR   R2, R0, #0x55      ; now, R2=0xFF
```

Example 2:
```
LDR   R0, =0x00010203     ; R0=00010203
LDR   R1, =0x30303030
ORR   R2, R0, R1          ; R2=0x30313233
```

Example 3:
```
LDR   R0, =0x55555555     ; R0=0x55555555
LDR   R1, =0xAAAAAAAA     ; R0=0xAAAAAAAA
ORR   R2, R1, R0          ; R1=0xFFFFFFFF
```

POP   POP register from Stack

Flags: Unaffected.

Format: POP {reg_list}   ; reg_reg = words off top of stack

Function: Copies the words pointed to by the stack pointer to the registers indicated by the reg_list and increments the SP by 4, 8, 12, 16, ... depending on the number of registers in the reg_list.

Example:
```
POP   {R1}     ; POP the top word of stack to R1
POP   {R1, R4, R7}       ; POP the top 3 words of stack to R1, R4, R7
POP   {R2-R6}      ; POP the top 5 words of stack to R2-R6
```

```
POP  {R0, R5}    ; POP the top 2 words of stack to R0 and R5
POP  {R0-R7}     ; POP the top 8 words of stack to R0-R7
```
The POP instruction is synonyms for LDMIA.

PUSH   PUSH register onto stack

Flags: Unaffected.

Format: PUSH {reg_list}   ; PUSH reg_list onto stack

Function: Copies the contents of registers stated in reg_list onto the stack and decrements SP by 4, 8, 12, 16, ... depending on the number of registers in reg_list.

```
Example:
  PUSH {R1}      ; PUSH the R1 onto top of stack
  PUSH {R1, R4, R7}        ; PUSH R1, R4, R7 onto top of stack
  PUSH {R2-R6}      ; PUSH the R2, R3, R4, R5, R6 onto top of stack
  PUSH {R0, R5}     ; PUSH the R0 and R5 onto top of stack
  PUSH {R0-R7}      ; PUSH the R0 through R7 onto top of stack
```
The PUSH instruction is synonyms for STMDB.

RBIT   Reverse Bits

Flags: Unaffected.

Format: RBIT Rd, Rn   ; Reverse the bit order of Rn and place in Rd

Function: Reverses the bit position order of the 32-bit value in Rn register and place the result in Rd.

Example:
```
  MOV  R1, #0x5F
  RBIT R2, R1        ; now, R2=0xF5000000
```

REV   Reverse byte order in a word

Flags: Unaffected

Format: REV Rd, Rn   ; Reverse the byte of Rn and place it in Rd

Function: Reverses the byte position order of the 32-bit value in Rn register and places the result in Rd. This can be used to convert from little endian to big endian or from big endian to little endian.

Example:
```
  LDR  R1, =0x12345678
  REV  R2, R1        ; now, R2=0x78564312
```

RV16   Reverse byte order in 16-bit

Flags: Unaffected

Format: REV16 Rd, Rn   ; Reverse the bits if Rn and place it in Rd

Function: Reverses the 16-bit position order of the 32-bit value in Rn register and places the result in Rd. This can be used to convert 16-bit little endian to big endian or from 16-bit big endian to little endian.

Example:
```
LDR  R1, =0x559922FF
RV16 R2, R1      ; now, R2=0x22FF5599
```

REVSH Reverse byte order in bottom halfword and sign extend
Flags: Unaffected

Format: REVSH Rd, Rn ; Rd=Reverse the byte and sign extend Rn

Function: Reverses the 16-bit position order of Rn register and after sign extending to 32-bit it is placed in Rd. This can be used to convert a signed 16-bit little endian to 32-bit signed big endian or from signed 16-bit big endian to 32-bit signed little endian.

Example:
```
LDR   R1, =0x559922FF
REVSH R2, R1               ; now, R2=0x22FF5599
```

ROR   Rotate Right
Flags: Unaffected.

Format: ROR Rd, Rm, Rn ; Rd=rotate Rm right Rn bit positions

Function: As each bit of Rm register shifts from left to right, they exit from the right end (LSB) and enter from left end (MSB). The number of bits to be rotated right is given by Rn and the result is placed in Rd register. The ROR does not update the flags.

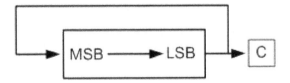

Example 1:
```
LDR   R2, =0x00000010
ROR   R0, R2, #8  ; R0=R2 is rotated right 8 times
         ; now, R0 = 0x10000000, C=0
```

Example 2:
```
LDR   R0, =0x00000018
MOV   R1, #12
ROR   R2, R0, R1  ; R2=R0 is rotated right R1 number of times
         ; now, R2 = 0x01800000, C=0
```

Example 3:
```
LDR   R0, =0x0000FF18
MOV   R1, #16
ROR   R2, R0, R1  ; R2=R0 is rotated right R1 number of times
```

```
                  ; Now, R2 = 0xFF180000, C=0
```

RRX   Rotate Right with extend
Flags: Unaffected.

Format: RRX Rd, Rm   ; Rd=rotate Rm right 1-bit position

Function: Each bit of Rm register is shifted from left to right one bit. The RRX does not update the flags.

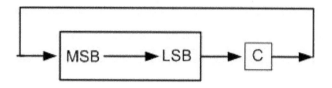

Example:
```
   LDR   R2, =0x00000002
   RRX   R0, R2  ; R0=R2 is shifted right one bit
           ; now, R0=0x00000001
```

RSB   Reverse Subtract
Flags: Unaffected

Format: RSB Rd, Rn, Op2   ; Rd = Op2 - Rn

Function: Subtracts the Rn from the Op2 and puts the result in the Rd. The RSB has no effect on flags. The steps for subtraction performed by the internal hardware of the CPU are as follows:

1. Takes the 2's complement of the Rn
2. Adds this to the Op2
3. Places the result in Rd

The Op2 and Rn operands remain unchanged by this instruction.

Example:
```
   LDR   R0, =0x55555555     ; R0=0x55555555
   LDR   R1, =0x99999999     ; R1=0x99999999
   RSB   R2, R0, R1  ; R2=R1-R0
     ; For "RSB R2, R0, R1" we have:
     ; R2=R1-R0=0x99999999 - 0x55555555 =
     ; R2=0x99999999 + 2's comp of 0x55555555
     ; R2=0x99999999 + 0xAAAAAAAB = 0x44444444
     ;    0x99999999
     ; - 0x55555555
     ;    ---------------
     ;    0x44444444
```

SBC   Subtract with Carry (Borrow)
Flags: Unaffected

Format: SBC Rd, Rn, Op2   ; Rd = Rn − Op2 − (1− C)

512

Function: Subtracts the Op2 operand from the Rn, placing the result in Rd. If C = 0, it subtracts 1 from the result; otherwise, it operates like SUB. The SBC has no effect on flags. This is used widely for multiword (64-bit) subtraction.

Example:
```
LDR  R0, =0x55555555    ; R0=0x55555555
LDR  R1, =0x99999999    ; R1=0x99999999
SUBS R2, R0, R1  ; R2=R0 - R1
MOV  R3, #0x09   ; R3=0x09
SBC  R4, R3,#03  ; R4=R3 - 0x3
     ; For SUBS we have:
     ; R2=R1 - R0 = 0x55555555 - 0x99999999 =
     ; R2=0x55555555 + 2's comp of 0x99999999
     ; R2=0x55555555 + 0x66666667 = 0xBBBBBBBC  C=0
     ; For SBC we have:
     ; R4=R3-0x3=0x09 - 0x3 -(1 - C) = 9 - 3 - 1
     ; R4= 0x9 +2'comp. of -4 = 0x9 + 0xFFFFFFFC = 0x05
     ;    0x0000000955555555
     ; - 0x0000000399999999
```

SBFX   Sign Bit Field extract

Flags: Unaffected

Format: SBFX Rd, Rn, #LSB, #Width

Function: Extracts the bit field from the Rn register and then after sign extending it is placed in Rd. The #LSB indicates which bit and #Width indicates how many bits.

Example 1:
```
LDR  R0, =0x00000543    ; R0=0x00000543
SBFX R2, R0, #8, #4     ; now, R2=0x00000005
```

Example 2:
```
LDR  R0, =0x00000C43    ; R0=0x00000C43
SBFX  R2, R0, #4, #8    ; now, R2=0xFFFFFFC4
```

SDIV   Signed Divide

Flags: Unaffected

Format: SDIV Rd, Rn, Rm   ; Rd= Rn/Rm

Function: Divides a signed integer word in Rn by another signed integer word in Rm. The quotient result is placed in Rd. If value in Rn register is not divisible by the value in Rm register, the result is rounded to zero and placed in Rd. Divide by zero causes interrupt type 3.

Example:
```
LDR  R0, =-20000 ; R0=-20000
LDR  R1, =-1000  ; R1=-1000
SDIV R2, R0, R1  ; now, R2 = -2000/-1000= 2
```

513

SEV   Send Event
Flags: Affected.

Format: SEV

Function: Sends signal to all the processors in the multiprocessors system. See the ARM Cortex manual.

SMLAL   Signed Multiply Accumulate Long
Flags: Unaffected

Format: SMLAL Rdlo, Rdhi, Rn, Rm ; Rdhi:Rdlo=(Rm × Rn) + (Rdhi:Rdlo)

Function: Multiplies signed words in Rn and Rm register, adds the 64-bit result to Rdhi:Rdlo register, and saves the final result in Rdhi:Rdlo. The Rdlo (low) and Rdhi(high) are the lower word and higher word of a 64-bit value.

Example 1:
```
LDR   R0, =0
LDR   R1, =0x23
LDR   R2, =-5000
LDR   R3, =-4000
SMLAL   R0, R1, R2, R3  ; now, R3:R2= (R3:R2)+ (R1 × R0)
        ; = 0x2300000000 + (-5000 × -4000)
        ; = 0x2300000000 + 20000000
        ; = 0x23000000 + 0x1312D00 = 0x2301312D00
        ; => R0 = 0x1312D00 and R1 = 0x23
```

SMULL   Signed Multiply Long
Flags: Unaffected

Format: SMULL Rdlo, Rdhi, Rn, Rm ; Rdhi:Rdlo = Rm × Rn

Function: Multiplies signed words in Rn and Rm register, and saves the result in Rdhi:Rdlo. The Rdl (low) and Rdh(high) are the lower word and higher word of a 64-bit value.

Example:
```
LDR   R0, =-20000 ; R0=-20000 (signed 2's comp)
LDR   R1, =-1000000      ; R0=-100000 (signed 2's comp)
SMLAL R2, R3, R0, R1     ; now, R3:R2= R1 × R0 = -20000 × -1000000 =
      ; 20000000000 =0x4A817C800 => R3 = 0x4 and
      ; R2 = 0xA817C800
```

SSAT   Sign Saturate
Flags: Unaffected.

Format: SSAT Rd, #n, Rm, shift#

Function: Used for saturation operation. See ARM Cortex manual.

STM   Store Multiple

Flags: Unaffected.

Format: STM Rn, {Rx, Ry, ...}

Function: Stores registers Rx, Ry, ... into consecutive memory locations. The starting address of memory location is given by Rn register. The source registers are separated by comma and placed in braces. In the ARM Cortex, the default stack is descending meaning that as information are pushed onto stack the stack pointer is decremented. This IA (Increment the address After each access) is the default. This instruction is widely used for Pushing (storing) multiple registers into ascending stack.

Example:
```
LDR   R7, =0x12000
LDR   R0, =0x82381046    ; R0=0x82381046
LDR   R2, =0x15585056    ; R2=0x15585056
LDR   R4, =0x39686063    ; R4=0x39686063
STM   R7, {R0, R2, R4}           ; now, R2=0x15585056, ..
; The contents of registers R0, R2, and R4 are stored into
; consecutive memory locations starting at an address given by R7.
; The R0 contents are stored into memory locations 0x12000-0x12003,
; the R2 contents are stored into memory locations 0x12004 through
; 0x12007, and so on. This is shown below.
; 12000=(46)
; 12001=(10)
; 12002=(38)
; 12003=(82)
; 12004=(56)
; 12005=(50)
; 12006=(58)
; 12007=(15)
; 12008=(63)
; 12009=(60)
; 1200A=(68)
; 1200B=(39)
```

STMDB Store Multiple register and Decrement Before

Flags: Unaffected.

Format: STMDB Rn, {Rx, Ry, ...}

Function: Stores registers Rx, Ry, ... into consecutive memory locations. The starting address of memory location is given by Rn register. The source registers are separated by comma and placed in braces. In the ARM Cortex, the default stack is descending meaning that as information are pushed onto stack the stack pointer is decremented. Since IA (Increment the address After each access) is the default we need to use DB (Decrement the address Before each access) is to overwrite the default. This instruction is widely used for Pushing (storing) multiple registers into Descending stack.

Example:
```
LDR   R7, =0x12000
LDR   R0, =0x39686063    ; R0=0x39686063
LDR   R2, =0x15585056    ; R2=0x15585056
```

```
    LDR   R4, =0x82381046    ; R4=0x82381046
    STMDB R7, {R0, R2, R4}
    ; The contents of registers R0, R2, and R4 are stored into
    ; consecutive memory locations starting at an address given by R7.
    ; The R0 contents are stored into memory locations 0x11FFF-0x11FFC,
    ; the R2 contents are stored into memory locations 0x11FFB
    ; through 0x11FF8, and so on. This is shown below.
    ; 11FF4=(46)
    ; 11FF5=(10)
    ; 11FF6=(38)
    ; 11FF7=(82)
    ; 11FF8=(56)
    ; 11FF9=(50)
    ; 11FFA=(58)
    ; 11FFB=(15)
    ; 11FFC=(63)
    ; 11FFD=(60)
    ; 11FFE=(68)
    ; 11FFF=(39)
```

STMEA Store Multiple register Empty Ascending

Flags: Unaffected.

Format: STMEA Rn, {Rx, Ry, ...}

Function: This is same as STM.

STMIA Store Multiple register Empty Ascending

Flags: Unaffected.

Format: STMIA Rn, {Rx, Ry, ...}

Function: This is same as STM.

STMFD   Store Multiple register Full Descending

Flags: Unaffected.

Format: STMFD Rn, {Rx, Ry, ...}

Function: This is another name for STMDB. The FD is for pushing onto Full Descending stacks

STR   Store Register

Flags: Unaffected.

Format: STR Rd, [Rx]   ; Store Rd into memory location pointed to be Rx

Function: Stores Rd register into four consecutive memory locations. The [Rx] points to starting address of memory location. This is widely used to store 32-bit register into memory locations.

Example:
```
    LDR   R1, =0x82381046    ; R1=0x82381046
    LDR   R0, =0x12000       ; R0=0x12000
```

```
   STR   R1, [R0]              ; now,
                   ; 12000=(46)
                   ; 12001=(10)
                   ; 12002=(38)
                   ; 12003=(82)
```

STRB   Store Register Byte

Flags: Unaffected.

Format: STRB Rd, [Rn]

Function: Stores the lowest byte of the Rd register into a single memory location indicated by Rn.

Example:
```
   LDR   R1, =0x82381046    ; R1=0x82381046
   LDR   R0, =0x12000           ; R0=0x12000
   STRB R1, [R0]               ; now, 12000=(46)
```

STRD   Store Register Double (two words)

Flags: Unaffected.

Format: STRD Rd, [Rn]

Function: Stores two registers of Rd and Rd+1 into 8 consecutive memory locations indicated by Rn. Rd can be R0, R2, R4, R6, R8, R10, or R12.

Example:
```
   LDR   R2, =0x12000
   LDR   R0, =0x82381046    ; R0=0x82381046
   LDR   R1, =0x15585056    ; R1=0x15585056
   STRD R0, R1, [R2] ; store R0 and R1 into memory locations starting
           ; at an address given by R2. Now, we have:
   ; 12000=(46)
   ; 12001=(10)
   ; 12002=(38)
   ; 12003=(82)
   ; 12004=(56)
   ; 12005=(50)
   ; 12006=(58)
   ; 12007=(15)
```

STRH   Store Register Halfword

Flags: Unaffected.

Format: STRH Rd, [Rn]

Function: Stores the lower 2 bytes of the Rd register into two consecutive memory locations indicated by Rn.

Example:
```
   LDR   R1, =0x82381046    ; R1=0x82381046
   LDR   R0, =0x12000       ; R0=0x12000
```

```
    STRB  R1, [R0]              ; now, 12000=(46), and  12001=(10)
```

STRT   Store Register

Flags: Unaffected

Format: STRT Rx, [Rn]

Function: Stores Rx register into memory location pointed to by Rx. This is the same as STR but is used for unprivileged memory access. See ARM Cortex manual.

Example:
```
    LDR   R1, =0x82381046    ; R1=0x82381046
    LDR   R0, =0x12000       ; R0=0x12000
    STRT  R1, [R0]
                ; now, 12000=(0x82381046)
```

SUB   Subtract

Flags: Unaffected

Format: SUB Rd, Rn, Op2   ; Rd = Rn – Op2

Function: Subtracts the Op2 from the Rn and puts the result in the Rd. Has no effect on flags. The steps for subtraction performed by the internal hardware of the CPU are as follows:

1. Takes the 2's complement of the Op2
2. Adds this to the Rn
3. Place the result in the Rd

The Rd and Op2 operands remain unchanged by this instruction.

Example:
```
    LDR   R0, =0x55555555    ; R0=0x55555555
    LDR   R1, =0x99999999    ; R1=0x99999999
    SUB   R2, R1, R0         ; R2=R1-R0
            ; For "SUB R2, R1, R0" we have:
            ; R2=R1-R0=0x99999999 - 0x55555555 =
            ; R2=0x99999999 + 2's comp of 0x55555555
            ; R2=0x99999999 + 0xAAAAAAAB = 0x44444444
            ;    0x99999999
            ; -  0x55555555
            ;    ---------------
            ;    0x44444444
```

SVC   supervisor Call (Software Interrupt)

Flags: Unaffected.

Format: SVC #imm_value

Function: It is used by application software to get services from operating systems (OS). This is like the SWI (software interrupt) instruction in ARM7.

SXTB   Sign Extend byte

Flags: Unaffected.

Format: SXTB Rd, Rm

Function: Converts a signed byte in Rm into a signed word by copying the sign bit (D7) of Rm into all the bits of Rd. Used widely to convert a signed byte in Rm to a signed word to avoid the overflow problem in signed number arithmetic.

Example:
```
  MOV  R1, #0xFB   ; R1=0xFB which is 2's complement of -5
  SXTB R0, R1  ; now, R0=0xFFFFFFFB
           ; R1= 0000 0000 0000 0000 0000 0000 1111 1011
           ; now R0=0xFFFFFFFB
           ; R0 = 1111 1111 1111 1111 1111 1111 1111 1011
```

SXTH   Sign Extend Halfword

Flags: Unaffected.

Format: SXTH Rd, Rm

Function: Converts a signed halfword in Rm into a signed word by copying the sign bit (D15) of Rm into all the bits of Rd. Used widely to convert a signed halfword (16-bit) in Rm to a signed word.

Example:
```
  ; assume R1=0xFFFB (which is -5)
  SXTH  R0, R1 ; now, R0=0xFFFFFFFB
  ; R1 = 0000 0000 0000 0000 1111 1111 1111 1011
  ; R0 = 1111 1111 1111 1111 1111 1111 1111 1011
```

TBB   Table Branch Byte

Flags: Unaffected.

Format: TBB [Rn, Rm]

Function: Branches forward using table of single byte offset using PC-relative addressing mode. Rn has starting address of the table and Rm is an index into the table. See ARM Cortex M3 manual.

TBH   Table Branch halfword

Flags: Unaffected.

Format: TBH [Rn, Rm, LSL #1]

Function: Branches forward using table of halfword offset using PC-relative addressing mode. Rn has starting address of the table and Rm is an index into the table. The "LSL # 1" shifts left the address once to make it halfword aligned address. See ARM Cortex M3 manual.

TEQ   Test Equivalence

Flags: Affected: N and Z

Format: TEQ Rn, Op2   ; performs Rn Ex-OR Op2

Function: Performs a bitwise logical Ex-OR on Rn and Op2, setting flags but leaving the contents of both Rn and Op2 unchanged. While the EORS instruction changes the contents of the destination and the flag bits, the TEQ instruction changes only the flag bits. This is widely used to see if two registers are equal.

Example 1:
```
   TEQ   R1, R2  ; check to see if R1=R2. If so Z=1. R1 and R2
              ; remain unchanged
```

Example 2:
```
   TEQ   R2, #0x01   ; check to see if D0 of R2 is 1, if so Z=1. R2
              ; remains unchanged
```

Example 3:
```
   TEQ   R1, #0xFF   ; check to see if D7_D0 of R1 are 1s,
              ; if so Z=1. R1 remains unchanged
```

TST   Test

Flags: Affected: N and Z

Format: TST Rn, Op2   ; performs Rn AND Op2

Function: Performs a bitwise logical AND on Rn and Op2, setting flags but leaving the contents of both Rn and Op2 unchanged. While the ANDS instruction changes the contents of the destination and the flag bits, the TST instruction changes only the flag bits. To test whether a bit of Rn is 0 or 1, use the TST instruction with an Op2 constant that has that bit set to 1 and all other bits cleared to 0.

Example 1:
```
   TST   R1, #0x01   ; check to see if D0 of R1 is zero, if so Z=1.
              ; R1 remain unchanged
```

Example 2:
```
   TST   R1, #0xFF   ; check to see if any bits of R1 is zero, if so
              ; Z=1. R1 remain unchanged
```

UBFX   Unsigned Bit field extract
Flags: Unaffected.

Format: UBFX Rd, Rn, #LSB, #Width

Function: Extracts the bit field from the Rn register and then zero extends it and places in Rd. The #LSB indicates from which bit and #Width indicates how many bits.

Example 1:
```
   LDR   R0, =0x00077555   ; R0=0x00077555
   UBFX R2, R0, #8, #4     ; now, R2=0x00000005
```

Example 2:
```
LDR   R0, =0x12345678   ; R0=0x12345678
UBFX R2, R0, #8, #12     ; now, R2=0x00000456
```

UDIV   Unsigned Divide

Flags: Unaffected

Format: UDIV Rd, Rn, Rm   ; Rd= Rn/Rm

Function: Divides an unsigned integer word in Rn by another unsigned integer word in Rm. The quotient result is placed in Rd. If value in Rn register is not divisible by the value in Rm register, the result is rounded to zero and placed in Rd. Divide by zero causes exception interrupt.

Example 1:
```
LDR   R0, =100    ; R0=100
LDR   R1, =2000
UDIV R2, R1, R0  ; now, R2=R1/R0=2000/100=20
```

Example 2:
```
LDR   R0, =20000  ; R0=20000
UDIV R2, R0, #100         ; now, R2=2000/100=20
```

UMLAL Unsigned Multiply with Accumulate

Flags: Unaffected

Format: UMLAL RdLo, RdHi, Rn, Rm ; RdHi:RdLo=(Rm × Rn) + (RdHi:RdLo)

Function: Multiplies unsigned words in Rn and Rm register, adds the 64-bit result to RdHi:RdLo registers, and saves the final result in RdHi:RdLo. The RdLo (low) and RdHi(high) are the unsigned lower word and higher word of the 64-bit value.

Example:
```
LDR   R0, =20000  ; R0=20000
LDR   R1, =1000
LDR   R2, =5000
LDR   R3, =4000
UMLAL R2, R3, R0, R1      ; now, R3:R2= R1 × R0 + R3:R2
```

UMULL Unsigned Multiply Long

Flags: Unaffected

Format: UMULL RdLo, RdHi, Rn, Rm ; RdHi:RdLo = Rm × Rn

Function: Multiplies unsigned words in Rn and Rm registers, and saves the result in RdHi:RdLo. The RdLo (low) and RdHi(high) are the lower word and higher word of a 64-bit value.

Example:
```
LDR   R0, =20000  ; R0=20000
LDR   R1, =10000  ; R1=10000
LDR   R2, =50000  ; R2=50000
```

```
LDR   R3, =40000   ; R3=40000
UMULL R2, R3, R0, R1    ; now, R3:R2= R1 × R0
```

UXBT   Zero extend a byte

Flags: Unaffected

Format: UXBT Rd, Rm

Function: Zero extends a byte in Rm and places in Rd. Used widely to convert a byte in Rm to word for signed number operations.

Example:
```
MOV   R1, #0xFB   ; R1=0xFB
UXBT  R0, R1  ; now, R0=0x00000000FB
      ; R1= 0000 0000 0000 0000 0000 0000 1111 1011
      ; now R0=0x000000FB
      ; R0 = 0000 00000 0000 0000 0000 0000 1111 1011
```

UXTH   Unsigned zero extend halfword

Flags: Unaffected

Format: UXTH Rd, Rm

Function: Zero extends a halfword in Rm and places in Rd. It extracts bits 0-15 and fills bits 16-31 with zeros.

Example:
```
         ; assume R1=0x1234FFFB
  UXTH   R0, R1  ; now, R0=0x00000FFFB
```

WFE   Wait for event

Flags: Unaffected

Format: WFE

Function: Used by power management. See ARM Cortex M3 manual.

WFI   Wait for interrupt

Flags: Unaffected

Format: WFI

Function: Suspends execution until one of the following events occurs:

   1.   a non-masked interrupt occurs and is taken,
   2.   an interrupt masked by PRIMASK becomes pending,
   3.   a Debug Entry request.

See ARM Cortex manual.

Appendix B: ARM Assembler Directives

Section B.1: List of ARM Assembler Directives

ALIGN

AREA

DCB directive (define constant byte)

DCD directive (define constant word)

DCW directive (define constant half-word)

ENDP or ENDFUNC

ENTRY

EQU (Equate)

EXPORT or GLOBAL

EXTRN (External)

FUNCTION or PROC

INCLUDE

RN (equate)

Section B.2: Description of ARM Assembler Directives

Directives, or as they are sometimes called, pseudo-ops or pseudo-instructions, are used by the assembler to translate Assembly language programs into machine language. Unlike the microprocessor's instructions, directives do not generate any opcode; therefore, no memory locations are occupied by directives in the final hex version of the assembly program. To summarize, directives give directions to the assembler program to tell it how to generate the machine code; instructions are assembled into machine code to give instructions to the CPU at execution time. The following are descriptions of the some of the most widely used directives for the ARM assembler. They are given in alphabetical order for ease of reference.

ALIGN
Format:

```
ALIGN   n ; n is any power of 2 from 2^0 to 2^31
```
This is used to make sure data is aligned in 32-bit word or 16-bit half word memory address. If n is not specified, ALIGN sets the current location to the next word (four byte) boundary. The following uses ALIGN to make the data word and half word aligned:

```
ALIGN   4     ; The next instruction is word (4 bytes) aligned
ALIGN         ; The next instruction is word (4 bytes) aligned
ALIGN   2     ; The next instruction is half word (2 bytes) aligned
```

Notice that, this ALIGN directive should not be confused with the ALIGN attribute of the AREA directive.

AREA
Format:

```
AREA    sectionname    attribute, attribute, ...
```

The AREA directive tells the assembler to define a new section of memory. The memory can be code or data and can have attributes such as ReadOnly, ReadWrite, and so on. This is widely used to define one or more blocks of indivisible memory for code or data to be used by the linker. Every assembly language program has at least one AREA.

The following line defines a new area named MY_ASM_PROG1 which has CODE and READONLY attributes:

```
AREA  MY_ASM_PROG1    CODE, READONLY
```

Among widely used attributes are CODE, DATA, READONLY, READWRITE, COMMON, and ALIGN. The following describes these widely used attributes.

CODE is an attribute given to an area of memory used for executable machine instruction. Since it is used for code section of the program it is by default READONLY memory. In ARM Assembly language we use this area to write our instructions.

DATA is an attribute given to an area of memory used for data and no instruction (machine instructions) can be placed in this area. Since it is used for data section of the program it is by default a READWRITE memory. In ARM Assembly language we use this area to set aside SRAM memory for scratch pad and stack.

READWRITE is an attribute given to an area of memory which can be read from and written to. Since it is READWRITE section of the program it is by default for DATA. In ARM Assembly language we use this area to set aside SRAM memory for scratch pad and stack.

READONLY is an attribute given to an area of memory which can only be read from. Since it is READONLY section of the program it is by default for CODE. In ARM Assembly language we use this area to write our instructions for machine code execution.

COMMON is an attribute given to an area of DATA memory section which can be used commonly by several program codes. We do not initialize the COMMON section of the memory since it is used by compiler exclusively. The compiler initializes the COMMON memory area with all zeros.

ALIGN is another attribute given to an area of memory to indicate how memory should be allocated according to the addresses. When the ALIGN is used for CODE and READONLY it aligned in 4-bytes address boundary by default since the ARM instructions are all 32-bit (4-bytes) word. The ALIGN attribute of AREA has a number after like ALIGN=3 which indicates the information should be placed in memory with addresses of 2^3, that is 0x50000, 0x50008, 0x50010, 0x50020, and so on. This ALIGN attribute of the AREA should not be confused with the ALIGN directive.

DCB directive (define constant byte)
Format:

```
label   DCB  n       ; n between -128 to 256, byte or string
```
The DCB directive allocates a byte size memory and initializes the values for reading only.

```
MYVALUE DCB  5              ; MYVALUE = 5
MYMSAGE DCB  "HELLO WORLD"; string
```

DCD directive (define constant word)
Format:

```
label      DCD  n
```
The DCD directive allocates a word size memory and initializes the values for reading only. The data is 32-bit aligned.

```
MYDATA  DCD  0x200000, 0xF30F5, 5000000, 0xFFFF9CD7
```

DCW directive (define constant half-word)
Format:

```
label      DCB  n
```
The DCW directive allocates a half-word size memory and initializes the values for reading only.

```
MYDATA  DCW  0x20, 0xF230, 5000, 0x9CD7
```

END
The END directive tells the assembler that it has reached the end of the program (not the end of the source file). All the text beyond the END directive is ignored by the assembler.

ENDP or ENDFUNC
The ENDFUNC or ENDP directive informs the assembler that it has reached the end of a function. ENDFUNC and ENDP are the same. See FUNCTION or PROC directives.

ENTRY
The ENTRY directive shows the entry point of a program to the assembler. Each program must have one entry point. *The newer versions of ARM Linker have alternative methods of specifying the program entry point that overwrite this directive.*

EQU (Equate)
To assign a fixed value to a name, one uses the EQU directive. The assembler will replace each occurrence of the name with the value assigned to it.

```
DATA1 EQU   0x39      ; the way to define hex value
PORTB EQU   0xF0018000      ; SFR Port B address
SUM1  EQU   0x40000120      ; assign RAM location to SUM1
```
Unlike data directives such as DCB, DCD, and so on, EQU does not assign any memory storage; therefore, it can be defined at any time and at any place, and can even be used within the code segment.

EXPORT or GLOBAL

To inform the assembler that a name or symbol will be referenced by other modules (in other files), it is marked by the EXPORT or GLOBAL directives. If a module is referencing a name outside itself, that name must be declared as EXTRN (or IMPORT). Correspondingly, in the module where the variable is defined, that variable must be declared as EXPORT or GLOBAL in order to allow it to be referenced by other modules. See the EXTRN directive for examples of the use of both EXTRN and EXPORT.

EXTRN (External)

The EXTRN directive is used to indicate that certain variables and names used in a module are defined by another module. In the absence of the EXTRN directive, the assembler would search for the definition and give an error when it couldn't find it. The format of this directive is:

```
EXTRN    name
```
The following example shows how the EXPORT and EXTERN directives are used:

```
; from the main program:
EXTRN MY_FUNC
...
BL MY_FUNC
...
; ---------------------------------------
; MY_FUNC is located in a different file:
AREA  OUR_EXAMPLE,CODE,READONLY
EXTRN    DATA1
EXPORT    MY_FUNC
MY_FUNC FUNCTION
...
LDR   R1,=DATA1
...
ENDFUNC
```

Notice that the EXTRN directive is used in the main procedure to show that MY_FUNC is defined in another module. This is needed because MY_FUNC is not defined in that module. Correspondingly, MY_FUNC is defined as GLOABAL in the module where it is defined. EXTRN is used in the MY_FUNC module to declare that operand DATA1 has been defined in another module. Correspondingly, DATA1 is declared as GLOBAL in the calling module.

FUNCTION or PROC

Often, a group of Assembly language instructions will be combined into a procedure so that it can be called by another module. The FUNCTION and ENDFUNC directives are used to indicate the beginning and end of the procedure. *Some versions of Keil uVision debugger require the code to be enclosed in PROC/ENDP pair to be single stepped.* See the following example:

```
MY_FUNC FUNCTION
    ...
    ...
    ENDFUNC
```

INCLUDE

When there is a group of macros written and saved in a separate file, the INCLUDE directive can be used to bring them into another file.

RN (Register Naming)

This is used to define a name for a register. The RN directive does not set aside a separate storage for the name, but associates a register with that name. The following code shows how we use RN:

```
VAL1  RN R1 ; define VAL1 as a name for R1
VAL2  RN R2 ; define VAL2 as a name for R2
SUM   RN R3 ; define SUM as a name for R3

   AREA    PROG_2_1, CODE, READONLY
   ENTRY
   MOV   VAL1, #0x25         ; R1 = 0x25
   MOV   VAL2, #0x34         ; R2 = 0x34
   ADD   SUM, VAL1, VAL2     ; add R2 to R1 and place it in R3
HERE  B  HERE
   END
```

Appendix C: Macros

What is a macro and how is it used?

There are applications in Assembly language programming where a group of instructions performs a task that is used repeatedly. For example, you might need to add three registers together. So, it does not make sense to rewrite them every time they are needed. Therefore, to reduce the time that it takes to write these codes and reduce the possibility of errors, the concept of macros was born. Macros allow the programmer to write the task (set of codes to perform a specific job) once only and to invoke it whenever it is needed, wherever it is needed.

MACRO definition

Every macro definition must have three parts, as follows:

```
      MACRO
[$label]        macroName parameter1, parameter2, ..., parameterN
   ...   ...
   ...   ...
      MEND
```

The MACRO directive indicates the beginning of the macro definition and the MEND directive signals the end. What goes in between the MACRO and MEND directives is called the body of the macro. The name must be unique and must follow Assembly language naming conventions. The parameters are names, or parameters, or even registers that are mentioned in the body of the macro. After the macro has been written, it can be invoked (or called) by its name, and appropriate values are substituted for parameters. For example, you might want to have an instruction that adds three registers. The following is a macro for the purpose:

```
      MACRO
ADD3VAL     $DEST, $ARG1, $ARG2, $ARG3
ADD       $DEST, $ARG1, $ARG2
ADD       $DEST, $DEST, $ARG3
      MEND
```

The above code is the macro definition. Note that parameters $DEST, $ARG1, $ARG2, and $ARG3 are mentioned in the body of the macro. To distinguish parameters, they must start with $. In the following example, the macro is invoked by its name with the user's actual data:

```
AREA OURCODE, READONLY, CODE
MOV   R1, #5
MOV   R2, #2
ADD3VAL     R0, R1, R2, #5
```

The instruction "ADD3VAL R0, R1, R2, #5" invokes the macro.

The assembler expands the macro by providing the following code in the .LST file:

```
3 00000008 E0810002        ADD     R0, R1, R2
4 0000000C E2800005        ADD     R0, R0, #5
```

Default Values for parameters

We can define default values for parameters as shown below:

```
MACRO
ADD3VAL    $DEST, $ARG1=R3, $ARG2, $ARG3=#5
ADD     $DEST, $ARG1, $ARG2
ADD     $DEST, $DEST, $ARG3
MEND
```

To use the default value, we put a '|' instead of the parameter while invoking the macro:

```
ADD3VAL    R0, R1, R2, |
```

The above code uses the default value of $ARG3 which is set to #5.

Using labels in macros

In the discussion of macros so far, examples have been chosen that do not have a label or name in the body of the macro. This is because if a macro is expanded more than once in a program and there is a label in the label field of the body of the macro, the same label would be generated more than once and an assembler error would be generated. To address the problem, we can give a unique label to the macro when we invoke it, as shown below:

```
    MACRO
$lbl   OUR_MACRO
    CMP  R1, #5
    BEQ  $lbl
    MOV  R1, #1
$lbl
    MEND


    AREA   OURCODE, READONLY, CODE
    ENTRY
    MOV R1, #3
label1 OUR_MACRO
    MOV  R1, #5
label2 OUR_MACRO
HERE   B   HERE
```

The assembler expands the macro by providing the following code in the .LST file:

```
20 00000000                    AREA OURCODE, READONLY, CODE
21
22 00000000 E3A01003           MOV  R1, #3
23 00000004          label1    OUR_MACRO
 3 00000004 E3510005           CMP  R1, #5
 4 00000008 0A000000           BEQ  label1
 5 0000000C E3A01001           MOV  R1, #1
 6 00000010          label1
24 00000010 E3A01005           MOV  R1, #5
25 00000014          label2    OUR_MACRO
 3 00000014 E3510005           CMP  R1, #5
 4 00000018 0A000000           BEQ label2
 5 0000001C E3A01001           MOV R1, #1
 6 00000020          label2
26 00000020 EAFFFFFE
                     HERE   B   HERE
```

In cases that there is more than one label in a macro the lines can be labeled as shown below:

```
    MACRO
$lbl    OUR_MACRO
```

```
        CMP   R1, #5
        BEQ   $lbl.equal
        MOV   R1, #1
        B  $lbl.next
$lbl.equal
        MOV   R1, #2
$lbl.next
        MEND


        AREA  OURCODE, READONLY, CODE

        MOV   R1, #3
label1  OUR_MACRO
        MOV   R1, #5
label2  OUR_MACRO
HERE  B  HERE
```

The assembler expands the macro by providing the following code in the .LST file:

```
13 00000000                     AREA OURCODE, READONLY, CODE
14
15 00000000 E3A01003            MOV R1, #3
16 00000004            label1   OUR_MACRO
 3 00000004 E3510005            CMP R1, #5
 4 00000008 0A000001            BEQ label1equal
 5 0000000C E3A01001            MOV R1, #1
 6 00000010 EA000000            B   label1next
 7 00000014            label1equal
 8 00000014 E3A01002            MOV R1, #2
 9 00000018            label1next
17 00000018 E3A01005            MOV R1, #5
18 0000001C            label2   OUR_MACRO
 3 0000001C E3510005            CMP R1, #5
 4 00000020 0A000001            BEQ label2equal
 5 00000024 E3A01001            MOV R1, #1
 6 00000028 EA000000            B   label2next
 7 0000002C            label2equal
 8 0000002C E3A01002            MOV R1, #2
 9 00000030            label2next
19 00000030 EAFFFFFE
                      HERE    B   HERE
```

Conditional macros

We can pass condition into macros, as well:

```
    MACRO
$lbl   OurMacro$cond
    CMP  R1, #5
    B$cond      $lbl.equal
    MOV    R1, #1
$lbl.equal
    MEND

    AREA  OURCODE, READONLY, CODE

    MOV   R1, #3
label1  OurMacroEQ  ; in the macro check equality
    MOV   R1, #3
```

```
label2  OurMacroLO  ; in the macro check if is lower
HERE   B HERE
```

The assembler expands the macro by providing the following code in the .LST file:

```
10 00000000                    AREA OURCODE, READONLY, CODE
11
12 00000000 E3A01003           MOV   R1, #3
13 00000004         label1   OurMacroEQ
 3 00000004 E3510005           CMP   R1, #5
 4 00000008 0A000000           BEQ   label1equal
 5 0000000C E3A01001           MOV   R1, #1
 6 00000010         label1equal
14 00000010 E3A01003           MOV   R1, #3
15 00000014         label2   OurMacroLO
 3 00000014 E3510005           CMP   R1, #5
 4 00000018 3A000000           BLO   label2equal
 5 0000001C E3A01001           MOV   R1, #1
 6 00000020         label2equal
16 00000020 EAFFFFFE
                    HERE          B     HERE
```

Notice that the first B$cond is substituted with BEQ while the second B$cond is substituted with BLO since the conditions EQ and LO are used respectively.

INCLUDE directive

Assume that there are several macros that are used in every program. Must they be rewritten every time? The answer is no if the concept of the INCLUDE directive is known. The INCLUDE directive allows a programmer to write macros and save them in a file, and later bring them into any file. For example, assuming that some widely used macros were written and then saved under the filename "MYMACRO1.S", the INCLUDE directive can be used to bring this file into any ".asm" file and then the program can call upon any of the macros as many times as needed. In the following example the ADD3VAL macro is defined in the MyMACRO.s file and it is used in the example.asm file.

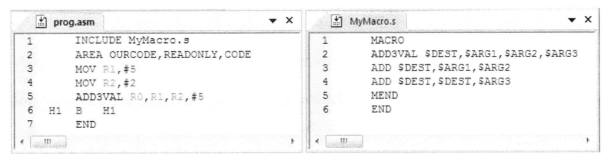

Figure C. 1: Defining a Macro in an Include File

Macros vs. subroutines

Macros and subroutines are useful in writing assembly programs, but each has limitations. Macros increase code size every time they are invoked. For example, if you call a 10-instruction macro 10 times, the code size is increased by 100 instructions; whereas, if you call the same subroutine 10 times, the code size is only that of the subroutine instructions. On the other hand, a function call takes 3 clocks and the return instruction takes 3 clocks to get executed. So, using functions adds around 6 clock cycles. The subroutines might use stack space as well when called, while the macros do not.

Appendix D: Passing Arguments into Functions

There are different ways to pass arguments (parameters) to functions. Some of them are:

- through registers
- through memory using references
- using stack

D.1: Passing arguments through registers

In the following program the BIGGER function gets two values through R0 and R1. After comparing R0 and R1, it returns the bigger value through R0.

Program D-1

```
        EXPORT  __main
        AREA    OUR_PROG, CODE, READONLY
__main
        MOV     R0, #5       ; R0 = 5
        MOV     R1, #7       ; R1 = 7
        BL      BIGGER       ; BIGGER(5, 7)
HERE    B       HERE   ; stay here


        ; =======================================
        ; BIGGER returns the bigger value
        ; Parameters:
        ;     R0 and R1: the values to be compared
        ; Returns:
        ;     R0: containing the bigger value
        ; =======================================
BIGGER
        CMP     R0, R1
        BHI     L1           ; if R0 > R1 go to L1
        MOV     R0, R1       ; R0 = R1
L1      BX      LR           ; return
        END
```

This is a fast way of passing arguments to the function.

D.2: Passing through memory using references

We can store the data in memory and pass its address through a register. In the following program the STR_LENGTH function gets the address of a zero-ended string through R0 and returns the length of the string through R1.

Program D-2

```
        EXPORT  __main
        AREA    OUR_PROG, CODE, READONLY
```

```
__main
        ADR    R0, OUR_STR ; R0 = addr. of OUR_STR
        BL     STR_LENGTH  ; STR_LENGTH(&OUR_STR)
HERE    B      HERE   ; stay here

OUR_STR DCB "HELLO!"
        ALIGN 4
        ; ======================================
        ; STR_LENGTH returns the length of string
        ; Parameters:
        ;     R0: address of the string
        ; Returns:
        ;     R0: the length of string
        ; ======================================
STR_LENGTH
MOV     R1, R0               ; move string pointer to R1
        MOV    R0, #0            ; use R0 as string length counter
L_BEGIN
        LDRB   R2, [R1]          ; fetch a character from string
        CMP    R2, #0
        BXEQ   LR            ; return if character is null (end of string)
        ADD    R1, R1, #1    ; point to next character in string
        ADD    R0, R0, #1       ; increment the counter
        B      L_BEGIN

        END
```

D.3: Passing arguments through stack

Passing through the stack is a flexible way of passing arguments. To do so, the arguments are pushed onto the stack just before calling the function and popped off after returning. In Program D-3, the BIGGER function gets two arguments through the stack and returns the bigger value in R0.

Program D-3

```
        EXPORT __main
        AREA   OUR_PROG, CODE, READONLY
  main
        MOV    R0, #5
        PUSH   {R0}       ; push Arg1
        MOV    R0, #7
        PUSH   {R0}       ; push Arg2

        BL     BIGGER     ; BIGGER(5, 7)
        ADD    SP, SP, #8  ; adjust the stack pointer to remove the arguments

HERE    B      HERE       ; stay here

        ; ======================================
```

```
; BIGGER returns the bigger value
; Parameters:
;     values to be compared on stack
; Returns:
;     R0: the bigger value
; ======================================
BIGGER
        LDR    R0, [SP, #8]      ; R0 = arg1
        LDR    R1, [SP, #4]      ; R1 = arg2

        CMP    R0, R1
        MOVLO  R0, R1            ; if R0 < R1 move R1 into R0
L1      BX     LR                ; return

        END
```

This method of passing arguments is used in x86 computers because they have very few general-purpose registers. In ARM CPU, the arguments are passed in the first four registers if there are four or fewer arguments. If there are more than four arguments, the first four are passed in the first four registers and the rest are passed on the stack.

It is important to remember that after returning from the call, the caller must clear the arguments on the stack.

D.4: AAPCS (ARM Application Procedure Call Standard)

The AAPCS provides a standard for implementing the functions and the function calls so that the codes made by different compilers and different programmers can work with each other. Some of the rules of the standard are:

- The arguments must be sent through R0 to R3. Each register cannot hold more than one argument. If there more than four words are needed, the first four words are sent in R0 to R3, the rest are passed on the stack.

- The return value must be returned in R0 (and R1 if the return value is 64-bit).

- The functions can use R4 to R8, R10 and R11 for temporary storage (Thumb code can only use R4 to R7). But their values must be saved upon entering the function and restored before returning. To do so, we push the registers before using them and pop them before returning from the function.

- The stack must be used as Full Descending

In Program D-4 the above rules are considered.

```
        EXPORT  __main
        AREA   OUR_PROG, CODE, READONLY
__main

        MOV    R0, #20
        BL     DELAY  ; DELAY(20)
HERE    B      HERE   ; stay here


        ; =======================================
        ; DELAY waits for a while
        ; Parameters:
        ;     R0: the amount of wait
        ; Returns:
        ;     none
        ; =======================================
DELAY
        CMP    R0, #0
        BXEQ   LR            ; return if zero

        PUSH   {R5}          ; save R5

        LDR    R5, =5000000     ; R5 = 5000000
L1      SUBS   R5, R5, #1  ; R5=R5-1
        BNE    L1            ; go to L1 if R5 is not zero

        POP    {R5}          ; restore R5
        BX     LR            ; return

        END
```

More information

For more information about AAPCS see the following article or search "AAPCS" on the Internet:

http://infocenter.arm.com/help/topic/com.arm.doc.ihi0042e/IHI0042E_aapcs.pdf

Appendix E: ASCII Codes

Dec	Hex	Ch		Dec	Hex	Ch		Dec	Hex	Ch		Dec	Hex	Ch
0	00			32	20			64	40	@		96	60	`
1	01	☺		33	21	!		65	41	A		97	61	a
2	02	☻		34	22	"		66	42	B		98	62	b
3	03	♥		35	23	#		67	43	C		99	63	c
4	04	♦		36	24	$		68	44	D		100	64	d
5	05	♣		37	25	%		69	45	E		101	65	e
6	06	♠		38	26	&		70	46	F		102	66	f
7	07	•		39	27	'		71	47	G		103	67	g
8	08	◘		40	28	(		72	48	H		104	68	h
9	09	○		41	29	)		73	49	I		105	69	i
10	0A	◙		42	2A	*		74	4A	J		106	6A	j
11	0B	♂		43	2B	+		75	4B	K		107	6B	k
12	0C	♀		44	2C	,		76	4C	L		108	6C	l
13	0D	♪		45	2D	–		77	4D	M		109	6D	m
14	0E	♫		46	2E	.		78	4E	N		110	6E	n
15	0F	☼		47	2F	/		79	4F	O		111	6F	o
16	10	►		48	30	0		80	50	P		112	70	p
17	11	◄		49	31	1		81	51	Q		113	71	q
18	12	↕		50	32	2		82	52	R		114	72	r
19	13	‼		51	33	3		83	53	S		115	73	s
20	14	¶		52	34	4		84	54	T		116	74	t
21	15	§		53	35	5		85	55	U		117	75	u
22	16	▬		54	36	6		86	56	V		118	76	v
23	17	↨		55	37	7		87	57	W		119	77	w
24	18	↑		56	38	8		88	58	X		120	78	x
25	19	↓		57	39	9		89	59	Y		121	79	y
26	1A	→		58	3A	:		90	5A	Z		122	7A	z
27	1B	←		59	3B	;		91	5B	[		123	7B	{
28	1C	∟		60	3C	<		92	5C	\		124	7C		
29	1D	↔		61	3D	=		93	5D	]		125	7D	}
30	1E	▲		62	3E	>		94	5E	^		126	7E	~
31	1F	▼		63	3F	?		95	5F	_		127	7F	⌂

Appendix F: Advanced C Programming

In this Appendix, first we introduce some preprocessor directives, and then we explain how to manipulate registers using the defined bits.

Section F.1: Preprocessor Directives

#define

#define can be used to define constant values like PI:

```
#define PI 3.1415
```

When the above line of code is written, the compiler replaces PI with 3.1415 at compile time. For example, after defining the PI constant if we write "float a = PI*2.0;", at compile time, the PI will be considered as 3.1415 and then it will be compiled.

Example F-1: Find the values for variables a and b in the following program:

```
#define NUM_OF_LINES        4
int main ( )
{
   int a = NUM_OF_LINES;
   int b = NUM_OF_LINES * 20;
   while(1);
}
```

Solution:

NUM_OF_LINES is replaced with 4 and the above program will be converted to:

```
int main ( )
{
   int a = 4;
   int b = 4 * 20;
   while(1);
}
```

So, a and b will contain 4 and 80, respectively.

Using #define we can define nickname for any expression. For example, in the following program, we name { and } as BEGIN and END respectively:

```
#define BEGIN   {
#define END     }

int main ()
BEGIN
   while(1)
   BEGIN
      while(1);
   END
END
```

Naming pins

It is a good practice to name pins of microcontrollers according to their usages. This makes the code more readable and also more reusable. For example, see the following snippet of code:

```
#define LED_PORT    GPIOC
#define LED_BIT     13
int main(){
   ...
   LED_PORT->ODR |= (1<<LED_BIT);
}
```

#define and Macros

#define can be used to define macros in C language. For example, in the following piece of code, a macro is defined that sets a bit of a variable:

```
#define SET_BIT(varName,bitNum) varName |= (1<<bitNum)
int main()
{
   ...
   SET_BIT(GPIOC->ODR,13); /* set bit 13 of GPIOC->ODR */
   while(1);
}
```

At compile time, SET_BIT(GPIOC->ODR,13) is replaced with "varName = varName|(1<<bitNum)" and the arguments varName and bitNum are substituted with GPIOC->ODR and 13, respectively. So, the line of code is replaced with "GPIOC->ODR |= (1<<13);".

See also the following snippet of code:

```
#define SQUARE(x)          x*x
int main ( )
{
  GPIOB->ODR = SQUARE(GPIOA->IDR);
   ...
}
```

In the above program, the compiler substitutes "GPIOB->ODR = SQUARE(GPIOA->IDR);" with "GPIOB->ODR = GPIOA->IDR*GPIOA->IDR;" and then converts it to machine code.

Note: macros vs. functions

 Macros do not have the function call overhead and do not use the stack space, neither. But they can increase the used program memory. It is good to use macros (or inline functions) when the function is too small.

#undef

#undef is opposite of #define. It undefines a symbol which was defined by #define:

```
#undef symbol
```

For example, in the following program we undefine MAX_VALUE which is defined in a few lines before:

```
#define MAX_VALUE 100
...
#undef MAX_VALUE
```

Conditional preprocessors

The conditional preprocessor has the following format:

```
#if condition
...  //The program that should be compiled when the condition is met
#endif
```

In the following program, the #if preprocessor checks the value of DISPLAY_CONFIG and the printf lines will be compiled only when DISPLAY_CONFIG is 1. Since DISPLAY_CONFIG is defined 0, the printf lines will not be compiled, now.

```
#define DISPLAY_CONFIG 0
int main ( )
{
   #if DISPLAY_CONFIG == 1
      printf("Hello World!\n\r");
      printf("Now Display config is set to 1.");
   #endif
}
```

Together with conditional preprocessor, we can use #else as well:

```
#if condition
 ...  /* The code that should be compiled when the condition is met */
#else
 ...  /* The code will be compiled when the condition is not met */
#endif
```

In the following program i is defined as uint8_t when ARRAY_LEN is defined less than 256:

```
#define ARRAY_LEN 64
int main ( )
{
   char ourArray[ARRAY_LEN];
#if ARRAY_LEN < 256
      uint8_t i = 0;
#else
      uint16_t i = 0;
#endif

   for (i = 0; i < ARRAY_LEN; i++)
      ourArray[i] = ' ';
}
```

Instead of #if we can use #ifdef or #ifndef directives, as well. We can check to see if a macro is defined or not using #ifdef and #ifndef respectively. For example, in the following code the printf line will be compiled only when DISPLAY_RESULTS is defined:

```
int main ()
{
   #ifdef DISPLAY_RESULTS
   printf("The result is as follows");
   #endif
}
```

Section F.2: Manipulating Registers Using Defined Bit Names

In Section 8-1, you learned to enable the clocks for different peripherals. For example, the following instruction enables the clock for GPIO port B:

```
RCC->APB2ENR |= (1<<3); /* set the IOPBEN bit to 1 (IOPBEN is bit 3 of APB2ENR) */
```

To write the above line of code, you have to check the book (or the reference manual of the chip) to see that bit 3 of the APB2ENR is for IOPBEN. Then, you can write the code.

In the "stm32f10x.h" header file, there are definitions for all bits of the I/O registers. For example, the bits of APB2ENR register are defined as shown below:

```
/***************** Bit definition for RCC_APB2ENR register  ****************/
#define RCC_APB2ENR_AFIOEN   ((uint32_t)0x00000001)   /*!< Alternate Function I/O clock
enable */
#define RCC_APB2ENR_IOPAEN   ((uint32_t)0x00000004)   /*!< I/O port A clock enable */
#define RCC_APB2ENR_IOPBEN   ((uint32_t)0x00000008)   /*!< I/O port B clock enable */
#define RCC_APB2ENR_IOPCEN   ((uint32_t)0x00000010)   /*!< I/O port C clock enable */
#define RCC_APB2ENR_IOPDEN   ((uint32_t)0x00000020)   /*!< I/O port D clock enable */
#define RCC_APB2ENR_ADC1EN   ((uint32_t)0x00000200)   /*!< ADC 1 interface clock enable
*/
...
```

So, to enable the clock for GPIO port B you can also write the following instruction:

```
RCC->APB2ENR |= RCC_APB2ENR_IOPBEN;
```

Similarly, the following line of code, enables the clock for ADC1:

```
RCC->APB2ENR |= RCC_APB2ENR_ADC1EN;
```

The bit definition naming follows the following convention:

peripheralName_registerName_bitName

For example, the BR5 bit for GPIOA->BSRR is named as GPIO_BSRR_BR5. So, the following line of code, sets PA4 and clears PA5:

```
GPIOA->BSRR = GPIO_BSRR_BS4|GPIO_BSRR_BR5;
```

As another example, the following snippet of code toggles PC13:

```
while(1)
{
  GPIOC->BRR = GPIO_BRR_BR13; /* make the pin low */
  delay_ms(500);              /* wait 0.5 sec. */
  GPIOC->BSRR = GPIO_BSRR_BS13; /* make the pin high */
  delay_ms(500);              /* wait 0.5 sec. */
}
```

In cases that a field of a register is bigger than 1 bit, the field is named as peripheralName_registerName_FieldName. To name the bits of the field, the bit number is added to the end of field name. For example, in the CRL and CRH registers, the MODEx fields are 2-bit. So, the fields are named as GPIO_CRL_MODEx and the bits are named as GPIO_CRL_MODEx_0 and GPIO_CRL_MODEx_1. For example, in STM32F10x.h the followings are defined for MODE6:

```
#define  GPIO_CRL_MODE6   ((uint32_t)0x03000000)   /*!< MODE6[1:0] bits (Port x mode
bits, pin 6) */
#define  GPIO_CRL_MODE6_0  ((uint32_t)0x01000000)             /*!< Bit 0 */
#define  GPIO_CRL_MODE6_1  ((uint32_t)0x02000000)             /*!< Bit 1 */
```

The following piece of code sets MODE6 to 2 using the bit definitions. It clears the MODE6 bits using AND and then, sets the MODE6_1 bit:

```
GPIOA->CRL &= ~GPIO_CRL_MODE6;
GPIOA->CRL |= GPIO_CRL_MODE6_1;
```

See the following programs.

Program F-1: Rewrite Program 8-5 using bit definitions (Toggle PC13 using BSRR and BRR)

```
#include <stm32f10x.h>

void delay_ms(uint16_t t);

int main()
{
   RCC->APB2ENR |= RCC_APB2ENR_IOPCEN;          /* Enable clocks for GPIOs */
   /* PC13 as output */
   GPIOC->CRH &= (GPIO_CRH_MODE13|GPIO_CRH_CNF13); /* clear MODE13 and CNF13 fields */
   GPIOC->CRH |= GPIO_CRH_MODE13_1|GPIO_CRH_MODE13_0; /* set MODE13 to 3 (Output) */

   while(1)
   {
      GPIOC->BRR = GPIO_BRR_BR13; /* make the pin low */
      delay_ms(500);              /* wait 0.5 sec. */
      GPIOC->BSRR = GPIO_BSRR_BS13;    /* make the pin high */
      delay_ms(500);              /* wait 0.5 sec. */
   }
}
// copy delay_ms from Program 8-1
```

Program F-2: Rewrite Program 8-9 using bit definitions (It toggles PC13 if PA2 is low)

```
#include <stm32f10x.h>

void delay_ms(uint16 t t);

int main()
{
   RCC->APB2ENR |= RCC_APB2ENR_IOPAEN|RCC_APB2ENR_IOPCEN; /* Enable clocks for PB and
PC*/
   /* PC13 as output */
   GPIOC->CRH = (GPIOC->CRH&~GPIO_CRH_CNF13)|GPIO_CRH_MODE;

   /* PA2 as input with pull-up */
   GPIOA->CRL &= ~(GPIOA->CRL&GPIO_CRL_CNF2);
   GPIOA->CRL |= GPIO_CRL_CNF2_1;
   GPIOA->ODR |= (1<<2);    /* pull-up PA2 */
```

```
   while(1)
   {
      if((GPIOA->IDR&GPIO_IDR_IDR2) == 0) /* is PA2 low? */
         GPIOC->ODR ^= GPIO_ODR_ODR13; /* toggle PC13 */
      else
         GPIOC->ODR &= ~GPIO_ODR_ODR13;
      delay_ms(500);
   }
}
//copy delay_ms from Program 8-1 to here
```

Program F-3: Rewrite Program 8-6 using bit definitions (Reading from port B and writing to port A)

```
#include <stm32f10x.h>

int main()
{
   RCC->APB2ENR |= RCC_APB2ENR_IOPAEN|RCC_APB2ENR_IOPBEN; /* Enable clocks for PA and
PB */

   GPIOA->CRL &= ~(GPIO_CRL_CNF|GPIO_CRL_MODE); /* PA0-PA7 as outputs */
   GPIOA->CRL |= GPIO_CRL_MODE;
   /* PA8-PA15 as outputs */
   GPIOA->CRH &= ~(GPIO_CRH_CNF|GPIO_CRH_MODE);
   GPIOA->CRH |= GPIO_CRH_MODE;
   /* PB0-PB7 as inputs */
   GPIOB->CRL &= ~(GPIO_CRL_CNF|GPIO_CRL_MODE);
   GPIOB->CRL |=
GPIO_CRL_CNF0_0|GPIO_CRL_CNF1_0|GPIO_CRL_CNF2_0|GPIO_CRL_CNF3_0|GPIO_CRL_CNF4_0|GPIO_C
RL_CNF5_0|GPIO_CRL_CNF6_0|GPIO_CRL_CNF7_0;
   /* PB8-PB15 as inputs */
   GPIOB->CRH &= ~(GPIO_CRH_CNF|GPIO_CRH_MODE);
   GPIOB->CRH |=
GPIO_CRH_CNF8_0|GPIO_CRH_CNF9_0|GPIO_CRH_CNF10_0|GPIO_CRH_CNF11_0|GPIO_CRH_CNF12_0|GPI
O_CRH_CNF13_0|GPIO_CRH_CNF14_0|GPIO_CRH_CNF15_0;

   while(1)
   {
      GPIOA->ODR = GPIOB->IDR; /* read from port B and write to port A */
   }
}
```

www.ingramcontent.com/pod-product-compliance
Lightning Source LLC
LaVergne TN
LVHW060132070326
832902LV00018B/2758